World Sustainability Series

Series Editor

Walter Leal Filho, Wissenschaften Hamburg, Hochschule für Angewandte, Hamburg, Hamburg, Germany

Due to its scope and nature, sustainable development is a matter which is very interdisciplinary, and draws from knowledge and inputs from the social sciences and environmental sciences on the one hand, but also from physical sciences and arts on the other. As such, there is a perceived need to foster integrative approaches, whereby the combination of inputs from various fields may contribute to a better understanding of what sustainability is, and means to people. But despite the need for and the relevance of integrative approaches towards sustainable development, there is a paucity of literature which address matters related to sustainability in an integrated way.

Notes on the quality assurance and peer review of this publication

Prior to publication, the works published in this book are initially assessed and reviewed by an in-house editor. If suitable for publication, manuscripts are sent for further review, which includes a combined effort by the editorial board and appointed subject experts, who provide independent peer-review. The feedback obtained in this way was communicated to authors, and with manuscripts checked upon return before finally accepted. The peer-reviewed nature of the books in the "World Sustainability Series" means that contributions to them have, over many years, been officially accepted for tenure and promotion purposes.

Rubee Singh · Walter Leal Filho · Vikas Kumar
Editors

Data-Driven Sustainability: Harnessing Technology for A Greener Future and Environmental Resilience

Springer

Editors
Rubee Singh
GLA University
Mathura, Uttar Pradesh, India

Walter Leal Filho
Hamburg University of Applied Sciences
Hamburg, Hamburg, Germany

Vikas Kumar
University of Portsmouth
Portsmouth, UK

ISSN 2199-7373 ISSN 2199-7381 (electronic)
World Sustainability Series
ISBN 978-3-032-13411-0 ISBN 978-3-032-13412-7 (eBook)
https://doi.org/10.1007/978-3-032-13412-7

This Springer imprint is published by the registered company Springer Nature Switzerland AG
The registered company address is: Gewerbestrasse 11, 6330 Cham, Switzerland

Preface

The twenty-first century stands at a crossroads where data, technology, and sustainability converge to shape humanity's shared destiny. The global community now faces the dual challenge of fostering technological progress while restoring environmental equilibrium. The rapid growth of digital ecosystems, artificial intelligence, and big data analytics offers immense possibilities to drive sustainable development and build environmental resilience. At the same time, these very technologies demand ethical frameworks and responsible use to ensure that innovation serves the planet rather than exploits it. It is in this context that *Data-Driven Sustainability: Harnessing Technology for a Greener Future and Environmental Resilience* has been conceived—a collaborative intellectual effort to explore how data can guide humanity toward a balanced, equitable, and greener future. This volume brings together a constellation of researchers and practitioners from more than 20 countries, including India, Sri Lanka, Bangladesh, China, Malaysia, Oman, Romania, Germany, Namibia, Chile, Canada, Ecuador, Iran, Uganda, Viet Nam, Indonesia, the Maldives, Serbia, Brunei, the United Arab Emirates, and the United Kingdom. The book captures a global voice for sustainability, reflecting both the technological aspirations of developed economies and the adaptive resilience of developing nations. It stands as a testament to the power of collaboration, where diverse geographies, disciplines, and ideologies unite under a common purpose: to make technology a force for ecological restoration.

The book contains 26 chapters, thoughtfully organized into 5 interrelated parts that unfold the evolving relationship between data and sustainability. The opening chapters establish the conceptual foundation of data-driven sustainability, describing how data analytics and digital transformation are revolutionizing environmental understanding. They discuss how data enables more precise measurement of carbon emissions, improved monitoring of natural resources, and evidence-based environmental policymaking. Contributors from India, Malaysia, and Romania present frameworks that link technology with ecological awareness, demonstrating how inclusive data governance and open-access platforms can strengthen decision-making for sustainable outcomes.

As the discussion progresses, the focus turns toward technological innovation and its capacity to reimagine the future of sustainability. Chapters contributed by scholars from China, Australia, Oman, and Sri Lanka explore how smart technologies such as the Internet of Things, blockchain, and artificial intelligence can address complex ecological problems. Case studies highlight applications ranging from precision agriculture and smart energy systems to digital waste management and real-time environmental surveillance. Through these examples, the authors collectively argue that technology, when guided by ethical design and social inclusion, can transform the sustainability discourse from reactive preservation to proactive innovation. The subsequent chapters shift attention to business transformation, public policy, and governance. Sustainability in the modern era cannot exist in isolation—it must integrate with markets, organizations, and communities. Contributors from Bangladesh, India, Serbia, and the United Kingdom examine how data analytics and predictive modeling are reshaping environmental, social, and governance (ESG) frameworks. They investigate how digital transparency, circular economy models, and sustainable financing mechanisms are helping industries and governments achieve carbon neutrality. By illustrating the convergence of technology, management, and policy, these chapters emphasize that sustainability is not merely an environmental concern but a strategic foundation for future economic stability. Further along, the book turns its gaze toward climate resilience and energy futures. Chapters from Namibia, Chile, Iran, and Uganda present data-based approaches to understanding and mitigating climate risks. The use of artificial intelligence in climate modeling, remote sensing for land management, and predictive analytics for disaster preparedness demonstrate how data serves as both a diagnostic and preventive instrument. These works illuminate the need for a holistic approach—combining scientific precision with community awareness—to design infrastructures that are not only efficient but also adaptable to the realities of a changing climate. The idea of resilience, as discussed across these contributions, extends beyond resistance to shock; it embodies the capacity to evolve intelligently in the face of uncertainty.

In its final part, the volume envisions future pathways that link digital transformation with global cooperation. Contributions from Germany, Canada, Viet Nam, and the Maldives reflect on the ethical, educational, and cultural dimensions of technological sustainability. They argue that while data empowers, it also obligates—demanding fairness, inclusivity, and transparency. The authors propose frameworks for open data sharing, digital education, and interdisciplinary collaboration to ensure that sustainability remains a shared endeavor rather than a fragmented pursuit. This concluding segment of the book reaffirms that the future of sustainability depends as much on the democratization of data as on the sophistication of technology. What distinguishes this book is the breadth of its authorship and the depth of its interdisciplinary engagement. The chapters encompass fields as varied as environmental science, business management, engineering, computer science, and social policy. Each contribution offers a different lens on how digital tools can be harmonized with sustainable practices. Together, they create a comprehensive narrative of transformation—one that recognizes technology not as an end in itself but as a means to restore balance between human progress and planetary well-being.

This compilation is also a celebration of collective knowledge. The contributors, representing universities, research centers, and professional institutions across continents, have worked collaboratively to address some of the most urgent sustainability issues of our era. From data-based decision models in agriculture and industry to the application of digital ethics in environmental governance, every chapter builds on the conviction that sustainable change emerges through shared learning and mutual responsibility. The inclusion of perspectives from both developed and emerging economies ensures that this volume remains globally inclusive while contextually grounded.

As editors, we express our heartfelt gratitude to every contributor for their intellectual generosity, rigor, and dedication. The process of compiling, refining, and integrating these diverse works into a unified vision has been a deeply rewarding experience. This book reflects not only the authors' research excellence but also their collective optimism for a sustainable world where data and technology serve humanity rather than dominate it. We hope that this volume will guide researchers, educators, policymakers, and practitioners to rethink how information systems can be aligned with ecological wisdom, and how innovation can be anchored in ethical sustainability. In an age defined by data, the challenge is not the scarcity of information but the scarcity of responsibility. This book aspires to bridge that gap—to encourage evidence-based choices, responsible innovation, and a culture of global collaboration. The pages that follow are an invitation to rethink progress, not as exploitation of nature but as partnership with it. Through data-driven insight and human empathy, we can together build a greener, smarter, and more resilient future.

Mathura, India — Rubee Singh
Hamburg, Germany — Walter Leal Filho
Portsmouth, UK — Vikas Kumar

Acknowledgements

We would like to express our heartfelt gratitude to all those who have contributed to the creation of this book from across the globe. Firstly, we thank the researchers and experts who have shared their knowledge and insights with us, providing valuable information on the current state of book titled: *Data-Driven Sustainability: Harnessing Technology for a Greener Future and Environmental Resilience.* We also extend our appreciation to the organizations and institutions that have supported our work, including government agencies, non-governmental organizations, and academic institutions.

We further acknowledge the invaluable contributions of our colleagues and editors, who have worked tirelessly to ensure that this book is informative, engaging, and accessible to a wide range of readers. We hope that this book will inspire further research and action toward data-driven sustainability and a greener future, promoting environmental resilience and sustainable technological advancement.

We also express our sincere gratitude to the experts and stakeholders who contributed their valuable insights and perspectives to this volume. Their contributions have been instrumental in shaping our understanding of how technology, data analytics, and innovation can be effectively leveraged to drive sustainability and strengthen environmental resilience in an increasingly digital world. We hope that this book will serve as a valuable resource for policymakers, practitioners, and researchers working toward a zero-carbon and sustainable future for our planet.

Finally, we are immensely thankful to the Book Series Editors, **Prof. Walter Leal Filho**, the Editorial Director/Publishing Editor, **Dr. Christian Witschel**, and the entire team at **Springer Nature** for bringing out this publication in a highly professional and timely manner.

Rubee Singh
Walter Leal Filho
Vikas Kumar

About This Book

This book explores and focuses on how data-driven innovation and technological intelligence are reshaping the global pursuit of sustainability and environmental resilience. In an age defined by information, *Data-Driven Sustainability: Harnessing Technology for a Greener Future and Environmental Resilience* investigates how data has become the foundation of informed decision-making, ethical governance, and sustainable progress across societies and economies.

The volume reflects a truly international and multidisciplinary effort, uniting researchers, innovators, and practitioners from more than twenty countries across Asia, Europe, Africa, Australia, and the Americas. Together, they examine how data science, artificial intelligence, machine learning, and digital ecosystems can be leveraged to confront urgent ecological challenges—ranging from climate change and energy transitions to urban management and green innovation.

Comprising 26 chapters organized into 5 thematic sections, the book integrates perspectives from environmental science, management, economics, engineering, public policy, and data analytics. Each contribution blends theory with practice, illustrating how digital technologies can support sustainable production, optimize resource use, and foster climate adaptation. The chapters highlight practical examples such as smart agriculture, renewable energy solutions, circular economy models, and ethical applications of AI for environmental governance.

At its core, the book emphasizes that sustainability is not only a technological challenge but a collective moral responsibility. By merging digital intelligence with ecological awareness, it envisions a future where innovation serves both people and the planet. *Data-Driven Sustainability* is an essential reference for academics, policymakers, industry professionals, and sustainability leaders seeking a multidisciplinary understanding of how technology and data can work together to build a greener, smarter, and more resilient world.

Contents

Editors and Contributors

About the Editors

Dr. Rubee Singh (D. Litt, Ph.D) is an Assistant Professor of Human Resources at the Institute of Business Management, GLA University, India, and a Post-Doctoral Researcher (D.Litt) at Kumaun University, Nainital, India. She holds a Ph.D. in Management from Noida International University, India, and an MBA in HR from Dr. A.P.J. Abdul Kalam Technical University, Lucknow, India. With over 12 years of teaching and research experience, and recognized internationally for her social contributions, she was awarded the Chartered Global Peace Building Professional (CGPP) by The George Washington University of Peace, USA, and the Centre for Peace Studies, Sri Lanka. She is a United Nations Volunteer and a contributing author for Voices of Youth (UNICEF). Dr. Singh is an honorary life member of the Women Economic Forum and serves as the National Secretary of IATM (Youth Wing)—International Anti-Terrorism Movement, India. As a prolific academic, Dr. Singh is the Managing Editor of IJARSH, India, Associate Editor for Circular Economy and Sustainability (Springer—Q1), and serves on the editorial boards of over 30 international journals. She is an Advisory Board Member for Heliyon (Elsevier) and Research in Globalization (Elsevier), and a reviewer for leading journals including Technological Society, Social Responsibility Journal, BSE (Wiley), Sustainable Development, European Journal of Innovation Management, British Food Journal, and Sustainability. She is a life member of several global professional associations, including the International Association of Engineers (Thailand), IMRTC (USA), and ISRD (UK).

Dr. Singh has published more than 100 research papers in SCI, Scopus, ABDC-A, and UGC-indexed journals, along with 30 academic books with renowned publishers such as Emerald, Springer, Elsevier, and De Gruyter. She has presented numerous papers at national and international conferences, including those held at IIT Madras, IIT Bombay, IIM Calcutta, and Mardin Artuklu University (Turkey) and Vietnam. Her research has earned over 1,800 citations on Google Scholar, with an H-index of 25. Her areas of expertise include Sustainability, Digital and Sustainable HRM, Circular

Economy, Net Zero, Industry 4.0, Artificial Intelligence, Blockchain Technology, Machine Learning, Sustainable Innovation, and Eco/Digital Innovation.

Her books are internationally recognized and included in the libraries of Ohio Christian University, University of California, UMass Chan Medical School, Aston University (UK), Dow University of Health Sciences (Pakistan), and University of Port Harcourt (Nigeria). One of her books, Training and Development, has been adopted in the MBA syllabus at The Institute of Aeronautical Engineering, Hyderabad. Among her numerous accolades are the Most Influential Woman Leader Award (2022), Third Consecutive Best Researcher Awards (2022–2024, GLA University), Child Ambassador Award (2022), Iconic Faculty Award (2019), International Young Researcher Award (2021), Young Scientist Award (2020), Nari Ratan Award (2023), India Book of Records (2020—confirmatory), and Asia Book of Records Holder (2021—confirmatory). Dr. Singh prioritizes impactful work over recognition, making her an eminent researcher, educator, and social activist.

Prof. Walter Leal Filho (D.Lit., Ph.D) holds the Chairs of Climate Change Management at the Hamburg University of Applied Sciences (Germany), and Environment and Technology at Manchester Metropolitan University (UK). He directs the Research and Transfer Centre "Sustainability Development and Climate Change Management". His main research interests are in the fields of sustainable development and climate change, also including aspects of climate change and health. Professor Leal has been author, co-author, editor or co-editor in many publications on matters related to environment, sustainable development and climate change issues. He has in excess of 700 publications to his credit, which includes 193 books, 102 book chapters and over 430 papers. His field experience in over 70 countries, involves missions undertaken on behalf of various international organizations (e.g., European Union, World Bank, OECD, UNESCO, UNEP) and attendance to specialized events in North America, Latin America/Caribbean, Europe, Africa, Asia and Pacific Region, and the Middle East. Among the many awards he has received, it can be mentioned the 1999 "Leading Editor of the Year Award" and the 2000 "Editor of the Year Award", by MCB University Press, the "Eco-Citizen Award" (2001) in Rio de Janeiro and the Aurelio Giuseppe Prize awarded by the Italian Council of Ministers for his environmental work.

Moreover, he was awarded the "Golden Page Award" (April 2003) in London for his editorial work and the North Sea Star Award (2006) by the EU-Interreg Secretariat for his project work in the North Sea Region. In the field of sustainable development, Prof. Walter Leal is the founder of the European School of Sustainability Science and Research and of the Inter-University Sustainable Development Research Programme (IUSDRP). He holds a record of publications on sustainability issues, some of which are available here: https://www.haw-hamburg.de/en/ftz-nk/programmes/iusdrp/ (please scroll down and click on "sharing"). He is also editor of the World Sustainability Series with Springer. In addition, he is the founding editor of the International Journal of Sustainability in Higher Education (IJSHE) launched in the year 2000. IJSHE is the only journal which fully focuses on sustainability in a higher education context. Professor Walter Leal is also Editor-in-Chief of the

Encyclopedia of the UN Sustainable Development Goals currently being prepared. With 17 volumes, one devoted to each SDG, more than 1.700 chapters and in excess of 2.300 authors, this is the largest editorial project on matters related to sustainable development, ever undertaken.

Prof. Vikas Kumar (Ph.D.) is the Deputy Vice-Chancellor for Research and Innovation at the University of Portsmouth, UK. He has held prominent leadership positions in higher education, including Associate Dean for Research, Innovation, and Enterprise at the Faculty of Business, Law and Social Sciences, Birmingham City University, and Director of Research and Scholarship at Bristol Business School, University of the West of England. Prof. Kumar is a distinguished Professor of Operations and Supply Chain Management with extensive experience spanning over a decade in teaching, research, and industry engagement. His academic journey includes a Ph.D. in Management from the University of Exeter, UK, and a B. Tech from the National Institute of Advanced Manufacturing Technology, Ranchi (formerly NIFFT), India. He has also contributed to academia in India, Ireland, and Hong Kong.

Professor Kumar has an outstanding research profile, having authored over 350 journal articles and conference papers, and published more than 15 books with leading publishers including Emerald, Springer, De Gruyter, and Taylor & Francis. His work has been widely cited, with over 21,000 citations on Google Scholar and an H-index of 78. In recognition of his research impact, he was ranked among the Top 1% of Highly Cited Researchers in 2024, and for the third consecutive year, acknowledged by Stanford University and Elsevier among the Top 2% of scientists globally. He serves on the editorial boards of multiple international journals, has guest-edited numerous special issues, and acts as a reviewer for more than 15 top-tier journals, including ABS 3* and 4* rated publications. Professor Kumar has secured over £1 million in competitive research funding from prestigious agencies such as Innovate UK, EPSRC, the British Council, the British Academy, the Newton Fund, and the Science Foundation of Ireland. His research focuses on Net Zero Emissions, Circular Economy, Sustainability, Industry 4.0, digital and sustainable supply chains, and operational excellence.

Contributors

Siti Suraya Abd Razak Faculty of Management, Universiti Teknologi Malaysia, Johor Bahru, Malaysia

Azamat Ali School of Liberal and Creative Arts, Lovely Professional University, Punjab, India

Dilanthi Amaratunga School of Applied Sciences, University of Huddersfield, Huddersfield, UK;
School of Architecture, Building and Civil Environment, Loughborough University, Loughborough, UK

A. Arun Kumar ICFAI Law School, ICFAI Foundation for Higher Education (Deemed to be University Under Sec 3 of UGC Act,1956), Hyderabad, India

Ir. Bambang Sugiyono Agus Purwono Mechanical Engineering Department, State Polytechnic of Malang, Malang, Indonesia

Derrick Frimpong Buabeng Department of Land and Spatial Sciences, Namibia University of Science and Technology, Windhoek, Khomas Region, Namibia

Benny Godwin J. Davidson School of Business, University of the Fraser Valley, Abbotsford, BC, Canada

Narayanage Jayantha Dewasiri Department of Accountancy and Finance, Faculty of Management Studies, Sabaragamuwa University of Sri Lanka, Belihuloya, Sri Lanka;
Qasim Ibrahim School of Business, Villa College, Maldives;
University Center for Research and Development (UCRD), Chandigarh University, Ajitgarh, India

Hiranya Dissanayake Department of Accountancy, Faculty of Business Studies and Finance, Wayamba University of Sri Lanka, Kuliyapitiya, Sri Lanka

Jacinta Dsilva SEE Institute, The Sustainable City, Dubai, United Arab Emirates

Lam Thuy Duong TNU—University of Education, Thai Nguyen, Vietnam

Kristina Encinas Bartos Institute of Sustainable Corporate Management, Ulm University, Ulm, Germany

R. Felista Sugirtha Lizy A.P.C. Mahalaxmi College for Women, Thoothukudi, Tamil Nadu, India

Jossy P. George Christ University, Bangalore, India

Moaz Gharib Department of Management, College of Commerce and Business Administration, Dhofar University, Salalah, Oman

Nitin S. Goje Department of Computer and Information Sciences, Webster University, Tashkent, Uzbekistan

Ernesto DR Santibanez Gonzalez Department of Industrial Engineering and CES 4.0, Faculty of Engineering, University of Talca, Curico, Chile

Liu Guangling Northwest University of Political Science and Law, Xi'an, China

Richard Haigh School of Applied Sciences, University of Huddersfield, Huddersfield, UK

Sameer Jain Faculty at NICMAR Business School, NICMAR University, Pune, India

Dilshad Jeffery Qasim Ibrahim School of Business, Villa College, Malé, Maldives

Amit Joshi Department of Management Studies, Kumaun University, Nainital, Uttarakhand, India

Som Aditya Juyal School of Management Studies, Swami Rama Himalayan University, Dehradun, India

Vinay Kandpal Department of Management Studies, Graphic Era Deemed to Be University Dehradun, Dehradun, India

Phi Dinh Khuong TNU—University of Education, Thai Nguyen, Vietnam

Rishi Kumar Department of Economics and Finance, BITS Pilani, Hyderabad Campus, Hyderabad, India

Vikas Kumar University of Portsmouth, Portsmouth, UK

Walter Leal Faculty of Life Sciences, Hamburg University of Applied Sciences, Hamburg, Germany

Songyue Li Faculty of Finance and Accounting, Zhejiang Business College, Hangzhou, China

Ying Tuan Lo XJTLU Entrepreneur College (Taicang), Xi'an Jiaotong-Liverpool University, Suzhou, China

Wei Shen Low Department of Hospitality and Information Technology, Sunway College Johor Bahru, Johor Bahru, Malaysia

Otilia Manta School of Advanced Studies of the Romanian Academy (SCOSAAR), Bucharest, Romania

S. Mojtaba Matinkhah Computer Engineering Department, Yazd University, Yazd, Yazd Province, Iran

Philip Mensah Department of Land and Spatial Sciences, Namibia University of Science and Technology, Windhoek, Khomas Region, Namibia

Florah Sewela Modiba Nelson Mandela University, Gqeberha, South Africa

Jabu Mtsweni Council for Scientific and Industrial Research (CSIR), Pretoria, South Africa

Peter V. Muttungal Department of Psychology, Christ University, Bangalore, India

Martin Müller Institute of Sustainable Corporate Management, Ulm University, Ulm, Germany

Deshika Nainanayake School of Business, Western Sydney University, Sydney, Australia

Ali Najeeb Qasim Ibrahim School of Business, Villa College, Male, Maldives

Pranab Kumar Panday Department of Public Administration, University of Rajshahi, Rajshahi, Bangladesh

Amrita Pani Department of Economics and Finance, BITS Pilani, Hyderabad Campus, Hyderabad, India

Kamalpreet Kaur Paposa Mittal School of Business, Lovely Professional University, Phagwara, Punjab, India

Sukhvinder Singh Paposa Division of Career Services, Lovely Professional University, Phagwara, Punjab, India

Paulina Permatasari Parahyangan Catholic University, Bandung, Indonesia

Nayana Prabhash School of Commerce and Management Studies, Dayananda Sagar University, Bengaluru, Karnataka, India

Nishi Priya Department of Psychology, Christ University, Bangalore, India

Asanga Ranasinghe Postgraduate Institute of Management, University of Sri Jayewardenepura, Nugegoda, Sri Lanka

Nilantha Randeniya School of Applied Sciences, University of Huddersfield, Huddersfield, UK

Seema Rani Chandigarh School of Business, Chandigarh Group of Colleges, Jhanjheri, India

Mananage Shanika Hansini Rathnasiri Department of Marketing Management, Faculty of Management Studies, Sabaragamuwa University of Sri Lanka, Belihuloya, Sri Lanka;
College of Society and Professions, London South Bank University, London, UK

Muhandiramge Vasana Maduwanthi Rodrigo Ministry of Industry and Entrepreneurship Development, Colombo, Sri Lanka

Julia Schwarzkopf HTW Berlin Business School, Berlin, Germany

Kavya Shabu Faculty of Management Studies, CMS Business School, Jain (Deemed-to-Be University), Bengaluru, India

Wasswa Shafik Dig Connectivity Research Laboratory (DCRLab), Kampala, Uganda;
School of Digital Science, Universiti Brunei Darussalam, Gadong, Brunei

Mohamed Irfan Shaikh SEE Institute, The Sustainable City, Dubai, United Arab Emirates

Akshay Singh School of Management Studies, Indra Gandhi National Open University, New Delhi, India

Rubee Singh Institute of Business Management, GLA University, Mathura, India; Department of Management Studies, Kumaun University, Nainital, Uttarakhand, India

Yogeeswari Subramaniam Universiti Teknologi Malaysia, Johor Bahru, Malaysia

Lalit Mohan Tewari Department of Botany, DSB Campus, Kumaun University, Nainital, Uttarakhand, India

Brankica Todorovic Faculty of Contemporary Arts, University of Business Academy, Belgrade, Serbia

Apurvaa Trivedi School of Management Studies, Swami Rama Himalayan University, Dehradun, India

Vibha Trivedi O.P. Jindal Global University, Sonipat, India

Hassan Tumwiine Department of Civil and Environmental Engineering, Monash University, Clayton, Australia

Himanshu Vaishnav School of Management, O.P. Jindal University, Raigarh, Chhattisgarh, India

Eric Yankson Department of Architecture, Planning and Construction, Namibia University of Science and Technology, Windhoek, Khomas Region, Namibia

Manuel A. Zambrano-Monserrate Universidad Espíritu Santo, Samborondón, Ecuador

Shufan Zheng School of Management, Zhejiang University of Finance and Economics, Hangzhou, China

Foundations of Data-Driven Sustainability and Industry Transformation

Global Best Practices in Tech Driven Sustainability: Smart CE and Manufacturing Sector

Rubee Singh, Walter Leal, Vikas Kumar, and Azamat Ali

Abstract The future of the circular economy (CE) and the manufacturing industry are being redefined by the use of innovative, smart solutions based on technology driven sustainability. Artificial intelligence (AI), digital twins, IoT enabled monitoring, blockchain are some of the tools that are redefining production systems through optimization of resource use, waste minimization and closed loop processes. Authority Global best practice shows a clear move towards the abandonment of linear take make dispose models toward regenerative and resilient ecosystems. As an illustration, predictive maintenance and energy analytics minimize operational downtime and emissions, whereas additive manufacturing and resource recovery make a process more efficient and resource circular. Online collaboration also enhances inter country supply chain traceability and accountability, and harmonizes industrial operations with sustainable development objectives. In addition to positive ecological effects, the innovations encourage the new business models and creation of social values. The incorporation of smart CE concepts into manufacturing, therefore, becomes not just an operationally required but also a strategic direction of attaining competitiveness, resilience, and long term sustainability in a fast changing global environment.

Keywords Technology driven sustainability · Circular economy · Smart manufacturing · Artificial intelligence · Blockchain · Sustainable development goals

---

R. Singh (✉)
Institute of Business Management, GLA University, Mathura, India
e-mail: dr.rubeerajput@gmail.com

W. Leal
Faculty of Life Sciences, Hamburg University of Applied Sciences, Hamburg, Germany
e-mail: walter.leal2@haw-hamburg.de

V. Kumar
University of Portsmouth, Portsmouth, UK
e-mail: Vikas.Kumar@port.ac.uk

A. Ali
School of Liberal and Creative Arts, Lovely Professional University, Punjab, India
e-mail: azamatcug@gmail.com

R. Singh et al. (eds.), *Data-Driven Sustainability: Harnessing Technology for A Greener Future and Environmental Resilience*, World Sustainability Series,
https://doi.org/10.1007/978-3-032-13412-7_1

1 Introduction

The manufacturing industry has been the focus of the sustainability challenges in the world. It is also among the biggest consumers of raw materials and energy and it also contributes a lot to the greenhouse gas emission, depletion of resources and generation of waste. The dominance of the world industrial growth is provided by the traditional production systems founded on the linear model of take, make and dispose during decades. Although this model helped in economic growth, it also posed a severe environmental pressure. These issues have been exacerbated by increasing global demands on products coupled with consumer demands of speed, quality and low cost (Zhang et al. 2024). Simultaneously, the climate change, the depletion of biodiversity, and the lack of resources are making industries rethink the organization of production and consumption. As a requirement rather than an option, these challenges have now come to be the key to the resilience of economies and societies (Nguyen et al. 2023).

The shortcoming of the linear model of production is becoming more evident. This system involves extraction of resources, processing them into goods, consumption, and they are discarded. These practices do not take into account that the availability of natural resources is limited and that wastes are of high ecological costs. The strain in the ecosystems is not sustainable with the fast population growth and economic growth (Shamaee et al. 2024). Landfills become full, toxic gases are spreading, and precious resources are wasted. The linear model is no longer able to guarantee long-lasting sustainability even though the model previously contributed to the short term growth. A new era of speed to regenerative systems with less waste and reuse of resources is urgently needed (Rashid and Malik 2023).

A structured alternative is provided by the CE. It encourages the use of processes in which the products are recycled into production; instead of disposing of the same as waste. It is based on the concepts of reuse, repair, remanufacturing, and recycling to create closed loops that close to natural ecosystem. In the case of manufacturing, it entails the redesigning of products, production processes and supply chain management. Through implementation of CE, industries are able to minimize the environmental footprint as well as create new value. To illustrate this, the recycling of metals, plastics, and electronic wastes not only conserves the raw materials but also energy than extracting the virgin resources (Burke et al. 2023).

In spite of the fact that CE offers the conceptual framework, technology allows large scale application. The main enablers of a circular manufacturing are artificial intelligence, digital twins, the Internet of Things (IoT), and blockchain. AI enhances efficiency by means of predictive maintenance and process optimization that lessens wastes and resource consumption. Digital twins form virtual models of machines and processes that enable industries to test the changes before actual implementation. IoT sensors can offer real-time information about energy consumption, emissions, and machine efficiency and can respond to the inefficiencies quickly. The blockchain will provide the supply chain with its transparency and accountability since it is able to trace all the material sources and sustainability across the borders (Rakshit et al. 2024).

The need to adopt these technologies demonstrates that sustainability is not only a moral responsibility, but a business benefit. Technology oriented sustainability will help companies to become more efficient, cost effective and to fulfil emerging regulatory and consumer demands. They also enhance resistance to shortages of resources and market changes. Technology in this way is not a burden but a force of competitiveness and innovation. The manufacturing sector has an opportunity to react to the global environmental issue by adopting the principles of the circular economy and integrating them with the introduction of innovations and smart technologies and achieve a sustainable, resilient future (Nyquist et al. 2025).

2 Smart Circular Economy: Concepts and Principles

2.1 Defining Sustainability and Circular Economy in Manufacturing

In manufacturing, sustainability is defined as the production of goods in such a manner that upholds the environmental wellbeing, social wellbeing and long term economic stability. It compels industries to reduce waste, decrease emissions and utilize energy and raw materials in a sensible manner. Environmentally friendly manufacturing is not a mere case of lessening the damage but also about establishing systems that can be sustained by the generations that will come. It usually implies short term changes in efficiency, but results in a more resilient scenario in the long term, where resources are used sparingly and production processes are made to be less risky to the environment (Khan et al. 2024). Circular economy (CE) offers an organised sustainability methodology. It pushes the line of traditional linear model of take, make and dispose where resources are extracted, consumed and discarded. Rather, CE encourages the constant movement of products and materials in industrial systems. The products are made in such a way that they are durable, repairable, upgradeable and recyclable. Waste is considered as a resource in this model. To manufacturers, implementation of CE results in decreased reliance on virgin resources, decreased cost of production, and increased resiliency to scarcity of resources (Dan et al. 2024).

2.2 Core Principles of the Circular Economy

Regeneration

Regeneration is concerned with repaying natural systems. It is concerned with the restoration of resources and minimizing harm to industrial activity. This is achievable through recycling of water, use of renewable energy or the choice of biodegradable material. Regeneration also promotes design of products which have long service life and are easily recycled. To give an example, rockers and other furniture made of sustainably sourced wood or naturally biodegradable materials helps with regeneration. With the concept, production stops being the process of exploiting resources and turns into one of renewing the resources (Edwards et al. 2024).

Efficiency

CE efficiency refers to the maximum utilization of the value of each unit of energy, water or raw material. It is not only related to money saving but also ecological effects. The improvement of efficiency can be achieved with the help of sophisticated equipment, energy control devices, and garbage recovery measures. As an example, sensor based real time data can indicate where energy or material is lost and take corrective action in a short time. Lean production also involves a cutting back on unnecessary processes and conservation of resources. Good practices boost competitiveness and reduce emissions and costs (Elroi et al. 2023).

Closed Loop Systems

Closed loop systems strive to maintain the materials in a continuous circulation as opposed to disposing of them. In these systems, the products are used to collect, repair or recycle such that the parts can be used back into the production. Product take backs in the electronics industry, such as recycling scrap metal in the automobile industry are examples. Closed loop systems may also be in the form of industrial symbiosis whereby the waste of one company is the resource of another. In one instance, steel manufacturing can use waste heat to drive facilities in the area. These systems lessen the raw material mining and restrict production of waste thus making industries more sustainable (Singh et al. 2025a).

2.3 Technology as an Enabler of Circular Economy

With the help of modern technologies, the principles of CE can be used more effectively. Some of the most significant tools in this transition are AI, IoT, digital twins, and blockchain. AI has the capacity to process vast data to enhance efficiency, forecast equipment breakdowns, and maximize resource utilization. This decreases wastage and prevents high expenses of downtime. IoT sensors enable machines, energy, and material flows to be monitored in real time and they offer industries with accurate information to make decisions.

Digital twins develop virtual models of production systems or machines, and by that, companies can experiment with new circular strategies, and this can be tested in a virtual environment before it is put into practice physically. The trust and transparency of blockchain is created through the process of recording material and product flows in secure systems that cannot be tampered. It allows supply chains to demonstrate material recycled, ethically sourced or sustainable production (Rakshit et al. 2024). These technologies have a cumulative effect on making CE become smart. They enable industries to gauge their progress, build up their effective strategies, and hold them accountable. With the technology combined with the principles of CE, manufacturing will be more resource-efficient, less pollutant and more transparent.

2.4 Alignment with Sustainable Development Goals (SDGs)

The smart circular economy, too, is in compliance with the Sustainable Development Goals (SDGs) of the United Nations. It is also very similar to SDG 12 (responsible consumption and production) as it keeps materials in use and minimizes waste. It promotes SDG 13 (climate action) by reducing emissions and increasing energy efficiency. CE also supports SDG 7 (affordable and clean energy), by supporting renewable energy and smarter energy systems. Moreover, CE opens up green workplaces, which promotes SDG 8 (decent work and economic growth) (Singh et al. 2024a). Technology reinforces this alignment in the sense that contribution is measurable. As an illustration, AI and IoT will be able to monitor energy consumption or emission reductions, whereas blockchain will be able to demonstrate adherence to sustainability criteria. This simplifies the process of industries reporting their successes and accountability of the stakeholders. By so doing, smart CE does not only benefit companies but also promotes global sustainability agendas (Ajakwe et al. 2025).

3 Key Enabling Technologies for Smart Manufacturing

3.1 AI, IoT, and Digital Twins for Optimization and Monitoring

AI allows industries to utilize data in a more efficient way. It is able to process massive amounts of data and identify trends that enhance decision making. In the manufacturing sector, AI is used as a predictive maintenance tool, allowing the detection of machines at risk of breaking down, which saves time, money and resources. The other way AI can improve quality control is by identifying flaws at an early stage of production, thus minimizing waste and guaranteeing a steady flow of output (Ucar et al. 2024). AI is used together with the Internet of Things (IoT) to deliver real time data. IoT is an application of sensors that are installed on machines to measure the temperature, energy consumption, and emissions. This knowledge assists managers to see where they are inefficient, and act accordingly. As an example, the alert system of IoT can inform managers in case energy use is higher than it should be which makes it possible to intervene quickly. In such a manner, IoT is a direct contributor to becoming green through increased efficiency and minimized waste (Turskis and Sniokiene 2024). Digital twins are AI-IoT systems that generate virtual machines, systems, or factories. They enable manufacturers to model the changes, determine the effects on energy or emissions, and create safer and more sustainable processes. AI, IoT, and digital twins are a combination to create a smart system that will save resources and reduce waste in the production process (Bhambri et al. 2024).

3.2 Blockchain for Supply Chain Transparency and Accountability

The supply chains of manufacturing are generally complex and worldwide. There are numerous suppliers of materials before they are converted to the final product. It is challenging to make sure that the resources are obtained in a responsible manner or that recycled materials are used at all. Blockchain aids in solving this issue. Blockchain is a virtual register which documents transactions in an impermanent and safe manner. Logging of all stages in the supply chain can be done, starting with the extraction of the raw materials, to delivery of the finished product. Record is unalterable therefore giving transparency and trust. To illustrate, a company that recycles metals can utilize blockchain in order to demonstrate what materials are original and their origin (Sunmola and Burgess 2023). It is a form of transparency that fosters confidence among the consumers, regulators and business partners. It minimizes the chances of fraud and unethical behaviour, including extraction of resources which are illegal or the use of labour force through exploitation. The companies also find it easier to meet the sustainability guidelines. Blockchain is thus enhancing accountability and the principles of the circular economy (Singh et al. 2024b).

3.3 Additive Manufacturing and Material Recovery for Circular Production

Additive manufacturing or 3D printing is a manufacturing method that involves production by adding layers. In contrast to the conventional manufacturing method when the material is cut or carved out, additive manufacturing utilizes the required amount. This saves on waste and makes it possible to use recycled materials or bio-based materials. The procedure also facilitates personalization and productivity. The equipment can be made to be less heavy, tougher and accessible to rehabilitate. By printing only the needed section it is possible to replace the broken parts and save time and cost. It also minimizes the necessity of the large inventories and reduces the transportation requirements that are a part of sustainability (Hegab et al. 2023). Material recovery is strictly connected with additive manufacturing. This entails the gathering and recycling of end of life materials of products. Metals, plastics and composites can be recovered using advanced recycling technologies. The recovered inputs can be re-used in new products. As an example, plastic waste can be managed into printing filament and this will be a closed loop system. Combined, additive manufacturing and material recovery eliminate waste, decrease the reliance on virgin raw materials, as well as facilitate innovative product design (Dumee 2022).

4 Global Best Practices in Tech-Driven Sustainability

4.1 Case Highlights from Europe

Europe is regarded as a world pioneer of sustainability through technology. The combination of powerful regulations and innovative technologies has been used in many European countries to facilitate circular practices. Industry 4.0 strategies have been embraced in Germany, with the implementation of AI, the IoT and digital twins in the production sector. The tools will be used to maximize energy consumption, control emissions, and aid predictive maintenance. German industries are also the best recycling systems with high metals and plastics recovery competition (Hagelüken and Goldmann 2022). The Nordic countries, in turn, are concerned with renewable energy and intelligent waste management (Sweden and Finland). Sweden is one example of creating a waste-to-energy program that heats whole cities and decreases landfills. Finland has put money in bio-based products and online platforms that facilitate the sharing and reuse in production. The European Union has a significant role as well since it establishes the policies regarding the circular economy which compels businesses to implement sustainable technologies (Nowak-Marchewka et al. 2025).

4.2 Case Highlights from Asia

Both manufacturing and sustainability practices have been growing at a high rate in Asia. Japan has a reputation of efficient usage of its resources and is recycling. Robotics and AI help Japanese manufacturers enhance precision and minimise wastage. Eco-design also is fostered in the country where products are designed to be durable and to easily recycle. South Korea has been concentrating on intelligent factories, and the wide usage of IoT sensors and digital twins is common in production monitoring. These factories minimize the consumption of energy and resource efficiency. As the largest manufacturer in the world, China has implemented national policies that promote a national policy of innovation on green. Chinese firms are putting money into renewable energy, electric vehicles, and blockchain based supply chains. The Chinese industrial park pilot projects demonstrate the ability of digital platforms to monitor the flows of resources and enhance the rates of recycling (Ullah et al. 2022).

4.3 Case Highlights from North America

North America has joined digital innovation and business based sustainability. The companies of the United States are applying AI and IoT to energy analytics that can be used to reduce emissions and cost. Aerospace and automotive uses of digital twins are used to create products that are lighter and more efficient. The U.S also leads in additive manufacturing where firms employ the 3D printing process in order to minimize waste of material and facilitate product personalization (Prashar et al. 2023). Canada has been concerned with clean integration of energy and recovery of resources. Industries in Canada apply blockchain in maximizing the visibility of supply chains particularly in the mining and forestry industries. This guarantees that the materials are sourced in a responsible manner and in a manner that is up to date with the standards of sustainability. In North America, it is common to find that private firms and universities collaborate to design technology solutions to facilitate the intent of circular economy (Farid 2025).

4.4 Examples from Developing Economies and Emerging Markets

Technology is also being embraced by developing economies to aid in the sustainability but mostly in smaller scale. Digital platforms are adopted to enhance the collection and recycling of waste in cities in India. Start-ups are developing mobile applications that would enable households to connect with recyclers, enhancing recovery of materials. The Indian manufacturers are also trying solar powered production systems to cut down the cost of energy and emission (Kamble et al. 2024). In Africa, Kenya and South African nations are the ones incorporating blockchain to enhance transparency in agriculture and manufacturing supply forms. Such systems are used to assure ethical production of products and minimize chances of fraud. Bio based materials and circular agricultural practices have been invested by Latin American countries, including Brazil. Affordable technologies used by small and medium enterprises (SMEs) in these regions to track the use of energy and water tend to include IoT devices of a low cost (Elkoraichi et al. 2025).

4.5 Comparative Insights and Lessons Learned

The worldwide experiences indicate that the sustainability frameworks informed by technology are localized. In Europe, government has been very supportive through stringent regulations that have seen massive use of advanced technologies. In Asia, the fast industrialization has promoted the investments in robotics, smart factories, and renewable systems. In North America, university industry partnerships and initiative by the private sector have already developed additive manufacturing breakthroughs and energy analytics. In developing economies, low cost solutions and local needs have resulted in innovative, low cost solutions, which nevertheless meet the targets of the circular economy (Dost et al. 2025). One of the lessons is that not all models are created equal as far as success is concerned. Every region has to make adaptation of technology to available resources, policies and the economic conditions. Nevertheless, there are certain common themes that are presented, namely, the significance of digital monitoring, supply chain transparency, and material recovery. International collaboration may also help in speeding up development as through cross border sharing of knowledge countries may learn to do things differently.

5 Benefits and Strategic Opportunities

5.1 Operational, Environmental, and Economic Advantages

Technology based sustainability has direct returns in the form of efficiency, environmental safety, and economic performance. The use of intelligent tools, including artificial intelligence, the Internet of Things, and digital twins, can enable industries to streamline production. Predictive maintenance avoids machine failures, therefore, minimizing repair prices and unnecessary waste of raw materials (Yazdi 2024). IoT sensors provide real-time data regarding the energy and resources consumption, which allows managers to detect wastefulness and minimize unproductive usage. With electricity and water usage smart meters, the factory may report huge utility cost savings. Digital twins also enhance decision making as they enable managers to experiment with strategies virtually as opposed to testing them in a trial and error costly manner.

Technologies that help in sustainability also generate considerable environmental benefits. The principles of the circular economy contribute to the preservation of the forests, water, and minerals due to lessening the use of virgin raw materials. Closed-loop systems promise to be able to reuse waste, and additive manufacturing reduces the amount of materials used by creating items in layers. The recycling and material recovery will ensure that plastics, metallic and electronic waste does not enter landfills. Smart monitoring is used to improve energy efficiency, and the integration of renewable energy helps decrease the need to use fossil-based fuels. Predictive analytics enable companies to align production to the energy demand, reducing unwarranted emissions (Joshi et al. 2024). These actions are not only aimed at the compliance with climate policies and international agreements like the Paris Agreement, but also enhance reputation. Companies with lower carbon footprint are credible to their customers, investors and governments.

5.2 Innovation, Social Impact, and Community Value

Sustainability that is technologically based promotes innovation and new business models. One such example is the product as a service model where companies lease products instead of selling products. The system keeps the resources in circulation and establishes long term relationships with the customers. An example of this is in the case of electronics companies, which can lease equipment and then reuse it to do remanufacturing. Additive manufacturing supports mass customization whereby they offer special solutions to each customer. Digital platforms also facilitate cooperation as it allows manufacturers to share excess materials or energy. With these innovations, new markets of green products are open, and the reputation of the brand is improved (Kandpal et al. 2024).

Social benefits are also of great significance. The blockchain enhances transparency in the supply chain and makes materials sourcing to be ethical. This minimizes chances of child labour, poor working conditions or unlawful extraction of resources. New employment opportunities in recycling and remanufacturing and renewable energy industries also benefit the communities. New skills have to be developed in these industries, thereby promoting investment in education and training. Quality of life also is enhanced by sustainability practices. Cleaner technologies imply fewer emissions, better waste collection and healthier living conditions of communities surrounding industrial centres. On the international level, emerging economies can jump to circular practices with the use of cost effective technologies. This will avoid the recurrence of intensive industrialization that is resource based and encourage inclusive development which will minimize disparities among the regions (Kaftan et al. 2023).

6 Challenges and Policy Perspectives

6.1 Barriers: Financial, Technological, and Organizational

The initial obstacle is the financial one. Most of the technologies are applied to smart manufacturing, including artificial intelligence, digital twins, or blockchain, are costly to establish. SMEs do not always have the resources to buy such tools. Moreover, ROI is not always certain and is more so in developing economies where advanced systems may not be supported through infrastructure. In fact, many industries will not achieve the scale of use of these technologies without having the affordable financing models. Second is technological barrier. The smart manufacturing relies on the highly developed digital infrastructure, good data, and qualified personnel. In other areas, there remains poor digital infrastructure, and internet connectivity is low. The process of deploying various technologies into the existing systems may be complicated even in developed countries. A lot of factories have old technology that cannot be connected to the IoT or AI applications. This poses technical difficulties to adoption (Dhanaraj et al., 2024). The third impediment is organizational. Linear to Circular model means significant cultural and structural changes in companies. There are chances that employees will be resistant to change unless they are trained to utilize new tools. The management will be reluctant to invest in new uncertain innovations. In other organizations sustainability is perceived as an expense and not as an opportunity. At worst, in the absence of a strong leadership and vision, organizational resistance may block progress (Mauss et al. 2023).

6.2 Risks: Data Privacy, Interoperability, and Workforce Readiness

Smart manufacturing is very dependent on computerized information. This poses new privacy and security threats. Sensitive information is collected and stored by IoT sensors, AI systems, and blockchain platforms all. Unless such information is secured, it will be abused or stolen. Cyber shock on manufacturing systems may interfere with the activities, cause financial, and destroy confidence. Privacy of data and cybersecurity is thus one of the key considerations (Dodiya et al. 2024). The other is the risk of interoperability. Various providers tend to develop various technologies. In case these systems fail to collaborate, the advantages of smart manufacturing are diminished. As an example, IoT systems of a certain company may not interoperate with blockchain systems of another. Absence of common standards complicates and makes integration quite expensive. Multivendor compatibility has to be enhanced with common protocols and industry wide interoperability (Zhuk 2025). It is also a major challenge in workforce preparation. Smarter manufacturing demands new data analysis skills, knowledge about managing digital systems and knowledge about advanced engineering. These skills are not yet developed in many workers and such training programs are not always present. This may cause loss of jobs on the usual positions and lack of skilful labour on the new ones. The fair and smooth transition of the workers should be ensured that the workers are reskilled and supported (Gabsi 2024).

6.3 Policy Frameworks and Governance for Global Sustainability Transitions

In order to get past these obstacles and risks then effective policy frameworks and structures of governance are required. Governments can be instrumental when it comes to funding innovation. These may take the form of subsidies, tax breaks or loans with low interest rates to firms investing in sustainable technologies. SMEs can be assisted by special programs, which in most cases are victims to the most financial problems. Research and development should also be encouraged by the policy structures. Advanced tools can be adopted faster through investments in innovation centres and through public private partnerships. Simultaneously, governments are able to establish infrastructural facilities like high speed internet and renewable power systems that facilitate digital and sustainable production (Tomai et al. 2024). The other significant aspect of governance is regulation. There must be clear data privacy, cybersecurity and interoperability rules. This is particularly significant in the international cooperation since supply chains are international. The common standards facilitate the cross-border interoperability of technologies. The forms of governance should thus be in a way that leads to co-operation among nations, businesses and organizations. Lastly, it is necessary to have workforce readiness policies. Governments and businesses ought to invest in digital and sustainable job preparation programs like education and training of workers. This encompasses technical expertise in AI, IoT, and additive manufacturing, and the familiarity with the practice of the circular economy. The workers who can lose traditional jobs during the transition should also be supported under social policies (Singh et al. 2025b).

7 Implementation of Tech-Driven Sustainability in Manufacturing

Technology driven sustainability in the manufacturing process should have a set of steps to be implemented. The areas of focus are the following:

- **Strategic Planning**: The first step is to have a clear and realistic plan of implementation. Businesses have to set sustainability objectives like the elimination of emissions, decrease in energy usage, or the improvement of recycled materials consumption. Such objectives are expected to be in line with business objectives to make sure that sustainability is part of long term strategies. Making a road map with timeframes, tasks and quantifiable results makes it structured and accountable.
- **Investment in Technology and Infrastructure**: The second one is investing in the empowering technologies that include artificial intelligence, IoT, digital twins, blockchain, and additive manufacturing. These technologies need funding and infrastructure facilities. Their integration into large firms is easier, whereas the small and medium enterprises are likely to be challenged with financial issues. This can be facilitated by state stimulus, subsidies and common centres of technology to lower the barriers and increase the adoption.
- **Workforce Training and Readiness**: Skilled employees are the key to intelligent and green production. It has to be realized through training to develop digital expertise and awareness of the practices of a circular economy. Capacity building may be assisted with the help of universities and technical institutes. Inclusion of employees in decision making also contributes to the minimization of resistance and a culture of innovativeness is established.

- **Supply Chain Collaboration**: Implementation is not limited to the individual companies and involves cooperation with supply chains. Materials can be monitored by using blockchain and digital platforms, making them transparent and traceable. There can be industrial collaborations that facilitate closed loop relationships whereby waste produced in one process finds its way as an input to another process enhancing efficiency and strength.
- **Monitoring and Continuous Improvement**: Implementation is a continuous process which needs monitoring and adaptation. The IoT and AI systems show real-time energy consumption and emissions and waste minimization data. Periodic reviews assist companies to change strategies and enhance performance with time.

8 Conclusion and Future Directions

Technology and circular economy are changing the manufacturing industry to a more sustainable and resilient industry. International best practice indicates how sectors can minimize waste, maximize resources, and generate value with the help of digital innovation. AI, IoT, digital twins, blockchain, additive manufacturing are some of the technologies that have shown a definite effect on efficiency, transparency, and circular production. Such practices can reduce costs and emissions as well as promote new business and social values. In the future, future trends will determine the future of sustainable manufacturing. The production will improve efficiency and lessen errors through automation. Greens energy, e.g., solar, wind and hydrogen will be more involved in powering factories. Closed loop systems will be formed with help of smart materials that are designed to be durable and recyclable. All these trends will render circular manufacturing more feasible and scalable. There are strategic suggestions that are aimed at cooperation between industries, governments, and societies. Businesses need to spend on technology and employee development. Policy-makers are supposed to offer enabling structures, financial rewards, and directives. A common mission statement of a strong and sustainable manufacturing will make it competitive and save the earth. The future of the manufacturing industry lies in the inclusion of smart technology in the practice of sustainability in order to establish long-term wealth.

9 Limitations

Regardless of the advancement, the implementation of tech-driven sustainability has a number of constraints to the broader application. Financial burdens are still an obstacle especially to small and medium enterprises that might not be able to afford sophisticated technologies. The inability to adopt is also delayed by technological issues like inability to have interoperability among the systems and ineffective infrastructure. There are additional issues of data privacy and security threats in the digital manufacturing environment. The other constraint is workforce readiness because most industries lack skilled personnel who can work to operate and manage smart systems. Lastly, there is also unequal support of policies to different countries, which implies that progression is quicker in certain areas and slower in others. These constraints also demonstrate that the transition towards sustainable manufacturing should be effective and fair and coordinated, as well as persistent and innovative.

References

Ajakwe IU, Ajakwe SO, Lee JM, Kim DS (2025) Internet-of-things-blockchain integration in environmental pollution monitoring data management: trends and techniques. Int J Environ Sci Technol 1–20. https://doi.org/10.1007/s13762-025-06615-x

Bhambri P, Rani S, Khang A (2024) AI-driven digital twin and resource optimization in industry 4.0 ecosystem. In: Intelligent techniques for predictive data analytics, pp 47–69. https://doi.org/10.1002/9781394227990.ch3

Burke H, Zhang A, Wang JX (2023) Integrating product design and supply chain management for a circular economy. Prod Plan Control 34(11):1097–1113. https://doi.org/10.1080/09537287.2021.1983063

Dan A, Raj A, Kumar P, Singh RK, Ahmed F (2024) Risk analysis of adopting the circular economy practices: a perspective of resource-dependent theory. Supply Chain Forum: Int J 25(4):504–521. Taylor & Francis. https://doi.org/10.1080/16258312.2023.2296387

Dhanaraj RK, Balusamy B, Samuel P, Bashir AK, Kadry S (eds) (2024) Digital twins in industrial production and smart manufacturing: an understanding of principles, enhancers, and obstacles. Wiley

Dodiya KR, Jha M, Jha S (2024) Fortifying the digital forge: unleashing cybersecurity in the interconnected world of digital manufacturing. In: Emerging technologies in digital manufacturing and smart factories. IGI Global Scientific Publishing, pp 230–256. https://doi.org/10.4018/979-8-3693-0920-9.ch014

Dost M, Lee M, Umrani WA (2025) Implementing circular economy through bricolage, frugal innovation and technological turbulence. Sustain Dev 33(2):1610–1622. https://doi.org/10.1002/sd.3200

Dumée LF (2022) Circular materials and circular design—review on challenges towards sustainable manufacturing and recycling. Circ Econ Sustain 2(1):9–23. https://doi.org/10.1007/s43615-021-00085-2

Edwards MG, Lindberg A, Larsson M, Angel J (2024) Regenerative business voices: values-based entrepreneurship for sustainable enterprises. Routledge. https://doi.org/10.4324/9781003330660

Elkoraichi Y, Elfezazi S, Belhadi A (2025) Analysis of barriers to blockchain technology adoption in the African agri-food supply chain. Discov Sustain 6(1):289. https://doi.org/10.1007/s43621-025-01125-9

Elroi H, Zbigniew G, Agnieszka WC, Piotr S (2023) Enhancing waste resource efficiency: circular economy for sustainability and energy conversion. Front Environ Sci 11:1303792. https://doi.org/10.3389/fenvs.2023.1303792

Farid O (2025) Climate-first innovation ecosystems: entrepreneurship and sustainability in the transition to a regenerative economy. https://doi.org/10.33774/coe-2025-7fk67

Gabsi AEH (2024) Integrating artificial intelligence in industry 4.0: insights, challenges, and future prospects–a literature review. Ann Oper Res 1–28. https://doi.org/10.1007/s10479-024-06012-6

Hagelüken C, Goldmann D (2022) Recycling and circular economy—towards a closed loop for metals in emerging clean technologies. Miner Econ 35(3):539–562. https://doi.org/10.1007/s13563-022-00319-1

Hegab H, Khanna N, Monib N, Salem A (2023) Design for sustainable additive manufacturing: a review. Sustain Mater Technol 35:e00576. https://doi.org/10.1016/j.susmat.2023.e00576

Joshi A, Singh R, Kumar V, Rani S (2024) Net zero strategy: exploring opportunities, confronting challenges for developing countries. In: Transition towards a sustainable future: net zero policies and environmental sustainability. Singapore: Springer Nature Singapore, pp 233–252. https://doi.org/10.1007/978-981-97-5756-5_11

Kaftan V, Kandalov W, Molodtsov I, Sherstobitova A, Strielkowski W (2023) Socio-economic stability and sustainable development in the post-COVID era: lessons for the business and economic leaders. Sustainability 15(4):2876. https://doi.org/10.3390/su15042876

Kamble AD, Das S, Vijaya Das B, Bordoloi U, Hazarika P, Kalita P (2024) Role of solar energy in the development of the Indian economy. In: Challenges and opportunities of distributed renewable power. Singapore: Springer Nature Singapore, pp 489–535. https://doi.org/10.1007/978-981-97-1406-3_18

Kandpal V, Jaswal A, Santibanez Gonzalez ED, Agarwal N (2024) Circular economy principles: shifting towards sustainable prosperity. In: Sustainable energy transition: circular economy and sustainable financing for environmental, social and governance (ESG) practices, pp 125–165. Cham: Springer Nature Switzerland. https://doi.org/10.1007/978-3-031-52943-6_4

Khan S, Singh R, Alnahas J, Abbate S, Centobelli P (2024) Navigating the smart circular economy: a framework for manufacturing firms. J Clean Prod 480:144007. https://doi.org/10.1016/j.jclepro.2024.144007

Mauss NA, Bühner D, Fottner J (2023) Applicability and limitations of change management for circular economy in manufacturing companies. Procedia Comput Sci 217:998–1007. https://doi.org/10.1016/j.procs.2022.12.298

Nguyen TT, Grote U, Neubacher F, Do MH, Paudel GP (2023) Security risks from climate change and environmental degradation: implications for sustainable land use transformation in the Global South. Curr Opin Environ Sustain 63:101322. https://doi.org/10.1016/j.cosust.2023.101322

Nowak-Marchewka K, Osmólska E, Stoma M (2025) Progress and challenges of circular economy in selected EU countries. Sustainability 17(1):320. https://doi.org/10.3390/su17010320

Nyquist AM, Farshid M, Brown T (2025) Employing digital twin technology in the pursuit to avert sustainable marketing myopia. J Res Mark Entrep 27(2):277–293. https://doi.org/10.1108/JRME-01-2024-0007

Prashar G, Vasudev H, Bhuddhi D (2023) Additive manufacturing: expanding 3D printing horizon in industry 4.0. Int J Interact Des Manuf (IJIDeM) 17(5):2221–2235. https://doi.org/10.1007/s12008-022-00956-4

Rakshit P, Saha N, Nandi S, Gupta P (2024) Artificial intelligence in digital twins for sustainable future. In: Transforming industry using digital twin technology. Cham: Springer Nature Switzerland, pp. 19–44. https://doi.org/10.1007/978-3-031-58523-4_2

Rashid S, Malik SH (2023) Transition from a linear to a circular economy. Renewable energy in circular economy. Springer International Publishing, Cham, pp 1–20

Shamaee SH, Yousefi H, Zahedi R (2024) Assessing urban development indicators for environmental sustainability. Discov Sustain 5(1):341. https://doi.org/10.1007/s43621-024-00563-1

Singh R, Gupta A, Kandpal V (2025a) ESG excellence in the digital world: Circular economy e-business solutions. In: Fintech for ESG and the circular economy, pp 193–204. https://doi.org/10.1002/9781394234035.ch10

Singh R, Gupta A, Bajpai A, Kandpal V (2024a) Exploring challenges of circular economy initiatives for smart cities. In: Smart cities and circular economy: the future of sustainable urban development. Emerald Publishing Limited, pp 59–69. https://doi.org/10.1108/978-1-83797-957-820241005

Singh R, Joshi A, Dissanayake H, Iddagoda A, Khan S, Félix MJ, Santos G (2025b) Integrating industry 4.0, circular economy, and green HRM: a framework for sustainable transformation. Sustainability 17(7):3082. https://doi.org/10.3390/su17073082

Singh R, Ravesangar K, Narayanan S, Singh KKH (2024b) HRM practices in creating net zero: supporting organizational sustainability. In: Net zero economy, corporate social responsibility and sustainable value creation: exploring strategies, drivers, and challenges. Cham: Springer Nature Switzerland, pp 1–18. https://doi.org/10.1007/978-3-031-55779-8_1

Sunmola F, Burgess P (2023) Transparency by design for blockchain-based supply chains. Procedia Comput Sci 217:1256–1265. https://doi.org/10.1016/j.procs.2022.12.324

Tomai M, Ramani SV, Papachristos G (2024) How can we design policy better? Frameworks and approaches for sustainability transitions. Sustainability 16(2):690. https://doi.org/10.3390/su16020690

Turskis Z, Šniokienė V (2024) IoT-driven transformation of circular economy efficiency: an overview. Math Comput Appl 29(4):49. https://doi.org/10.3390/mca29040049

Ucar A, Karakose M, Kırımça N (2024) Artificial intelligence for predictive maintenance applications: key components, trustworthiness, and future trends. Appl Sci 14(2):898. https://doi.org/10.3390/app14020898

Ullah S, Khan FU, Ahmad N (2022) Promoting sustainability through green innovation adoption: a case of manufacturing industry. Environ Sci Pollut Res 29(14):21119–21139. https://doi.org/10.1007/s11356-021-17322-8

Yazdi M (2024) Maintenance strategies and optimization techniques. In: Advances in computational mathematics for industrial system reliability and maintainability. Cham: Springer Nature Switzerland, pp 43–58. https://doi.org/10.1007/978-3-031-53514-7_3

Zhang Q, Liu G, Liu Y (2024) Sustainable practices in steel manufacturing: Substitution elasticities between waste, raw materials, and energy for greenhouse gas emission reduction. J Clean Prod 477:143845. https://doi.org/10.1016/j.jclepro.2024.143845

Zhuk A (2025) Beyond the blockchain hype: addressing legal and regulatory challenges. SN Soc Sci 5(2):11. https://doi.org/10.1007/s43545-024-01044-y

Smart Cities in Action: Integrating IoT, AI, and Cloud Infrastructure for Sustainable Urban Living

Hiranya Dissanayake, Otilia Manta, Deshika Nainanayake, and Lalit Mohan Tewari

Abstract The acceleration of urbanization presents complex challenges for sustainability, governance, and citizen well-being. Smart cities emerge as adaptive solutions, grounded in the convergence of Internet of Things (IoT), Artificial Intelligence (AI), and cloud computing. IoT sensors embedded across infrastructures generate vast, real-time data streams, while AI algorithms provide predictive insights for mobility, energy, and safety management. Cloud infrastructures enable interoperability, secure storage, and scalable integration across heterogeneous systems. Theoretically, this integration aligns with socio-technical systems theory, highlighting the interdependence of technological and social layers; resilience theory, emphasizing adaptive capacities for coping with urban uncertainties; and sustainability transitions theory, framing how innovations drive systemic change in urban environments. These perspectives reveal how smart cities can function not merely as technological solutions but as socio-technical ecosystems that enhance sustainability, inclusivity, and resilience. However, ethical concerns regarding privacy, security, and governance necessitate robust regulatory frameworks. This study contributes to understanding the transformative potential and theoretical underpinnings of IoT–AI–cloud synergies for future urban living.

H. Dissanayake (✉)
Department of Accountancy, Faculty of Business Studies and Finance, Wayamba University of Sri Lanka, Kuliyapitiya, Sri Lanka
e-mail: hiranya@wyb.ac.lk

O. Manta
School of Advanced Studies of the Romanian Academy (SCOSAAR), Bucharest, Romania
e-mail: otilia.manta@rgic.ro

D. Nainanayake
School of Business, Western Sydney University, Sydney, Australia
e-mail: deshika.thilini@gmail.com

L. M. Tewari
Department of Botany, DSB Campus, Kumaun University, Nainital, Uttarakhand, India
e-mail: l_tewari@rediffmail.com

R. Singh et al. (eds.), *Data-Driven Sustainability: Harnessing Technology for A Greener Future and Environmental Resilience*, World Sustainability Series,
https://doi.org/10.1007/978-3-032-13412-7_2

Keywords Smart city · Internet of Things (IoT) · Artificial Intelligence (AI) · Cloud computing · Sustainability · Cyber-physical systems · Urban governance · Predictive maintenance · Data-driven decision-making

1 Introduction

Urbanization is changing the spatial, social and economic demarcation of the twenty-first century. As the estimates by the United Nations imply the bulk of the population in the world today resides in urban settlements and urban concentration will continue to grow until the middle of the century (United Nations 2023). This concentration brings with it both solid innovation, cultural exchange, and economic productivity prospects, and, at the same time, it generates the environmental strains, infrastructure strains, social disparity, and vulnerability to climatic and health shocks. The extent and rate of urban change demands a different order of government, and of instruments of action, which can manage the movements of persons and of commodities, energy and communications, on a far more responsive and prospective scale than most conventional bureaucracies can usually admit of.

A solution to these challenges of governance has become a very popular solution by the smart city (He et al. 2022). Layered Pragmatically, a contemporary smart city refers to a type of system that consists of three layers of technology that interact to systematically form an urban nervous system (Kashef et al. 2021; Ramaprasad et al. 2017; Uribe-Pérez and Pous 2017). Internet of Things (IoT) offers omnipresent sensing of streets, buildings, vehicles and utilities to generate dense streams of space and time-based data (fine-grained) (Dou et al. 2023). Cloud computing is provided with scalable storage, interoperability (cross domain integration) which allows the heterogeneous datasets to be amalgamated, maintained and disseminated amongst the different stakeholders (Rozony et al. 2024). Artificial intelligence provides the pattern recognition of analytic capability, predictive modelling and decision support which would enable the conversion of data into operational decisions highly proximate to real time (Dissanayake et al. 2024). With these layers, coupled together with institutional capacity and ethical government, pretentiousness-restraint uses are allowed to predict and stabilize energy provision and aimed at maintenance intercession before the sequential breakdowns.

But, the ability of technology (IoT), and artificial intelligence (AI) of cloud systems do not offer the socially desirable outcomes. Introduction of digital systems into the city is a socio-technical change: seemingly technical choices concerning what to measure, where to place sensors, how to state the goals of optimization and who is in control of information stores introduce values and have distributive effects. Algorithms are not neutral instruments, and they represent prioritization and trade-offs, and their operation can render some neighborhoods, demographics or daily practices very visible or very invisible to the agencies of the state. In the case of the lack of clear governance, researchers caution that algorithmic decision-making will break down into new forms of surveillance and discrimination as opposed to promoting democratic responsibility (Hollands 2008; Kitchin 2014). In short, the promise of technical efficiency will be offset by the problems of equity, privacy and civic legitimacy.

With the implementation under appropriate managerial and regulatory schemes, the paper will come up with a more integrative socio-technical perspective to explain how the IoT-AI-cloud convergence can help in achieving city sustainability, resilience and justice. Two urban spheres in which digital convergence is already transforming flows are mobility and logistics, the empirical focus of the study. Mobility systems organize access to jobs, education and services and contribute to significant portion of greenhouse gas emissions in the cities. Mobility innovations that are driven by AI, such as connected and autonomous vehicles, dynamic signal control and Mobility-as-a-Service platforms, can lessen congestion and enhance safety but will have little value to the public unless they are regulated, interoperable, and designed with inclusiveness that avoids uneven distributions of services (Nikitas et al. 2020). The final testbed is urban logistics—the last-mile movement of products which supports urban economies, where IoT telemetry, cloud orchestration, and autonomous delivery vehicles can be used to implement strategies of vehicle-kilometers and emissions reduction, though this must be accompanied by standardized protocols, cybersecurity, and governance that introduce social goals into optimization logic (Mohsen 2024).

In addition to these interventions on the domain level, smart city transitions need to be placed into wider theoretical contexts. Socio-technical systems (STS) theory assists in sketching the co-evolution of technologies and institutions and determining what is possible and acceptable; resilience thinking underlines what cities require to endure any shock and ensure the continuation of fundamental functions in times of crisis; and sustainability transitions scholarship revolves around the ability of niche experiments to scale to a larger system through policy mixes, market creation and cultural transformations in times of crisis. By combining these viewpoints, it is possible to understand that integrating micro-level design-based decisions (where sensors are placed, algorithmic goals) with meso-level management decisions (procurement, KPIs, public- Private coordination) and macro-level transition processes (standards, incentives and supply-chain governance).

This expressly integrative focus, and foreshadowing distributive results, is the novelty of the paper. The current literature is inclined to address technology, governance or social impacts separately; in comparison, the given study interweaves them and makes questions of visibility, access and fairness its core. It addresses three gaps: lack of analytic frameworks that can both explain socio-technical design, resilience needs and transition conditions in urban digitalization; scanty empirical evidence on how AI pilots in mobility and logistics can yield different outcomes based on socio-economic groups; and a lack of actionable advice on how municipal managers need to align technical procurement with social performance measures.

The research follows the following manner. The first is that it expounds the technological architecture and its political economy. Second, it discusses empirical examples of both advantages and pitfalls of mobility and logistics. Third, it shows an interdependent heuristic, socio-technical fit, resilience profile and transition potential and a conceptual figure, which illustrates the way IoT, AI and cloud interact within normative and institutional limits. Finally, it offers a research agenda and managerial recommendations. The primary contribution is both analytical and prescriptive: the paper's synthesis, along with the examples of real-world situations and managerial guidance, gives city leaders guidance on how digitalization should proceed to produce effective, resilient, inclusive, and ecologically friendly outcomes. The analysis is written with policy relevance in mind, allowing towns to use it when making choices.

1.1 Smart Cities and the Promise of Technological Convergence

The smart city narrative has evolved substantially since its origins in late twentieth-century aspirations for improved urban administration through information and communications technologies (Batty 2018). Early conceptions centered on centralized, proprietary "city operating systems" designed to orchestrate municipal services. Those early visions often presumed a technologically deterministic relationship in which innovations would naturally resolve urban problems. Subsequent critiques exposed the political and normative assumptions embedded in such an approach and stressed that technology alone is insufficient (Hollands 2008; Kitchin 2014). The contemporary discourse on smart urbanism therefore emphasizes integration: the alignment of sensors, platforms, analytic capabilities, institutions, and civic practices that together generate adaptive capacities.

The three-layered technical architecture at the core of modern smart cities—IoT, cloud computing and AI—each contributes distinct capacities and limitations. IoT provides continuous environmental and operational data; cloud systems offer capacity for storage, integration, and cross-domain interoperability; and AI converts raw data into predictive and prescriptive outputs. When these layers are designed to work together, urban systems can shift from rule-bound, reactive administration toward anticipatory, feedback-driven governance. Examples include dynamic traffic-signal coordination that responds to live flows, energy systems that balance distributed generation and demand through predictive control, and waste-collection routes that evolve with occupancy patterns and sensor inputs.

But convergence in technology creates new distributional and political issues. Data never remains neutral, as it is premised on the agendas of the instrument developers, resource overseers and business or government financiers. Dependencies that create the asymmetries of control and access of the past are reintroduced by the centralization of cloud and analytical capabilities to a few large technology companies. The term purposeful inclusion infrastructures have been proposed by the researchers as ensuring connectivity, digital literacy and redistributive forces in such a way that the gains of smart cities do not accrue to those already empowered (Mishra 2025). In addition, there is a set of convergence opportunities between digitalization and the notion of the circular economy: material movement enabled by IoTs, analytics of lifecycle made with the help of AI, and the coordination enabled by the cloud could help transform the linear city metabolism into the circular one under the condition of governance and management system changes (Khan et al. 2024). Nevertheless, in order to move pilot projects to systemic circularity, policy incentives, standards and institutional capacity are required.

The evidence of both geographies varies. Such a move as the creation of so-called highly instrumented cities such as Singapore is a sign that coordinated platforms may coordinate traffic, energy and resource movement, but it is also catalyzing problems of surveillance and data sovereignty that individuals are talking about. An alternative approach to data governance co-design by citizens can be found on the open-source civic platforms in Barcelona. African cities have explored low-cost sensors, mobile networks and community-driven designs to handle energy monitoring and transport inefficiencies, and IoT and AI may be tailored to local capabilities in low-income settings in creative, low-cost adaptations (Cinderby et al. 2024; Moumen 2024). These experiences highlight the main argument: technical capacities provide socially positive effects in case they are supported by inclusive institutions and reflexive governance.

1.2 Reimagining Urban Futures: Sustainability and Community

Sustainability offers an epistemic and moral framework regarding the rethink of urban futures. Urban centers and policies Urban policies and technologies as well as structures of governance are the center of the world's environmental processes and will be critical in ensuring that cities contribute either to the acceleration or reducing stresses on the planet. The sustainable city does not see development as an unchecked growth but a more comprehensive capacity to not only sustain the life of human beings and the environment but also within a sustainable biosphere. Smart technologies may be added to that ability by making that process of utilizing the resources easier through data-based interventions and developing new forms of communication. Nevertheless, the shift in power balance, the re-evaluation of expertise, and ethical limitations that govern extractive processes must be brought in the technological twist as well.

Sustainable urbanism is based on the ethics of the community. The most appropriate type of technology is one that has enhanced group thinking and democracy. Participatory platforms, neighborhood digital twins, open-data portals may make the situation more transparent, in which new platforms of co-production are generated but they will only add to the already existing democracy when institutional reforms are provided which would allow citizens to voice and power. Lacking the redistribution of the decision-making powers, the digital tools are likely to be the tools of managerialism, rather than the inclusion tool.

Sustainability also concerns the transition to local optimisation to address the global supply chain and global life-cycle externalities. Digitalization must consider the effect it has on the environment since digitalization depends on cloud infrastructures and AI, which needs data centers, resource elicitation, and computation-intensive techniques. It involves green cloud architecture, renewable energy-powered data centers, and electronic waste cycle to complement the effectiveness on the urban scale (Khan et al. 2024; Mishra 2025). These managers and policymakers need to adopt a global-system thinking which needs to be linked between local smart interventions and supply-chain custodianship and supply-chain circular-economy incentives. Finally, the non-material aspects of sustainability were highlighted during the pandemic years. Neighborhood solidarity, social cohesion, and mental health are important elements of strong urban life. Digital technology may make social communication and awareness more accessible, and it will not make the community out of nothing. The technological deployments must then be intended to add to the trust, reciprocity and civic capacity but not to replace them.

Smart Mobility and Artificial Intelligence Transport

Mobility determines the physical form of urban space, economic chance, and everyday living. Formal transport planning based on fixed forecasting and capacity expansion has not managed to deal with the rise of congestion, air pollution and unfair treatment. Intelligent mobility is based on an alternative approach and combines sensor networks, predictive analytics, and real-time coordination across modes. Examples of such technologies include connected and autonomous vehicles (CAVs), Mobility-as-a-Service platforms, and aerial urban systems that can be combined to design efficiency and responsiveness in mobility systems (Nikitas et al. 2020).

There are multiple physical contributions of AI to mobility systems (Wolniak and Stecuła 2024). Machine learning algorithms can be used to optimize traffic-signal timing and minimize idling; predictive maintenance can be used to reduce downtime of public fleets; and shared-mobility algorithms can be used to reduce empty vehicle miles through dynamic pooling. MaaS integration to streamline ticketing, multimodal routing, and autonomous shuttles and advanced coordination have also been tested in cities like Helsinki and Singapore, respectively. However, the socio-technical conditions leading to successful intelligent mobility are quite massive. The public needs to trust automation and data sharing, and the regulatory environment and clarity and engagement by the populace. The system of sensors and vehicle telemetry produces very sensitive information: in the absence of a well-developed policy, such information sets can be sold or tracked.

Mobility transitions are pressing issues of equity. Automation will threaten the employment of drivers and logistics personnel, and the algorithmic routes may concentrate the benefits in lucrative urban routes and townships and leave the marginalized or low-income areas underserved. The potential of the environment is in place but conditional: AI has the potential to decrease total vehicle kilometers and promote modal shifts towards low carbon could a set of optimization objectives explicitly combine carbon, equity and accessibility goals. Emerging economies and low-income settings have also shown creative ways of smart mobility through the low-cost sensors, mobile payment infrastructure and community-based governance, implying that clever mobility may be inclusive when it is locally based and participatory.

1.3 AI-Driven Urban Logistics: Integrating Autonomous Vehicles and IoT for Sustainable Delivery

Logistics is the circulatory system of the city. The growth of e-commerce, densification, and on-demand expectations have exerted significant pressure on last-mile delivery, remaking small streets into loading bays and increasing emissions. An AI-driven logistics architecture that combines IoT telemetry, cloud coordination, and autonomous or semi-autonomous delivery vehicles can reduce redundant travel, improve vehicle utilization, and enable consolidation strategies that materially reduce urban congestion and emissions (Mohsen 2024).

In this architecture, IoT sensors on vehicles and in urban infrastructure feed positional and environmental data into cloud platforms. AI then analyzes routes, forecasts demand and dynamically allocates fleet resources. Practical pilots indicate substantial potential: consolidation centers, combined deliveries across carriers, and electric micro-vehicles for last-mile distribution can reduce vehicle kilometers travel and energy use while maintaining service levels. However, realizing these benefits at scale requires standardized data protocols, cybersecurity measures and governance models that prevent monopolistic control over data and infrastructure.

The social implications of logistics transformations are profound. If optimization criteria privilege efficiency alone, algorithms may route services toward affluent, high-demand zones and away from marginalized neighborhoods. Similarly, automation without a policy for retraining and social protection risks displacing workers. The governance challenge is therefore to embed social and ecological objectives into the very logic of optimization. Regulatory frameworks, public procurement rules, and performance metrics must align incentives such that logistical efficiency advances equity and climate goals.

Another urgency is to develop logistics systems that directly incorporate social and environmental externalities instead of viewing them as the side effects of making efficiency. This, in practice, implies that cities need to mix infrastructural interventions (urban consolidation centres, special micro-distribution points, and low-emission delivery zones) and regulatory and market instruments, which would influence algorithmic objectives toward public objectives. Multi-carrier consolidation may be facilitated by data governance arrangements including shared data trusts or mandatory open APIs, and the rules applied in public procurement can compel carriers contracted to be socially and environmentally scored, which will incentivise operations towards electrification and equal coverage of service (Khan et al. 2024; Singh Gupta and Kandpal 2024a, b, c). The change of workforce should be a strand of policy on par: reasons to automate conditionally should be coupled with retraining programs, wage guarantees, and access to higher value jobs in urban logistics systems. Lastly, accountability and monitoring rely on clear indicators: vehicle-kilometres travelled, emissions per parcel, percentage coverage of underserved neighbourhoods, displacement and re-employment of workforce than must be published publicly and utilised to re-calibrate optimisation goals in deployed systems (Mohsen 2024).

Socio-Technical, Resilience and Sustainability Transition Perspectives

Understanding the interplay of IoT, AI and cloud computing in urban systems requires the integration of three theoretical lenses: socio-technical systems theory, resilience thinking, and sustainability transitions. Each lens highlights different dynamics, and together they offer a comprehensive approach to both analysis and practice.

Socio-technical systems theory underscores that technologies are co-constructed with social institutions. In urban contexts, sensor networks, analytic models and cloud platforms are embedded in specific governance arrangements, contractual relations, and civic cultures. The political nature of design is evident in seemingly technical choices, such as sensor placements or data retention policies, which determine whose experiences are made visible and who benefits from optimization. Therefore, technical deployment must be accompanied by inclusive design processes, participatory procurement and mechanisms for public contestation of embedded values. Research on circular economy frameworks has shown that technology alone is insufficient; institutional redesign and cross-sector governance are required to scale circular business models and avoid perverse outcomes (Khan et al. 2024).

Resilience theory focuses attention on a city's capacity to anticipate, absorb and recover from shocks. IoT and AI can strengthen resilience through real-time situational awareness and predictive coordination. Yet these same technologies can create brittle dependencies if systems are designed with single points of failure or without fallback mechanisms. Resilience requires diversity in provisioning, redundancy in critical systems, local edge capabilities that maintain core functions during network or cloud outages, and regular stress-testing of socio-technical responses. Urban logistics experiments illustrate that decentralized, interoperable architectures with shared data standards improve failure tolerance relative to proprietary, monolithic platforms (Mohsen 2024).

Sustainability transitions scholarship looks at how niche innovations may disrupt the regimes of incumbent systems and redesign socio-technical systems to be more sustainable. Pilots with digital capabilities are valuable segments, but the transition to success relies on policy combinations, market motivation, and cultural transformation. The spread of circular-economy activities that IoT and AI will enable will thus need regulatory alleviation, procurement levers, and safeguarding and increasing successful pilots. Cities that foster such innovations, offer regulatory sandboxes, and create markets of green services through procurement stand a better chance of niche innovations coalescing into regime change.

A framework incorporating the socio-technical fit, resilience profile, and transition potential gives useful advice to policymakers. Socio-technical fit aims to establish the inclusion of the marginalized stakeholders in an intervention and availability of institutional structures to actualize the technology. The profile of resilience identifies the redundancy of the system, edge processing along with recovery protocol. Transition potential determines the intervention and its capability of scaling the intervention without difficulty, market incentives or change in the policy expectations. This heuristic creates useful managerial recommendations when applied to projects such as AI-based consolidation platforms: demand open APIs to provide interoperability, demand fallback and edge-processing, demand social metrics of performance, and design procurement channels to make pilot success to be mainstream.

The integrated point of view implies numerous core activities of the managers. The managers will be required to institutionalize participatory governance in the design and procurement process, require interoperability to reduce vendor lock-in, invest in resilience testing and cybersecurity as well as require specification of procurement and KPI frameworks to entrench equity and sustainability goals. The work to the policymakers entails the creation of policy mixes that comprise the use of standards, incentives and public procurement that determines the market conditions to scale sustainable digital infrastructures.

1.4 Ethics, Privacy and Governance Problems

The success of smart city initiatives is impacted by existential concerns about privacy, ethics, and governance brought up by the growth of IoT-AI-cloud platforms. Vulnerable groups may receive disproportionate benefits and services as a result of algorithms' ability to perpetuate and reinforce bias. These ethical conundrums can also be caused by socio-technical choices that dictate data collection and problem-solving methods, in addition to algorithmic opaqueness. These ethical concerns can be addressed with the aid of established models, algorithmic impact testing, and stakeholder involvement in the design process.

Data security and privacy are the main concerns. Fine spatiotemporal profiles of citizens' activities are provided by continuous sensing. Because single cloud platforms are centralized, they concentrate critical data and are therefore vulnerable to misuse or hacks. Cities should hence enact strict data regimes of governance that include data minimization, limitation of purposes, anonymization in case it is needed, explicit consent systems and data stewardship frameworks that specify explicit custodianship and access policies. Investment in urban infrastructure should consider cybersecurity as one of the top priorities.

Multi-actor ecosystems also complicate governance and accountability because they involve the production of services by municipal governments, private vendors, and civil society. Public procurement which transfers the risk of control of critical data to private firms without any protective measures prevents the importance of public accountability. To maintain the trust and accountability of the people, more transparency of the terms of the contract, and open-data requirements, citizen consultations and legally binding norms of liability in the automated systems are required. The implication of social equity cannot be an accident. The inequalities can be worsened by digital divides, uneven connectivity and low digital literacy. Inclusive deployment approaches should therefore be an integration of infrastructural commitment on connection and digital literacy programs and directive designs that entrench accessibility and customizations as ultimate goals and not accessories.

1.5 Theoretical and Practical Implications

The analysis provides several theoretical implications that can be utilized in debates in the fields of urban studies, STS, resilience scholarship and transition research. To begin with, the idea of smart cities as socio-technical ecosystems makes purely technological or managerial explanations more complicated as it focuses on co-constitution: technologies create and are created by institutional structures, regimes of power and civic values. This point of view builds upon STS by demonstrating how urban scale magnifies the possibilities and dangers of algorithmic mediation and it challenges scholars to understand how visibility, measurability and algorithmic goals refreeze urban politics. Second, the concept of resilience and transitions scholarship is useful in explaining how cities can cope with shocks together with the STS which is helpful in showing how digital experimentation can become embedded with sustainable urban metabolisms and how digital tools can be embedded in the pathway of regime shifts out of niche pilot projects. Third, the article is one of the first to anticipate the governance parameters in which IoT and AI become resources of lifecycle decreases instead of novel kinds of extraction and requires theory to relate design decisions with supply-chain outcomes (Khan et al. 2024).

In practice, the results suggest that urban administrators and policymakers should use a comprehensive strategy that would focus on socio-technical fit, resilience design and pathways to transition. This in the operational sense would mean the appending of weight to the participatory procurement processes that involve social specifications; imposing interoperability standards in order to avoid vendor lock-in; investing in edge-processing and redundancy in order to allow core services to be sustained even in the event of an outage; integrating equity and environmental measurements into performance measures; and the designing of procurement and policy tools that would offer scaling cues toward successful pilots. Data governance is also to be redesigned on stewardship principles that clarify ownership of who has what and is to give responsible access and make sure that the data is used in the best of interests of the civilians. Finally, workforce policies should assist in automation transitions by facilitating the risk of displacement through retraining and social protection and distributing the benefits of efficiency gains more broadly.

1.6 Conclusion and Recommendations

IoT, AI and cloud computing offer a wider array of tools than ever before to cities to make them more efficient, responsive and sustainable. However, technology will not always be fair and robust. One should consider smart cities as socio-technical ecosystems where the design of technologies is co-created with civic institutions, regulatory systems and community-based practices. Managers and policymakers have the task of coming up with people-friendly systems that are shock resistant and systems that can assist in systemic sustainability transitions. Part of these actions are in design of participation governance processes to get citizens and marginal groups involved in design and procurement, interoperability and open standards to prevent vendor lock-in and redundancy, edge-processing capacity, cybersecurity to scale successful pilots to larger scale regime transformation, social and environmental metrics in procurement and KPIs, and policy instruments that can be used to scale successful pilots to larger regime change. Finally, each phase of smart city delivery must be grounded on a pledge of transparency, accountability, and even-handedness. The success of the smart city will not be evaluated according to the number of sensors it encompasses to the sophistication of the algorithm used, but on whether the digital systems will assist civic agency, privacy, fair sharing of the benefits and recovery of relationship between city and planetary frontiers.

References

Batty M (2018) Artificial intelligence and smart cities. Environ Plan b: Urban Anal City Sci 45(1):3–6

Bijker WE, Hughes TP, Pinch T (eds) (1987) The social construction of technological systems: new directions in the sociology and history of technology. MIT Press

Cinderby S, Haq G, Opiyo R, Muhoza C, Ngabirano A, Wasike Y, Mwamba D, Cambridge H (2024) Inclusive climate resilient transport challenges in Africa. Cities 146:104740

Dissanayake H, Manta O, Iddagoda A, Palazzo M (2024) AI applications in business: trends and insights using bibliometric analysis. Int J Manag Educ 22(3):101075

Dou F, Ye J, Yuan G, Lu Q, Niu W, Sun H, Guan L, Lu G, Mai G, Liu N, Lu J, Liu Z, Wu Z, Song W (2023) Towards artificial general intelligence (AGI) in the internet of things (IOT): opportunities and challenges. arXiv:2309.07438

He W, Li W, Deng P (2022) Legal governance in the smart cities of China: functions, problems, and solutions. Sustainability 14(15):9738

Hollands RG (2008) Will the real smart city please stand up? City: Anal Urban Trends Cult Theory Policy Action 12(3):303–320. https://doi.org/10.1080/13604810802479126

Kashef M, Visvizi A, Troisi O (2021) Smart city as a smart service system: human-computer interaction and smart city surveillance systems. Comput Hum Behav 124:106923

Khan S, Singh R, Alnahas J, Abbate S, Centobelli P (2024) Navigating the smart circular economy: a framework for manufacturing firms. J Clean Prod 480, Article 144007. https://doi.org/10.1016/j.jclepro.2024.144007

Kitchin R (2014) The data revolution: big data, open data, data infrastructures and their consequences. SAGE Publications

Meerow S, Newell JP, Stults M (2016) Defining urban resilience: a review. Landsc Urban Plan 147:38–49. https://doi.org/10.1016/j.landurbplan.2015.11.011

Middelstadt BD, Allo P, Taddeo M, Wachter S, Floridi L (2016) The ethics of algorithms: mapping the debate. Big Data Soc 3(2). https://doi.org/10.1177/2053951716679679

Mishra A (2025) Smart cities project: a catalyst to inclusive economic growth (in reference to Uttar Pradesh). Int J Multidiscip Res 7(4). https://doi.org/10.36948/ijfmr.2025.v07i04.54355

Mohsen BM (2024) AI-driven optimization of urban logistics in smart cities: integrating autonomous vehicles and IoT for efficient delivery systems. Sustainability 16(24), Article 11265. https://doi.org/10.3390/su162411265

Moumen N, Radoine H, Nahiduzzaman KM, Oulidi HJ (2024) Smartainity: a comprehensive framework for urban performance assessment in African smart cities with key performance indicators. In: Ezziyyani M, Kacprzyk J, Balas VE (eds) International conference on advanced intelligent systems for sustainable development (AI2SD' 2023) (Lecture Notes in Networks and Systems), vol 931. Springer, Cham. https://doi.org/10.1007/978-3-031-54288-6_12

Nikitas A, Michalakopoulou K, Njoya E, Karampatzakis D (2020) Artificial intelligence, transport and the smart city: definitions and dimensions of a new mobility era. Sustainability 12(7):2789. https://doi.org/10.3390/su12072789]

Pliatsios A (2024) An ontology-based middleware for enhancing semantic interoperability in IoT-enabled smart city applications (Doctoral dissertation, Ph. D. dissertation, Department of Information and Communication Systems Engineering)

Ramaprasad A, Sánchez-Ortiz A, Syn T (2017) A unified definition of a smart city. In: International conference on electronic government. Cham: Springer International Publishing, pp 13–24

Rozony FZ, Aktar MNA, Ashrafuzzaman M, Islam A (2024) A systematic review of big data integration challenges and solutions for heterogeneous data sources. Acad J Bus Adm Innov Sustain 4(04):1–18

Singh R, Khan S, Kumar A, Kumar V (eds) (2024) Artificial intelligence enabled management: an emerging economy perspective. Walter de Gruyter GmbH & Co. KG

Singh R, Gupta A, Kandpal V (2024) ESG excellence in the digital world: circular economy e-business solutions. In: Santibanez-Gonzalez EDR, Kandpal V, Ozili PK, Chatterjee P (eds) Fintech for ESG and the circular economy. Wiley, pp 193–204. https://doi.org/10.1002/9781394234035.ch10

Singh R, Gupta A, Bajpai A, Kandpal V (2024) Exploring challenges of circular economy initiatives for smart cities. In: Kandpal V, Santibanez-Gonzalez EDR, Chatterjee P, Nallapaneni MK (eds) Smart cities and circular economy: the future of sustainable urban development. Emerald Publishing, pp 59–69. https://doi.org/10.1108/978-1-83797-957-820241005

Singh R, Ravesangar K, Narayanan S, Singh KKH (2024) HRM practices in creating net zero: supporting organizational sustainability. In: Singh R, Khan S, Kumar A, Luthra S, Chokshi H (eds) Net zero economy, corporate social responsibility and sustainable value creation. Springer, pp 1–18. https://doi.org/10.1007/978-3-031-55779-8_1

Singh A, Singh R, Katragadda R (2025) Leveraging artificial intelligence for climate neutrality: human-centric mitigation strategies to combat global warming. In: Singh R, Leal Filho W (eds) Climate neutrality through smart eco-innovation and environmental sustainability. Springer, pp 99–114. https://doi.org/10.1007/978-3-031-83250-5_7

Uribe-Pérez N, Pous C (2017) A novel communication system approach for a Smart City based on the human nervous system. Futur Gener Comput Syst 76:314–328

United Nations, Department of Economic and Social Affairs (2023) World urbanization prospects: The 2023 revision. United Nations. https://population.un.org/wup/

Xu H et al (2024) An IoT-based low-cost architecture for smart libraries and related applications. Sci Rep. https://doi.org/10.1038/s41598-024-57484-2

Wolniak R, Stecuła K (2024) Artificial intelligence in smart cities—applications, barriers, and future directions: a review. Smart Cities 7(3):1346–1389

Smart Workforce Metrics for a Zero Carbon Future: A Conceptual Framework for IoT-Driven Human Capital Performance

Ernesto DR Santibanez Gonzalez, Apurvaa Trivedi, Som Aditya Juyal, and Vinay Kandpal

Abstract Organizations are rapidly adopting to digital technologies as an integral part of their sustainability strategies for achieving net-zero carbon targets, though a significant progress is made in areas such as energy monitoring and operational efficiency, human capital as a direct contributor to carbon reduction goals is an important factor that often remain overlooked. To evaluate the performance of the human capital in an IoT-enabled environment which is designed to support a zero-carbon future, the chapter proposes a conceptual framework, that has Smart Workforce Metrics (SWMs) at its core which utilizes variety of real time data ranging from occupancy sensors and energy-tracking wearables to digital mobility records in an IoT enabled ecosystem, this will hep in quantifying the environmental footprint of employees at the individual, team, and organizational levels. The framework draws from key theoretical perspectives including Green Human Resource Management, Socio-technical Systems theory, and Performance Management. By merging data-driven insights with behavioural analysis, the model establishes a multi-dimensional approach that directly connects employee actions to sustainability outcomes. Ultimately, the proposed model contributes to the ongoing discussion on linking human capital development with environmental accountability. It does so by embedding carbon-sensitive key performance indicators (KPIs) into performance management systems, thereby ensuring that workforce evaluation and sustainability goals are aligned. The chapter also discusses ethical considerations in IoT-based monitoring, the potential for digital ESG dashboards, and the strategic implications for HR

E. D. S. Gonzalez
Department of Industrial Engineering and CES 4.0, Faculty of Engineering, University of Talca, Curico, Chile

A. Trivedi (✉) · S. A. Juyal
School of Management Studies, Swami Rama Himalayan University, Dehradun, India
e-mail: apurvaa.trivedi@gmail.com

V. Kandpal
Department of Management Studies, Graphic Era Deemed to Be University Dehradun, Dehradun, India

R. Singh et al. (eds.), *Data-Driven Sustainability: Harnessing Technology for A Greener Future and Environmental Resilience*, World Sustainability Series,
https://doi.org/10.1007/978-3-032-13412-7_3

leaders and policymakers aiming to operationalize sustainability goals. The chapter addresses a key gap in current literature by unifying digital performance tracking and workforce sustainability metrics and by offering a forward-looking tool for smart, resilient, and responsible organizations.

Keywords Smart workforce metrics · IoT · Human capital performance · Zero carbon · Green HRM · ESG · Sustainability analytics

1 Introduction

Reaching net-zero greenhouse-gas emissions has emerged as a test of definition for governments and companies alike, with the IPCC (2023) stressing that near-term, systemic mitigation is needed to limit warming below 1.5–2 °C. Concurrently, companies are leveraging digital technologies in particular the Internet of Things (IoT), data analytics, and AI to track, optimize, and decarbonise operations (e.g., smart buildings, industrial energy management, and circular flows). These technologies generate high-frequency, rich-context data to inform sustainability decisions at scale. However, even with these improvements in technical infrastructure, the human drivers of decarbonisation workforce behaviours, abilities,

and choices remain under-valued and seldom measured as part of formal sustainability performance frameworks. Researchers and practitioners increasingly demonstrate that individuals and organisations do count digital technologies can unleash value for sustainability when coupled with fresh habits and skills; IoT within buildings and facilities can lower energy consumption if the behaviour and working practices of occupants evolve; and employees' green behaviour (EGB) is consistently associated with improved environmental performance.

Yet, mainstream carbon accounting and sustainability reporting primarily concern assets, processes, and footprints rather than human capital performance. This creates a quantifiable gap between "what organisations measure" and "how people in reality drive decarbonisation" in everyday work. To close this gap, we propose a conceptual framework for Smart Workforce Metrics (SWMs) that leverages real-time data from IoT ecosystems (e.g., occupancy sensors, energy-tracking wearables, mobility and process logs) to quantify employees' environmental impact at individual, team, and organizational levels. SWMs translate digital traces into carbon-aware Key Performance Indicators (KPIs) aligned with net-zero pathways and operational realities.

The framework integrates three complementary lenses. First, Green Human Resource Management (GHRM) clarifies how HR practices (recruitment, training, incentives) shape pro-environmental behaviours and outcomes. Second, Socio-technical Systems (STS) theory explains how social structures (work design, norms) and technical systems (sensors, platforms) must be co-designed. Third, Performance Measurement & Management (PMM) provides the architecture for embedding carbon-aware indicators, targets, and feedback loops into everyday performance systems, including Industry 4.0 contexts.

This chapter advances the discourse on aligning environmental accountability with human capital development in four ways: (1) it defines a coherent construct Smart Workforce Metrics for linking IoT data streams with GHRM-driven behaviours and sustainability outcomes; (2) it specifies a multi-layer model spanning data inputs, behavioural insights, and carbon-linked KPIs; (3) it discusses ethical, privacy, and governance considerations for IoT-based monitoring; and (4) it outlines digital ESG dashboards and change-management strategies for HR leaders and policymakers to operationalize net-zero goals through people. In doing so, the chapter offers a forward-looking toolset for smart, resilient, and responsible organizations.

2 Literature Review

2.1 Sustainability Crisis and Digital Transformation

The climate crisis has increasingly matured from scientific prediction to operational urgency. Governments, corporations, and civil society have committed via net zero pledges, ESG frameworks, and the UN Sustainable Development Goals (SDGs) to reduce greenhouse gas emissions, shift toward renewable energy sources, and minimize ecological footprints of industrial and societal systems (Zhang et al. 2025; Baratta et al. 2023).

A latest report entitled "Global readiness for carbon neutrality: From targets to action" highlights that numerous countries have made carbon neutrality targets but the gap between policy implementation is wide, particularly in monitoring, enforcement, and setting targets into action. Another paper, "The Impact of ESG Practices in Industry with a Focus on Carbon Emissions: Insights and Future Perspectives" (Baratta et al. 2023), charts the convergence of industrial sectors toward adopting ESG practices and delineates strengths and shortcomings in how such actions combat emissions. Along with these policy and strategic commitments, there is growing momentum in digital transformation as both means and driver of sustainability.

IoT, big data, and AI are identified nowadays as the key technologies for monitoring real-time energy consumption, forecasting emissions, managing resources effectively, and enabling responsive systems. For example, "Comprehensive review of artificial intelligence applications in renewable energy systems: current implementations and emerging trends" (Ejiyi et al. 2025) overviews over 400 recent publications and concludes that AI is used in conjunction with IoT and data analytics to perform demand forecasting, predictive maintenance, integration of distributed energy resources, and improvement of grid stability. Similarly, "Artificial intelligence and IoT driven technologies for environmental pollution monitoring" (Popescu et al.

2024) discusses the utilization of sensor networks and ML models to track air, water, and land pollution, typically almost in real-time, and discusses the technical and governance concerns (quality of data, privacy, interpretability of models) associated with such deployment. Such research finds a range of emergent trends including decision making with data where large volume of sensor and IoT environmental data is helping in accurate measurement of energy usage and emissions. The integration of AI is used for prediction of equipment breakdown where AI is helping to in waste reduction as well as preventing energy loss due to faults, and inefficiency. Digital enabler technologies are insufficient in and of themselves; the regulatory environment, requirements for ESG reporting, and compliance substantially mediate their impact. Evidence suggests that the availability of climate policy uncertainty or weak ESG enforcement limits the translation of digital potential into reduced emissions.

Although there are gaps, much of the effort remains descriptive or review-based; few of them touch upon how human actor's organizational culture, managers, employees shape whether digital systems are being used to deliver sustainability impacts. Moreover privacy, ethics, data governance, and equity are mainly talked about as an afterthought. This calls for frameworks bringing human capital, behaviour, and performance to bear in the digital-sustainability relationship.

2.2 *Human Capital and Sustainability*

Human capital is now considered as an equally important driver of sustainability as advancements in technology and operations. While organizations have achieved notable improvements in areas such as energy efficiency and infrastructure, the attitudes, behaviours, and cultural values of the workforce remain at the heart of meaningful environmental performance (Amrutha and Geetha 2020).

An emerging approach in this filed is Green Human Resource Management (GHRM), as it integrates the sustainability principles into the core functions of HR. This includes linking the environmental goals within various HRM functions such as recruitment, training, performance evaluation, and reward systems to encourage the employees to adopt eco-friendly practices at work (Tang et al. 2018). By incorporating green practices into HR strategies alignment between employee contributions and broader sustainability objectives can be ensured by the organizations (Ansari et al. 2021).

Studies have consistently shown that individual employee behaviours play a critical role in shaping an organization's sustainability outcomes. Everyday actions such as reducing energy use, cutting down on waste, and participating in green innovation initiatives—collectively referred to as Employee Green Behaviour (EGB)—have been linked with measurable improvements in environmental performance (Zacher 2023). However, whether these behaviours translate into significant results often depends on the surrounding organizational context. Factors such as workplace culture, leadership commitment, ethical standards, and external stakeholder expectations create the conditions that determine how effectively employee actions support sustainability (Guerci et al. 2016; Pinzone et al. 2016; Sarwar et al. 2025).

Current research affirms the connection between Green HRM and the well-being of employees, with findings indicating sustainable HR practices benefit not only the environment but also happiness, satisfaction, and engagement. For instance, Martínez-Falcó et al. (2024) illustrate Green HRM within the wine sector increases well-being and work engagement, whereas Islam et al. (2025) illustrate the mediating effect of green innovation and engagement towards stimulating productivity and sustainability in industrial sectors.

Some other studies emphasize the role of psychological and knowledge-based mediators. For example, Rahman et al. (2025) observed that workers' green concerns influence green behaviour via green knowledge and effectiveness as mediators, especially in consumer product industries. Likewise, Carbone et al. (2024) assert that employees themselves can be key agents for change, and green workplace behaviour can contribute measurably to sustainability performance.

Cumulatively, the research base underpinning this paper lends credibility to the proposal that Smart Workforce Metrics (SWMs) can operationalize the people dimension of sustainability by quantifying workforce value to carbon neutrality. By expanding the Green HRM lineage and adding digital tracking and IoT-enabled data, SWMs offer organisations measurement and leverage to drive the sustainability value of their people.

2.3 *IoT and Performance Management*

The concept of organizational performance is transformed by Internet of Things (IoT), as compared to the conventional systems that used indicators such as supervisors rating or productivity metrics, the IoT based technologies offer a robust realtome data that can link the workers' behaviors with organizational efficiency and sustainability outcomes (Mohamed et al. 2025).

IoT based devices are the core of this transformation, stress, fatigue and ergonomic risks can be traced with the help of wearables, the office usage and energy consumption can be monitored through occupancy sensors, while the mobility logs can keep a track of commuter and traveller conduct to encourage low-carbon mobility initiatives (Popescu et al., 2024). While offering the capabilities to measure the employee's contributions toward environmental performance, these technologies also maximize operational efficiency by integrating workforce analytics with net-zero strategies.

The application of technologies can extend to industrial maintenance and operation. Scheffer et al. (2021) highlighted how the IoT-supported augmented reality is useful in the diagnosis of Information Technology (IT) and Operational Technology (OT) systems faults in a digitalized environment, reducing the downtime hence leading to reduce wastage of of resources and link human action, technical infrastructure, and sustainable performance to each other directly. Similarly, Deen and Asselin (2020) highlighted how the IoT powered monitoring of electrochemical processes (copper leaching and zinc electrowinning) raises energy efficiency, highlighting the IoT contribution to co-optimization of human and technical systems.

From the viewpoint of socio-technical systems, these developments offers both opportunities and risks. While the IoT data support more precise sustainability-related Key Performance Indicators (KPIs) it also raises concerns about surveillance, autonomy, and ethics (Ajunwa 2020). Thus, integrating IoT analytics with HR dashboards requires careful governance to ensure that performance management systems are not only effective but also dependable and ethical. Together the workforce analytics enabled by IoT extend the scope of performance management to include sustainability measures alongside productivity, creating a framework wherein human capital performance is automatically tied to carbon reduction targets.

2.4 *Ethical and Governance Considerations in IoT-Based Monitoring*

While IoT-driven workforce analytics offer unprecedented visibility into employee performance and contribution to sustainability, they simultaneously generate deep ethical and governance issues. The same technologies that enable organizations to optimize energy consumption, track productivity, and minimize carbon prints can also pose threats of over-surveillance, misuse of data, and undermining employee autonomy (Ajunwa 2020).

One of the first concerns is privacy. IoT sensors like wearables, occupancy sensors, and mobility trackers gather extremely detailed personal data. Left without strong data governance structures, this data can be used in manners that staff finds intrusive, resulting in resistance, mistrust, or even ethics breaches (Bodie et al. 2017). Transparency in data collection with respect for staff dignity must therefore be balanced.

A second challenge is bias and fairness. Algorithms interpreting IoT data may embed systemic biases, leading to inequitable evaluations of employee performance. This issue resonates with broader debates in algorithmic management, where automation has been shown to unintentionally reinforce discrimination in employment decisions (Kellogg et al. 2020).

From a governance perspective, organizations are increasingly encouraged to integrate IoT monitoring within broader Ethics, Social, and Governance (ESG) frameworks. This entails not only compliance with data protection laws such as GDPR but also the adoption of ethical principles like proportionality, explainability, and employee participation in system design (Floridi and Cowls 2019). Such approaches align socio-technical systems theory with human-centric innovation, ensuring that technology enhances both organizational outcomes and employee well-being.

Finally, the emergence of digital sustainability dashboards that integrate IoT-based workforce metrics with ESG reporting presents both an opportunity and a responsibility. These dashboards can enable managers and policymakers to track human capital contributions to carbon goals more transparently. Yet they also require strict governance to prevent the reduction of employees to "data points" in sustainability reporting (Benlian et al. 2022).

The ethical and governance considerations are not peripheral but foundational to the deployment of IoT-based performance systems. By embedding safeguards for privacy, fairness, and accountability, organizations can build trust and legitimacy while advancing sustainability objectives.

2.5 Strategic Implications for HR Leaders and Policymakers

The integration of IoT-driven workforce analytics into sustainability strategies has direct implications for both HR leaders and policymakers. Organizations seeking to achieve net-zero commitments must go beyond monitoring energy systems to include the human dimension of sustainability performance. Embedding Smart Workforce Metrics (SWMs) into performance management systems allows HR leaders to align employee behavior with carbon reduction goals (Zacher 2023).

For HR practitioners, the challenge is to translate Green HRM principles into measurable outcomes. IoT-enabled devices and digital dashboards provide an opportunity to track how employee behaviours such as energy-conscious practices, mobility choices, or remote work adoption contribute to sustainability. However, HR leaders must ensure that the use of such data is not merely extractive but integrated into developmental processes, including training, recognition, and career progression (Pinzone et al. 2016). Embedding sustainability-linked Key Performance Indicators (KPIs) into appraisal systems can foster accountability while reinforcing organizational culture around environmental responsibility (Renwick et al. 2016).

For policymakers, the expansion of IoT-based workforce analytics raises questions of standardization, compliance, and regulation. Global reporting frameworks such as the Global Reporting Initiative (GRI) and the Task Force on Climate-related Financial Disclosures (TCFD) are beginning to emphasize the human capital dimension of ESG reporting (Douglas et al. 2022). Policymakers can accelerate this trend by mandating disclosure of workforce-related sustainability metrics, ensuring that human contributions to carbon reduction are systematically recognized.

Moreover, there is a strong case for incentivizing responsible innovation. Tax incentives, subsidies, or recognition schemes for organizations that deploy IoT ethically and transparently can encourage widespread adoption without compromising privacy or employee trust (Ajunwa 2020; Floridi and Cowls 2019). Public–private partnerships can also play a role in establishing industry-wide ethical data governance frameworks, balancing innovation with accountability.

Lastly, HR leaders and policymakers need to vision the work of the future in the context of sustainability imperatives. With digital surveillance increasingly woven into organizational existence, training programs for the workforce should prioritize data literacy, ethical literacy, and sustainability skills. Through preparing workers to interact purposefully with IoT-capable systems, organizations can create not only compliance but also worker agency for realizing carbon–neutral futures.

Effectively, strategic human capital performance alignment to sustainability objectives calls for a multi-level governance solution: HR executives need to incorporate IoT insights in equitable and enabling management systems, whereas policymakers need to establish regulatory and incentive systems that guarantee accountability, inclusiveness, and ethical responsibility.

2.6 Research Gaps and Conceptual Opportunities

Together, these streams reveal three interrelated gaps. First, digital sustainability research places emphasis on technological efficacy but neglect the human capital as a measurable factor in carbon outcomes. Second, HRM research depicts the significance of employees' involvement but lacks mechanisms for integrating real-time data into performance systems. Third, ethical scholarship reveals vulnerabilities in IoT-enabled monitoring but provides few mechanisms for embedding protection within sustainability measures.

This fragmented geography justifies the development of Smart Workforce Metrics (SWMs) as a theoretical construct. Integrating IoT-facilitated data, HRM theory, and PMM infrastructures, SWMs implement human capital contributions to sustainability. SWMs enrich GHRM by integrating carbon responsibility into evaluation systems, drive STS by initiating participatory policymaking, and consolidate PMM by integrating employee-level KPIs into ESG dashboards. By doing so, SWMs reconcile technological, human, and ethical views and place the workforce as a core driver of decarbonization.

3 Conceptual Framework: Smart Workforce Metrics (SWMs)

3.1 Framework Overview

The Smart Workforce Metrics (SWMs) framework conceptualizes how human capital performance can be systematically aligned with sustainability objectives in IoT-enabled environments. Unlike traditional performance systems that emphasize efficiency and productivity, SWMs embed carbon-awareness and environmental accountability into multi-level indicators across individual, team, and organizational levels. *Inputs* consist of IoT-enabled data streams such as wearables, occupancy

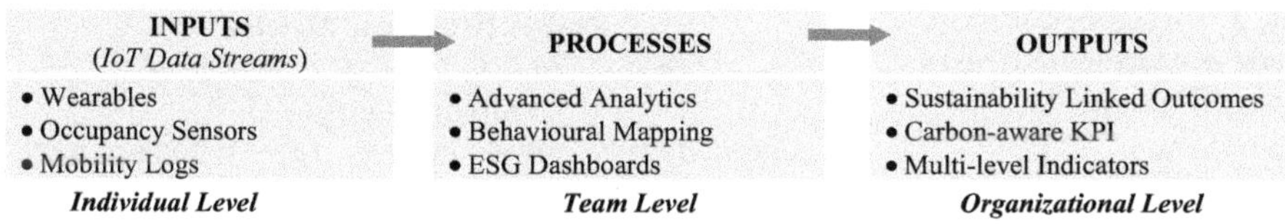

Fig. 1 Conceptual framework of smart workforce metrics (SWMs). Source Developed by the authors

sensors, and mobility logs, which provide granular, real-time insights into workforce behaviour. *Processes* involve advanced analytics, behaviour mapping, and integration with digital ESG dashboards, transforming raw data into interpretable sustainability indicators that align with HR processes. *Outputs* include sustainability-linked outcomes, particularly carbon-aware Key Performance Indicators (KPIs), enabling organizations to quantify workforce contributions to net-zero targets while balancing productivity and well-being. This layered model positions employees as central actors in sustainability and demonstrates how IoT ecosystems can translate workforce practices into measurable carbon impacts, advancing Green HRM and socio-technical perspectives toward a novel paradigm of sustainability-focused performance management (Fig. 1).

3.2 Dimensions of Smart Workforce Metrics (SWMs)

The Smart Workforce Metrics (SWM) framework is operationalized through five dimensions that link IoT-enabled workforce data to sustainability outcomes, demonstrating how employee behaviours and organizational practices directly influence carbon goals and positioning human capital as a measurable driver of performance (Zacher 2023; Amrutha and Geetha 2020). Energy behaviour metrics use IoT wearables and sensors to monitor electricity use, HVAC adjustments, and device operation, identifying patterns such as idle time or thermostat changes that, when aggregated, support carbon accounting (Poyyamozhi et al. 2024). Mobility metrics capture commuting and travel data from logs and telematics, enabling interventions like public transport incentives, carpooling, or hybrid work policies (Popescu et al. 2024). Workspace utilization metrics apply occupancy sensors and booking systems to measure desk-sharing, occupancy per square metre, and energy costs, helping organizations optimize usage and move towards carbon–neutral facilities (Deen and Asselin 2020). Team collaboration metrics evaluate digital and physical interactions, balancing travel-related emissions from meetings with energy costs of data centres to ensure sustainability without productivity loss (Scheffer et al. 2021). Finally, well-being and productivity metrics integrate human outcomes, as wearables and surveys track stress, health, and engagement, showing how sustainable practices enhance both accountability and HR outcomes like satisfaction and retention (Martínez-Falcó et al. 2024; Islam et al. 2025). Collectively, these dimensions form a holistic toolkit

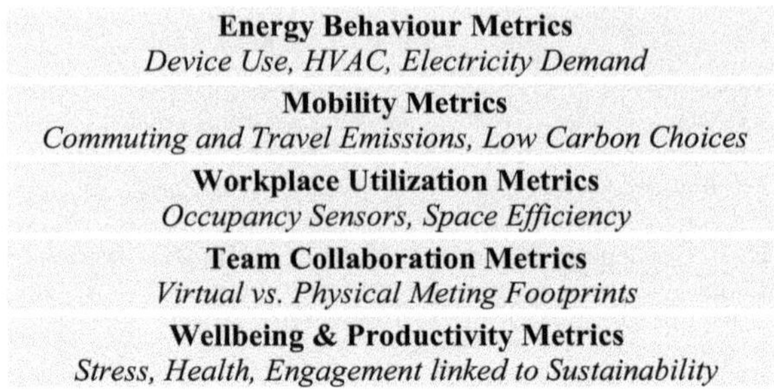

Fig. 2 Dimensions of smart workforce metrics (SWMs). Source Developed by the authors

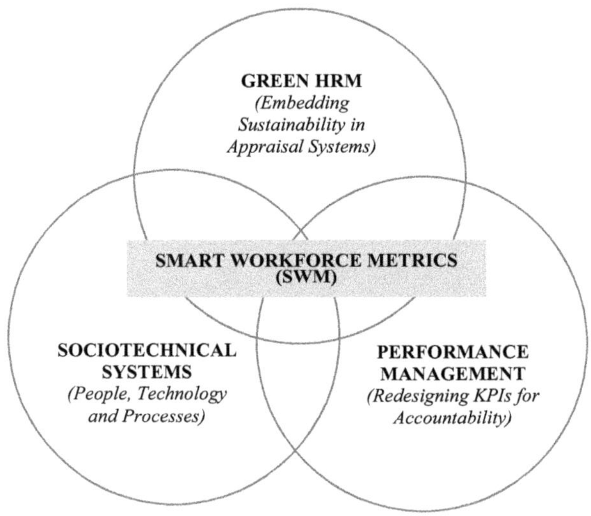

Fig. 3 Framework of smart workforce metrics (SWMs). Source Developed by the authors

aligning ecological accountability with organizational resilience and competitiveness (Figs. 2 and 3).

3.3 Alignment with Theories

The Smart Workforce Metrics (SWM) framework is grounded in three theoretical traditions that provide conceptual strength and practical applicability. Green Human Resource Management (Green HRM) integrates sustainability into HR functions such as recruitment, training, appraisal, and rewards, with carbon-aware indicators in appraisal systems encouraging pro-environmental behaviours like energy-conscious device use or sustainable commuting (Tang et al. 2018; Ansari et al. 2021). Socio-technical Systems (STS) theory highlights that technological solutions succeed when co-designed with social structures, emphasizing that IoT devices and ESG dashboards must operate within cultures of trust and participatory governance to ensure sustainable outcomes (Baxter and Sommerville 2011). Finally, Performance Measurement and Management (PMM) extend traditional systems by embedding environmental accountability, as sustainability-linked KPIs in real-time dashboards allow employees and teams to directly observe the ecological impact of their actions, thereby institutionalizing carbon accountability alongside financial and operational goals (Demartini and Taticchi 2022).

3.4 Managerial Applications of Smart Workforce Metrics (SWMs)

The practical relevance of Smart Workforce Metrics (SWMs) lies in their ability to transform abstract sustainability goals into actionable managerial practices by embedding IoT-enabled indicators into everyday processes and operationalizing sustainability at multiple decision-making levels. SWMs provide managers with tools for diagnostics and benchmarking; enabling them to assess workforce-related sustainability practices and compares them with internal targets or industry standards. For example, the number of employees using low-carbon modes can be identified through mobility logs, while occupancy sensors can help in detecting office energy efficiencies, leading to evidence-based policy intervention. The integration of sustainability into HR systems can be done through SWMs by embedding carbon-conscious indicators in recruitment, appraisal, training, and rewards aligning individual performance with shared environmental goals.

SWMs can also influence incentive mechanism design, where employees demonstrating low-carbon behaviours such as energy efficiency or hybrid work can be rewarded through gamified dashboards or reward schemes, once more ensuring accountability and motivation, this will lead to a higher possibility of demonstrating such behaviours. SWMs enable real-time decision support, as ESG dashboards provide real-time insight into energy, mobility, and workspace behaviour, informing adaptive choices such as re-arranging layouts, scheduling hybrid workdays, or regulating HVAC controls to meet actual occupancy, thus reducing carbon consumption. The managerial value of SWMs is illustrated using a multinational information technology company that used mobility logs to identify excessive solo driving; carpooling incentives and discounted public transportation passes applied as interventions reduced single-occupancy vehicle travel by 20% in six months. Collectively, these applications highlight SWMs as a diagnostic, integrative, and decision-making toolkit for embedding sustainability into organizational practice (Figs. 4 and 5).

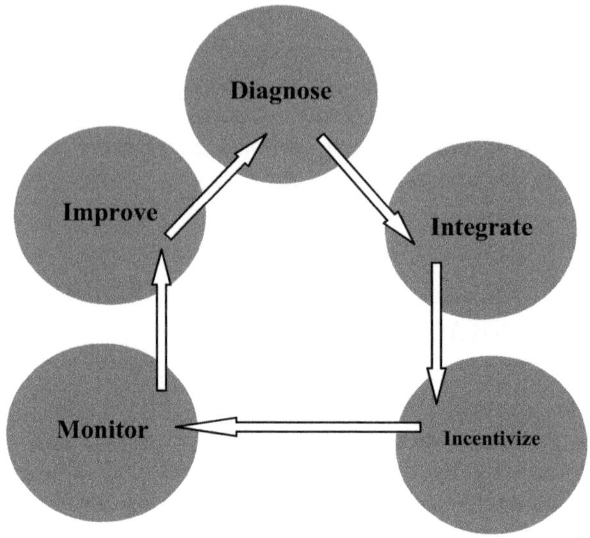

Fig. 4 Managerial application cycle of smart workforce metrics (SWMs). Source Developed by the authors

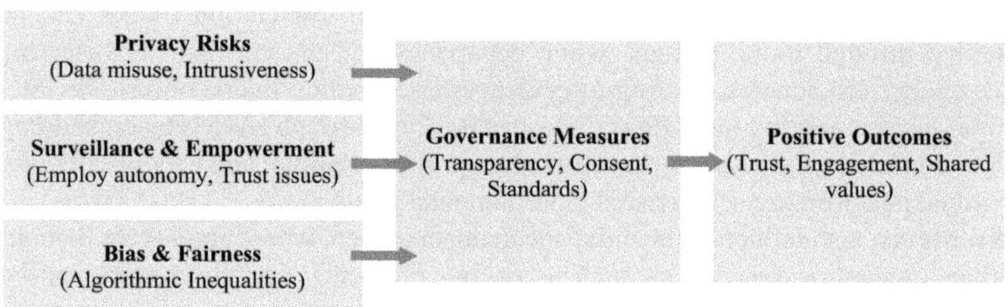

Fig. 5 Ethical and managerial considerations in smart workforce metrics (SWMs). Source Developed by the authors

4 Ethical and Managerial Considerations

The integration of IoT-driven Smart Workforce Metrics (SWMs) promises advances in transparency, accountability, and carbon-aware decision-making, yet it introduces critical ethical and managerial challenges. Privacy remains a primary concern, as wearables, mobility trackers, and occupancy sensors collect highly granular data that may be perceived as intrusive surveillance (Bodie et al. 2017). To build legitimacy and trust, the organizations must ensure transparent communication, consent, anonymization, and compliance with data protection laws such as GDPR (Floridi and Cowls 2019). A further tension lies in surveillance versus empowerment: while excessive monitoring risks undermining autonomy and psychological safety (Ajunwa 2020), SWMs designed as developmental tools can foster engagement and ownership by providing employees with feedback and recognition. Algorithmic fairness also requires attention, as IoT data may embed systemic biases, privileging some groups (e.g., remote workers) over others; ensuring algorithmic transparency, conducting bias audits, and adopting proportionality principles in KPI design are essential (Jarrahi et al. 2021; Kellogg et al. 2020). Managerially, ESG dashboards that consolidate IoT-derived metrics offer actionable insights but must integrate indicators with HR practices such as training, recognition, and career progression to avoid reducing employees to "data points" (Benlian et al. 2022). At the governance level, aligning SWMs with ESG frameworks such as the GRI and TCFD is critical, while policymakers should mandate disclosure and incentivize ethical practices (Douglas et al. 2022).

5 Strategic Implications

The operationalization of Smart Workforce Metrics (SWMs) carries wide-ranging strategic implications for organizations, HR leaders, and policymakers, as it integrates IoT-enabled workforce analytics with sustainability strategies to align human capital performance with carbon reduction targets while fostering trust, engagement, and innovation. In an organization, SWMs enable businesses to infuse sustainability into the DNA of business-as-usual by measuring energy behaviours, mobility decision-making, and workspace usage, aligning workforce practices with net-zero goals and strengthening resilience in industries facing ESG scrutiny. By uncovering inefficiencies, such as carbon-emitting commuting or unused workplaces, SWMs facilitate focused interventions such as carpool incentives or real estate optimization, turning sustainability into aspiration into quantifiable ecological and economic results. For HR professionals, SWMs bring Green HRM principles to action by integrating sustainability-linked results into appraisals, training, and reward systems. Dashboards that identify sustainable actions like energy conservation or low-carbon travel promote accountability while fostering a culture of environmental stewardship, equity, and inclusion. Furthermore, HR can foster engagement by positioning sustainability as a shared organizational value and by equipping employees with data literacy and ethical awareness to participate actively in the net-zero transition. For policymakers, SWMs highlight the urgency of establishing standards for human capital sustainability reporting. While frameworks like GRI and TCFD acknowledge workforce dimensions, stronger mandates are needed, including compulsory disclosure of workforce-related metrics, tax incentives for ethical IoT adoption, and public–private partnerships to co-create governance frameworks. Collectively, these measures ensure that workforce sustainability is treated as a systematic requirement of global climate commitments. Strategically, the integration of SWMs positions organizations to achieve carbon neutrality and workforce development simultaneously, offering competitive advantage and ESG legitimacy for firms, engagement and developmental tools for HR, and regulatory innovation pathways for policymakers, thereby advancing smart, resilient, and responsible organizations (Fig. 6).

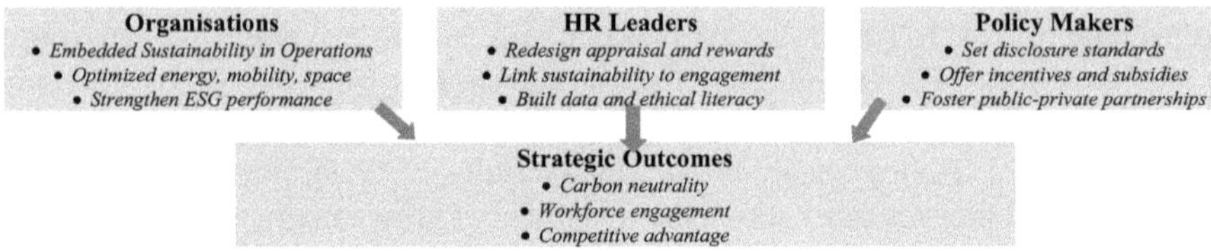

Fig. 6 Strategic implications of smart workforce metrics (SWMs). Source Developed by the authors

6 Implications

The Smart Workforce Metrics (SWM) framework advances workforce accountability in sustainability by repositioning employees from passive actors to active agents of decarbonisation, integrating human behaviour into carbon accounting through quantification of energy use, mobility choices, and workspace utilization. Its multi-level design spanning individuals, teams, and organizations ensures distributed accountability, enabling employees to receive feedback on their contributions, teams to benchmark practices, and leadership to assess overall impact. This granularity bridges the gap between sustainability reporting, which often emphasizes physical assets, and the lived realities of workforce practices. By integrating IoT-enabled monitoring with HRM systems, SWMs provide organizations with a comparative advantage, not only achieving ecological outcomes but also strengthening employee engagement, retention, and innovation. Firms adopting SWMs enhance employee well-being and organizational citizenship behaviours while meeting investor and consumer expectations for ESG performance, thus creating a dual dividend of ecological accountability and organizational legitimacy. Beyond measurement, SWMs enable predictive sustainability modelling by linking IoT data with machine learning to forecast carbon outcomes, support proactive interventions, and facilitate scenario planning on HR policies, workplace designs, or mobility programs. This shifts SWMs from reactive monitoring tools to instruments of strategic foresight. However, challenges remain, including ethical concerns around privacy, autonomy, and fairness, as well as sectoral differences that influence applicability across industries. Future research should explore cross-sectoral applications in manufacturing, healthcare, and services; examine cultural variations in perceptions of IoT monitoring; and assess technological integration, particularly blockchain and digital twins, to refine SWMs for broader adaptability and legitimacy (Fig. 7).

7 Conclusion and Future Directions

Achieving net-zero carbon goals requires rethinking how sustainability is measured and managed. This chapter introduced Smart Workforce Metrics (SWMs) as a framework integrating IoT-enabled monitoring with Green HRM, socio-technical systems theory, and performance management, positioning employees as active agents of

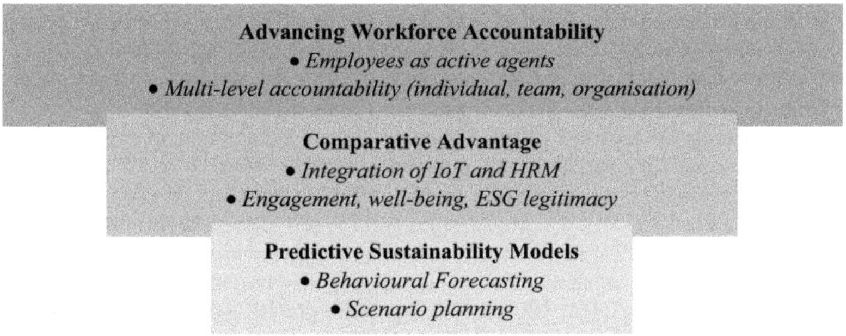

Fig. 7 Key contributions of smart workforce metrics (SWMs). Source Developed by the authors

decarbonisation rather than peripheral actors. The study contributes conceptually by developing a multi-layered framework that links IoT inputs, behavioural processes, and sustainability-linked outputs; theoretically by aligning Green HRM, STS, and PMM to embed carbon accountability in workforce systems; and practically by outlining ethical, managerial, and strategic pathways for application. In practice, SWMs enable organizations to embed sustainability into HR systems, leverage IoT dashboards for feedback and ESG reporting, foster employee engagement by aligning well-being with environmental accountability, and enhance competitiveness through carbon-conscious workforce management. Future research should examine AI-enhanced analytics for behavioural forecasting, explore adoption across diverse sectors, consider cultural and regional variations, assess the role of emerging technologies such as blockchain and digital twins, and conduct longitudinal studies on long-term impacts. Overall, SWMs provide a forward-looking toolset to align human capital with sustainability goals, and as they evolve, they hold the potential to become a practical standard for building smart, resilient, and responsible organizations in a zero-carbon future.

References

Ajunwa I (2020) The paradox of automation as anti-bias intervention. Cardozo Law Rev 41(5):1671–1714. https://doi.org/10.2139/ssrn.2746078

Amrutha VN, Geetha SN (2020) A systematic review on green human resource management: implications for social sustainability. J Clean Prod 247:119131. https://doi.org/10.1016/j.jclepro.2019.119131

Ansari NY, Farrukh M, Raza A (2021) Green human resource management and employees' pro-environmental behaviours: Examining the underlying mechanism. Corp Soc Responsib Environ Manag 28(1):229–238. https://doi.org/10.1002/csr.2044

Baratta A, Cimino A, Longo F, Solina V, Verteramo S (2023) The impact of ESG practices in industry with a focus on carbon emissions: insights and future perspectives. Sustainability 15(8):6685. https://doi.org/10.3390/su15086685

Baxter G, Sommerville I (2011) Socio-technical systems: from design methods to systems engineering. Interact Comput 23(1):4–17. https://doi.org/10.1016/j.intcom.2010.07.003

Benlian A, Wiener M, Cram WA, Rossi M, Sein MK, Simon JC (2022) Algorithmic management. Bus Inf Syst Eng 64(6):825–839. https://doi.org/10.1007/s12599-022-00764-w

Bodie MT, Cherry MA, McCormick ML, Tang J (2017) The law and policy of people analytics. Univ Color Law Rev 88(4):961–1042. https://ssrn.com/abstract=2769980

Carbone E, Chierici R, D'Onofrio I (2024) Green workplace behaviors: can employees make the difference? Sustainability 16(24):11188. https://doi.org/10.3390/su162411188

Deen KM, Asselin E (2020) Integration of Cu extraction and Zn electrowinning processes for energy storage. J Clean Prod 253:119779. https://doi.org/10.1016/j.jclepro.2019.119779

Demartini M, Taticchi P (2022) Performance measurement and management: a literature review focused on the role of management theories with a deep dive into the Industry 4.0 environment. Int J Prod Perform Manag 71(4):1008–1033. https://doi.org/10.1108/IJPPM-02-2021-0063

Douglas MA, Mollenkopf DA, Castillo VE et al (2022) Journeys, not destinations: theorizing a process view of supply chain integrity. J Bus Ethics 181(1):195–220. https://doi.org/10.1007/s10551-021-04906-0

Ejiyi CJ, Cai D, Thomas D, Obiora S, Zhang QX, Bamisile OO, Osei-Mensah E (2025) Comprehensive review of artificial intelligence applications in renewable energy systems: current implementations and emerging trends. J Big Data 12, Article 169. https://doi.org/10.1186/s40537-025-01178-7

Floridi L, Cowls J (2019) A unified framework of five principles for AI in society. Harv Data Sci Rev 1(1). https://doi.org/10.1162/99608f92.8cd550d1

Guerci M, Longoni A, Luzzini D (2016) Translating stakeholder pressures into environmental performance: the mediating role of Green HRM practices. Int J Hum Resour Manag 27(2):262–289. https://doi.org/10.1080/09585192.2015.1065431

IPCC (2023) Climate change 2023: synthesis report. https://doi.org/10.59327/IPCC/AR6-978929 1691647

Islam MF, Al Masud A, Emon M et al (2025) Integrating green HRM for productivity and sustainability: green innovation, engagement and pro-environmental behavior as key mediators. Futur Bus J 11:24. https://doi.org/10.1186/s43093-025-00433-w

Jarrahi MH, Sutherland W, Nelson SB, Sawyer S (2021) Algorithmic management in a work context. Big Data Soc 8(2):20539517211020332. https://doi.org/10.1177/20539517211020332

Kellogg KC, Valentine MA, Christin A (2020) Algorithms at work: the new contested terrain of control. Acad Manag Ann 14(1):366–410. https://doi.org/10.5465/annals.2018.0174

Li S, Ao X, Zhang M, Pu M (2025) ESG performance and carbon emission intensity: examining the role of climate policy uncertainty and the digital economy. Front Environ Sci 12, Article 1526681. https://doi.org/10.3389/fenvs.2024.1526681

Martínez-Falcó J, Sánchez-García E, Marco-Lajara B, Millán-Tudela LA (2024) Enhancing employee wellbeing and happiness management in the wine industry: Unveiling the role of green human resource management. BMC Psychol 12:203. https://doi.org/10.1186/s40359-024-017 03-y

Mohamed A, Ismail I, AlDaraawi M (2025) IoT-driven intelligent energy management: leveraging smart monitoring applications and artificial neural networks (ANN) for sustainable practices. Computers 14(7):269. https://doi.org/10.3390/computers14070269

Pinzone M, Guerci M, Lettieri E, Redman T (2016) Progressing in the change journey towards sustainability in healthcare: the role of "Green" HRM. J Clean Prod 122:201–211. https://doi.org/10.1016/j.jclepro.2016.02.031

Popescu SM, Mansoor S, Wani OA, Kumar SS, Sharma V, Sharma A, Arya VM, Kirkham MB, Hou D, Bolan N, Chung YS (2024) Artificial intelligence and IoT driven technologies for environmental pollution monitoring and management. Front Environ Sci 12:1336088. https://doi.org/10.3389/fenvs.2024.1336088

Poyyamozhi M, Murugesan B, Rajamanickam N, Shorfuzzaman M, Aboelmagd Y (2024) IoT-a promising solution to energy management in smart buildings: a systematic review, applications, barriers, and future scope. Buildings 14(11):3446. https://doi.org/10.3390/buildings14113446

Rahman MH, Amin MB, Hasan MN, Yasmin N, Rahaman MA, Oláh J (2025) The nexus among employees' green concerns, green effectiveness, and green behavior through the mediating role of employees' green knowledge: evidence from Bangladeshi consumer goods industry. Environ ChallS 19:101161. https://doi.org/10.1016/j.envc.2025.101161

Renwick DWS, Jabbour CJC, Muller-Camen M (2016) Contemporary developments in Green (environmental) HRM scholarship. Int J Hum Resour Manag 27(2):114–128. https://doi.org/10.1080/09585192.2015.1105844

Sarwar U, Baig W, Rahi S, Sattar S (2025) Fostering green behavior in the workplace: the role of ethical climate, motivation, and environmental knowledge. Sustainability 17(9):4083. https://doi.org/10.3390/su17094083

Scheffer S, Martinetti A, Damgrave R, van Dongen L (2021) Augmented reality for IT/OT failures in maintenance operations of digitized trains: current status, research challenges and future directions. Procedia CIRP 100:816–821. https://doi.org/10.1016/j.procir.2021.05.038

Tang G, Chen Y, Jiang Y, Paillé P, Jia J (2018) Green human resource management practices: scale development and validity. Asia Pac J Hum Resour 56(1):31–55. https://doi.org/10.1111/1744-7941.12147

Zacher H (2023) Employee green behavior as the core of environmentally sustainable work behavior. Annu Rev Organ Psych Organ Behav 10:353–373. https://doi.org/10.1146/annurev-orgpsych-120920-050421

Zhang S, Cai W, Zheng X, Lv X, An K, Cao Y, Cheng HS, Dai J, Dong X, Fan S, Gao Y, Gong Z, Guan Y, Hon, C, Li J, Li M, Li Y, Liang S, Liao W, Ma Z, He K (2025) Global readiness for carbon neutrality: from targets to action. Environ Sci Ecotechnology 25:100546. https://doi.org/10.1016/j.ese.2025.100546

Fostering the Applications of Industry 4.0 in Sustainable Mobility in Smart Cities: An Alternative-Fuel Adoption Perspective

Amrita Pani and Rishi Kumar

Abstract The transportation sector significantly influences an economy's development. The ongoing energy transitions and urgency to address an alternative to fossil fuels have simultaneously led to innovation-led growth. The stance of transport and research on sustainable mobility has become critical for 'Smart Cities'. Climate change complexities put forward the need to adopt "Net-Zero" by 2050. Further, as the transport sector significantly contributes to total emissions, the 'human–computer interaction' aspect of Industry 5.0 can contribute to devising mitigation techniques, especially in the Global South. This chapter explores the infrastructural and technological barriers from a developed and developing nations context. Through the lens of alternative fuels such as electric vehicles, it critically evaluates the applications of Industry 4.0 in terms of cleaner technology and sustainable transport in smart cities through a comparative analysis and a detailed case study on sustainable transport. As Industry 5.0 entails intelligent machines coexisting with people, aiming for increased sustainability and resilience, it acts as a catalyst to the pre-existing pillars of Industry 4.0 to mitigate climate change. Hence, this chapter highlights the role of Industry 4.0 in the transport sector through the development of alternative fuel vehicles to aid in sustainable mobility solutions in Smart Cities.

Keywords Sustainable and SMART mobility · Transport · Emissions · Industry 4.0 and industry 5.0 · Alternative fuels · Electric vehicles · Sustainable Development Goals (SDGs)

A. Pani (✉) · R. Kumar
Department of Economics and Finance, BITS Pilani, Hyderabad Campus, Hyderabad, India
e-mail: p20210026@hyderabad.bits-pilani.ac.in

R. Singh et al. (eds.), *Data-Driven Sustainability: Harnessing Technology for A Greener Future and Environmental Resilience*, World Sustainability Series,
https://doi.org/10.1007/978-3-032-13412-7_4

1 Introduction

The territory of the Global South has been the subject of several debates across the globe. The dualistic hierarchies categorised them as low-income, poor infrastructure, dense population, and political or cultural marginalisation victims (Mignolo 2011). The Global South has been dealing with its political stance in international power relations. In the present day, the geographical boundaries of the Global South consist of Asia, Africa (excluding South Korea, Japan and Israel), and the nations of Latin America, the Caribbean and Oceania (without New Zealand and Australia) (Chant 2009). The geographical diversity of the Global South highlights its wealth inequality due to unequal power relations among them (Mahler 2017). In contrast, the G77 has unity, complementarity, cooperation and solidarity between marginalised countries in the global order. Despite the arguments around the characteristics, the adaptation of digitalisation is on the rise. So is urbanisation with the acknowledgement of the role of geography in economic parameters. Further, the second phase of the twentieth century highlighted economic development regarding smart city models, sustainable societal transitions and modernised plans for sustainable futures (Bach 2021).

The Fourth Industrial Revolution, or Industry 4.0, refers to the "manufacturing renaissance" based on the pillars of the Internet of Things (IoT), artificial intelligence (AI) and big data to enhance the connectivity of the existing technologies by ensuring cybersecurity (Bag et al. 2021), fuelling a diverse adaptation of technology. The developed countries such as Singapore, the United States of America (USA), and Germany have a high potential to adapt to new technologies, production structures, and product allocation (Bhunia 2018). The adaptation scenario, especially in the Global South, is rather nascent.

Several existing literatures have claimed that rapid technological advancements and the availability of the latest technology act as catalysts for growth and development (Hekkert et al. 2007; Deng and Nelson 2011) critical in evaluating an economy's growth. This phenomenon has been keenly observed in developed countries. However, the countries in the Global South have been put behind due to a lack of infrastructure, investment and accessibility of existing skills towards efficient production options (Lin 2011). Hence, the concept of SMART (Specific, Measurable, Achievable, Relevant and Time Bound) cities that aim at inclusion, safety, efficiency and quality of life within a targeted time frame was popularised in the 20th century in the urban planning of several developing nations. Sustainable mobility, which includes alternative fuels, non-motorised transport and providing adequate infrastructure to transfer costs, air and noise pollution, improved transfer speed, and reduce traffic congestion. Thus, aiming at building resilient urban spaces with the help of technology (Venezia and Vergura 2015).

Transition, be it in terms of technological adaptation or otherwise, entails certain uncertainties. In the Global South context, the characteristics of income disparity, poverty and accessibility have been the major reasons for their delay in adaptation to technological advancement compared to the rest of the world (Chant 2009). Therefore, this chapter deals with analysing the stance of the countries of the Global

South, especially achieving the adoption of Industry 4.0 and outlining prospects for them to adapt to the upcoming Industry 5.0, shortly in sustainable mobility in smart cities. This chapter has been further subdivided based on key features of Industry 4.0, prospects and characteristics of Industry 5.0 and the difference between Industry 4.0 and 5.0. It also discusses the global leaders of technological advancements and their stance on Industry 4.0, followed by a review of the literature that states adaptation and improvements from a Global South perspective. Further, it discusses the scenario of GHG emissions, how Industry 4.0 and 5.0 could help mitigate them, and the energy security concerns. This is further substantiated by a relevant case study followed by a policy framework highlighting the optimum usage of the Industrial Revolution. Thus, aiming to provide a roadmap to accelerate the adoption of Industry 5.0 to achieve 'Net-Zero' in the Global South.

2 Understanding Industry 4.0 and Industry 5.0 and Their Differences

2.1 Industry 4.0 and Its Key Features

The Industrial Revolution has played a quintessential role in developing the manufacturing industries and their functions. Mechanisation using steam and water power, mass production in large assembly lines using electricity and automated production using computers and robotics were the contributions of the First (1750–1850), Second (1850–1915) and Third (1970-Early 2000s) Industrial Revolutions (Kearney 2018). Further, the fourth Industrial Revolution, which originated in 2011 as an advanced strategy by the government of Germany to promote computerisation, was introduced as Industry 4.0 at the 'Hannover Fair'. Presently, Industry 4.0 is referred to as a combination of the existing structure of manufacturing along with modern digital technology such as cloud computing, robots, drones, autonomous cars, 3D printing, AI, and the IoT (Abdelmajied 2022). Distinguished with the ability to network, analyse, act and communicate, these technologies enable their adapters to make data-driven, intelligent and responsive decisions.

Furthermore, the salient features of Industry 4.0 include interconnection, transparency of information, technical aid and decentralised decision-making through the Internet of People (IoP), making the production process involving technological systems that can support individuals in making decisions, solving problems, and performing risky or difficult tasks.

2.2 *Prospects of Industry 5.0:*

Industry 5.0 refers to an escalation to mass production and an industrial approach to growth and development. It is the mechanism characterised by the "human-centric" approach wherein humans, with the help of robotics and artificial intelligence, can increase their production and efficiency by resilient and sustainable methods of production (Adel 2022).

Industry 5.0 includes 'human centrism', 'resilience', and 'sustainability' as its three pillars. These pillars aim to support human intuition and intellect with utmost productivity, increase production by ensuring enhancement of production capacity across diverse sectors, and improvise technology and its production to cater to the rising side effects of climate change.

2.3 *How Can One Differentiate Between Industry 4.0 and 5.0?*

Both Industry 4.0 and 5.0 highlight the technological push in industrial practice. With the rise in population and subsequent increasing demands, the twenty-first century has delivered the need to include socio-economic dynamics such as income, educational qualification, social structure, and employment status of a society (Bach 2021). Hence, with the increasing human–computer interaction came the risk of surrounding oneself with technology such as cybersecurity.

Industry 4.0 is based on big data analytics, artificial intelligence, simulation, and the Industrial Internet of Things (IIOT). On the other hand, the Fifth Industrial Revolution or Industry 5.0 further leads the relay of productivity by considering environmental sustainability, human-centric customer experience, and resilience. (Martini et al. 2024).

Further, in terms of approaches to production, Industry 4.0 functions in repetitive movements to achieve reliability through human–computer interaction. However, in the context of Industry 5.0, this idea of human-capital interaction extended to the safety of employees and management decisions on organizational behaviour in terms of training and skill development of employees (Bag et al. 2021).

Investing in Industry 4.0 ensures waste prevention through data analytics and system optimization, but it leads to increased material consumption and energy usage. Industry 5.0 addresses these limitations by promoting waste prevention and recycling, investing in renewable energy sources, and integrating smart and energy-autonomous sensors into the existing industry structure for energy efficiency.

3 Objective of the Study

The chapter highlights the 'Net-Zero emission' goals in the Global South in the context of their abilities to adapt to Industrial Revolution 4.0 and outlines the prospects of Industry 5.0 in these nations with respect to alternative fuels and smart cities.

4 Review of Literature

One of the distinct characteristics of a developing country could be gauged by its income disparity with that of a developed country. As technology expands, production increases, thus ensuring increased employment, earnings, and returns to the investment (Basu et al. 2006). Initially, the approach to adaptation of Industry 4.0 was adopting smart technology to increase production and digitalise the manufacturing process (Pozzi et al. 2023). The context of Industry 4.0 in the present day is rather an extension of technology applications in the social space.

The manufacturing sector in the present day has included sustainability as one of its factors. However, only twenty-five nations from East Asia, Europe, and North America involve Industry 4.0 in their production process, and these nations alone are responsible for global manufacturing value added (Wade 2018). The involvement of Industry 4.0 in the existing operational technologies expands to include all industries, such as oil and gas, automobiles, pharmaceuticals, and electronics. Digitalisation, AI, and the industrial IoT have addressed the limitations of efficiency, access, and remote monitoring (Javaid et al. 2021).

In its fourth edition, the IR (Industrial Revolution) has contributed to growth and development in the economies of the Global South (Manda and Ben Dhaou 2019). Along with providing multisided platforms for inclusive business environments, the involvement of robotics and big data analytics has helped to reshape the production hierarchy of the existing industries in the Global South. Latin American nations like Brazil and Mexico have demonstrated great innovation and influence worldwide (Herrero-Solana and Piedra-Salomón 2022). In addition, the travel and tourism sector in the Global South is embracing the idea of Industry 4.0 and investing in its ramifications and potential future trends (Henama and Sifolo 2021).

Industry 4.0, characterised by digital transformation and revolutionary technologies, is vital to the Global South's attempt to attain net-zero emissions. It offers prospects for intelligent and sustainable production, encompassing not only smart factories but the entire value chain, inclusive of logistics and product life cycles (Hofmann and Marco 2017). Moreover, Industry 4.0's potential for sustainability is consistent with the Sustainable Development Goals (SDGs) of the United Nations (UN), emphasising the significance of establishing an autonomous, interoperable system that can process data in real-time. Thus, substantiating the need of adapting the need of the Global South to adapt to Industry 4.0.

4.1 SMART Cities and Sustainable Mobility

The concept of SDG 11 of sustainable cities aims at making cities inclusive, safe, resilient and sustainable. 'Sustainable mobility' is an integrated approach that involves sustainable and safe transport networks, Intelligent and cooperative vehicle technologies, and alternative fuel adoption (Pallonetto 2023).

Further, the Sustainable Urban Mobility Plan (SUMP) is inclusive of the SMART Mobility strategy and SMART Mobility measures (Müller-Eie and Kosmidis 2023). The adoption of alternative fuel vehicles, such as Electric Vehicles, hydrogen, and biofuels, plays a major role in enhancing energy efficiency. While the developing economies are shifting their dependence on vehicles from privately owned vehicles to public transport (Pallonetto 2023). The developed nations, for instance, the European Union Alternative Fuels Infrastructure Regulation (AFIR), have taken initiatives such as expansion of public charging and hydrogen refuelling stations, which have been an encouragement for the consumer adoption of alternative fuels. Further implementing carbon–neutral policies in the US is a combination of 'clean fuel' tax incentives and investment in EV infrastructure to attract investment, consumers, and innovation towards sustainable mobility in society (Auriemma 2025).

4.2 Climate Change, Industry 4.0 and Sustainable Mobility

The twenty-first century stands as a witness to the fact that anthropogenic activities are the primary cause of climate change. One of the most viable indicators of climate change is Greenhouse Gas (GHG) emissions (Bruhwiler et al. 2021). The greenhouse effect observably traps the heat on the Earth's surface for longer periods. Thus, contributing to the rise of surface temperatures and climate change.

Further, 'Climate Change' has been integral to several debates in the past decade (Ajmi et al. 2015). However, in 2015, the Paris Agreement highlighted the notion of 'Net-Zero'. Net Zero was targeted to limit global emissions to not more than 1.5 °C, reduce emissions by 43% in 2030, and reach 'net zero' by 2050 (International Energy Agency 2023).

Industry 4.0 has been proven efficient in reducing carbon footprint in several ways, such as promoting energy efficiency, reducing water usage, and recycling liquid waste. Thus, assuring optimal usage of resources. However, the applications of Industry 4.0 are diverse and vary across manufacturing, oil and gas, food and hospitality, logistics, and transport (Kaul 2013).

A few applications of Industry 4.0 in curbing climate change include explicit usage of big data, IoT integration, and blockchain technology that could be extended to international entities such as the International Monetary Fund (IMF) and the United Nations (UN) for financial and strategic support. Hence, simulation can aid in gauging the potential hindrances in the manufacturing process in advance by utilizing the available data from the existing machinery. Moreover, as cybersecurity enhances

Table 1 Role of industry 4.0 in application in sustainable mobility

Industry 4.0 technology	Application in sustainable mobility	Impact on climate change mitigation
Internet of things (IoT)	Real-time monitoring of vehicle emissions, traffic flow, and energy usage	Reduces emissions through optimised routes and energy-efficient operations
Artificial intelligence (AI)	Traffic management systems, predictive maintenance for vehicles, and EV optimisation	Minimises congestion and fuel consumption, reducing greenhouse gas emissions
Big data analytics	Analyzing travel patterns to improve public transit and shared mobility systems	Encourages efficient transit solutions, lowering private vehicle dependency
Digital twins	Virtual simulation of urban transport systems for optimal planning and development	Reduces resource wastage and ensures sustainable urban mobility solutions
Blockchain	Securing transactions in shared mobility and EV charging infrastructure	Promotes trust in shared services, reducing vehicle ownership and emissions
Autonomous vehicles (AVs)	Deployment of electric AVs for first- and last-mile connectivity	Lowers emissions by optimizing energy-efficient routes and reducing idle time
3D printing	Producing lightweight vehicle components	Improves vehicle fuel efficiency and reduces material waste
Smart grids	Integration of renewable energy with EV charging infrastructure	Supports a shift to clean energy, cutting fossil fuel reliance
Drones and robotics	Automating freight delivery in urban areas	Reduces emissions by replacing traditional delivery vehicles with drones

Source Authors' own creation

the reliability of technical advancements, the transition and adaptation to modern methods and technical advancements under Industry 4.0 can be made more feasible.

Table 1 Role of Industry 4.0 in Application in Sustainable Mobility Author's own creation:

4.3 *World Stance on Adaptation of Industry 4.0*

Digitalisation and automation have prompted developed countries to embrace technological changes. Industry 4.0 highlights the significance of digitalisation as a key factor for development. Countries can be categorised based on their drivers and structure of production into four types, as presented in Table 1: Leading, High-Potential, Legacy, and Nascent (Bhunia 2018; Kearney 2018) (Table 2).

Table 2 Classification of countries based on their structure and drivers of production

	Large/Complex structure of production	Remarks	Small/Simple structure of production	Remarks
Favourable drivers of production	Leading	Strong current Base Successful future prospects	High-Potential	Restricted current Base Delivers hopeful future prospects
Unfavourable drivers of production	Legacy	Strong Current Base Considerable risks for the future	Nascent	Strong Current Base Holds risk for future

Source Authors' own creation

In the pretext of the given characteristics of the countries based on their drivers and structure of production, Singapore, Switzerland, the United Kingdom and Germany, are some of the leading nations. Further, Lithuania, the Russian Federation and India, are categorised as legacy nations. Meanwhile, Hong Kong SAR, Australia, and Portugal present their characteristic 'High-potential' characteristics. Last but not least, South Africa, Brazil, Indonesia, Argentina, Kenya and Nigeria are nascent in terms of Industry 4.0 and its related advancement (Kearney 2018).

4.4 The Stance of the Global South in the Adaptation of Industry 4.0.

In recent years, with the increase in population, the Global South has witnessed a significant increase in its dependence on technology (Chant 2009). The Global South has seen considerable penetration of smartphones in the economy. Countries like Vietnam, Tunisia, Colombia, South Africa, Mexico, Brazil, the Philippines, Venezuela, China and India rank among the world's leading nations in terms of smartphone penetration. Similarly, Internet penetration is also vastly rising in this region (Ling 2011). China, Indonesia, Brazil, Nigeria, Mexico and Pakistan are the Global South nations with the most internet users worldwide (Duarte 2023). Interestingly, about 1.05 million people, or 75% of the Chinese population, and almost half (49.15% or 692 million individuals) of the Indian population depend on internet usage (Fraske 2022; Duarte, 2023). Further, it is observed that South Africa records an all-time high global average of 9 h and 38 min of internet usage per day (Duarte 2023).

With the advancement in technology and the accumulation of several countries in the Global South, they have taken measures to adapt to Industry 4.0 (Bhunia 2018). However, the key issues among these countries lie in their legal frameworks, consumer protection and investments. For instance, Vietnam has made significant attempts to strengthen its e-commerce base, taxation policy, origin regulation, and product flow control. It has also identified significant barriers across its regions to enhance its digital economy and become a pioneer in Industry 4.0 adaptation. On the other hand, Indonesia aims to improve its existing artificial intelligence base and further strengthen its legal framework (Bogoviz 2019).

India has marked its name as one of the leading economies in the world. Along with its increase in mass production, it has also taken initiatives for cybersecurity by revisiting legal frameworks to ensure sustainable adaptation of Industry 4.0 with Artificial Intelligence and Blockchain technology. Similarly, Thailand has put forward its focus on developing cloud strategy and Artificial Intelligence adoption (Mihardjo 2019; Fraske 2022).

4.5 Emissions and Energy Security

The indication of depleting resources, especially fossil fuels, has urged economies to research and develop their renewable energy sources. Renewable energy has been variedly addressed in the literature due to its direct impact on climate variables such as carbon emission reduction and curbing the increasing surface temperature, thus contributing to a sustainable future (Adams and Nsiah 2019). Energy transition refers to a sectoral shift from a non-renewable source such as oil, coal or gas to a renewable source such as wind, solar or lithium alternatives. The adoption of non-renewable resources has played a critical role in the energy sector (Furlan and Mofurrtarino 2018). In the past century, a few countries like the U.S.A., China, Denmark and Germany have enabled their economies to adapt to renewable energy alternatives (Wang et al. 2021).

Today, power generation, vehicular emissions, and industrial production account for more than the emissions. Meanwhile, 60% of emissions in the Global South are from India and China S (Fuhr 2021). Several studies have established that an energy-secure economy can be efficient in creating a ripple effect in other policies, including environmental preservation, social equity, and poverty alleviation (Scheyvens and Momsen 2020). Hence, involving technological advances in the Fourth and Fifth Industrial Revolutions shall be a prosperous initiative, individually in all sectors and assist in minimising carbon emissions and reducing greenhouse gas emissions across the Global South.

Limitations of Global South to Achieve "Net-Zero":

Slower adaptation of technology and low investment in research and development in the Global South nations is one of the major reasons for the delay in the adoption of Industry 4.0. The Global South is also a victim of undefined cybersecurity and data privacy laws. During 2005–2009, eleven out of twelve nations most attacked by cybercrime were from the Global South including Colombia, Romania, Bangladesh and Vietnam, to name a few. Thus, the rising trends of adaptation of Industry 4.0 severely indicate the need to define cybersecurity and data privacy laws

(Kirgel 2016). But, due to low literacy rates, technological advances and adaptation applications are rather restricted (Hammer and Ness 2021). Moreover, poor working conditions, monotonous work structure, and limited accessibility to technology are existing limitations of labour surplus economies. Due to its Geopolitical Diversity the Global South nations are affected by border issues, political tensions, and security investment. However, the share of trade from regional cooperation among these nations is less.

5 Usage of Technology to Solve Climate Change: A Case Study on Singapore's Sustainable Mobility Plan

The decreasing air quality levels have put forward the urgency of addressing climate action. In an attempt to reduce the repercussions of climate change, several countries of the Global South have resorted to technological advancements, but keeping up with the environmental facet is still far in sight (Banister 2011). However, countries like Singapore and India could represent themselves as excellent examples that incorporating modern technology and increasing investments in research could accelerate the adaptation of Industry 5.0 in our societies. Singapore's Smart Nation Program, launched in 2014, was a multidimensional vision to incorporate technology into society (Ho 2017). Initially, its aim was digitalisation and the expansion of SMART governance. However, at present, this initiative successfully forms Intelligent networks and digital platforms for urban management in real-time and optimises policy implementation. Further, their Smart mobility agenda involves the use of AI, promotes EV infrastructure, and enhances the smart traffic systems. Apart from the above functions, the government has also taken up literacy initiatives to educate the population about the benefits of technology. Thus, enabling and ensuring sustainability and resilience in mobility through climate-smart infrastructure (Frana 2024).

In the Indian context, the transition to Electric Vehicles from traditional Internal Combustion Engine (ICE) Vehicles was initially regulated in the form of the National Council for Electric Mobility (NCEM) in 2011 to make policy recommendations to boost electric mobility in the country. This was followed by a National Automotive Board (NAB) in 2012 to accelerate the adoption of Electric Mobility through Faster Adoption and Manufacturing of Electric and Hybrid Vehicles (FAME) (NITI Aayog 2015). A similar attempt at EV adoption in Pune, India, was initiated through the FAME scheme. However, the unequal infrastructure distribution, geographical barriers and investment limitations have a negative impact on digitalisation and EV adoption in this Indian city (Patil and Thopate 2025). Thus, highlighting the stark difference between the two nations.

5.1 Insights from the Case Study

The involvement of Industry 4.0 in developing economies involves implementation at the grassroots level. These procedures involve much strategy, planning, and implementation. The stance of the Global South in all these scenarios is unequal. Hence, this procedure involves an in-depth analysis of skills, availability of economic and non-economic resources, and the risk-taking characteristics of a state.

Therefore, state policies are expected to invest in research to ensure the feasibility and scalability of human–computer interactions like the one in the above-mentioned case study. Further, leading nations like Singapore can act as a unique reference to other nations by helping them become guides to start small businesses and enabling them to create a technology-oriented production ecosystem. Last but not least, if the Global South nations individually strategize their technological ecosystem regarding the guidelines of Sustainable Development Goals (SDGs) by the United Nations Framework Convention on Climate Change (UNFCCC), they should make 'climate-specific' policy decisions to achieve 'Net-Zero' in an efficient manner.

5.2 Conclusion

The advancements in technology and their impact on human adaptation are the modern-day transition. While the world awaits the growth of the Global South, it has become quintessential for these nations to ensure sustainability and leadership. The member nations of the Global South contribute to more than half of the world's population and are victims of almost similar socio-economic conditions. Hence, these regions require an exhaustive policy framework incorporating the interregional advantages and benefits. Thus, benefit from the advantages of Industrial 4.0 and achieve faster adoption of mitigating measures for the Net-Zero target at the earliest.

The implications of Industry 4.0 and 5.0 are large. Therefore, the Global South should take references from member nations in terms of trade, skills and policy regimes. For instance, the educational promotion initiative of Bangladesh's '10 min School' and Sri Lanka's 'Layup'. Similarly, India's initiative to go local globally through schemes like 'Make in India' and 'Start-up India'.

This chapter highlights and puts forward the favourable solutions to adapt to Industry 4.0 to increase production, promote development and achieve 'net-zero'. With the use of artificial intelligence, the Internet of Things, and big data, the complexities of the world, especially the Global South, could be ironed out as they provide energy efficiency and a boost in production and thus ensure sustainability. One such instance wherein the integration of Industry 4.0 and Industry 5.0 is illustrated with the help of a case study in the chapter. It highlights that with the help of Industry 4.0 and 5.0, it is possible to reduce the adverse effects of climate change.

Currently, most countries in the Global South are in their nascent stage of adoption of Industry 4.0 due to the need for more finance, organisational structure and resistance to change. In such a scenario, attempts at international collaboration, a broader 'end-to-end' value chain, informed and updated digital platforms, and sustainability-driven education, consumption, and resource management campaigns can not only foster growth and development in the underprivileged countries of the Global South; but also pave prospects for achieving Industry 5.0 by enabling sustainable adoption of sustainable development goals through Society 5.0.

5.3 Future Implications of the Study

The study highlights the challenges and barriers related to the Global South Nations in achieving 'Net-Zero' in the pretext of Industry 4.0 adoption. Thus, adding itself to the literature for the prospects of Industry 5.0. Further, the insights from the chapter shall help adapt the 'smart' approach and aid in policymaking, production decisions and research awareness in the context of the Global South. Thus, paving the way for Industry 5.0, along with Energy Security.

References

Abdelmajied FY (2022) Industry 4.0 and its implications: concept, opportunities, and future directions. In: Tamás Bányai ÁB (ed) Supply chain—recent advances and new perspectives in the industry 4.0 era. https://doi.org/10.5772/intechopen.102520

Adams S, Nsiah C (2019) Reducing carbon dioxide emissions; does renewable energy matter? Sci Total Environ 693:133288

Adel A (2022) Future of industry 5.0 in society: human-centric solutions, challenges and prospective research areas. J Cloud Comput 11(1). https://doi.org/10.1186/s13677-022-00314-5

Ajmi AN, Hammoudeh S, Nguyen DK, Sato JR (2015) On the relationships between CO2 emissions, energy consumption and income: the importance of time variation. Energy Econ 49:629–638. https://doi.org/10.1016/j.eneco.2015.02.007

Auriemma V, Nardi L (2025) New models of sustainable mobility in smart cities. Acad Int Sci J 16(32):84–101

Bach C (2021) Leadership in the context of a new world: digital leadership and industry 4.0.. In: Managing global transitions, vol 19. https://doi.org/10.26493/1854-6935.19.209-226

Bag S, Yadav G, Dhamija P, Kataria KK (2021) Key resources for industry 4.0 adoption and its effect on sustainable production and circular economy: an empirical study. J Clean Prod 281:125233. https://doi.org/10.1016/j.jclepro.2020.125233

Banister D (2011) Cities, mobility and climate change. J Transp Geogr 19(6):1538–1546. https://doi.org/10.1016/j.jtrangeo.2011.03.009

Basu S, Fernald JG, Kimball, MS (2006) Are technology improvements contractionary? Am Econ Rev 96(5):1418–1448. https://doi.org/10.1257/aer.96.5.1418

Bekhet HA, Othman NS (2018) The role of renewable energy to validate dynamic interaction between CO2 emissions and GDP toward sustainable development in Malaysia. Energy Econ 72:47–61. https://doi.org/10.1016/j.eneco.2018.03.028

Bhunia P (2018) Only 25 countries well-positioned to benefit from Industry 4.0 according to new World Economic Forum report. OPEN GOV. https://opengovasia.com/only-25-countries-well-positioned-to-benefit-from-industry-4-0-according-to-new-world-economic-forum-report/
Bogoviz AV (2019) Comparative analysis of formation of industry 4.0 in developed and developing countries. In: Industry 4.0: industrial revolution of the 21st century, pp 155–164. https://doi.org/10.1007/978-3-319-94310-7_15
Bruhwiler L, Basu S, Butler JH, Chatterjee A, Dlugokencky E, Kenney MA, McComiskey A, Montza SA, Stanitski D (2021) Observations of greenhouse gases as climate indicators. Clim Chang 165(1):12. https://doi.org/10.1007/s10584-021-03001-7
Chant SH (2009) In: Geographies of development in the 21st century: an introduction to the global South. Edward Elgar Publishing
Deng T (2011) Recent developments in bus rapid transit: a review of the literature. Transp Rev 31(1):69–96
Duarte F (2023) Countries with the highest number of internet users. Retrieved from exploding topics. https://explodingtopics.com/blog/countries-internet-users
Frana PL (2024) Assessing smart nation Singapore as an international model for AI responsibility. Int J Responsib 7(1):2
Fraske T (2022) Industry 4.0 and its geographies: a systematic literature review and the identification of new research avenues. Digit Geogr Soc 3:100031. https://doi.org/10.1016/j.diggeo.2022.100031
Fuhr H (2011) The rise of the global south and the rise in carbon emissions. Third World Q 42(11):2724–2746. https://doi.org/10.1080/01436597.2021.1954901
Fuhr H (2021) The rise of the global south and the rise in carbon emissions. Third World Q 42(11):2724–2746. https://doi.org/10.1080/01436597.2021.1954901
Furlan C, Mortarino C (2018) Forecasting the impact of renewable energies in competition with non-renewable sources. Renew Sustain Energy Rev 81:1879–1886
Goldstone JA (2002) Efflorescences and economic growth in world history: rethinking the rise of the west and the industrial revolution. J World Hist 323–389
Hammer A, Ness I (2021) Informal and precarious work: Insights from the global south. J Labor Soc 24(1):1–15
Hekkert MP (2007) Functions of innovation systems: a new approach for analysing technological change. Technol Forecast Soc Chang 74(4):413–432
Henama US, Sifolo PP (2021) Fourth industrial revolution (4IR) in the global south: the future of work in tourism–challenges, opportunities, and future prospects. In: Future of work, work-family satisfaction, and employee well-being in the fourth industrial revolution, pp 259–278
Herrero-Solana V, Piedra-Salomón Y (2022) Industry 4.0 in Latin America: countries and institutions with the greatest innovation and global impact. Preprints. https://doi.org/10.20944/preprints202208.0403.v1
Ho E (2017) Smart subjects for a smart nation? Governing (smart) mentalities in Singapore. Urban Stud 54(13):3101–3118
Hofmann E, Marco R (2017) Industry 4.0 and the current status as well as future prospects on logistics. Comput Ind 89:23–34. https://doi.org/10.1016/j.compind.2017.04.002
Industrallai (2023) What are the differences between industry 4.0 and industry 5.0? https://industriall.ai/blog/what-are-the-differences-between-industry-4-0-and-industry-5-0#:~:text=Industry%204.0%20has%20as%20one,be%20considered%20individually%20and%20personalized
International Energy Agency (2023) Global EV outlook 2023- catching up with climate ambitions. Paris, France: International Energy Agency. https://www.iea.org/reports/global-ev-outlook-2023
Javaid M, Haleem A, Singh RP, Rab S, Suman R (2021) Upgrading the manufacturing sector via applications of industrial internet of things (IIoT). SensS Int 2:100129

Kaul I (2013) The rise of the global South: implications for the provisioning of global public goods. United Nations Development Programme. UNDP-HDRO Occasional Papers. https://hdr.undp.org/system/files/documents/hdro1308kaulpdf.pdf
Kearney A (2018) Readiness for the future of production report 2018. Cologny, Switzerland: World Economic Forum. https://www3.weforum.org/docs/FOP_Readiness_Report_2018.pdf
Kirgel A (2016) Cyber Crime nation typologies: k-means clustering of countries based on Cyber Crime rates. Int J Cyber Criminol 10(2)
Kloß ST (2017) The Global South as subversive practice: challenges and potentials of a heuristic concept. Glob South 11(2):1–17
Lin JY (2011) New structural economics: a framework for rethinking development. World Bank Res Obs 26(2):193–221
Ling R (2011) Mobile communication in the global south. New Media Soc 13(3):363–374. https://doi.org/10.1177/1461444810393899
Liu D, Guo X, Xiao B (2019) What causes growth of global greenhouse gas emissions? Evidence from 40 countries. Sci Total Environ 661:750–766. https://doi.org/10.1016/j.scitotenv.2019.01.197
Martini B, Bellisario D, Coletti P (2024) Human-centered and sustainable artificial intelligence in industry 5.0: challenges and perspectives. Sustainability 16(13):5448
Mahler AG (2017) Global south. Retrieved from Oxford bibliographies in literary and critical theory ed. Eugene O'Brien. http://spanitalport.as.virginia.edu/people/agm7b
Manda MI, Ben Dhaou S (2019) Responding to the challenges and opportunities in the 4th Industrial revolution in developing countries. In: Proceedings of the 12th international conference on theory and practice of electronic governance, pp 244–253
McKinsey & Company (2022) What are industry 4.0, the fourth industrial revolution, and 4IR? McKinsey & Company. https://www.mckinsey.com/featured-insights/mckinsey-explainers/what-are-industry-4-0-the-fourth-industrial-revolution-and-4ir
Mignolo WD (2011) The global south and world dis/order. J Anthropol Res 67(2):165–188
Mihardjo LS (2019) Digital leadership role in developing business model innovation and customer experience orientation in industry 4.0. Manag Sci Lett 9(11):1749–1762. https://doi.org/10.5267/j.msl.2019.6.015
Müller-Eie D, Kosmidis I (2023) Sustainable mobility in smart cities: a document study of mobility initiatives of mid-sized Nordic smart cities. Eur Transp Res Rev 15(1):36
Niti Aayog, Government of India (2021) State Level Policies, from e-Amrit State Level Policies: https://e-amrit.niti.gov.in/state-level-policies. Accessed on 23 June 2023
NITI Aayog, Government of India (2015) Status quo analysis of various segments of electric mobility and low carbon passenger road transport in India. New Delhi. https://www.niti.gov.in/sites/default/files/2021-04/FullReport_Status_quo_analysis_of_various_segments_of_electric_mobility-compressed.pdf. Accessed on 20 June 2023
Pallonetto F (2023) Towards a more sustainable mobility. Handbook of computational social science for policy. Springer International Publishing, Cham, pp 465–486
Patil A, Thopate K (2025) Barriers to EV adoption and sustainable energy. In: Modern computing technologies for EV efficiency and sustainable energy integration. IGI Global Scientific Publishing, pp 251–288
Patel, V. (2023, May 8). *Reducing Carbon Footprint with Industry 4.0: A Pathway to Net-Zero Emissions.* Retrieved from LinkedIn: https://www.linkedin.com/pulse/reducing-carbon-footprint-industry-40-pathway-net-zero-vish-patel
Pozzi R, Rossi T, Secchi R (2023) Industry 4.0 technologies: critical success factors for implementation and improvements in manufacturing companies 34(3):139–158. https://doi.org/10.1080/09537287.2021.1891481
Scheyvens R, Momsen JH (2020) Tourism and poverty reduction: issues for small island states. In: Tourism and sustainable development goals. Routledge, pp 111–129
Venezia E, Vergura S (2015) Transport issues and sustainable mobility in smart cities. In: 2015 international conference on clean electrical power (ICCEP). IEEE, pp 360–365

Wade R (2018) Governing the market: economic theory and the role of government in East Asian industrialization

Wang WH, Moreno-Casas V, Huerta de Soto J (2021) A free-market environmentalist transition toward renewable energy: the cases of Germany, Denmark, and the United Kingdom. Energies 14(15):4659

Harnessing Namibia's National Spatial Data Infrastructure for Data-Driven Sustainability: Sectoral Evidence, Technological Innovations, and Pathways to Environmental Resilience

Philip Mensah, Eric Yankson, and Derrick Frimpong Buabeng

Abstract Namibia has made important progress in deploying digital infrastructure to advance sustainable development, with the Namibia Statistics Agency spearheading the establishment of a National Spatial Data Infrastructure (NSDI). Conceived as a framework for improving real-time data accessibility and coordination across government institutions, the NSDI was designed to strengthen evidence-based decision-making in national development planning. Yet, its potential has not been fully realized, owing to weak sectoral integration, limited performance monitoring, and slow adoption of advanced digital tools. This study responds to the lack of empirical insight into how Namibia's NSDI supports sustainability outcomes, particularly its role in tracking progress toward the Sustainable Development Goals. Drawing on a qualitative review of institutional reports, national frameworks, and available geospatial platforms, the research examines the extent to which the NSDI enables data-driven governance in sectors such as urban planning, environmental management, water and land use, and disaster preparedness. Findings indicate that, while Namibia's NSDI provides a solid institutional and technical base, its application to real-time monitoring and inter-sectoral data sharing remains uneven. The study concludes by recommending actionable strategies—such as integrating mobile geographic information systems, interactive dashboards, and automated reporting systems—to optimize NSDI's effectiveness, offering lessons for countries seeking resilient, data-driven pathways to sustainability.

P. Mensah (✉) · D. F. Buabeng
Department of Land and Spatial Sciences, Namibia University of Science and Technology, Windhoek, Khomas Region, Namibia
e-mail: pm319834@gmail.com

D. F. Buabeng
e-mail: derrickfrimpongbuabeng@gmail.com

E. Yankson (✉)
Department of Architecture, Planning and Construction, Namibia University of Science and Technology, Windhoek, Khomas Region, Namibia
e-mail: eyankson@nust.na; ericyankson22@gmail.com

R. Singh et al. (eds.), *Data-Driven Sustainability: Harnessing Technology for A Greener Future and Environmental Resilience*, World Sustainability Series,
https://doi.org/10.1007/978-3-032-13412-7_5

Keywords Data-driven governance · Environmental resilience · Geospatial intelligence · National spatial data infrastructure · Sustainable development goals · Technological innovation

1 Background Orientation

The accelerating pace of environmental change, compounded by climate disruption, unsustainable resource consumption, and socio-economic inequality, underscores the urgent need for innovative mechanisms that align data, technology, and policy in tackling sustainability crises. Worldwide, governments and development partners are increasingly recognizing the central role of spatial data infrastructures (SDIs) in providing reliable, interoperable geospatial information to inform decision-making in environmental governance, disaster risk management, urban planning, and resource allocation (Scott and Rajabifard 2017). National Spatial Data Infrastructures (NSDIs) in particular represent institutional frameworks that integrate policies, technologies, standards, and human resources to facilitate spatial data collection, sharing, and use. By enabling collaboration and reducing duplication, they support sustainable development pathways and help bridge sectoral divides in environmental management (Bello 2024; Scott and Rajabifard 2017). Namibia's NSDI, more than a decade in the making, reflects this global trend yet faces critical challenges and opportunities that must be addressed if it is to fulfill its role in sustainability transitions.

Namibia, one of the most arid nations in sub-Saharan Africa, remains highly vulnerable to environmental and socio-economic pressures. Recurring droughts, land degradation, biodiversity loss, and persistent water scarcity intersect with rapid urbanization and growing inequality (Spear et al. 2018). Climate change threatens to intensify extreme events, further stressing agriculture, water management, energy supply, and settlement systems. In this fragile context, data-driven strategies are essential for resilience and long-term planning. Namibia's NSDI was conceived to provide a coherent framework for managing and disseminating geospatial data that can guide policy and sectoral interventions. However, despite institutional progress, including the establishment of the Namibia Statistics Agency (NSA) and data standardization efforts, major gaps remain in integrating sectors, adopting technological innovations, and translating data into actionable sustainability outcomes (Ogunbiyi 2021).

Globally, sustainable development efforts highlight the need for interconnected data ecosystems that transcend national boundaries. Initiatives such as the United Nations Global Geospatial Information Management (UN-GGIM) and the Group on Earth Observations (GEO) have emphasized the value of harmonized geospatial frameworks in supporting the Sustainable Development Goals (SDGs). The SDGs, adopted in 2015, rely heavily on spatially disaggregated data for tracking progress on poverty reduction, climate action, environmental sustainability, and urban growth. Yet many developing countries, Namibia included, still lack robust infrastructures to generate such evidence (Mariathasan et al. 2019). This underscores a crucial gap: while Namibia's NSDI exists as a policy framework, the extent of its use in producing sectoral evidence, driving innovation, and supporting resilience remains underexplored.

Emerging technologies are reshaping how societies manage urban and environmental systems. Smart city initiatives, supported by the Internet of Things (IoT), artificial intelligence (AI), cloud computing, and blockchain, are already optimizing mobility, reducing emissions, improving energy use, and enhancing service delivery (Alahi et al. 2023; Van Hoang 2024). In environmental management, tools such as digital twins, predictive analytics, and waste intelligence platforms now generate real-time insights into ecosystems and resource flows, strengthening responsiveness to climate shocks. For Namibia, embedding such technologies into its NSDI could allow the country to bypass traditional constraints and accelerate sustainability planning. Yet empirical evidence on the integration of these technologies within Namibia's key sectors—agriculture, water, energy, and urban development—remains scarce, representing another gap this study seeks to address.

Urbanization and consumption trends further complicate Namibia's sustainability outlook. Cities such as Windhoek are expanding rapidly, driving demand for land, housing, energy, and transport. This growth often leads to unsustainable consumption of natural resources and increasing carbon footprints (Bebi and Iyambo 2025). Without geospatial frameworks that capture consumption dynamics, urban expansion can deepen socio-environmental inequalities. Namibia's NSDI could serve as the architecture for mapping these patterns, identifying vulnerability hotspots, and informing sustainable production and consumption strategies. However, no systematic mechanism currently links consumption-related data with environmental monitoring, leaving an important gap in both policy and practice.

Ethical considerations also accompany digital sustainability. As Namibia strengthens its NSDI, issues of data privacy, security, accessibility, and equity cannot be ignored. Global debates stress the importance of transparent governance, inclusivity, and fair access to geospatial information (Micheli et al. 2022). For Namibia, it is vital that rural populations, informal settlement dwellers, and smallholder farmers benefit from NSDI innovations. Addressing these ethical dimensions requires frameworks for equitable access and digital literacy; yet such safeguards are still limited in Namibia's current NSDI strategy, further highlighting the need for deeper investigation.

Sector-specific applications illustrate both potential and shortfalls. In agriculture, spatial data could advance precision farming, drought forecasting, and sustainable land-use planning, thereby supporting food security. For water resources, NSDI applications could improve integrated planning and balance competing demands. In urban development, spatial intelligence can inform strategies that harmonize expansion with sustainability, while in conservation, it can guide biodiversity protection and ecosystem valuation (Mwange et al. 2018; Ogunbiyi 2021). Despite these opportunities, Namibia lacks comprehensive studies evaluating NSDI's performance across these sectors or identifying emerging innovations that strengthen resilience.

Against this background, the present study investigates how Namibia's NSDI can be harnessed for data-driven sustainability through sectoral evidence, technological

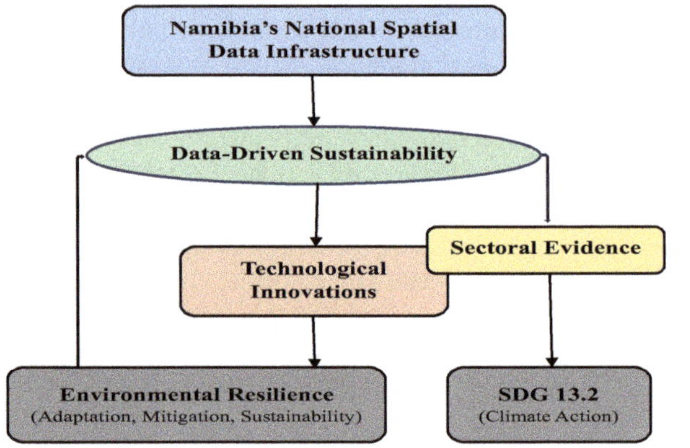

Fig. 1 Conceptual framework of harnessing Namibia's NSDI for sustainability. *Source* Authors' research (2025)

innovations, and pathways to resilience. Specifically, it examines NSDI's current sectoral performance, explores its integration of emerging technologies, and identifies strategies for strengthening geospatial intelligence in sustainability transitions. The study fills a critical knowledge gap by linking Namibia's NSDI to global sustainability discourses and situating its experience within wider international frameworks. Crucially, it contributes to SDG 13—particularly Target 13.2, which calls for integrating climate measures into national policies and planning. By illustrating how geospatial data infrastructures can embed climate resilience into development strategies, this research positions Namibia's NSDI as a catalyst for transformative adaptation and mitigation (see Fig. 1).

The study's broader significance lies in its ability to inform national policy and global debates alike. Nationally, it provides evidence-based insights to guide Namibia's NSDI toward greater impact in sustainability transitions. Globally, it contributes lessons for resource-constrained countries seeking to use NSDIs in advancing resilience and equity. By integrating digital innovations and ethical perspectives, the study expands the discourse beyond infrastructure to the broader dimensions of environmental resilience and digital sustainability.

In summary, Namibia's NSDI embodies both opportunities and constraints in advancing data-driven sustainability in a context of climate vulnerability and social inequality. By explicitly aligning the study's objectives with SDG 13.2 and situating the Namibian case in a comparative global framework, this research offers actionable insights for policymakers, practitioners, and scholars committed to leveraging geospatial infrastructures for sustainability transitions.

2 Materials and Methods

The methodological orientation of this study was anchored in a qualitative design aimed at examining how Namibia's NSDI supports data-driven governance across sustainability-related sectors. A qualitative approach was most appropriate as it enabled contextual depth and interpretive analysis of institutional frameworks, governance structures, and technological adoption in Namibia's policy environment (Cantelmi et al. 2021). Unlike quantitative models, which emphasize statistical generalizations, this study prioritized nuanced insights into how institutional arrangements shape the performance and implementation of NSDI. This was crucial because NSDI, though designed to standardize and streamline geospatial data, is deeply embedded in governance and institutional structures that require interpretive exploration to reveal both opportunities and constraints (Masser and Crompvoets 2018; Georgiadou et al. 2015). To achieve this, the study analyzed institutional reports, national planning frameworks, and existing geospatial platforms to build a comprehensive understanding of NSDI's sectoral impact.

An exploratory case study design was employed, with Namibia's NSDI constituting a bounded system of policies, institutions, and technologies. This methodology allowed for a holistic examination of how the NSDI functions within the Namibian context (Ogunbiyi 2021). The NSA, as the lead implementing authority, served as the institutional entry point. Perspectives were also drawn from line ministries, municipal authorities, and sectoral agencies involved in urban planning, land and water management, environmental protection, and disaster risk preparedness. This triangulation enabled the research to capture diverse experiences of how NSDI was applied in practice and aligned with both national priorities and global sustainability agendas (Nunes and Camboim 2025). By focusing on Namibia as a deeply investigated case, the study generated insights relevant to other developing countries building spatial data infrastructures.

Primary data came from institutional reports by the NSA and other government agencies, as well as key national frameworks including Namibia's Fifth National Development Plan (NDP 5), the Harambee Prosperity Plans, and the National Spatial Development Framework (NSDF). Sectoral policies on water, land, and environment were examined to assess the integration of geospatial data into strategic objectives (Ahmad and Ali 2023). Technical assessment was also carried out on geospatial platforms such as the Namibia Geoportal and associated metadata catalogues to evaluate functionality, accessibility, and capacity to support decision-making. To situate Namibia's NSDI in a wider perspective, reports from international organizations such as the United Nations and African Union were included.

Institutional documents and policy frameworks were analyzed through qualitative content analysis. Texts were systematically coded to identify recurring themes related to governance, coordination, sectoral applications, and technological integration of NSDI (Nunes and Camboim 2025). Both inductive and deductive coding were applied, which allowed the study to capture emergent issues while staying aligned with the research questions. This process identified challenges around data accessibility, metadata standards, reporting mechanisms, and alignment with SDG indicators.

Sectoral integration was a focal point, given Namibia's pressing challenges in urban development, water scarcity, environmental management, and disaster preparedness. Urban planning was assessed through municipal frameworks and settlement growth reports, with emphasis on Windhoek. Environmental management was examined in relation to biodiversity conservation and environmental impact assessments. Water and land governance were reviewed through Ministry of Agriculture, Water, and Land Reform documents, focusing on water allocation and tenure security. Disaster preparedness was considered by evaluating early warning systems, drought monitoring, and flood mapping. Across these sectors, NSDI's contributions and performance gaps were evaluated.

To enhance empirical grounding, a preliminary assessment of sectoral performance against selected SDG indicators was undertaken. While not a full quantitative analysis, the study linked NSDI-supported processes to progress on global targets (Sachs et al. 2023; 2024; 2025). Urban planning performance was tied to SDG 11.3 on sustainable urbanization, environmental management to SDG 15.1 on terrestrial ecosystems, water governance to SDG 6.4 on water-use efficiency, and disaster preparedness to SDG 13.1 on resilience to climate hazards. These connections structured the assessment of how NSDI underpins sustainability efforts.

The performance review synthesized available statistics with geospatial outputs. For example, urban land use data from geoportals were compared with municipal reports, while biodiversity monitoring data were cross-checked against conservation reports. Water resource mapping was analyzed alongside water access indicators, and disaster assessments were compared with early warning outputs. Although limited by data gaps, this triangulated analysis revealed the enabling role of NSDI in evidence-based SDG monitoring.

A technical review of Namibia's Geoportal and metadata catalogues complemented this work. The Geoportal was assessed for accessibility, user interface, and inter-sectoral sharing capabilities, while catalogues were reviewed for compliance with international data standards. Special consideration was given to advanced technologies, including cloud-based systems, artificial intelligence in environmental monitoring, and mobile GIS tools for field data. This highlighted both achievements and gaps in NSDI's technological application.

Triangulation was used throughout for validation, cross-referencing institutional reports, platform data, and regional benchmarks (Vivek 2023). For instance, environmental monitoring findings were compared across national, geoportal, and international sources, strengthening reliability.

The study recognized several limitations, including uneven availability of sectoral data, the evolving nature of NSDI, and the inability to generalize qualitative findings beyond Namibia. Nonetheless, the approach ensured robust insights that inform both national and global debates. Ethical standards were maintained by relying on publicly available data, respecting confidentiality, and citing institutional sources responsibly.

In summary, the methodological workflow combined qualitative content analysis with technical platform review to investigate NSDI's role in supporting governance for sustainability. Sectoral performance was aligned with selected SDG indicators, and triangulation enhanced reliability. Table 1 summarizes the methodological stages, showing how the study examined NSDI's rationale, effectiveness, and optimization pathways, situating Namibia's experience within broader global discourses on geospatial infrastructures and sustainable development.

Table 1 Summary of methodological approach

Stage	Description	Key elements
Research design	Qualitative, exploratory case study of Namibia's NSDI	Institutional processes, governance frameworks, technological integration
Case study focus	NSDI as a bounded system	NSA entry point; ministries, municipalities, sector agencies
Data sources	Institutional reports, national frameworks, geospatial platforms, international reports	NDP5, Harambee Plans, NSDF, Geoportal, UN/AU reports
Data analysis	Qualitative content analysis	Coding, thematic categorization, SDG alignment
Sectoral integration	Application of NSDI in sustainability-related sectors	Urban planning, environment, water/land, disaster preparedness
SDG alignment	Linking performance to SDG indicators	SDG 11.3, SDG 15.1, SDG 6.4, SDG 13.1
Validation	Triangulation across sources	Reports, platforms, international benchmarks
Limitations	Constraints acknowledged	Uneven data, evolving NSDI, limited generalizability
Ethical considerations	Responsible practices	Confidentiality, appropriate citation
Findings and implications	Contribution to sustainability	NSDI as enabler of governance and sustainable development

Source Authors' interpretive method (2025)

3 Results and Discussion

This section presents the results and discussions, thematically aligned with the research objectives. It highlights Namibia's NSDI success stories, empirical evidence, and critical insights, while also exposing challenges, gaps, and opportunities. The analysis underscores its transformative role in sustainable development and tracking SDG progress.

3.1 Data-Driven Governance in Key Sectors

The first step in addressing the broader aim of this study is to explore the question: *To what extent does Namibia's NSDI enable data-driven governance across sectors such as urban planning, environmental protection, water and land management, and disaster preparedness?* This is central to evaluating whether the institutional and technical investments in Namibia's NSDI are shaping governance outcomes in critical sectors and whether these contributions are sufficient to support sustainable development trajectories.

Findings from institutional reports, national frameworks, and geospatial platforms suggest that the NSDI has created a technical foundation for sectoral integration, largely coordinated by the Namibia Statistics Agency. In urban planning, spatial data has informed settlement upgrading strategies and land allocation policies. Municipalities such as Windhoek have used improved cadastral mapping to streamline property registration and expansion planning. Yet, many local authorities lack the technical and financial capacity to fully embed these datasets into real-time planning systems, limiting the transformative potential of NSDI-driven governance. Similar challenges of uneven adoption have been noted in other African contexts, where national geospatial systems exist but are weakly localized (Ballo 2024; Ogunbiyi 2021).

In environmental protection, NSDI-enabled platforms have supported monitoring of land degradation, biodiversity loss, and conservation outcomes. Land cover change data, for instance, has been used in reporting under the United Nations Convention to Combat Desertification and in managing wildlife corridors and national parks. These contributions demonstrate that the NSDI provides a basis for evidence-driven decision-making. However, the persistence of institutional silos—particularly between the Ministry of Environment, Forestry and Tourism and agencies handling land, water, and infrastructure—continues to weaken coordinated environmental governance. This mirrors findings elsewhere that highlight inter-agency fragmentation as a key barrier to the effective use of geospatial systems (Akingbemisilu 2024; Sarker 2025).

Water and land management is another sector where the NSDI has enhanced transparency and accountability. Geospatial datasets are now used by the Ministry of Agriculture, Water and Land Reform for irrigation planning, groundwater mapping, and land adjudication. Digitization of land tenure systems has reduced duplication of records and improved dispute resolution, which is critical for rural development. Nonetheless, the integration of real-time monitoring, such as satellite-based water assessments or drought surveillance, remains limited. The lack of cloud-based GIS systems capable of synchronizing regional water-use data emerged as a critical gap, reducing the ability of NSDI applications to respond to fast-changing climatic pressures.

In disaster preparedness, results point to a mixed performance. The Office of the Prime Minister, through its disaster risk management division, has begun using NSDI datasets to map flood-prone regions in northern Namibia and to improve early-warning systems for drought and flooding. These efforts have been important in targeting vulnerable communities. However, reliance on ad hoc data collection during emergencies persists, reflecting limited integration of NSDI into comprehensive disaster response frameworks. This weakens the timeliness and accuracy of interventions, a trend consistent with broader findings in sub-Saharan Africa, where disaster governance lags behind technical capacity (Bachmann et al. 2022; Van den Homberg and Susha 2018).

Taken together, the results show that Namibia's NSDI has provided the structural basis for data-driven governance, but the degree of impact varies by sector. Urban planning and environmental protection show stronger applications, though capacity gaps and institutional silos constrain their effectiveness. Water and land management highlight the importance of NSDI for dispute resolution and planning, but real-time integration is still underdeveloped. Disaster preparedness has benefited from geospatial mapping but remains too reliant on episodic rather than continuous data use. A synthesis of sectoral performance is presented in Table 2.

Overall, these results confirm that Namibia's NSDI contributes to data-driven governance, but uneven implementation, weak inter-sectoral collaboration, and

Table 2 Extent of NSDI support for data-driven governance in key sectors

Sector	Achievements	Limitations
Urban planning	Access to cadastral, demographic, land-use data; aids infrastructure planning	Slow updates; weak monitoring of informal settlement
Environmental protection	Biodiversity, land, and climate data support conservation planning	Limited remote sensing/AI use; weak predictive tools
Water and land management	Digitized water points, aquifers, parcels; boosts planning efficiency	Institutional fragmentation; poor system integration
Disaster preparedness	Hazard maps and vulnerability data available	Weak real-time data, early warning; and responder access

Source Authors' interpretive review (2025)

limited adoption of advanced geospatial technologies constrain its potential. As highlighted in global debates, SDIs act as enablers rather than guarantees of good governance, with effectiveness depending on the depth of integration and institutional ownership (Crompvoets et al. 2018). For Namibia, strengthening technical capacity, enhancing inter-agency collaboration, and expanding the use of AI and cloud-based GIS are critical steps toward realizing the NSDI's full role in sustainability transitions.

3.2 NSDI's Role in SDG Monitoring

Having analyzed the extent to which Namibia's NSDI supports data-driven governance in key sectors, this section now addresses the research question: *How does Namibia's NSDI contribute to monitoring and reporting progress toward selected SDG indicators?* The results point to both opportunities and limitations, reflecting the NSDI's role as a national mechanism for enhancing sustainability tracking, particularly where spatially disaggregated data is indispensable for accurate reporting.

Evidence from institutional reports and national planning frameworks shows that Namibia's NSDI provides an important backbone for spatial data integration, particularly in land use, environmental management, and socio-economic planning. The NSA, together with sectoral ministries, has increasingly embedded SDG reporting mechanisms in geospatial datasets, which enhance understanding of regional disparities. For example, data on land degradation (SDG 15.3), access to clean water (SDG 6.1), and urban expansion (SDG 11.3) are now represented on interactive geospatial platforms, allowing planners to visualize localized development trajectories rather than relying solely on national averages (Sachs et al. 2023; 2024; 2025). These initiatives demonstrate how the NSDI helps contextualize SDG indicators within Namibia's geographic realities.

Despite this progress, adoption of NSDI-supported tools is uneven across sectors. Environmental protection and disaster preparedness are comparatively advanced, having integrated remote sensing and geospatial monitoring into climate adaptation and disaster risk reduction strategies. Applications tracking drought, floods, and vegetation cover directly support SDG 13 and SDG 15. In contrast, urban planning and social service delivery have been slower to incorporate NSDI tools. Some efforts to map informal settlements in Windhoek and other towns to support SDG 11.1 on housing have emerged, but integration into national reporting remains inconsistent due to the absence of standardized geospatially aligned indicators (Bachmann et al. 2022; Sachs et al. 2024).

Another limitation is that Namibia's NSDI has not yet matured into a system capable of real-time SDG monitoring. Capacity gaps, weak uptake of cloud-based GIS, and reliance on manual reporting processes hinder timely updates. This is particularly evident in water and land management, where datasets often remain static and fail to capture seasonal or annual variations crucial for accurate assessments. The

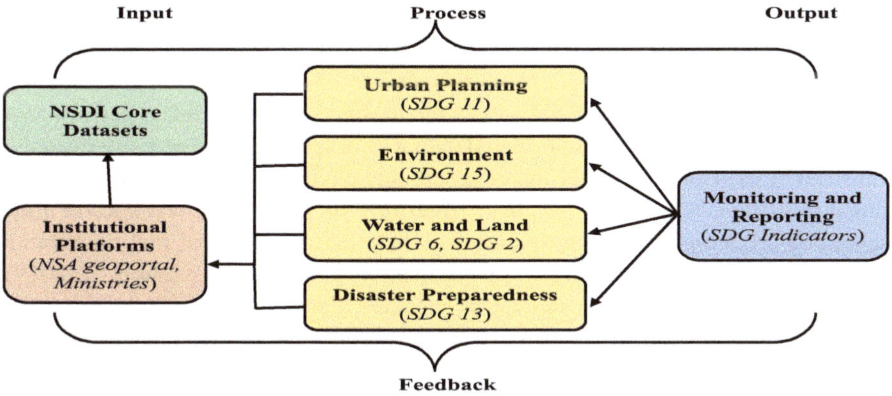

Fig. 2 Proposed conceptual framework of NSDI's role in SDG monitoring based on systems theory. *Source* Authors' research (2025)

delays also affect harmonization of geospatial evidence with Namibia's Voluntary National Reviews (VNRs) to the United Nations.

Nevertheless, pilot initiatives suggest promising pathways. Mobile GIS applications for community-level data collection have shown value in tracking SDG 6 (clean water and sanitation) and SDG 11 (sustainable cities and communities), enabling bottom-up participation and wider spatial coverage. Institutional efforts to align NSDI outputs with the African Union's Agenda 2063 and the UN Global Geospatial Information Management (UN-GGIM) framework further highlight the growing recognition of NSDI as a strategic asset for accountability (Ezigbalike et al. 2016).

In sum, Namibia's NSDI has established a foundation for integrating geospatial intelligence into SDG monitoring, with strong performance in environmental and disaster preparedness but weaker adoption in urban and socio-economic reporting. Expanding standardized indicators, automated dashboards, and real-time data systems would enhance the NSDI's role in generating actionable, timely, and globally comparable evidence for SDG tracking (see Fig. 2).

3.3 *Effectiveness of Performance Reporting*

To combat persistent barriers in sectoral performance, this section addresses the research question: *What are the current limitations in sectoral performance tracking and inter-sectoral data sharing within the NSDI?* Findings drawn from the review of institutional reports, planning frameworks, and geospatial platforms reveal that Namibia's sectoral performance reporting remains at an evolving stage, with significant gaps undermining its alignment with spatial data use. While the NSDI has established institutional and technical foundations through the NSA and a custodianship framework for geospatial data, these foundations have yet to translate into effective performance tracking across key sectors.

One of the most notable challenges is the absence of standardized performance indicators that explicitly integrate spatial data within sectoral reporting. For instance, ministries responsible for water management and environmental protection publish annual performance reviews, but these often rely on descriptive statistics or fragmented datasets rather than geospatial analytics that could reveal real-time trends in land degradation, water scarcity, or biodiversity risks (Mariathasan et al. 2019). The lack of spatially explicit indicators weakens the ability of reports to inform cross-sectoral planning, which remains a central objective of the NSDI framework. Interoperability between platforms is also limited. In agriculture, urban planning, and disaster preparedness, data-sharing practices are constrained, with urban planning especially affected. Reports on informal settlements are still presented in narrative terms, disconnected from geospatial mapping systems, despite their critical relevance for sustainable urban development (Scott and Rajabifard 2017).

Weak adoption of advanced geospatial technologies further constrains performance monitoring. Although Namibia has piloted GIS portals and land information systems, there is little systematic use of cloud-based GIS or artificial intelligence-enabled monitoring tools, which have become international best practices (Lifelo et al. 2024). The absence of such tools makes reporting slow, reactive, and reliant on static datasets that fail to capture dynamic environmental or socio-economic shifts. This finding is consistent with broader African assessments that identify institutional inertia and limited technical capacity as barriers to maximizing the effectiveness of NSDIs (Mwange et al. 2018).

Institutional silos present another obstacle. Agencies continue to produce separate reports with limited cross-validation of data. For example, agricultural performance assessments rarely incorporate spatial data on water availability or soil fertility maintained by environmental agencies. Similarly, disaster preparedness reports underutilize spatial risk maps available within NSDI repositories. These disconnects reflect broader governance issues where reporting is geared toward fulfilling internal accountability rather than promoting multi-sectoral learning and coordinated policymaking (Lee and Ospina 2022).

The formats used for reporting also limit their effectiveness. Most performance reports are still presented in text or table formats with little use of spatial visualizations such as thematic maps or interactive dashboards. This reduces accessibility for decision-makers unfamiliar with complex datasets but who could easily interpret visualized outputs. The lack of visualization also limits transparency and weakens opportunities for public engagement, since communities are unable to track spatial disparities in service delivery or exposure to environmental risks. Table 3 below summarizes the key reporting practices, limitations, and NSDI integration gaps across sectors.

Overall, the effectiveness of performance reporting under the NSDI is constrained by three interlinked weaknesses: the absence of spatially integrated indicators,

Table 3 Sectoral performance reporting practices, limitations, and NSDI integration gaps

Sector	Reporting practices	Key limitations	NSDI gaps
Urban planning	Municipal service and infrastructure data	Fragmented, siloed, non-standard	Weak aggregation; poor GIS integration
Environment	Biodiversity and climate reports	Static, limited spatial use	Low AI/remote sensing uptake; weak SDG link
Water and land	Allocation and land-use statistics	Mostly non-spatial; sector silos	Not harmonized with NSDI standards
Disaster preparedness	Response-oriented reports	Few maps; poor cross-sector exchange	No real-time dashboards; fragmented data
Cross-sectoral	Separate ministerial reporting	Overlap; no standard indicators	No binding data policy; hub role unclear

Source Authors' interpretive analysis (2025)

limited uptake of advanced geospatial tools, and persistent institutional fragmentation. These findings are consistent with global scholarship emphasizing the need to align reporting systems with spatial data infrastructures in order to strengthen accountability and monitor sustainable development more effectively (Neiroukh and Çağlar 2025; Scott and Rajabifard 2017). For Namibia, moving beyond technical foundations to embed geospatial intelligence into reporting systems is critical for enhancing both national and sectoral accountability.

By situating these findings within the African context, it becomes evident that Namibia's challenges are systemic, reflecting the wider regional struggle to shift from data availability toward effective data use. Policy priorities must therefore include harmonizing reporting formats, investing in geospatial technologies for real-time monitoring, and fostering cross-sectoral collaboration. Only then can performance reporting mechanisms fully leverage the NSDI to advance the SDGs and build environmental resilience.

3.4 Adoption of Advanced Digital Tools

Having examined the persistent barriers to sectoral performance, this subsection turns to the research question: *How effectively are advanced digital tools (e.g., AI, mobile GIS, cloud-based GIS) being integrated into Namibia's NSDI for sustainability monitoring?* Evidence from institutional assessments, planning frameworks, and existing geospatial platforms suggests incremental progress, but also reveals significant gaps in technology adoption across key sectors.

The foundational NSDI framework has improved accessibility and coordination of spatial data, yet the integration of cutting-edge tools such as artificial intelligence (AI), mobile GIS, and cloud-based systems remains limited. For example, the Namibia Statistics Agency has implemented GIS and remote sensing applications for land management and census mapping, but most operations are confined to conventional desktop platforms rather than fully mobile or cloud-enabled environments. This reflects a slow pace of digital transition, constrained by institutional readiness, limited technical expertise, and infrastructure bottlenecks (Franchi et al. 2023).

AI, with its potential for real-time monitoring, predictive modelling, and anomaly detection, remains almost absent from Namibia's operational frameworks. Although global practices highlight AI's value for drought forecasting, land degradation assessment, and disaster preparedness (Ziemba et al. 2024), Namibia still relies heavily on manual or semi-automated analyses. This reliance stems from limited investment in computing infrastructure and a shortage of skilled personnel able to deploy AI-driven solutions. Consequently, opportunities to improve early warning systems, track SDG indicators, and strengthen evidence-based policy remain largely underexploited.

Mobile GIS has shown promise but remains underdeveloped within the NSDI. Pilot projects, such as handheld data collection during the 2021 Population and Housing Census, demonstrated efficiency gains and enhanced data accuracy. However, these initiatives have not translated into a standardized framework for regular application across sectors like urban planning, environmental monitoring, and water resource management. This reflects a wider challenge: project-based innovations are rarely scaled into institutionalized practice (Coetzee et al. 2021; Putra et al. 2019).

Cloud-based GIS, widely regarded as a scalable and cost-effective tool for multi-stakeholder collaboration, also lags behind. Namibia's NSDI remains heavily reliant on centralized government-hosted servers, limiting interoperability and real-time analytics. Weak broadband connectivity, particularly in rural areas, further constrains deployment and undermines inclusivity in decision-making. Despite cloud GIS being adopted globally to facilitate cross-sector collaboration and dynamic monitoring (Musa et al. 2024; Mwungu 2017), Namibia has yet to mainstream such approaches into its NSDI.

Three structural factors emerge from these findings. First, institutional inertia and fragmented governance hinder innovation, with agencies often operating in silos rather than through coordinated platforms. Second, reliance on donor-driven projects leads to uneven adoption—pilot initiatives are rarely expanded at national scale. Third, human resource gaps persist: Namibia faces shortages of AI specialists, cloud engineers, and data scientists, leaving geospatial practice rooted in conventional methods (Coetzee et al. 2021; Ye et al. 2018).

Even so, momentum is building. Stakeholders are increasingly interested in cloud-based dashboards for SDG monitoring, particularly in urban resilience initiatives. Furthermore, partnerships with international organizations are slowly exposing Namibian institutions to AI-enabled applications. While integration is still at an early stage, these emerging trends point to gradual progress and the possibility of more advanced adoption in the coming years (see Table 4).

Table 4 Adoption of advanced digital tools in Namibia's NSDI

Tool	Global best practice	Namibia's NSDI status	Adoption gap
Artificial intelligence	Widely used in monitoring, modelling, automation	Minimal use, manual or semi-automated processes	High
Mobile GIS	Common in field data collection, real-time updates	Pilots (e.g., 2021 Census), no framework	Moderate
Cloud-based GIS	Standard for collaboration, real-time analytics	Centralized servers, low interoperability, weak broadband	High

Source Authors' interpretive assessment (2025)

In summary, Namibia's NSDI has provided a platform for coordinated geospatial governance, but its integration of AI, mobile GIS, and cloud-based systems remains embryonic. Without investments in digital infrastructure, institutional reform, and human capacity, the transformative potential of these tools will remain underutilized. To realize the NSDI's role in sustainability monitoring and SDG implementation, Namibia must prioritize scaling up pilot projects, improving inter-agency collaboration, and embedding advanced digital tools into national policy frameworks.

3.5 Strategies for Strengthening NSDI Effectiveness

Having established the foundations of Namibia's NSDI and the challenges hindering its sectoral performance, this section now proposes actionable strategies for enhancing its role in sustainable development. The guiding research question was: *What practical steps can be taken to enhance the institutional and technical effectiveness of the NSDI in achieving sustainability outcomes?*

Findings from the qualitative review of institutional reports, geospatial platforms, and planning frameworks indicate that although an institutional structure exists under the NSA, the NSDI has not yet evolved into a fully operational, real-time system. Data exchange across ministries, municipalities, and sectoral agencies remains fragmented, reducing the NSDI's potential to track and report on SDG progress. Stakeholders emphasized the absence of user-friendly interfaces and integrated reporting tools, which has constrained its wider adoption. For example, environmental and land management bodies still depend on parallel databases and manual reporting formats, delaying information flows critical for sustainability assessments. Likewise, disaster preparedness agencies noted difficulties in accessing harmonized spatial datasets during emergencies, hampering timely interventions (Tsoutsos and Vescoukis 2025).

One central strategy is the introduction of interactive dashboards. Dashboards offer a centralized platform to integrate, visualize, and analyze datasets from multiple sectors in near real-time. Experiences from the Kenya Open Data Initiative and Rwanda's Land Administration Information System illustrate how dashboards strengthen collaboration and public access to data, ultimately improving transparency and accountability (Jing et al. 2019; Lata et al. 2022). For Namibia, dashboards could be tailored to monitor indicators linked to the Harambee Prosperity Plan II and National Development Plans, allowing policymakers to align sectoral interventions with SDG targets more effectively.

Automated reporting tools also emerged as a key innovation. Current reporting processes are dependent on static PDF outputs and manual data entry, which are slow and prone to error. Automated systems embedded within cloud-based GIS could enable continuous validation, updating, and dissemination of data, thereby shortening reporting cycles. For instance, automated scripts could generate dashboards on SDG indicators such as water access, urban growth, or forest cover at regular intervals, providing decision-makers with dynamic and timely insights (Alahi et al. 2023; Jing et al. 2019). Beyond efficiency, such tools would standardize metrics across institutions and strengthen accountability.

However, technical innovations alone are insufficient without parallel institutional reforms. Capacity-building initiatives remain crucial, as many agencies lack skilled personnel in GIS, data science, and analytics. Training programs, knowledge-sharing workshops, and partnerships with academic institutions are necessary to build a professional base that can sustain and scale digital innovations. Additionally, formalized data-sharing agreements are needed to overcome entrenched practices of data hoarding. Establishing governance mechanisms that encourage cooperation between ministries, local governments, and private actors will ensure more equitable use of spatial data resources (Pluto-Kossakowska et al. 2022).

Ultimately, the strategies proposed—interactive dashboards, automated reporting, capacity-building, and improved governance—should be understood as systemic reforms rather than isolated fixes. Dashboards and automation provide the technological backbone, while training, governance alignment, and policy harmonization serve as the institutional scaffolding. Together, they can transform Namibia's NSDI from a passive repository into an active governance instrument that drives resilience, equity, and sustainability. Strengthening NSDI effectiveness, therefore, is not a one-off achievement but an ongoing, iterative process requiring sustained adaptation, institutional commitment, and multi-stakeholder collaboration.

4 Implications and Conclusion

The study has shown that Namibia's NSDI is a vital but underutilized mechanism for advancing sustainable development and tracking progress on the SDGs. While Namibia has established a solid institutional foundation through the NSA and supporting policy frameworks, its effectiveness remains constrained by limited sectoral integration, weak performance reporting, and the slow adoption of advanced geospatial technologies. By reviewing institutional reports, national planning documents, and existing geospatial platforms, this study was able to highlight both the achievements—such as technical infrastructure and governance structures—and the persistent gaps in data sharing, standardization, and stakeholder collaboration.

The findings suggest that Namibia's NSDI is at a transitional stage, with the basic systems in place but not fully operationalized for real-time monitoring or multi-sectoral use. Strengthening this capacity requires targeted interventions such as interactive dashboards, automated reporting mechanisms, and the integration of artificial intelligence for predictive environmental monitoring. These innovations could enable the NSDI to better support evidence-based governance in critical areas like disaster management, urban planning, environmental conservation, and water resource allocation. Importantly, the inclusion of civil society, local governments, and research institutions is essential to ensure that the NSDI fosters not only state-level decision-making but also community empowerment and accountability.

At the national scale, improving the NSDI will advance policy coherence, efficient resource allocation, and resilience to climate and socio-economic shocks. Regionally, aligning Namibia's geospatial framework with continental and global standards offers opportunities for cross-border collaboration, SDG monitoring, and positioning Namibia as a leader in digital geospatial governance in Africa. Future opportunities lie in embedding cloud-based GIS, machine learning, and participatory data models, including citizen science and community reporting.

In conclusion, Namibia's NSDI has the potential to evolve from a static database into a dynamic, integrated, and inclusive decision-support ecosystem. Achieving this transformation requires combining technological innovation, institutional reform, and stakeholder engagement, ensuring that the NSDI becomes both a driver of resilience and a benchmark of accountability in advancing sustainable development.

References

Ahmad M, Ali A (2023) A review of governance frameworks for national spatial data infrastructure. In: Innovation, strategy, and transformation frameworks for the modern enterprise, pp 186–207. https://doi.org/10.4018/979-8-3693-0458-7.ch008

Akingbemisilu TH (2024) A critical evaluation of government role in spatial data infrastructures for healthcare decision-making. J Public Policy 8(1):14–23. http://www.sciencepg.com/journal/545/archive/5450801

Alahi MEE, Sukkuea A, Tina FW, Nag A, Kurdthongmee W, Suwannarat K, Mukhopadhyay SC (2023) Integration of IoT-enabled technologies and artificial intelligence (AI) for smart city scenario: recent advancements and future trends. Sensors 23(11):5206. https://doi.org/10.3390/s23115206

Bachmann N, Tripathi S, Brunner M, Jodlbauer H (2022) The contribution of data-driven technologies in achieving the sustainable development goals. Sustainability 14(5):2497. https://doi.org/10.3390/su14052497

Bebi BB, Iyambo SN (2025) A study of spatial and temporal variation of urban population growth in Windhoek, Namibia. Environ Res Commun 7(5):055012. https://doi.org/10.1088/2515-7620/add3d2

Bello I (2024) Harnessing national geospatial data infrastructure and integrated geospatial information framework for good governance and national security. Int J Space Technol Manag Innov 1:140–163. https://orcid.org/0000-0001-6514-2294

Cantelmi R, Di Gravio G, Patriarca R (2021) Reviewing qualitative research approaches in the context of critical infrastructure resilience. Environ Syst Decis 41(3):341–376. https://doi.org/10.1007/s10669-020-09795-8

Coetzee S, Gould M, McCormack B, Mohamed-Ghouse ZS, Scott G, Kmoch A, Devarajan T (2021) Towards a sustainable geospatial ecosystem beyond SDIs. EUROGI: Druento, Italy, pp 1–33. https://doi.org/10.13140/RG.2.2.22555.39203

Crompvoets J, Vancauwenberghe G, Ho S, Masser I, De Vries WT (2018) Governance of national spatial data infrastructures in Europe. Int J Spat Data InfrastructS Res 13:253–285. https://doi.org/10.2902/1725-0463.2018.13.art16

Ezigbalike C, Njagi PK, Ngogang Wandji L, Chiliswa Z (2016) Convergence of spatial data infrastructure and data revolution. Smart World 15. https://www.academia.edu/download/96064800/

Franchi F, Graziosi F, Di Fina E, Galassi A (2023) A survey of cloud-enabled GIS solutions toward edge computing: challenges and perspectives. IEEE Open J Commun Soc 5:312–331. https://doi.org/10.1109/OJCOMS.2023.3344198

Georgiadou Y, Puri SK, Sahay S (2015) Towards a potential research agenda to guide the implementation of spatial data infrastructures—a case study from India. Int J Geogr Inf Sci 19(10):1113–1130. https://doi.org/10.1080/13658810500286950

Jing C, Du M, Li S, Liu S (2019) Geospatial dashboards for monitoring smart city performance. Sustainability 11(20):5648. https://doi.org/10.3390/su11205648

Lata K, Sood A, Kaur K, Benipal AK, Pateriya B (2022) Web-GIS based dashboard for real-time data visualization and analysis using open source technologies. J GeomatS 16(2):134–146. https://doi.org/10.58825/jog.2022.16.2.42

Lee S, Ospina SM (2022) A framework for assessing accountability in collaborative governance: a process-based approach. Perspect Public Manag GovAnce 5(1):63–75. https://doi.org/10.1093/ppmgov/gvab031

Lifelo Z, Ding J, Ning H, Dhelim S (2024) Artificial intelligence-enabled metaverse for sustainable smart cities: technologies, applications, challenges, and future directions. Electronics 13(24):4874. https://doi.org/10.3390/electronics13244874

Mariathasan V, Bezuidenhoudt E, Olympio KR (2019) Evaluation of earth observation solutions for Namibia's SDG monitoring system. Remote Sens 11(13):1612. https://doi.org/10.3390/rs11131612

Masser I, Crompvoets J (2018) Qualitative monitoring of information infrastructures: a case study of INSPIRE. Environ Plan B: Urban Anal City Sci 45(2):330–344. https://doi.org/10.1177/0265813516675871

Micheli M, Gevaert CM, Carman M, Craglia M, Daemen E, Ibrahim RE, Kotsev A, Mohamed-Ghouse Z, Schade S, Schneider I, Shanley LA, Vespe M (2022) AI ethics and data governance in the geospatial domain of digital earth. Big Data Soc 9(2):20539517221138767. https://doi.org/10.1177/20539517221138767

Musa M, Gasmalla O, Elmahal A, Ganawa E (2024) Distributed geospatial information systems challenges and opportunities. Explor Remote Sens-Methods Appl. https://doi.org/10.5772/intechopen.1007737

Mwange C, Mulaku GC, Siriba DN (2018) Reviewing the status of national spatial data infrastructures in Africa. Surv Rev 50(360):191–200. https://doi.org/10.1080/00396265.2016.1259720

Mwungu CM (2017) Spatial data infrastructure in Africa: a technical and institutional analysis. PhD thesis, University of Nairobi. http://hdl.handle.net/11295/101373

Neiroukh N, Çağlar D (2025) Information systems quality and corporate sustainability: unpacking the interplay of financial reporting, artificial intelligence, and green corporate governance. Systems 13(7):537. https://doi.org/10.3390/systems13070537

Nunes DM, Camboim SP (2025) Are national spatial data infrastructures adequate for achieving the 2030 agenda? A case study of Brazil's NSDI from the perspective of UN-GGIM fundamental themes. Trans GIS 29(1):e70011. https://doi.org/10.1111/tgis.70011

Ogunbiyi J (2021) Reviewing the status of national spatial data infrastructure: a case study in Southern African countries. PhD thesis, Technische Universität Wien. https://doi.org/10.34726/hss.2021.89045

Pluto-Kossakowska J, Fijałkowska A, Denis M, Jaroszewicz J, Krzysztofowicz S (2022) Dashboard as a platform for community engagement in a city development—a review of techniques, tools and methods. Sustainability 14(17):10809. https://doi.org/10.3390/su141710809

Putra TYD, Sekimoto Y, Shibasaki R (2019) Toward the evolution of national spatial data infrastructure development in Indonesia. ISPRS Int J Geo Inf 8(6):263. https://doi.org/10.3390/ijgi8060263

Sachs JD, Lafortune G, Fuller G (2024) The SDGs and the UN summit of the future. Sustainable Development Report 2024. Paris: SDSN, Dublin: Dublin University Press. https://doi.org/10.25546/108572

Sachs JD, Lafortune G, Fuller G, Drumm E (2023) Implementing the SDG stimulus. Sustainable Development Report 2023. Dublin: Dublin University Press. https://doi.org/10.25546/102924

Sachs JD, Lafortune G, Fuller G, Iablonovski G (2025) Financing sustainable development to 2030 and mid-century. Sustainable Development Report 2025. Paris: SDSN, Dublin: Dublin University Press. https://doi.org/10.25546/111909

Sarker S (2025) Role of management information systems in environmental risk assessment: a systematic review of geographic and ecological applications. Am J Interdiscip Stud 6(1):95–126. https://doi.org/10.63125/k27tnn83

Scott G, Rajabifard A (2017) Sustainable development and geospatial information: a strategic framework for integrating a global policy agenda into national geospatial capabilities. Geo-Spat Inf Sci 20(2):59–76. https://doi.org/10.1080/10095020.2017.1325594

Spear D, Zaroug M, Daron J, Ziervogel G, Angula M, Haimbili E, Hegga S, Baudoin MA, New M, Kunamwene I, Togarepi C, Davies J (2018) Vulnerability and responses to climate change in drylands: the case of Namibia. CARIAA-ASSAR Working Paper. University of Cape Town, Cape Town, South Africa. www.assar.uct.ac.za

Tsoutsos MC, Vescoukis V (2025) Natural hazards and spatial data infrastructures (SDIs) for disaster risk reduction. Eng Proc 87(1):101. https://doi.org/10.3390/engproc2025087101

Van den Homberg M, Susha I (2018) Characterizing data ecosystems to support official statistics with open mapping data for reporting on sustainable development goals. ISPRS Int J Geo Inf 7(12):456. https://doi.org/10.3390/ijgi7120456

Van Hoang T (2024) Impact of integrated artificial intelligence and internet of things technologies on smart city transformation. J Tech Educ Sci 19(1):64–73. https://doi.org/10.54644/jte.2024.1532

Vivek R (2023).A comprehensive review of environmental triangulation in qualitative research: methodologies, applications, and implications. J Eur Econ 22(4):157–532. http://dspace.wunu.edu.ua/handle/316497/51854

Ye H, Brown M, Harding J (2018) GIS for all: exploring the barriers and opportunities for underexploited GIS applications. OSGeo J 1–17

Ziemba EW, Duong CD, Ejdys J, Gonzalez-Perez MA, Kazlauskaitė R, Korzynski P, Mazurek G, Paliszkiewicz J, Stankevičienė J, Wach K (2024) Leveraging artificial intelligence to meet the sustainable development goals. J Econ Manag 46:508–583. https://doi.org/10.22367/jem.2024.46.19

Smart Cities, Data Infrastructure, and Environmental Resilience

Water Sustainability and Smart Monitoring: IOT and AI for Water Use

Peter V. Muttungal, Jossy P. George, Nishi Priya, and Benny Godwin J. Davidson

Abstract This chapter discusses the contributions of Artificial Intelligence (AI) and the Internet of Things (IoT) to sustainable water management. It considers how traditional water management, including reservoirs, qanats, and aqueducts, has provided important lessons in climate adaptation and sustainability, while contemporary technologies effectively address challenges of scarcity, leakage, and inefficient management. AI-supported predictive analytics, IoT-enabled smart sensors, together with precision irrigation, can support leak detection and water demand forecasting, detecting soil moisture through sensors, and water quality monitoring. The chapter also touches on trends such as blockchain capabilities, circular water systems, and the use of citizen science. It offers new opportunities to innovate, focus on global partnerships, and small-scale capabilities, to help combine ancient wisdom with innovation to promote equitable, efficient, and durable approaches to water sustainability.

Keywords Water management · Artificial Intelligence · Internet of Things · Water sustainability

---

P. V. Muttungal · N. Priya
Department of Psychology, Christ University, Bangalore, India
e-mail: peter.mv@christuniversity.in

N. Priya
e-mail: nishi.priya@christuniversity.in

J. P. George
Christ University, Bangalore, India
e-mail: frjossy@christuniversity.in

B. G. J. Davidson (✉)
School of Business, University of the Fraser Valley, Abbotsford, BC, Canada
e-mail: Benny.JeyekumarDavidson@ufv.ca

R. Singh et al. (eds.), *Data-Driven Sustainability: Harnessing Technology for A Greener Future and Environmental Resilience*, World Sustainability Series,
https://doi.org/10.1007/978-3-032-13412-7_6

1 Introduction

According to UNICEF, water scarcity has become a global issue, with 2.2 billion people lacking safe drinking water (UNICEF 2020). Even in countries with abundant water, shortages can happen for a variety of reasons, such as poor infrastructure, contaminated supplies, or poor management (Kumar and Yaashikaa 2019). Clearly, climate change, along with human activities, is compounding the challenges children face in obtaining clean water and sanitation. Increased pressures of industrialization, urban growth, and agricultural necessity also place additional demand on available water sources. If these trends are not reversed, they will have significant humanitarian, economic, and environmental consequences. Much of the research into water conservation has involved looking to the past, examining a wide range of practices from ancient to modern. Historically, societies came up with innovative ways to manage water resources. For example, in India, ends of ancient water conservation systems like rainwater harvesting with reservoirs, check dams, and step wells were crucial for survival. These systems tended to be local to the region and made adjustments based on the place's climate. In other regions of the world, such as the Middle East, qanats were developed to transport water across dry regions with little evaporation.

In the contemporary world, water conservation has advanced to encompass advanced technologies and strategies. Water conservation and wise utilization have given rise to a new concept known as water sustainability, which is the effective and sustainable management of water resources to satisfy current demands without sacrificing future generations' capability to satisfy their own demands. Conventional methods of water conservation, as useful as they are, tend to be inefficient and reactionary in the context of contemporary challenges such as climate change, demographics, and old systems. AI algorithms can monitor sensor data in water pipes to detect slight variations in pressure or flow. Small variations can forewarn of a leak long before it becomes a significant break. This technology is employed in Bangkok, Thailand, by the Metropolitan Authority of Bangkok, as a strategy to assimilate the amount of physical water lost from their distribution system (Vanijjirattikhan et al. 2022). Similarly, IoT-based smart meters allow consumers and utility companies to monitor water usage levels with granularity for a smarter use of water. AI-based systems can utilize information provided by weather forecasts, soil moisture sensors, and plant health indicators (Sinwar et al. 2019) to quantify the precise amount of water a crop needs at any given time. This practice of precision irrigation could be used to decrease water usage.

This chapter aims to explore applications for new technologies such as the IoT and AI to support sustainable water management and water use efficiency. It will examine how smart sensor systems could be designed to collect, analyze, and enable decision-making of in-time data, to effectively allocate, distribute, and conserve water. The chapter will discuss new applications of IoT sensors and AI-enhanced predictive analytics to detect leaks, track water consumption patterns, forecast demand, and improve water quality management. Also, it will evaluate how these technologies could facilitate policymaking, consumer participation, and environmental stewardship goals such as reducing waste, improving efficiency, and increasing water accessibility. Using a multidisciplinary perspective, this chapter aims to narrow the gap between water, sustainability, and technology, while providing a series of actionable recommendations for policymakers, researchers, and practitioners.

2 The Global Water Crisis: Drivers and Challenges

2.1 Scale of the Crisis

Only 0.26% of Earth's water is usable by humans. This limited supply, combined with a growing demand from population and economic growth, is creating a water crisis. Global water demand is increasing by about 1% annually due to rising populations, changing economies, and new consumption patterns (An et al. 2021). By 2050, the world population could reach 5 billion, worsening already serious shortages. The World Water Development Report 2018 estimates that nearly half the world already experiences at least one month of water scarcity each year. The security of freshwater supply is an escalating challenge, threatening millions of people. In the summer of 2018, Cape Town, South Africa, nearly faced Day Zero, a scenario where the city's water supply would run out (Ahmadi et al. 2020). This event was significant as Cape Town became the first major developed city to face such an acute water crisis, demonstrating that severe shortages are no longer exclusive to developing nations. Water consumption patterns are being significantly altered by factors such as population growth, socioeconomic development, and climate change. A consistent annual increase of approximately 1% in worldwide water consumption has been observed (Saleem et al. 2021). Should this trend persist, global water demand is anticipated to rise by 20–30% of current usage by the year 2050.

2.2 Historical Practices of Water Management

To boost crop yields, water productivity, energy efficiency, and soil fertility, various tools and systems for land and water management have been developed (Fyles and Madramootoo 2015). Water management involves a set of strategies to optimize water use (Brooks 2006). These include reducing the quantity or quality of water needed for a task and adapting the task itself to use less or lower-quality water. It also focuses on minimizing water loss from the source through use and disposal. Furthermore, good management shifts water usage to off-peak periods and improves a system's ability to operate during droughts. This comprehensive approach ensures more efficient and resilient water systems. Indigenous rainwater harvesting (RWH) techniques often serve multiple purposes (Musaed Aklan et al. 2024). For example, over 6,500 years ago in Iraq's Ur area, people used RWH for both domestic and agricultural needs. The Marib Dam in Yemen diverted floodwater for irrigation, while Syria's Resafa Dam provided drinking water. Similarly, ancient Mayan ponds in Central America, with capacities of up to 10,000 cubic meters, were used for both domestic and agricultural purposes. Ancient civilizations developed incredibly resourceful systems to store and conserve water. In India, they used reservoirs, check dams, and step wells. These structures captured seasonal rainfall and groundwater for year-round availability. Step wells also served as social and cultural community spaces. Check dams slowed water flow, allowing for groundwater recharge. Reservoirs supported agriculture and provided drinking water during dry periods. In the Middle East, qanats were underground channels. They transported water from aquifers in the mountains to arid plains. Their design minimized evaporation in hot, dry climates. This system enabled settlements to flourish in inhospitable environments. The Roman aqueducts were also iconic engineering examples. They transported water over long distances to support urban growth. These systems provided water for public baths and agriculture. Together, these traditional methods highlight climate-adaptive and localized approaches. They offer valuable lessons about sustainability, resilience, and community-centered management.

3 Overview of AI and IoT Technologies

Artificial intelligence can be defined as the ability of a computer program to think like its human counterpart and execute cognitive functions via its built-in Artificial Neural Network system (Zohuri and Behgounia 2023). AI has three distinct approaches, including symbolic AI systems that mimic human rationality and thinking (do Rosário et al. 2015). These cognitive models are based on logic and learning. The next approach is Connectionist AI, which focuses on systems that model the human brain's structure. The last approach is Distributed AI that focuses on the interactive processes, like LLM models that can converse like a human.

3.1 Artificial Intelligence in Resource Management

AI has emerged as a crucial tool in managing natural resources. Traditional methods of monitoring and managing natural resources often rely on historical records and observable trends. This approach may not appear significant during the timely decision-making process. With the help of AI, resource managers can analyse various variables like climate data, land use patterns, water consumption, and environmental indicators for prediction. Analysis of these factors can help us to predict resource availability, potential shortages, or areas at risk of degradation. AI-based modelling is becoming crucial for sustainable resource management. It is also enabling proactive planning and early intervention, which is crucial in today's times.

3.2 Internet of Things (IoT) Fundamentals

The Internet of Things (IoT) generally encompasses any device that connects to the Internet. This term is most often used to define objects that talk to one another (Atkinson 2021). For example, RFID and QR codes are two common types of these connected objects. When combined with automated systems, these devices can collect data, analyze it, and then perform an action. There are four key components that make up a typical IoT system. First, sensors, actuators, and smart meters gather essential data on physical and chemical parameters. Next, communication technologies like ZigBee and Wi-Fi connect all the system's elements. The data is then sent to an IoT cloud for storage, processing, and analysis. Finally, the system is powered by rechargeable batteries, which can also have energy-harvesting components. Under IOT, Real-time smart sensors make remote sensing applications incredibly powerful and engaging by sensing and analyzing data (Kadrolli and Kalnoor 2024). Navigation, surveillance, and communication are a few examples of these potential remote sensing applications. Remote sensing also uses many different types of sensors in fields like weather monitoring and prediction support.

3.3 AI and IoT Convergence

The integration of artificial intelligence and the Internet of Things is called AIoT. AI significantly enhances IoT by enabling better data interpretation and decision-making capabilities. This powerful combination allows for intelligent, automated feedback loops for real-time actions. For instance, in water networks, AI analyses sensor data for leaks or unusual usage. It then immediately adjusts valves or notifies maintenance teams. Similarly, AI in agriculture processes weather data from sensors to manage irrigation. This enhances efficiency and promotes sustainability by reducing resource waste. This integration enables predictive maintenance and proactive planning for changing conditions.

4 Applications in Sustainable Water Management

4.1 Leak Detection and Infrastructure Monitoring

In major urban areas, Pipe systems are the main method for underground power transmission. However, dielectric fluid leaks pose a major problem for high-pressure fluid-filled (HPFF) cable systems (Tylman et al. 2010). These leaks are often caused by general steel pipe corrosion or external mechanical damage. When a leak is extensive, the pumping plant may fail to maintain the required pressure. About 20–30% of water is lost in distribution systems before reaching consumers (Farah and

Isam Shahrour 2024). In some regions, this number can climb as high as 50%. These losses are known as non-revenue water. This is water that is produced but never sold to customers. NRW is a major inefficiency that stresses already-strained freshwater sources. These sources are stressed by over-extraction, population growth, and climate change. A recent revolutionary study has focused on identifying leaks in pipelines via one sensor (Rajasekaran et al. 2025). It uses a 1D Convolutional Neural Network and a modified AdaBoost algorithm. A 1D Convolutional Neural Network analyses sequential data to recognize patterns. This AI works by interpreting vibrations in a pipeline. A modified AdaBoost algorithm then uses these patterns to detect and locate leaks accurately. A recent study wanted to compare various machine learning models for leak detection (Vanijjirattikhan et al. 2022). The system compares machine learning models like Support Vector Machine (SVM), Convolutional Neural Network (CNN), and Deep Neural Network (DNN). The DNN proved to be the most accurate model, with over 94% accuracy. In field trials, novice operators using this system were as accurate as professional experts. The platform integrates acoustic rods, smartphones, and cloud servers, providing a very efficient approach to leak management. A systematic review of 47 studies explored research trends in leak detection (Islam et al. 2022). Findings show an increasing focus on experimental data and machine learning algorithms. IoT technology is also a prominent topic in recent research. Popular sensors include vibration, acoustic, and flow sensors, along with some novel ones. ML and threshold-based algorithms are the dominant leak detection methods. Wi-Fi, cellular IoT, and LoRa lead as communication technologies in this research. These communication methods represent a significant 80% of all research.

4.2 Smart Metering and Consumption Tracking

A water smart meter is an electronic device that records a user's water consumption. Unlike traditional meters, smart meters collect and send data in real-time (Sønderlund et al. 2014). This provides instant information on water usage to both the user and the utility. Studies show that data from smart water meters can provide a better understanding of household water use (Gurung et al. 2014). This enhanced understanding makes the meters a more effective infrastructure and management tool. While smart meters don't directly reduce water demand, they can inform both utilities and customers about usage patterns and potential leaks. This information can then help reduce demand through real-time usage portals and instant leak alerts. Such information has the strong potential to reduce peak demand and defer expensive infrastructure upgrades. The current top-down water infrastructure network modelling approach often uses out-of-date consumption data.

4.3 Agricultural Water Management

Agriculture is the world's largest consumer of freshwater resources. It accounts for about 70% of all freshwater usage (Ye et al. 2024). The global population is projected to reach ten billion by 2050. Consequently, the demand for agricultural products is expected to rise dramatically. 4.0 technologies are emerging as a key solution for improving agricultural sustainability (Preite and Vignali 2024). These innovations include blockchain, the Internet of Things (IoT), deep learning, and machine learning algorithms, along with other computer applications. Among these, predictive algorithms are especially valuable. They are straightforward solutions used widely because they offer highly scalable automation. This allows for more efficient and sustainable agricultural practices. Climate change has significantly disrupted rainfall, water availability, and irrigation needs. To address this, it's essential to develop efficient irrigation systems and water harvesting strategies (Naqvi et al. 2025). AI-based algorithms can help mitigate climate change by enabling water conservation techniques. Many researchers have already implemented AI-powered irrigation practices in various systems. One example is an IoT pivot irrigation system that uses an AI algorithm to analyze real-time images, detect issues, and respond to them, optimizing water usage (Sampath Boopathi 2024). Machine learning algorithms process data to predict water demand and optimize irrigation schedules. This leads to more efficient water usage and increased crop productivity. Additionally, AI can monitor water quality to detect contamination early. The primary application is smart irrigation, which uses IoT sensors to monitor key parameters.

4.4 Water Quality Monitoring and Safety

Water pollution is a significant concern that requires considerable effort from governments and researchers. Both marine and freshwater environments are frequently contaminated by intensive industrial and agricultural activities (Bria et al. 2020). This contamination can also affect tap water. According to the World Health Organization (WHO), roughly two billion people globally use polluted water. Manual methods for monitoring water quality are inherently inefficient, as they are based on a labor-intensive and potentially costly approach. A person, usually a farmer or health official, must travel to a site, obtain a water sample (Zulkifli et al. 2022), and transport it back to a laboratory for testing and analysis of certain water quality indicators. This is a lengthy and inefficient process due to the required time to determine which pollutants are present and their concentrations, along with any attempts to trace the sources of those pollutants. Water quality sensors have developed substantially in recent years. Proper placement of these sensors can improve detection efficiency and sensitivity (Shahra and Wu 2020). Sensors today can measure temperature, pH, and chlorine, and in addition, turbidity, conductivity, and dissolved oxygen (Shahra and Wu 2020). A recent empirical study examined K-Nearest Neighbor (KNN) and

Support Vector Machine (SVM). KNN is an algorithm that classifies an unknown new data point based on the most common class of its nearest neighbors. SVM is an algorithm that separates data points into various classes by determining an optimal hyperplane. Sensor values were sent via LPWAN and NodeMCU to a cloud server. These methods can replace the traditional indicators of water status classification. This also protects individuals from waterborne infections and diseases.

4.5 Flood and Drought Prediction Systems

Droughts and floods have significant implications for society, ecosystems, and the economy (Brunner et al. 2021). Water management of extreme events relies on understanding, modeling, and forecasting extreme events across different time horizons. For short-term forecasting, from hours to seasons, predictions about the likelihood of flash flooding, extreme precipitation, or drought are barometers for planning emergency events and issuing warnings. For long-term forecasts which can be leveraged for planning purposes (e.g., proposal of flood defences and reservoir design), estimates of the return period and magnitudes of events are needed. Climate change driven by human action is affecting global temperatures, including precipitation patterns (Adikari et al. 2021). This changing climate has led to unfavourable conditions such as sea-level rise and increased frequency of extreme weather. Urbanization and the expansion of infrastructure will further exacerbate the problem. Researchers have investigated, researched, and forecasted extreme events such as widespread flooding and regular occurrences of drought to mitigate the negative impacts caused by extreme events through in-depth study and analysis. Scientists have concluded that climate change, which is related to human activity, is causing changes in the patterns of temperature and precipitation. This leads to unfavourable conditions such as rising sea levels and increased frequency of extreme weather conditions. Urbanization and infrastructure expansion will only lead to an increased risk of these conditions. Researchers are actively studying and predicting possibilities such as national floods and droughts to reduce impacts (Sohn et al. 2012). A wide range of AI methods for forecasting droughts includes Artificial Neural Networks (ANN), Support Vector Machines (SVM), Extreme Learning Machines (ELM), and Deep learning (Kikon and Deka 2021). Because these models analyze complex input–output relationships to anticipate when a drought may start, the duration of drought, and the degree of severity, they are able to reduce error rates and provide more accurate, efficient, and robust prediction capabilities in the management of water resources than statistical methods. Drought is a serious natural disaster with important economic and social implications (Oyounalsoud et al. 2024). Many countries use drought indices for monitoring and management, but there is no global drought index due to the complexity of droughts and different hydroclimatic conditions around the world. A hybrid artificial intelligence model for predicting droughts was addressed in a study conducted by Azimi et al. (2022). The study used four models to predict droughts at time intervals of 1 month, 6 months, and 12 months based on observed data.

Forecasting skill increased when models combined precipitation, temperature, and drought data, especially in 6-month or 12-month time intervals.

5 Socio-Economic and Environmental Implications

5.1 Humanitarian Impacts

Water service provision is a core component of public health and quality of life. The global endeavor to provide water services evenly and sustainably is perhaps the most significant challenge for the world in the twenty-first century (Pereira and Marques 2021). Overall, the international system did succeed in attaining the Millennium Development Goal 7C, which called for the halving of the population without access to improved water supplies by 2015 (Huang et al. 2021). However, there are still approximately 1.8 billion people worldwide who do not have access to safe drinking water, and a further 144 million people who are reliant upon untreated surface water. As such, we have a new target, which is to achieve universal access to safe drinking water by 2030, using both existing tools and new evidence to meet that goal. Sustainable water management can have important economic implications. Utilities, such as water treatment plants, can potentially save millions of dollars each year due to decreased financial losses from leaks and other unnecessary inefficiencies in the pipelines. AI/IoT solutions can enhance agricultural output with precise irrigation techniques/technologies that use less water while maintaining the quantity and quality of crop yield, which can directly benefit farmers and farmers are key to food security. Increased interest in smart water management solutions has attracted investors to new water technology markets. This can lead to more opportunities focused on innovation, job creation, and the growth of new water markets. In the long term, these encourage resilience economically, making water sustainability an opportunity and a global financial priority. Sustainable water management can have important environmental implications. Water use that decreases groundwater depletion and minimizes stress on rivers and lakes helps better protect ecosystems from over-extraction of their water supply and long-term environmental damage. Rational water use ensures that a balance is achieved to share limited, valuable water resources responsibly across human needs and natural systems. This balance is important to support biodiversity, as wetlands, forests and aquatic habitats rely on regular access to water to support plant and animal life. Advanced and emerging technologies, like AI and IoT, provide opportunities to optimize water consumption, ensure waste is minimized, and preserve fragile ecosystems. A more efficient use of resources encourages environmental resilience to stressors and protects natural habitats and ecosystems for generations to come.

5.2 *Environmental Impacts*

Throughout the twentieth century, pressure has been placed on global water resources related to climate change and population growth, which may have even greater implications moving forward (Dangar et al. 2021). Groundwater represents an important source of fresh water across the world—it supplies groundwater for domestic purposes for approximately half of the world's population and meets roughly 38% of global water needs for irrigation purposes. Groundwater depletion is a long-term drop in groundwater levels and volume and occurs when groundwater is extracted at a greater rate over a sequence of years than it is replenished. When an aquifer is extremely dry or does not receive any natural recharge, groundwater pumping is termed groundwater mining because a resource is being administered like a non-renewable resource, which can present a near-permanent reduction of the groundwater supply. Good water management will reduce depletion of groundwater and stress on lakes or rivers and protect long-term sustainability; therefore, it can play a dual role in maintaining a balanced ecosystem and contribute to mitigating habitat and biodiversity loss. Good water management can also help foster resilience among ecosystems while reducing anthropogenic pressures on the environment.

6 Barriers, Risks, and Ethical Concerns

6.1 *Technological Barriers*

Despite being data hungry, AI models in hydrology face considerable challenges relating to data quality and availability. Data scarcity and inconsistencies are especially challenging in regions in development where monitoring equipment is limited (Gacu et al. 2025). Even areas rich with data provide new challenges if sensors malfunction or available data is historically or spatially fragmented, thus limiting model usability and dependability. Changes and shifts in hydrological systems find additional limitations through climate change or changing land use, while arrangements for data-informed models are more constrained with uncertainty. Moreover, these limitations are not isolated to AI, as traditional mechanistic, physical models also struggle with changing conditions and require adjustment through calibration that may not transfer to future conditions. Lastly, many AI models, such as complex neural networks, come with a significant drawback of being 'black boxes' without any transparency, which hinders the capacity to interrogate the cause-and-effect nature of the model. Lack of transparency serves as a significant barrier to entry and is generally unacceptable for more critical applications, such as dam safety and flood forecasting. The Internet of Things (IoT) is becoming more significant. It's penetrating sensitive areas like healthcare and finance, as well as our homes and cities. In 2020, more than 50 billion connected devices represented enormous amounts of data and are encouraging manufacturers to focus on security in light of the emergence of Artificial Intelligence (AlSalem et al. 2023). To ensure a trustworthy device, that device must be protected. The connection to the internet exposes IoT devices to numerous vulnerabilities and threats. Those threats include cybercrime, software piracy, malware, etc. And new threats are being developed around the globe every day, meaning our existing ways to overcome IoT risks often are not enough. We need to create new security frameworks and security solutions as our tech space continues to rapidly evolve in the IoT landscape.

6.2 *Institutional and Governance Barriers*

Local governments have much difficulty adopting Artificial Intelligence because of several primary areas. Local governments have financial factors at play, including the inability to buy into an initial investment and the ongoing deployment (Yigitcanlar et al. 2022). There is also a substantial gap in knowledge with staff who do not have the necessary skills and expertise to properly develop and employ AI. As AI becomes a larger part of local governments, it is crucial that these systems are deployed responsibly and ethically. This goes far beyond just compliance with the law (Yigitcanlar et al. 2021). It entails that decisions supported by AI are compatible with situations apparent to society's general standards, and that decision-making reflects upon their effect on persons, individual communities and the world.

6.3 *Ethical Issues*

Hydrological data may be seen as non-personal information; however, when scrutinized by intelligent algorithms, it may introduce a new type of sensitive information. Specifically, behaviors drawn from water use data may also relate to individual or industrial behaviors that introduce a new type of privacy risk. Data protection law does not fully address the new risks related to how AI draws all of these inferences and connects discrete data together. From a security angle, AI systems will be subjected to a new type of cybercrime, such as model poisoning attacks in which malicious data is introduced and the model becomes corrupted (Stahl 2021). Conversely, AI could also be used to identify vulnerabilities in other systems to exploit them. In summary, the complexity of security also points to a higher concern about the reliability and uncertain response of AI systems. Although profiling water users serves to enhance supply efficiency, it poses a significant threat to privacy. Water data aggregated with GIS and census records effectively provides a complete profile of a family's lifestyle, habits, and perhaps, their socioeconomic status. This information has the potential to be misused in ways that violate informational privacy (Doorn 2021). There is also potential for discriminatory practices stemming from the use of this information. Utilities may decide to segment users based on profiles, which could limit water accessibility, particularly to lower-income households, large households, or others that display unusual consumption patterns or reside in particular neighborhoods. This sets the stage for a two-tiered system where even written procedures may rank some users superior to others, and the fairness and equity of the system overall could be brought into question.

7 Future of AI and IOT in Water Sustainability

The future of Artificial Intelligence (AI) and the Internet of Things (IoT) holds tremendous promise for water sustainability. Current trends point to growing integration with blockchain technology that will promote water trading that is secure, transparent, accountable, and equitable. The use of AI applications in circular water systems, where wastewater is treated and reused, will lessen demand for freshwater. Citizen science, or using crowdsourced data from local communities with IoT devices and technologies, is another vital trend diversifying, democratizing, and making water monitoring participatory. Innovation pathways will also exist that blend ancient, climate-adapted customs, such as step wells and qanats, with modern technologies and practices to produce hybrid models that are sustainable and culturally responsive. Open-source platforms for water data sharing could provide access and streamline collaboration effectively for citizens, governments, and researchers to work together on pressing water issues. Global partnerships will be key to shaping this future. International partnerships can facilitate knowledge exchange across boundaries, while public and private partnerships can bring the capital, innovation, and scalability behind those partnerships. Collectively, these actions will help guarantee that AI and IoT can both optimize water management while serving long-term environmental stewardship, resilience, and equity in access.

8 Conclusion

The ability to discuss AI and IoT in the context of water sustainability showcases the urgent need for solutions while also illuminating potential technological value when used for good. Climate change, population growth, and lack of efficiency around water use have resulted in water scarcity, which requires projects that are not based on the traditional system of reservoirs, qanats, and aqueducts. Those systems offer essential lessons in localized climate-adaptive management; contemporary challenges require improved data-driven scaling and application. AI and IoT address those projects to facilitate precision irrigation, leak detection, smart meters, water quality monitoring, and disaster forecasting while eliminating waste and enhancing resources. However, we must also acknowledge the challenges associated with limited data, significant and long-term costs, cybersecurity, and ethical responsibilities. Robust governance, transparency of policy, and designing an inclusive process will support shifting the projected bias of the technologies, as a tool to enhance equity rather than deepen inequity. The next step in implementing AI and IoT across various mediums stems from ancient practices with modern systems, backed by open-source platforms and global thinking. When we enhance our mutually shared knowledge, international collaboration, and public–private partnerships, and utilize AI and IoT as the first-stage applications, they can advance from experimental tools into the foundational infrastructure of resilience. Therefore, the future of water sustainability will hinge on our ability to align a sense of human responsibility, cultural knowledge or wisdom, and new technologies concerning water secured for present and future generations.

References

Adikari KE, Shrestha S, Ratnayake DT, Budhathoki A, Mohanasundaram S, Dailey MN (2021) Evaluation of artificial intelligence models for flood and drought forecasting in arid and tropical regions. Environ Model Softw 144:105136. https://doi.org/10.1016/j.envsoft.2021.105136

Ahmadi MS, Sušnik J, Veerbeek W, Zevenbergen C (2020) Towards a global day zero? assessment of current and future water supply and demand in 12 rapidly developing megacities. Sustain Cities Soc 61:102295. https://doi.org/10.1016/j.scs.2020.102295

Aklan M, de Fraiture C, Hayde LG (2024) Why we should revitalize indigenous water harvesting systems: lessons learned. Int Soil Water Conserv Res. https://doi.org/10.1016/j.iswcr.2024.05.004

AlSalem TS, Almaiah MA, Lutfi A (2023) Cybersecurity risk analysis in the IoT: a systematic review. Electronics 12(18):3958. MDPI. https://doi.org/10.3390/electronics12183958

An M, Fan L, Huang J, Yang W, Wu H, Wang X, Khanal R (2021) The gap of water supply-demand and its driving factors: from water footprint view in Huaihe River Basin. PLoS ONE 16(3):e0247604–e0247604. https://doi.org/10.1371/journal.pone.0247604

Atkinson J (2021) Technology and change management in academic libraries: an overview and literature review. Technol, Chang Acad Libr 11–32. https://doi.org/10.1016/b978-0-12-822807-4.00002-6

Azimi SME, Sadatinejad SJ, Malekian A, Jahangir MH (2022) Application of artificial intelligence hybrid models for meteorological drought prediction. Nat Hazards. https://doi.org/10.1007/s11069-022-05779-w

Boopathi S (2024) Sustainable development using IoT and AI techniques for water utilization in agriculture. Pract, Prog, Profic Sustain 204–228. https://doi.org/10.4018/979-8-3693-1722-8.ch012

Bria A, Cerro G, Ferdinandi M, Marrocco C, Molinara M (2020) An IoT-ready solution for automated recognition of water contaminants. Pattern Recogn Lett 135:188–195. https://doi.org/10.1016/j.patrec.2020.04.019

Brooks DB (2006) An operational definition of water demand management. Int J Water Resour Dev 22(4):521–528. https://doi.org/10.1080/07900620600779699

Brunner MI, Slater L, Tallaksen LM, Clark M (2021) Challenges in modeling and predicting floods and droughts: a review. WIREs Water 8(3). https://doi.org/10.1002/wat2.1520

Dangar S, Asoka A, Mishra V (2021) Causes and implications of groundwater depletion in India: a review. J Hydrol 596:126103. https://doi.org/10.1016/j.jhydrol.2021.126103

De Rosário CR, Kipper LM, Frozza R, Mariani BB (2015) Modeling of tacit knowledge in industry: simulations on the variables of industrial processes. Expert Syst Appl 42(3):1613–1625. https://doi.org/10.1016/j.eswa.2014.09.023
Doorn N (2021) Artificial intelligence in the water domain: opportunities for responsible use. Sci Total Environ 755:142561. https://doi.org/10.1016/j.scitotenv.2020.142561
Farah E, Shahrour I (2024) Water leak detection: a comprehensive review of methods, challenges, and future directions. Water 16(20):2975–2975. https://doi.org/10.3390/w16202975
Fyles H, Madramootoo C (2015) Water management. Elsevier EBooks 117–134. https://doi.org/10.1016/b978-1-78242-335-5.00006-8
Gacu JG, Monjardin CEF, Mangulabnan RGT, Pugat GCE, Solmerin JG (2025) Artificial intelligence (AI) in surface water management: a comprehensive review of methods, applications, and challenges. Water 17(11):1707. https://doi.org/10.3390/w17111707
Gurung TR, Stewart RA, Sharma AK, Beal CD (2014) Smart meters for enhanced water supply network modelling and infrastructure planning. Resour Conserv Recycl 90:34–50. https://doi.org/10.1016/j.resconrec.2014.06.005
Huang Z, Nya EL, Cao V, Gwenzi W, Rahman MA, Noubactep C (2021) Universal access to safe drinking water: escaping the traps of non-frugal technologies. Sustainability 13(17):9645. https://doi.org/10.3390/su13179645
Islam MR, Azam S, Shanmugam B, Mathur D (2022) A review on current technologies and future direction of water leakage detection in water distribution network. IEEE Access 10:107177–107201. https://doi.org/10.1109/access.2022.3212769
Ismail S, Dawoud DW, Ismail N, Marsh R, Alshami AS (2022) IoT-based water management systems: survey and future research direction. IEEE Access 10:35942–35952. https://doi.org/10.1109/access.2022.3163742
Kadrolli V, Kalnoor G (2024) IoT and smart sensors for remote sensing healthcare and agriculture applications. Remote Sens Earth Syst Sci. https://doi.org/10.1007/s41976-024-00129-9
Kikon A, Deka PC (2021) Artificial intelligence application in drought assessment, monitoring and forecasting: a review. Stoch Env Res Risk Assess. https://doi.org/10.1007/s00477-021-02129-3
Oyounalsoud, Abdullah Gokhan Y, Abdallah M, Abdeljaber A (2024) Drought prediction using artificial intelligence models based on climate data and soil moisture. Sci Rep 14(1). https://doi.org/10.1038/s41598-024-70406-6
Pereira MA, Marques RC (2021) Sustainable water and sanitation for all: are we there yet? Water Res 207:117765. https://doi.org/10.1016/j.watres.2021.117765
Preite L, Vignali G (2024) Artificial intelligence to optimize water consumption in agriculture: a predictive algorithm-based irrigation management system. Comput Electron Agric 223:109126. https://doi.org/10.1016/j.compag.2024.109126
Rajasekaran U, Kothandaraman M, Hong Pua C (2025) Water pipeline leak detection and localization with an integrated AI technique. IEEE Access 13:2736–2745. https://doi.org/10.1109/access.2024.3524669
Saleem A, Mahmood I, Sarjoughian H, Nasir HA, Malik AW (2021) A water evaluation and planning-based framework for the long-term prediction of urban water demand and supply. Simulation 97(5):003754972098425. https://doi.org/10.1177/0037549720984250
Senthil Kumar P, Yaashikaa PR (2019) 1—Introduction—Water. In: Muthu SS (ed) ScienceDirect, Woodhead Publishing. https://www.sciencedirect.com/science/article/abs/pii/B9780081026335000014
Shahra EQ, Wu W (2020) Water contaminants detection using sensor placement approach in smart water networks. J Procedia Eng. https://doi.org/10.1007/s12652-020-02262-x
Sinwar D, Dhaka VS, Sharma MK, Rani G (2019) AI-based yield prediction and smart irrigation. Studies in Big Data 2:155–180. https://doi.org/10.1007/978-981-15-0663-5_8
Sohn S-J, Tam C-Y, Ahn J-B (2012) Development of a multimodel-based seasonal prediction system for extreme droughts and floods: a case study for South Korea. Int J Climatol 33(4):793–805. https://doi.org/10.1002/joc.3464

Sønderlund AL, Smith JR, Hutton C, Kapelan Z (2014) Using smart meters for household water consumption feedback: knowns and unknowns. Procedia Eng 89:990–997. https://doi.org/10.1016/j.proeng.2014.11.216

Stahl BC (2021) Ethical issues of AI. Artif Intell Better Futur 35–53. https://doi.org/10.1007/978-3-030-69978-9_4

Tylman W, Kolczyński J, Anders GJ (2010) Fully automatic AI-based leak detection system. Energy 35(9):3838–3848. https://doi.org/10.1016/j.energy.2010.05.038

UNICEF (2020) Water scarcity. UNICEF. https://www.unicef.org/wash/water-scarcity

Vanijjirattikhan R, Khomsay S, Kitbutrawat N, Khomsay K, Supakchukul U, Udomsuk S, Suwatthikul J, Oumtrakul N, Anusart K (2022) AI-based acoustic leak detection in water distribution systems. Results Eng 15:100557. https://doi.org/10.1016/j.rineng.2022.100557

Ye Z, Yin S, Cao Y, Wang Y (2024) AI-driven optimization of agricultural water management for enhanced sustainability. Sci Rep 14(1). https://doi.org/10.1038/s41598-024-76915-8

Yigitcanlar T, Corchado JM, Mehmood R, Li RYM, Mossberger K, Desouza K (2021) Responsible urban innovation with local government artificial intelligence (AI): a conceptual framework and research agenda. J Open Innov: Technol, Mark, Complex 7(1):71. https://doi.org/10.3390/joitmc7010071

Yigitcanlar T, Agdas D, Degirmenci K (2022) Artificial intelligence in local governments: perceptions of city managers on prospects, constraints and choices. AI & Soc. https://doi.org/10.1007/s00146-022-01450-x

Zaigham Abbas Naqvi SM, Hussain S, Awais M, Tahir MN, Saleem SR, Al-Yarimi FAM, Ashurov M, Saidani O, Khan M Ijaz, Wu J, Wei Z, Hu J (2025) Climate-resilient water management: leveraging IoT and AI for sustainable agriculture. Egypt Inform J 30:100691. https://doi.org/10.1016/j.eij.2025.100691

Zohuri B, Behgounia F (2023) Chapter 8—application of artificial intelligence driving nano-based drug delivery system. In: Philip A, Shahiwala A, Rashid M, Faiyazuddin Md (eds) ScienceDirect. Academic Press. https://www.sciencedirect.com/science/article/abs/pii/B9780323899253000071

Zulkifli CZ, Garfan S, Talal M, Alamoodi AH, Alamleh A, Ahmaro IYY, Sulaiman S, Ibrahim AB, Zaidan BB, Ismail AR, Albahri OS, Albahri AS, Soon CF, Harun NH, Chiang HH (2022) IoT-based water monitoring systems: a systematic review. Water 14(22):3621. https://doi.org/10.3390/w14223621

Exploring the Role of Artificial Intelligence and Smart Mobility Solutions in Reducing Transport Emissions: A Qualitative Exploration in South Asia

Narayanage Jayantha Dewasiri, Rubee Singh, and Dilshad Jeffery

Abstract Transport is a major contributor to greenhouse gas emissions in South Asia, where rapid urbanisation, rising vehicle use, and inadequate infrastructure create significant sustainability challenges. With the integration of artificial intelligence (AI) and intelligent mobility—including innovative traffic management, ride-sharing platforms, autonomous systems, and electric mobility—there is potential to transform emissions reduction and efficiency, as well as mobility operations. However, a lack of contextual, qualitative evidence—particularly in South Asia—limits understanding of the drivers, barriers, and enablers. This chapter addresses this gap by examining how AI and smart mobility solutions can help decarbonise transport in Sri Lanka, India, the Maldives, and Bangladesh. Data were gathered through 16 in-depth interviews with policymakers, transport authorities, technology providers, and mobility users. Thematic analysis identified four main themes: (1) optimisation of traffic flow and mobility systems through AI, (2) potential to adopt low-carbon transport alternatives, (3) ongoing barriers such as infrastructure gaps, affordability issues, and data governance, and (4) enablers: supportive policies and cross-country collaboration. Results suggest that, despite the promising potential of AI-enabled smart mobility to decarbonise transport, its adoption remains limited due to financial, institutional, and regulatory challenges. The chapter concludes by discussing practical, theoretical, and policy implications, emphasising inclusive and

N. J. Dewasiri (✉)
Department of Accountancy and Finance, Faculty of Management Studies, Sabaragamuwa University of Sri Lanka, Belihuloya, Sri Lanka
e-mail: jayantha@mgt.sab.ac.lk

University Center for Research and Development (UCRD), Chandigarh University, Ajitgarh, India

R. Singh
Institute of Business Management, GLA University, Mathura, India
e-mail: dr.rubeerajput@gmail.com

Department of Management Studies, Kumaun University, Nainital, Uttarakhand, India

D. Jeffery
Qasim Ibrahim School of Business, Villa College, Malé, Maldives
e-mail: dilshad@bristolinstitute.edu.lk

R. Singh et al. (eds.), *Data-Driven Sustainability: Harnessing Technology for A Greener Future and Environmental Resilience*, World Sustainability Series,
https://doi.org/10.1007/978-3-032-13412-7_7

sub-region-specific strategies to promote sustainable and climate-resilient transport futures in South Asia.

Keywords Artificial intelligence · Smart mobility · Transport emissions · Sustainable transport · South Asia · Qualitative exploration · Climate-resilient futures

1 Introduction

The transport sector remains a significant contributor to global greenhouse gas (GHG) emissions, accounting for approximately a quarter of the world's total carbon dioxide (CO_2) emissions (Kulagin et al. 2025). The heavy dependence on fossil fuel-based transport results from rapid urbanisation, population growth, and increasing mobility demands, raising environmental and public health issues. Challenges such as vehicle ownership privacy, poorly functioning transit systems, road congestion, and inadequate infrastructure that offer low-carbon alternatives have created a critical issue that requires innovative solutions to achieve industry decarbonization (Dewasiri et al. 2025).

Other forms of sustainability in the transport industry include not only emissions but also energy use, waste, and urban sprawl. Traditional approaches, such as fuel efficiency policies, carbon taxes, and modal shifts, while effective to some extent, are insufficient to meet the demands of rapidly evolving urban mobility. Consequently, there is a growing recognition that technological innovation—particularly the challenges of Artificial Intelligence (AI) and data-driven smart mobility—can unlock revolutionary possibilities for developing sustainable and low-carbon transport systems. These technologies can optimise traffic flow, enhance fuel efficiency, enable predictive maintenance, and facilitate the transition to electric and autonomous mobility as part of an integrated effort to reduce environmental impact.

1.1 Global and South Asian Perspectives

The rapid adoption of artificial intelligence and other transformative mobility programmes aimed at achieving carbon neutrality is occurring globally. The European Union Sustainable and Smart Mobility Strategy (2020) emphasises the utilisation of digital technologies as tools for decarbonising mobility (Kunytska et al. 2022). Developed economies, such as Japan, Singapore, and the United States, have integrated AI systems to develop intelligent transportation, optimise web traffic navigation, and enhance real-time transport network management. These international projects demonstrate that AI-based innovations can improve environmental resilience and support urban development.

This is not the case in South Asia. Inequality in infrastructure, economic restrictions, inconsistent policies, and varying levels of technological readiness can be observed in countries such as India, Sri Lanka, Bangladesh, and even the Maldives. These countries have experienced significant growth, with a large number of vehicle owners; however, pollution levels are also high. For example, transportation pollution is a primary contributor to air pollution and elevated carbon emissions in major cities such as Delhi, Dhaka, and Colombo. The shortage of transit services, increased use of two-wheelers, vehicle maturity, and limited investment in electric mobility exacerbate the situation.

However, despite these challenges, it shows promising signs of improvement. The National Electric Mobility Mission Plan of India (Juyal 2022), the National Transport Master Plan of Sri Lanka (Samarasekera et al. 2023), and Bangladesh's Sustainable Transport Policy also recognise that AI and smart mass transportation could be potential solutions for sustainable development in the capital. There is also a growing trend of regional cooperation and technology transfer, particularly in areas such as electric buses, intelligent transportation systems, and ride-sharing apps. However, limited empirical data exist on how these technologies are being adopted, their actual impact on reducing emissions, and the socio-political factors affecting their implementation. Therefore, it is vital to examine the context of South Asia qualitatively to understand how AI and smart mobility can effectively support transport decarbonisation in resource-limited settings (Rathnasiri et al. 2024).

The primary objective of this chapter is to explore how Artificial Intelligence and innovative mobility solutions can contribute to reducing transport emissions in South Asia from a qualitative perspective. It aims to produce specific findings about how AI-driven technologies are viewed, implemented, and adapted within the diverse transport systems of the region. This chapter is organised into seven sections. Following this introduction, Sect. 2 provides a detailed examination of the theoretical and conceptual foundations of AI, smart mobility, and sustainable transportation systems. Section 3 describes the study design, including data collection, sampling, and thematic analysis methods. In Sect. 4, the author presents the most important findings, organised by themes derived from qualitative interviews. Section 5 discusses the theoretical, practical, and policy implications of the study, highlighting opportunities for implementing AI within sustainable transport agendas. Finally, Sect. 6 summarises the chapter with key insights, limitations, and directions for future research.

2 Theoretical and Conceptual Background

2.1 Defining Artificial Intelligence (AI)

Artificial Intelligence (AI) is one of the most revolutionary technologies of the twentieth century, increasingly recognised as a key driver of sustainability and environmental resilience (Dewasiri and Choudhary 2025). Broadly, AI refers to the ability of machines and computer systems to perform tasks that were previously carried out by human intelligence, such as reasoning, learning, perception, and decision-making. Contemporary AI solutions utilise subdisciplines like machine learning (ML), deep learning (DL), natural language processing (NLP), and computer vision to analyse data, identify patterns, and deliver adaptive solutions. In the transportation sector, AI has proven to be immensely valuable for optimisation, traffic forecasting, congestion management, and enhancing energy efficiency (Kumar et al. 2024).

2.2 Understanding Smart Mobility

Smart mobility is a multidimensional term that involves innovative technologies, data analytics, and online connectivity to enhance the efficiency, accessibility, and sustainability of transportation networks (Porru et al. 2020). It extends beyond technological innovation to encompass new business models and behavioural shifts that enable the movement of people and goods more efficiently and cleanly. Included in smart mobility are intelligent transport systems (ITS), electric vehicles (EVs), autonomous and connected vehicles (CAVs), ride-sharing and carpooling services, mobility-as-a-service (MaaS), and integrated multimodal networks.

2.3 Concept of Sustainable Transport

A sustainable transport system supports society's mobility needs without harming the environment, while promoting social equity and social sustainability. The Sustainable Development Goals (SDGs), which primarily include SDG 9 (Industry, Innovation, and Infrastructure), SDG 11 (Sustainable Cities and Communities), and SDG 13 (Climate Action), represent global efforts to promote sustainability in the mobility sector (Jayarathne et al. 2025). A sustainable transport system aims to reduce the environmental impact of transport through energy efficiency, the utilisation of renewable energy, the minimisation of emissions, and the provision of modal, equitable, and affordable transport to all sections of society (Singh et al. 2025).

The solution to sustainable transport is complex and involves advanced solutions that require technology, policy change, and behavioural shifts. Traditional policies, such as fuel-efficiency standards, vehicle electrification, and transportation system improvements, are effective but can be further developed. There is still room to implement digital methods, such as AI, IoT, and big data analytics. These technologies can enhance monitoring, control, and prediction. Therefore, intelligent mobility, which utilises AI, is a key technological and policy measure to achieve sustainable transport. It reduces waste, promotes low-carbon travel, and improves user experience. A set of principles has been introduced at the United Nations (SDG 9: Industry, Innovation, and Infrastructure; SDG 11: Sustainable Cities and Communities; SDG 13: Climate Action) to promote sustainable mobility (Dewasiri 2025). Addressing environmental costs in operating sustainable transport networks should involve energy conservation, the use of renewable energy, emission reduction, and modal diversification. It must also be accessible to all societal groups and affordable.

2.4 Linking to Relevant Theoretical Foundations

This chapter connects AI and smart mobility to a sustainable transport system through two important theoretical frameworks, the Technology Acceptance Model (TAM) and the Innovation Diffusion Theory (IDT).

Technology Acceptance Model (TAM)

The Technology Acceptance Model (TAM), proposed by Davis (1989), posits that the adoption of new technology is influenced by its perceived ease of use and effectiveness. He views these factors as heavily influenced by whether a person is willing to try new technology. TAM explains how AI and smart mobility innovations are either adopted or rejected by users and stakeholders. For instance, policymakers and transportation officials are more likely to support AI solutions if they believe these will reduce emissions and improve operational efficiency. On the other hand, everyday users may oppose new mobility services because they seem complicated, raise privacy concerns, or lack trust in autonomous technologies.

Innovation Diffusion Theory (IDT)

Rogers' (2003) Innovation Diffusion Theory explains how technological objects are adopted within a social system over time, influenced by factors such as relative advantage, compatibility, complexity, trialability, and observability. When applied to AI mobility, IDT emphasises the crucial role of systematic preparedness, infrastructure, and social trust in facilitating the diffusion of AI. For instance, despite recognising the environmental advantages, electric vehicle adoption remains slow in India and Sri Lanka, mainly due to high costs, limited charging infrastructure, and challenges related to compatibility and trialability.

2.5 *Empirical Review*

Overview of AI Applications in Sustainable Transport

Artificial Intelligence (AI) is playing a crucial role in sustainable development across various sectors, and transportation is one of the most promising fields for minimising emissions and enhancing the efficiency of operating processes. On an international level, AI-based solutions in the transportation industry can be divided into four broad categories: traffic management, predictive maintenance, energy optimisation, and user-focused mobility systems.

Research by Ponnusamy et al. (2024) and Lv and Shang (2023) demonstrate that AI-driven traffic signal optimisation can cut congestion by over 25% and lower fuel use by 10–15%. These systems leverage reinforcement learning algorithms to adjust signal timings based on live traffic data. Similarly, AI-based predictive maintenance models analyse sensor data from vehicles to identify potential faults, saving time and reducing unnecessary resource consumption (Mahale et al. 2025). In logistics and public transportation, AI also plays a key role in route optimisation by using big data analytics to predict high-demand routes and improve fleet deployment (Fatorachian and Kazemi 2025).

Moreover, AI-based electric mobility studies indicate that grid integration and smart charging can enhance the effectiveness of EVs (Sarker et al. 2025). Planning charging to align with renewable energy production will help AI decarbonise the grid by reducing strain on the grid (Mishra et al. 2024). In general, the evidence provided above suggests that AI can potentially reduce emissions not only directly, due to increased energy efficiency, but also indirectly, by promoting behavioural change towards cleaner and more sustainable modes of transportation.

Global Trends in Smart Mobility and Emission Reduction

Smart mobility has become a key driver in the low-carbon transformation of the global transportation industry. The European Union (EU) has made significant progress in its efforts to reduce transport emissions by 90% by 2050, notably through the Sustainable and Smart Mobility Strategy (2020) and the European Green Deal in 2019 (Gkoumas et al. 2021).

In the United States, AI-based smart logistics have the potential to reduce urban freight emissions by 20 per cent, mainly through route optimisation and responsive delivery scheduling. Similarly, Japan and Singapore have successfully implemented AI-powered intelligent transport systems (ITS) to streamline urban mobility, minimise traffic congestion, and enhance public safety.

Nordic has adopted AI technologies to develop the so-called green corridors, integrating buses and autonomous vehicles to support carbon–neutral city strategies. These international case studies demonstrate that AI, when properly integrated into the transport sector—including supportive infrastructure and regulation—can lead to substantial emission reductions and environmentally sustainable urban growth (Singh et al. 2024a, 2024c).

Smart Mobility Developments in Emerging Economies

Asian, Latin American, and African emerging economies are increasingly adopting innovative mobility technologies, although some have been more successful than others. Artificial intelligence is also among the many elements associated with China's New Infrastructure campaign, which emphasises smart streets, autonomous vehicles, and the integration of electric vehicles (EVs).

Latin America: AI-enabled services in Mexico City (EcoBici) and in São Paulo (intelligent bus scheduling system) have demonstrated a positive environmental and social impact, particularly in promoting shared mobility. However, scalability is often limited by infrastructural and data governance issues. In Africa, pilot trials in Kenya and South Africa indicate that AI-driven ride-sharing could reduce fuel costs, with affordability and access to online technology being key obstacles.

Research Gaps and Need for Qualitative Inquiry

Although international literature highlights numerous technical and quantitative benefits of AI and smart mobility, significant research gaps remain, especially regarding qualitative and context-specific knowledge in South Asia. Firstly, most studies heavily depend on simulation practices, performance indicators, or pilot information from developed countries. Such approaches often overlook the social, cultural, and institutional dynamics that shape the use of technology in the developing world.

Secondly, stakeholder perceptions, including those of policymakers, transport operators, technology providers, and end-users, regarding the viability, trust, and ethical considerations of AI-structured transport solutions, are poorly understood. Techniques such as interviews and focus groups are qualitative, and their results may reveal subtle insights into the enablers and barriers to the diffusion of sustainable technology.

Third, there is limited research on how AI-driven innovation is reflected in regulations. Despite policy frameworks, their application tends to be inconsistent and lacks support from empirical evidence at the ground level. To achieve sustainable results, it is essential to understand the dynamic aspects of governance and develop a context-specific approach tailored to each situation.

Lastly, there is a lack of cross-country comparative research in South Asia. While social and political conditions vary across countries, cooperation in mobility technology remains the least developed. Most South Asian contexts can be examined qualitatively to identify shared success factors, challenges, and the potential for collective policy-making.

3 Methodology

The present study employs a qualitative and exploratory design to examine methods for reducing transport emissions in South Asia using AI and innovative mobility solutions. The qualitative approach was chosen based on its capacity to provide in-depth insights into the complex social, technological, and policy factors that purely quantitative, inside-out analyses cannot fully capture. The exploratory aim is to identify context-specific sensitivities, experiences, and drivers of AI-powered sustainability across diverse socio-economic and infrastructural settings.

This study adopts an interpretivist approach, focusing on the creation of meaning through participants' narratives and existing field data. This approach is suitable because AI application in South Asian transport is an emerging area that lacks empirical evidence regarding stakeholder perceptions and situational issues.

It targets four countries: Sri Lanka, India, Bangladesh, and the Maldives, each with varying levels of digital preparedness, transport infrastructure, and environmental policy. The countries illustrate the region's heterogeneity: in India, small-scale innovative city plans are gaining momentum; in Sri Lanka, smaller AI-enhanced traffic solutions are emerging; in Bangladesh, a city transport restructuring is underway; and in the Maldives, small island mobility innovations are taking place. Examining them together provides a comparative view of the development of AI and smart mobility solutions within the context of different policies and infrastructure.

Participants with expertise in AI, transportation, or sustainability were selected using a purposive sampling method. A total of sixteen in-depth market research interviews were conducted from March to August 2025, involving four participants in each of four stakeholder groups: Policy Makers and Regulators, Technology and Industry Experts, Transport Authorities and Planners, and Academics and Users. This diversity provided insights into the technological, institutional, and user-based aspects of AI adoption. The variety of subjects offered a comprehensive view of these aspects, initially gathered through semi-structured interviews conducted via virtual platforms (Zoom and MS Teams) due to geographic dispersion.

Interviews, ranging from 45 to 75 min in length, were conducted in detail to facilitate a thorough discussion. All the interviews were recorded for comprehension and subsequently transcribed verbatim. Confidentiality and anonymity were maintained through pseudonymization and the exclusion of organisational identifiers. The ethics committee of the university affiliated with the respective author granted ethical approval.

3.1 Data Analysis and Coding Process

Thematic analysis was used as part of the six-phase approach to data analysis detailed by Braun and Clarke (2006): (1) familiarisation with the data, (2) initial coding, (3) searching for themes, (4) reviewing themes, (5) defining and naming themes, and

(6) creating the report. The transcripts were organised with NVivo 14, coded, and patterns identified.

The analysis was conducted using an inductive approach, where the data is expected to reveal themes without the application of a pre-existing theory. A cross-country analysis identified both convergent and divergent patterns. The final thematic pattern included four predominant themes.

Theme 1: AI-centric rationalisation of traffic flows and mobility systems;
Theme 2: transport feasibility prospects of low-carbon alternatives;
Theme 3: Infrastructure-, affordability-, and governance-related obstacles;
Theme 4: The facilitating effects of positive policies and cooperation at the regional level.

These themes would draw on examples showing how AI-based mobility solutions can help reduce emissions and expose institutional and socio-technical constraints in South Asia.

4 Findings and Discussion

Our analysis of 16 detailed interviews revealed four key themes that explain how AI and smart mobility solutions are influencing efforts to reduce transport emissions in South Asia. One theme concerns values and norms, another focuses on global standards, and the last involves stories and memory. Below, we present decision-by-objective statements from participants (identified individually as P1-P16) and our analysis based on theories relevant to this study and the contextual factors influencing the results.

4.1 Theme 1: AI-Driven Optimisation of Traffic and Mobility Systems

In all four countries, participants observed that the most apparent environmental impact of AI is related to improved performance in existing traffic flows and mobility operations. Pilot projects involving policymakers and city engineers in India and Sri Lanka utilised AI-enabled adaptive signal controls to reduce average intersection delays and idling time, cut fuel consumption, and lower local emissions.

> *The smart-signal system has already reduced vehicle waiting lines in Colombo by nearly a third during rush hour*s
>
> (P3 Sri Lanka Transport Authority).

Likewise, Indian interviewees reported that AI-based predictive analytics at the logistic hubs in megacities eased freight scheduling and reduced empty-truck miles.

These findings align with previous literature, demonstrating that network efficiency, a key factor in achieving run-time reduction, is maximised by machine-learning algorithms.

Regarding the theoretical context, the core idea of the Technology Acceptance Model, which is perceived usefulness, is supported by these findings: actors are more likely to adopt AI when its benefits are apparent at both the environmental and operational levels. Bangladeshi and Maldivian users showed more positive acceptance of digital apps used for navigation and ride-sharing when they noticed a slight reduction in traditional time and fuel consumption.

> *Expanding on why GPS applications are more trusted, drivers in this region believed that GPS devices in cars were a tool used to make a quick buck, but now they realise it saves them a considerable amount of petrol*
>
> (P9—Bangladesh, Ride-Hailing Operator).

However, despite these profits, most cities have not yet matured. In the absence of city sensors or central data repositories, optimisation remains fragmented. Participants requested continuous data flows across agencies, power utilities, and individual operators, as well as the conditions of long-lasting diffusion described by Innovation Diffusion Theory, such as system connectivity and compatibility.

4.2 Theme 2: Opportunities for Integrating Low-Carbon Transport Alternatives

The theme highlights AI's role in integrating electric, shared, and multimodal transport to support low-carbon mobility, supporting the argument made by Singh et al. (2024b). In all surveyed countries, respondents noted AI's ability to coordinate various transportation systems, manage demand, and optimise EV charging stations.

> *Our EV buses are connected to an AI dashboard which forecasts the required power and charges at the off-peak renewable-energy time*
>
> (P6—India, Technology Provider).

Sri Lankan and Bangladeshi interviewees discussed emerging initiatives that integrate ride-hailing services with public transport timetables through mobile Mobility-as-a-Service (MaaS) applications. These are still at the experimental stage, but they encourage commuters to use buses, ride-sharing services, and walk to reduce their reliance on single-occupancy vehicles.

Renewable energy is also integrated with AI. The Maldives participants discussed microgrids that automatically charge electric scooters based on solar generation estimates. These ideas demonstrate that technology can simultaneously have a positive impact on economic performance, enhance social inclusion, and protect the environment, aligning with the Sustainable Development Framework. However, participants stated that technology alone is not enough. Behaviour change remains slow due to cultural attachment to personal vehicles, unreliable public transport, and ineffective policies.

> *People perceive that electrical mobility represents prestige and modernity; only, said one policymaker, do they have rules and pets they prefer to drive: we must have behavioural campaigning, not just technology*
>
> (P11—Bangladesh, Policy Official).

Therefore, AI and smart-mobility initiatives must be accompanied by awareness campaigns and incentive schemes that encourage users to adopt greener lifestyles. These findings support the observability aspect of IDT; visible improvements in the environment can drive environmental adoption, but cultural advocacy remains essential to turn perceptions into actions.

4.3 Theme 3: Persistent Barriers — Infrastructure, Affordability, and Governance

Even with the increase in enthusiasm, respondents consistently cited structural and institutional barriers that hinder AI from achieving its full potential for sustainability.

(a) Infrastructure Limitations

The most frequently mentioned constraint was the lack of a trusted digital infrastructure, characterised by limited sensor network coverage, inconsistent data standards, and inadequate cloud computing infrastructure. Urban planners in Bangladesh and Sri Lanka have noted that the reliability of traffic signal connectivity is compromised due to power outages and the hardware reaching the end of its lifespan.

> *You cannot do smart mobility by running it on dumb infrastructure*
>
> (P2—Sri Lanka, Urban Planner).

Without reliable data pipelines, AI systems will never be able to operate continuously and with accuracy, leading to partial optimisation and limited emissions reduction.

(b) Economic and Affordability Challenges

AI implementation incurs significant costs in both hardware and software, yet it also offers inherent advantages. Public-sector actors often feel that competing budget priorities hinder the adoption of new initiatives. Despite the long-term savings promised by AI solutions, the high initial costs discourage municipalities that lack fiscal autonomy. Similarly, operators in India and Bangladesh privately express that government subsidies for electric and autonomous mobility are still insufficient to cover these initial expenses.

(c) Governance and Institutional Fragmentation

Another central theme was on governance issues. Transport, energy and environment duties in most cities in South Asia are divided among different agencies that have staggered mandates. Rachel, gay respondents stated that they experienced challenges in sharing and the coordination of data:

It really is that transport data is housed in one ministry, and emission data in the other; AI must have an interface to communicate with each other

(P12—India, Transport Consultant).

The participants recommended national AI-for-Mobility frameworks to synchronise standards and promote bilateral partnerships between the public and private sectors, as well as regional cooperation to share experiences and draw lessons beyond national boundaries. Together, these obstacles demonstrate that the potential of AI as a sustainable trend depends not only on technology but also on the capacity of institutions and the willingness to act politically—a crucial factor to consider in the developing world.

4.4 *Theme 4: The Enabling Role of Supportive Policies and Regional Collaboration*

The final theme is how AI-driven emission reductions can be realised through policy frameworks, capacity building, and the sharing of regional knowledge. Respondents highlighted the new policy initiatives, including the Digital India initiative in India, the National AI Strategy in Sri Lanka, and the Sustainable Transport Policy in Bangladesh, but also pointed out the issue of uncoordinated implementation.

We have good policies on paper, but the practice and monitoring fall short.

(P15—Bangladesh, Academic Researcher).

There has always been a demand among stakeholders to develop stronger institutional systems that align digital transformation with sustainability goals. Three priority actions were identified:

1. **Policy Integration:** Integrate AI and smart-mobility strategies into country-climate-action plans and city-transport plans.
2. **Capacity Development:** To develop technical literacy among government officials and transport managers through training programmes.
3. **Regional Collaboration:** Data-sharing Partnerships, Pilot projects, and digital platform standardisation. Form South-Asian partnerships on data sharing, pilot projects, and standardisation of digital platforms.

The participants indicated that regional issues, such as air pollution, energy security, and infrastructure shortages, are problems that require collective action. As one respondent from the Maldives stated:

Our islands are tiny, but we learn through the wisdom of India to fit all things to the proper scale. The South-Asian knowledge network we need.

(P16—Maldives, Sustainability Expert)

These understandings reinforce the collaborative spirit of the sustainable development paradigm: Sustainability depends on cooperation across spatial and institutional boundaries. Along with the Innovation Diffusion Theory, the findings suggest that policy support and inter-governmental efforts act as barriers to technological diffusion, preventing early adoption from transitioning into widespread regional success.

4.5 Cross-Theme Interpretation

The four themes discussed in this paper emphasise AI as both a technological resource and a socio-institutional driver of low-carbon mobility, as shown in Fig. 1. Further analysis reveals that emission reduction results from a combination of AI algorithms improving efficiency (technological optimisation) and behavioural adoption (user acceptance and shared mobility participation). Institutional coordination (such as policies and alliances among various actors) also plays a crucial role. This multilevel interaction aligns with the conceptual framework outlined earlier (optimisation - integration - innovation - governance). Interviews have empirically confirmed that these four dimensions must be synchronised to ensure success. For instance, India is innovative, but its government is dysfunctional, whereas Sri Lanka and Bangladesh have policy-driven governments, yet lack adequate infrastructure. Despite its small size, the Maldives has achieved high levels of integration in renewable energy and mobility planning.

Fig. 1 Cross-theme interpretation. *Source* Authors' own

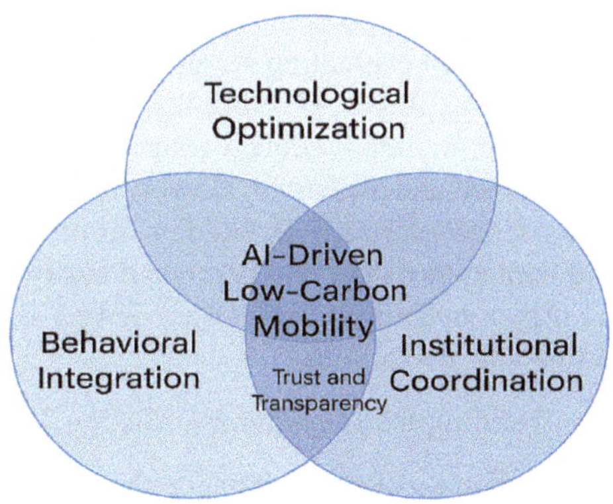

Source: Authors' own

Further analysis indicates that people's overall view of AI tends to be positive, though transparency and trust are key factors influencing these attitudes. Ethical concerns raised by participants—such as digital bias, data ownership, and internet safety—highlight the importance of establishing ethical standards to build societal trust. It is crucial to address these issues to prevent neglecting the perceived ease-of-use element of TAM, which suggests that simplification and transparency might pose challenges to sustained adoption.

5 Implications

The results of this qualitative review of Sri Lanka, India, Bangladesh, and the Maldives indicate that Artificial Intelligence (AI) and new mobility technologies can play a significant role in reducing emissions, provided they are supported by harmonious governance, institutional preparedness, and social acceptance. This study has implications across theoretical, practical, and social-policy domains, and overall, it enriches the discussion on sustainable transport transitions in developing countries.

5.1 Theoretical Implications

This study contributes to the theoretical development of sustainable transport and digital innovation by situating AI and intelligent mobility within the broader frameworks of Technology Acceptance, Innovation Diffusion, and Sustainable Development.

(a) Extending the Technology Acceptance Model (TAM)

The paper supports the applicability levels of perceived usefulness and perceived ease of use in evaluating technology adoption among different stakeholder groups. Respondents found AI relevant when the environmental and operational benefits of AI, which included reduced congestion and decreased gasoline consumption, could be observed and measured. However, trust and transparency emerged as new determinants of acceptance. Stakeholders raised issues such as algorithm bias, cybersecurity, and data authorship, and perceived trustworthiness—considered a key factor in the AI-driven mobility trend—was viewed as essential within the TAM framework to define new economies.

(b) Refining the Innovation Diffusion Theory (IDT)

The findings further enhance the Innovation Diffusion Theory by demonstrating that systemic readiness, rather than technological factors, primarily drives the spread of AI innovations in South Asia. Relative advantage and compatibility are influenced not only by functionality but also by institutional elements, such as coordination and regulation, as well as socio-cultural acceptance.

For example, AI-driven traffic management has demonstrated significant advantages in Indian cities. However, widespread adoption remains limited due to a lack of practical inter-agency cooperation. This highlights the importance of prioritising institutional interoperability as a key factor for successful implementation in emerging economies.

5.2 *Practical and Managerial Implications*

This study provides pragmatic insights for policymakers, transportation authorities, and industry practitioners involved in developing sustainable mobility systems in South Asia.

(a) Institutional Integration and Governance Reform

The study highlights the importance of integrating policies across ministries and agencies responsible for transportation, environment, and energy. Fragmented political structures hinder the sharing of real-time data, which is crucial for the effectiveness of AI. It is recommended that governments establish National AI-to-Mobility Councils or inter-ministerial task forces to improve coordination, standardise data formats, speed up dissemination, and incorporate emission reduction targets into wider transport and urban planning strategies.

(b) Investment in Digital and Physical Infrastructure

Implementing AI deployment relies on the readiness of the infrastructure. Policymakers should prioritise investments in digital infrastructure, including cloud computing, sensors, and communication networks, along with physical infrastructure like electric vehicle charging stations and smart intersections. Public–private collaborations can facilitate resource coordination, leverage private sector expertise, and share the risks associated with these technologies.

(c) Capacity Building and Technical Literacy

It reveals that the lack of technical expertise within government departments is a critical issue, hindering the scaling of AI projects. Specific capacity-building initiatives, designed to train engineers, planners, and regulators, should be launched by governments and universities to enhance their expertise and capabilities. These schemes must enhance computer skills and information analysis. Close cooperation between international technology companies and development organisations can also strengthen knowledge ecosystems at the local level.

(d) Incentivising Innovation and Private Participation

Partnering with the private sector is both an impetus and a bottleneck. The interviewees highlighted that monetary and non-monetary benefits, such as financial considerations including tax incentives, grants, or innovation vouchers, can motivate local start-ups to develop AI-based solutions tailored to the region's specific needs. Governments can also support regulatory sandboxes to test out new technologies safely before their mass application. These rooms facilitate experimentation and learning.

(e) Data Governance and Ethical Frameworks

Data governance systems must be accurate to ensure a high level of trust among the populace. Transparent regulations regarding data security, algorithm accountability, and cybersecurity must accompany the implementation of AI. National definitions of AI in the ethical implementation of mobility would protect human rights and allow progress in innovation, in harmony with the current UNESCO 2021 AI ethics recommendation. These procedures address issues related to the misuse of data and ensure that AI-based versions are legitimate and accessible.

5.3 Social and Policy Implications

This paper demonstrates that, beyond technology and institutions, broader social issues must be addressed to achieve equitable and sustainable transitions.

(a) Promoting Public Awareness and Behavioural Change

Sustainable mobility will require technology, and humans are often reluctant to change due to status, habits, and convenience. Attitudes can be shifted through campaigns that highlight the environmental and cost savings of smart mobility. Promoting sustainability in schools and universities is also likely to lead to lasting behavioural changes.

(b) Ensuring Equity and Accessibility

AI-driven smart mobility can help eliminate digital and economic divides. Services should remain affordable and inclusive, particularly for low-income users and those in rural areas. Access can be facilitated through subsidies on digital access or isolating AI-capable services offered to public transportation, which will help prevent tech-based marginalisation.

(c) Fostering Regional Collaboration

South Asia faces ongoing environmental challenges, and local cooperation can enhance AI-driven sustainable initiatives. The South Asia Digital Sustainability Network aims to share knowledge, coordinate collaborative research, and establish unified standards across the region. National progress can be supported by initiatives like regional low-emission corridors and shared information systems. Additionally, local education programs, exemplified by the Maldives and Sri Lanka, help reduce redundancy, improve technological literacy, and align policies across different levels.

6 Conclusion and Future Directions

The study concludes that AI can significantly assist in decarbonising transport systems and enhancing environmental resilience, but this requires integrated meta-systems linking technology, governance, and society. Technologically, AI applications like traffic optimisation and fleet management can deliver quick, measurable emission reductions, especially when data infrastructures are reliable and shared across agencies. Institutional coherence is crucial, as inconsistent leadership can undermine the impact of AI. Achieving digital transformation for sustainability depends on the coordination of policies and effective collaboration among relevant agencies. Socially, factors such as trust, awareness, and inclusiveness are crucial for societal acceptance of AI-enabled mobility; technology must be supported by social legitimacy and ethical governance. Lastly, regional cooperation is crucial in South Asia, where shared technological standards, collective capacity-building, and cross-border learning can address domestic limitations and turn AI's role in sustainable transport into a regional, rather than solely national, goal.

6.1 Quantitative Validation

Future studies will need to be mixed or purely quantitative to assess the reduced emissions from specific AI technologies, based on real-life validation through the transport systems of South Asian cities.

6.2 Comparative Regional Studies

Artificial intelligence and the regulatory issues surrounding it seem to be characterised by the same level of hype as the issue of the Internet of Things (IoT) creates regarding its relationship.

6.3 AI Ethics and Governance Models

To foster trust among the people and ensure fair access to potential, researchers should develop ethical systems that preserve data privacy, reduce algorithm bias, and make mobility systems transparent to the public.

6.4 Integration with Renewable Energy Systems

Research into the interaction of AI-powered transport and renewable energy sources and storage can support the use of veridical climate-mitigation strategies.

6.5 Behavioural and Cultural Dimensions

Further research is needed on how the adoption of mobility technologies based on AI can be influenced by societal sociability, cultural identity, and user psychology.

By conducting this study, researchers and professionals can gain a deeper understanding of the complex interaction between digital technologies and social systems, thereby supporting the acceleration of low-carbon initiatives.

References

Braun V, Clarke V (2006) Using thematic analysis in psychology. Qual Res Psychol 3(2):77–101

Davis FD (1989) Technology acceptance model: TAM. Al-Suqri, MN, Al-Aufi, AS: Inf Seek Behav Technol Adopt 205(219):1–33

Dewasiri NJ, Choudhary M (2025) Harnessing artificial intelligence (AI) for sustainable innovation: case studies and practical applications. In: Innovation Ecosystems And Sustainable Technologies: Enhancing Competitiveness And Sustainability. Emerald Publishing Limited, pp 169–188. https://doi.org/10.1108/978-1-83662-368-720251008

Dewasiri NJ, Gangodawilage D, Singh R (2025) Pathways to climate neutrality: integrating sustainability in transport, energy, and communication for the indo-pacific region. In: Climate neutrality and sustainable ecosystems: practical approaches to addressing global warming. Singapore: Springer Nature Singapore. pp 95–114. https://doi.org/10.1007/978-981-96-7181-6_6

Dewasiri NJ (2025) Highlights of circular economy actions in the climate change policies: a call for action for the Asia-pacific region. Clim Chang Soc Responsib 23–41. https://doi.org/10.1108/S2043-052320250000025002

Fatorachian H, Kazemi H (2025) Sustainable optimisation strategies for on-demand transportation systems: enhancing efficiency and reducing energy use. Sustain Environ 11(1):2464388

Gkoumas K, Marques dos Santos FL, Stepniak M, Pekár F (2021) Research and innovation supporting the European sustainable and smart mobility strategy: a technology perspective from recent European Union projects. Appl Sci 11(24):11981

Jayarathne PGSA, Dewasiri NJ, Khan S (2025) Challenges of integrating social responsibility and climate change for the sustainable development goals: experience from the South Asian context. In: Climate change and social responsibility. Emerald Publishing Limited, pp 205–226. https://doi.org/10.1108/S2043-052320250000025011

Juyal S (2022) Electric mobility and electric vehicles management in India. In: Jadhav P, Choudhury RN (eds) Infrastructure planning and management in India. Studies in Infrastructure and Control. Springer, Singapore. https://doi.org/10.1007/978-981-16-8837-9_9

Kulagin V, Kuklina A, Grushevenko D (2025) The impact of logistics and political decisions on greenhouse gas emissions in market transformation. Carbon Research 4(1):e8. https://doi.org/10.1007/s44246-024-00172-7

Kumar A, Singh R, Khan S, Kumar V (2024) Artificial intelligence-enabled management: an emerging economy perspective. De Gruyter, Retrieved December 23, 2024

Kunytska O, Persia L, Gruenwald N, Datsenko D, Zakrzewska M (2023) The sustainable and smart mobility strategy: country comparative overview. In: Arsenyeva O, Romanova T, Sukhonos M, Tsegelnyk Y (eds) Smart technologies in urban engineering. STUE 2022. Lecture Notes in Networks and Systems, vol 536. Springer, Cham. https://doi.org/10.1007/978-3-031-20141-7_59

Lv Z, Shang W (2023) Impacts of intelligent transportation systems on energy conservation and emission reduction of transport systems: a comprehensive review. Green Technol Sustain 1(1):100002

Mahale Y, Kolhar S, More AS (2025) A comprehensive review on artificial intelligence-driven predictive maintenance in vehicles: technologies, challenges and future research directions. Discov Appl Sci 7(4):e243

Mishra U, Rathnasiri MSH, Dewasiri NJ (2024) Net-zero economy and energy sustainability: a Nepalese perspective. In: Singh R, Khan S, Kumar A, Luthra S, Chokshi H (eds) Net zero economy, corporate social responsibility and sustainable value creation. CSR, Sustainability, Ethics & Governance. Springer, Cham. https://doi.org/10.1007/978-3-031-55779-8_9

Ponnusamy S, Chourasia H, Rathod SB, Patil D (2024) AI-driven traffic management systems: Reducing congestion and improving safety in smart cities. In: Smart cities. CRC Press, pp 96–121. https://doi.org/10.1201/9781003442660

Porru S, Misso FE, Pani FE, Repetto C (2020) Smart mobility and public transport: Opportunities and challenges in rural and urban areas. J Traffic Transp Eng (Engl Ed) 7(1):88–97

Rathnasiri MSH, Dewasiri NJ, Kumar AA (2024) Policy framework and implementation strategies for Sri Lanka's transition to a net-zero economy. In: Singh R, Khan S, Kumar A, Luthra S, Chokshi H (eds) Net zero economy, corporate social responsibility and sustainable value creation. CSR, Sustainability, Ethics & Governance. Springer, Cham. https://doi.org/10.1007/978-3-031-55779-8_3

Rogers E (2003) The diffusion of innovation, 5th edn. The Free Press, New York, NY

Samarasekera G, Sivakumar T, Nanayakkara S, Jayasinghe A, Liyanage T (2023) Policy-driven strategic transport planning for the Western Province, Sri Lanka. J South Asian Logist Transp 3(2):129–152

Sarker MT, Al Qwaid M, Shern SJ, Ramasamy G (2025) AI-driven optimisation framework for smart EV charging systems integrated with solar PV and BESS in high-density residential environments. World Electr Veh J 16(7):385

Singh R, Gupta A, Bajpai A, Kandpal V (2024b) Exploring challenges of circular economy initiatives for smart cities. In: Smart cities and circular economy: the future of sustainable urban development. Emerald Publishing Limited, pp 59–69. https://doi.org/10.1108/978-1-83797-957-820241005

Singh R, Crowther D, Ravesangar K (2024) Harnessing human capital for a sustainable future: HR's pivotal role toward net zero policies. In: Singh R, Crowther D (eds) Transition towards a sustainable future. approaches to global sustainability, markets, and governance. Springer, Singapore. https://doi.org/10.1007/978-981-97-5756-5_2

Singh R, Ravesangar K, Narayanan S, Singh KKH (2024) HRM practices in creating net zero: supporting organisational sustainability. In: Singh R, Khan S, Kumar A, Luthra S, Chokshi H (eds) Net zero economy, corporate social responsibility and sustainable value creation. CSR, Sustainability, Ethics & Governance. Springer, Cham. https://doi.org/10.1007/978-3-031-55779-8_1

Singh R, Gupta A, Kandpal V (2025) ESG excellence in the digital world: circular economy e-business solutions. Fintech ESG Circ Econ 193–204. https://doi.org/10.1002/9781394234035.ch10

Co-Evolutionary Pathways of Digital-Green Integration for Industrial Upgrading in Southern Xinjiang

Songyue Li, Shufan Zheng, Ying Tuan Lo, Wei Shen Low, and Siti Suraya Abd Razak

Abstract Southern Xinjiang, historically dependent on traditional industries, faces dual challenges of environmental stress and economic stagnation. Leveraging digital technologies alongside green innovation offers a transformative pathway, yet little is known about how these trajectories interact in peripheral, culturally distinctive regions. This study develops a co-evolutionary framework combining Innovation Diffusion Theory and the Multi-Level Perspective to explain the mechanisms and conditions enabling industrial upgrading. Using fuzzy-set Qualitative Comparative Analysis (fsQCA) on 12 enterprises across textiles, agriculture, and mineral processing, supplemented by embedded case studies, the research identifies three dominant pathways: (1) digital acceleration, (2) green-led structural response, and (3) digital-green synergy. Findings reveal that digital capability acts as a necessary but insufficient condition; the most robust upgrading occurs when digital infrastructures integrate with ecological technologies and institutional incentives. These interactions enhance real-time transparency, operational flexibility, and market legitimacy. The study contributes theoretically by contextualizing socio-technical transition models for frontier economies and provides practical implications for policymakers and managers, emphasizing targeted infrastructure, policy alignment, and socio-cultural adaptability for sustainable industrial transformation.

S. Li
Faculty of Finance and Accounting, Zhejiang Business College, Hangzhou, China

S. Zheng
School of Management, Zhejiang University of Finance and Economics, Hangzhou, China

Y. T. Lo (✉)
XJTLU Entrepreneur College (Taicang), Xi'an Jiaotong-Liverpool University, Suzhou, China
e-mail: lyingtuan@hotmail.com

W. S. Low
Department of Hospitality and Information Technology, Sunway College Johor Bahru, Johor Bahru, Malaysia

S. S. Abd Razak
Faculty of Management, Universiti Teknologi Malaysia, Johor Bahru, Malaysia

R. Singh et al. (eds.), *Data-Driven Sustainability: Harnessing Technology for A Greener Future and Environmental Resilience*, World Sustainability Series,
https://doi.org/10.1007/978-3-032-13412-7_8

Keywords Digital economy · Green technology · Industrial upgrading · Southern Xinjiang · Sustainable development

1 Introduction

Southern Xinjiang combines rich cotton, oil, and solar resources with chronic peripheral constraints such as long transport chains, limited modern logistics, and complex ethnic governance. Uyghur and Hui majority communities shape production rhythms through religious calendars, clan-based trust networks, and bilingual labour markets. Conventional studies on digital or green upgrading seldom incorporate these sociocultural variables, relying instead on evidence from China's coastal clusters or European eco-industries. Although some research has examined technology-driven experiences in other sectors, such as smart dining environments (Lo et al. 2024; Sun et al. 2025), manufacturing-focused analyses in culturally complex, peripheral economies remain scarce. The resulting knowledge gap obscures why identical technologies diffuse unevenly across regions and how environmental targets interact with local institutions in a Muslim-majority setting. This study closes that gap through three contributions. First, it fuses innovation-diffusion theory with the multi-level perspective on socio-technical transitions (MLP) to model a co-evolutionary mechanism in which digital capability, green-technology integration, and institutional embedding reinforce one another. Second, it introduces the concept of contextual complementarity, arguing that religious norms, clan trust, and resource heterogeneity jointly moderate both technology uptake and environmental upgrading. Third, it empirically tests this mechanism in a peripheral border economy, offering the first systematic account of dual transformation pathways in Southern Xinjiang and demonstrating the portability and limits of mainstream transition theory beyond core regions. By specifying the how and for-whom of digital green convergence, the paper moves beyond generic integration calls and provides a structured explanation of boundary conditions that shape industrial upgrading in culturally distinct, resource-rich frontier areas.

2 Literature Review

2.1 Digital Economy and Industrial Upgrading

Digital economy refers to the production, allocation, and consumption of value that relies on data-intensive infrastructure, platform coordination, and algorithmic intelligence. Core technologies include cloud computing, industrial IoT, blockchain-based traceability, and AI-driven analytics (Lund-Thomsen et al. 2019). In peripheral regions such as Southern Xinjiang, digital tools function under four structural constraints: long-haul logistics, multilingual labour pools, limited broadband coverage, and variable institutional capacity. These constraints shape both the pace of adoption and the form of upgrading.

Empirical studies in minority regions report that e-commerce penetration lifts smallholder income by compressing intermediary tiers and expanding market reach. Digital payment ecosystems embedded in WeChat and Alipay reduce cash-flow volatility for textile micro-enterprises operating under informal credit norms (Sun and Qu 2025). Similar findings in consumer-facing sectors show how AI-mediated experiences enhance engagement and sustainability perception, ultimately influencing purchase intentions (Jing et al. 2025). These insights underscore the dual role of digital tools in shaping both operational efficiency and consumer trust. Cross-border platforms supported by the Khorgos dry port integrate Uyghur-run agribusinesses into Central Asian value chains, enhancing export price realisation (Chen et al. 2025). These findings confirm that digital infrastructure can mitigate geographic remoteness when complemented by local social capital.

Upgrading outcomes depend on the interaction between data capability and organisational learning. Firms possessing bilingual ERP dashboards and sensor-based production monitoring display higher defect-reduction and process flexibility than purely manual systems (Chen et al. 2022). AI-enabled predictive maintenance shortens downtime in cotton-spinning mills where machine vintage averages fifteen years, illustrating a leapfrogging effect that bypasses incremental automation stages (Sun et al. 2025). Similar relationships have been observed in entrepreneurial and accounting contexts, where AI integration mediates the impact of strategic management systems on performance outcomes, highlighting its transformative role across sectors (Yang et al. 2025) Platform data analytics support bespoke design services for small-batch Islamic apparel, moving producers into design-branding segments of the value chain.

In 2022, the Aksu Textile Industrial Park initiated a pilot using membrane-based closed-loop dyeing systems across 8 major processing units. These upgrades resulted in a 42.1% reduction in freshwater intake and a 36.8% decline in wastewater discharge compared to 2020 baselines. According to locally available monitoring records, COD concentrations reportedly decreased from about 85 mg/L to 61 mg/L over two production cycles. Firms reported average annual savings of CNY 1.5 million in water and chemical treatment costs. These outcomes are consistent with Zhao et al (2018) findings in similar textile clusters in northwest China, where closed-loop systems generated substantial environmental and cost benefits for SMEs.

In Kashgar's Shule County, solar-powered drip irrigation was deployed over 39,700 hectares of cotton fields in 2021 to 2023. Remote sensor-linked control systems optimized irrigation timing, reducing diesel consumption by 68% and improving water-use efficiency by 44.9%. Empirical studies from Kong and Chen (2025) further confirm that solar-powered irrigation improves yield stability and water precision in arid Xinjiang, especially when linked to remote monitoring infrastructure.

Biomass valorization projects in Hotan also demonstrate how low-grade agricultural byproducts contribute to energy substitution. In 2023, the Qira Bioenergy Cooperative converted 280,000 tons of cotton stalks into compressed briquettes and biochar, replacing 23.5% of coal inputs for nearby brick kilns. Emissions audits estimated a reduction of 120,800 tons of CO_2. Similar effects were reported by (Li et al. 2024), who found that biomass reuse in Hotan's rural industries reduces particulate pollution while generating stable rural employment.

Digitalization and green technology integration rarely unfold as isolated processes. In structurally constrained regions like Southern Xinjiang, their convergence emerges through recursive feedback loops, institutional intermediaries, and dual-function infrastructures. Rather than two parallel shifts, digital and green transformations become co-constitutive when they reshape organizational incentives, data visibility, and compliance capacities in tandem.

Case evidence from the Aksu textile sector shows that real-time water quality sensors, originally installed to monitor dyeing processes, were later linked to municipal environmental platforms, automating regulatory reporting and reducing inspection frequency by 26%, according to local environmental authority records. Similar dual-purpose infrastructure emerged in the Shache agricultural zone, where satellite-linked irrigation systems were simultaneously used for evapotranspiration control and compliance with local groundwater quotas. These systems reduced manual reporting burdens and enabled access to CNY 4.2 million in green equipment subsidies. The synergistic value was not in the individual technologies but in their shared data interface and modular coordination.

2.2 *Theoretical Framework*

This study integrates Innovation Diffusion Theory (IDT) and the Multi-Level Perspective (MLP) to construct a processual and context-sensitive model of dual transformation in Southern Xinjiang. Rather than treating digital and green transitions as independent trajectories, this framework conceptualizes them as co-evolving dynamics within a layered socio-technical system, where adoption depends not only on technology attributes but also on alignment with institutional structures, community norms, and spatial constraints.

Innovation Diffusion Theory, as formulated by (Rogers et al. 2019), centers on how new practices spread through communication channels over time within a social system. It emphasizes perceived relative advantage, compatibility with local values, and observability as key determinants of adoption. In Southern Xinjiang, religious norms, trust-based social capital, and clan-based decision networks shape what innovations are seen as legitimate, feasible, and sharable. Digital dashboards and closed-loop water systems are not diffused merely on economic grounds but also on how well they fit into community-based governance and interethnic cooperation norms (Yang and Xu 2025).

The Multi-Level Perspective, derived from socio-technical transition theory (Geels 2002; Geels and Schot 2007), explains systemic transformation through interactions across three levels: niche innovations, socio-technical regimes, and macro landscapes. In this model, digital tools and green technologies function as niche experiments that challenge established production routines and ecological baselines. Regional policy initiatives such as the "Green Silk Road Corridor" and Xinjiang's "Digital-Rural Integration Pilot Plan" operate as regime destabilizers, creating windows for niche-regime realignment. Local ethnic governance structures, climate resource constraints, and infrastructure asymmetries constitute elements of the macro landscape that condition the speed and direction of change.

3 Research Methodology

3.1 Case Selection and Data Sources

Twelve enterprises were purposively selected across three industries central to the Southern Xinjiang economy: textiles (5 cases), agricultural processing (4 cases), and mineral-based manufacturing (3 cases). Selection criteria included evidence of recent digital or green technology adoption, operational transparency, and representativeness of varying ownership forms (state-owned, private, cooperative). Enterprises were drawn from Aksu, Hotan, Korla, and Kashgar. The sampling strategy aimed at maximum configuration heterogeneity while ensuring cross-sectoral comparability.

3.2 Variable Construction and Calibration

The measurement model assessment for the operationalized core conditions variables is as shown in Table 1. Each condition was calibrated into fuzzy sets (range: 0 to 1) using direct method of calibration (Rogers et al. 2019). For example, a firm with fully integrated IoT-based monitoring across production stages was coded as DC = 1, partial adoption with digital dashboards only = 0.67, sporadic tool use = 0.33, and no digital system = 0. Case-based thresholds were defined using empirical anchors

Table 1 Measurement model assessment for operationalized core conditions variables (author's own creation)

Condition	Code	Description	Source
Digital capability	DC	Integration of ERP, IoT, or digital supply chain platforms	Interviews, site records
Green integration	GI	Adoption of closed-loop water, solar, biomass reuse, or low-carbon labeling	Audit reports, CSR disclosures
Institutional support	IS	Participation in green/digital subsidy programs, pilot zones	Local gov. policy archives
Structural response	SR	Evidence of value chain repositioning, product diversification	Interview narratives, sales data
Industrial upgrading	IU	Joint improvement in profitability, eco-compliance, and innovation output	Internal KPIs, export records

and theoretical cutoffs from recent studies (Feng et al. 2022). Calibration details are provided in Appendix A.

3.3 fsQCA Procedure

The fsQCA 3.1 software was used to analyze the dataset. The following steps were followed:

- Necessity analysis for each individual condition to test whether any single factor was consistently required for industrial upgrading (threshold: consistency $\geq$0.9).
- Truth table construction, listing all possible combinations of DC, GI, IS, and SR across 12 cases.
- Frequency threshold set at 2 (minimum 2 cases per configuration) and consistency threshold at 0.80 based on Ragin's guidelines and empirical testing.
- Boolean minimization using the Quine-McCluskey algorithm to extract parsimonious, intermediate, and complex solutions.
- Coverage and consistency of each solution path assessed to ensure explanatory strength.
- Robustness checks conducted by varying consistency thresholds (0.78 0.85) and recalibrating one condition to test model sensitivity.

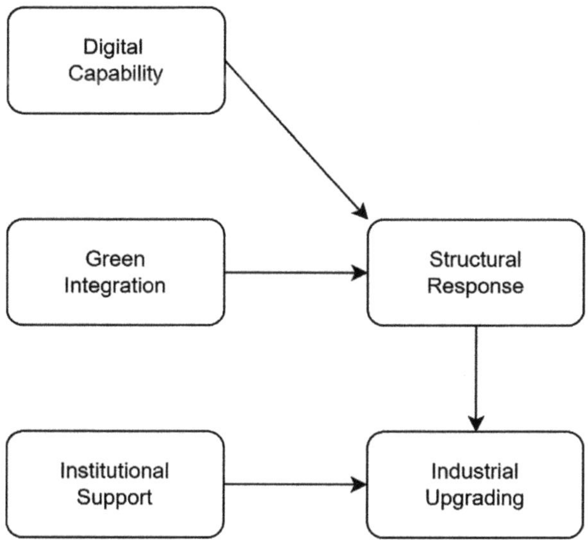

Fig. 1 Conceptual model of digital-green-policy interactions in industrial upgrading. *Note* Structural response refers to organizational and process-level adjustments within firms, enabling the effective assimilation of digital and green inputs

3.4 Integration with Case Study Logic

The case study component provides within-case process tracing for each of the high-consistency configurations. For instance, two agricultural processors (Cases B and E) matched the path {DC GI IS → IU}, and their internal investment memos, staff interviews, and product repositioning strategies were examined in detail to validate causality. This allowed theoretical insights to move beyond correlation into mechanism-level explanation. The two methods thus work in tandem: fsQCA identifies sufficiency conditions across the sample, while case studies unpack how and why those configurations succeed or fail under real-world constraints.

3.5 Model Diagram

The model diagram is as shown in Fig. 1 (author's own creation).

4 Findings and Discussion

The Enterprise Upgrade Pathways Analysis is as shown in Table 2. fsQCA results show that high digital capability appears in all configurations associated with successful industrial upgrading. In particular, the configuration {DCGIIS → IU} emerged in four cases with consistency ≥ 0.91 and raw coverage of 0.42, suggesting that digital infrastructure operates not as a standalone trigger but as a necessary enabler for the successful alignment of green practices and institutional support. This condition also passed the necessity test for IU (consistency = 0.93), making it the only factor consistently present across all success paths.

Case A, a state-holding cotton processing enterprise in Aksu, illustrates this pathway. Since 2021, the firm deployed an integrated IoT-based monitoring system across its scouring and dyeing lines, paired with ERP modules that track water and energy usage in real time. These systems reduced average production cycle times by 17.2% and defect rates by 14.5% compared to pre-adoption levels, according to internal company records. Sensor logs also improved coordination between dye scheduling and wastewater treatment, enabling the plant to meet provincial COD discharge targets without retrofitting its effluent unit. Managers noted that automation had enhanced interdepartmental communication, reducing paperwork dependency in a bilingual workforce.

Case E, a private pomegranate juice exporter in Kashgar, adopted a cloud-based cold-chain management platform and blockchain-enabled farm-to-factory traceability in 2022. These tools stabilized delivery intervals by ±3 h despite road delays and reduced product loss by 22.4% over two harvest seasons. Export documentation was streamlined using the Xinjiang Customs digital portal, enabling the firm to complete halal and organic certification applications entirely online. As a result,

Table 2 Enterprise upgrade pathways analysis (authors own creation)

Pathway type	Condition combination	Representative cases	Key mechanism description	Upgrade results
Pathway A	DC + GI (± IS)	Case A/B/E	Real-time data transparency; Platform-based supply chain collaboration; Digital certification driving market entry	Improved response speed; Enhanced compliance efficiency; Export expansion
Pathway B	GI + IS + ~DC	Case C/G	Structural reconstruction through energy-saving equipment; Policy subsidies driving cost transfer; Environmental compliance strengthening market positioning	Reduced energy/ water consumption; Expanded green channels; Improved resource efficiency
Pathway C	DC + GI + IS/SR	Case B/D	Sensor and closed-loop system integration; Blockchain and green labeling overlay; Digital platform coordinating certification and production	Enhanced collaboration; Access to high-standard markets; Better brand and financing performance

the firm entered new markets in the UAE and Malaysia, posting a 31% year-on-year increase in export revenue from 2022 to 2023.

The comparative case F, a traditional textile firm in Yecheng without any digital system, failed to meet export growth targets or qualify for regional green finance despite installing solar boilers. Interviews revealed poor documentation of environmental metrics and low responsiveness to market fluctuations due to manual inventory and order tracking. fsQCA confirmed that absence of DC was present in all failure configurations.

Three mechanisms explain how digitalization accelerates upgrading:

- Real-time transparency: Digital dashboards and IoT sensors reduce information asymmetry within firms, enabling faster and more coordinated decisions across procurement, production, and compliance functions.
- Operational flexibility: Predictive analytics tools help firms anticipate demand shifts and resource constraints, allowing quicker adjustment of production plans.
- Market legitimacy: Digital traceability platforms support participation in green or ethical value chains where documentation, certification, and response speed are prerequisites. This is consistent with evidence from other green technology markets, such as electric vehicles, where consumer purchase intentions hinge on perceived innovation legitimacy and trust in domestic brands (Thoo et al. 2024; Yang and Liu 2024).

Contrary to linear assumptions, digital adoption does not guarantee upgrading unless integrated with complementary capacities. Case D, a mining machinery firm in Hotan, adopted QR-based order management but lacked green technology or policy support. The firm's overall performance remained stagnant, confirming fsQCA's finding that DC is a necessary but not sufficient condition for upgrading.

This section supports the broader argument that digital capability functions not as an isolated driver but as a structuring condition that enhances a firm's ability to absorb complementary innovations, respond to institutional expectations, and enter regulated markets. In peripheral regions with multilingual labor, dispersed operations, and policy volatility, digital systems provide coordination scaffolding that analog infrastructures cannot match.

5 Conclusions

This study explored how traditional industries in Southern Xinjiang undergo structural upgrading through the dual pathways of digitalization and green transformation. By integrating innovation diffusion theory and the multi-level perspective, the research developed a co-evolutionary framework that explains how technologies, institutions, and organizational behaviors interact under peripheral constraints. Using fuzzy-set Qualitative Comparative Analysis (fsQCA) and multiple embedded case studies, the study identified three distinct upgrading mechanisms: digital acceleration, green-led structural response, and digital green synergy.

Findings show that digital capability functions as a necessary condition across most success paths, providing coordination, traceability, and adaptive learning. Green technology, when supported by institutional incentives or embedded in restructured operations, can also independently drive upgrading. The most robust outcomes arise from digital green integration, where data systems and ecological tools form mutually reinforcing systems. These synergistic firms not only improve performance but also reconfigure their organizational logic to meet new policy and market demands.

The framework introduced in this study captures both adoption behavior and systemic change. It illustrates how technology diffusion is shaped by cultural compatibility, institutional mediation, and infrastructural gaps. Southern Xinjiang, with its multilingual labor base, dispersed value chains, and religious compliance regimes, presents conditions that challenge conventional transition models. This research demonstrates that peripheral regions are not passive adopters but can serve as laboratories for hybrid upgrading strategies under constraint.

Theoretical contributions include a mechanism-based reconciliation of IDT and MLP, clarification of when and how green transformation proceeds without digital support, and an empirical demonstration of co-evolutionary upgrading in a structurally disadvantaged yet strategically important region. These insights extend the generalizability of socio-technical transition frameworks beyond industrial cores to frontier economies.

By distinguishing functional complementarity, structural compensation, and institutional leverage, this study provides a typology of pathways that contribute to sustainable industrial renewal. It suggests that effective upgrading depends less on the presence of individual technologies and more on their alignment with the region's infrastructural, regulatory, and socio-cultural context.

While the study advances theoretical and practical insights, its scope is limited by the small sample of 12 enterprises, a cross-sectional design, and the methodological constraints of fsQCA, which may not capture long-term dynamics. Future research should broaden the geographical scope to include other frontier and coastal regions, adopt longitudinal designs to trace evolving pathways, and evaluate the effectiveness of specific policy interventions such as subsidy mechanisms, infrastructure investment, and governance integration to strengthen the evidence base for sustainable industrial transformation. At the same time, the findings highlight how Uyghur and Hui cultural governance, including clan-based trust networks and halal compliance, shapes the uptake of institutional support and can either reinforce or constrain digital–green pathways. For policymakers, this suggests the importance of tailoring subsidies and training programs to bilingual and multicultural enterprises, developing trust-building mechanisms between ethnic minority firms and green finance providers, and promoting halal-certified digital platforms to integrate local producers into export markets.

Appendix A: Condition and outcome calibration logic

Variable	Calibration Logic	Fully In (1.0)	Crossover Point (0.5)	Fully Out (0.0)
Digital Capability	Extent of integration of digital tools (ERP, IoT, platforms, AI) across core operations	Full integration across production, sales, logistics	Partial application in one or two segments	No use of digital tools
Green Integration	Adoption of resource-efficient or eco-friendly technologies and circular practices	Closed-loop production, zero emissions, solar adoption	Some energy-saving technologies, no systemic integration	Traditional inputs, no green adoption
Institutional Support	Degree of formal support from government or policy environments	Policy pilot zone participation, green subsidies, formal incentives	Indirect policy exposure, occasional funding	No visible government interaction
Structural Response	Internal changes in firm structure/ processes reflecting technological adaptation and environmental goals	Process reengineering, dedicated digital-green teams	Incremental improvements, fragmented changes	No organizational change, rigid structures
Industrial Upgrading	Composite of three indicators: value chain movement, market expansion, eco-performance	Evidence of all three upgrading markers	Improvement in one or two dimensions	No clear upgrading trajectory

References

Chen G, Wu P, Wang J, Zhang P, Jia Z (2022) Ridge–furrow rainfall harvesting system helps to improve stability, benefits and precipitation utilization efficiency of maize production in loess plateau region of China. Agric Water Manag 261:107360. https://doi.org/10.1016/j.agwat.2021.107360

Chen Y, Wang Q, Zhang J (2025) The role of cross-border E-commerce platforms in the digital economy: Empower firms to gain global market insights to increase global competitiveness. J Digit Manag 1(1):4. https://doi.org/10.1007/s44362-024-00003-0

Feng H, Wang F, Song G, Liu L (2022) Digital transformation on enterprise green innovation: effect and transmission mechanism. Int J Environ Res Public Health 19(17):10614. https://doi.org/10.3390/ijerph191710614

Geels FW (2002) Technological transitions as evolutionary reconfiguration processes: a multi-level perspective and a case-study. Res Policy 31(8–9):1257–1274. https://doi.org/10.1016/S0048-7333(02)00062-8

Geels FW, Schot J (2007) Typology of sociotechnical transition pathways. Res Policy 36(3):399–417. https://doi.org/10.1016/j.respol.2007.01.003

Jing H, Thoo AC, Lo YT, Lau TC, Gan HS (2025) Green threads and digital dreams: exploring the influence of AI-mediated experiences on purchase intentions in sustainable apparel markets. In: Chen W, et al. (eds) Selected proceedings from the 2nd international conference on intelligent manufacturing and robotics, ICIMR 2024, 22–23 August, Suzhou, China. ICIMR 2024 2024. Lecture Notes in Networks and Systems, vol 1316. Springer, Singapore. https://doi.org/10.1007/978-981-96-3949-6_10

Kong L, Chen J (2025) Impact of digital transformation on green and sustainable innovation in business: a quasi-natural experiment based on smart city pilot policies in China. Environ Dev Sustain 27(10):24629–24657. https://doi.org/10.1007/s10668-024-05643-w

Li H, Liu Z, Hachard V (2024) Digital transformation driving green innovation: Evidence from chinese a-share firms. Int Rev Econ Financ 95:103487. https://doi.org/10.1016/j.iref.2024.103487

Lo YT, Ma W, Tan AHP, Majeed APPA, Lau TC, Razak SSA (2024) Predicting the impact of restaurant automation and food safety in china: identifying key factors for smart dining experience. In: Tan A, et al (eds) Advances in intelligent manufacturing and robotics. ICIMR 2023. Lecture Notes in Networks and Systems, vol 845. Springer, Singapore. https://doi.org/10.1007/978-981-99-8498-5_14

Lund-Thomsen P, Wendelboe Hansen M, Lindgreen A (eds) (2019) Business and development studies: issues and perspectives, 1st ed. Routledge. https://doi.org/10.4324/9781315163338

Rogers EM, Singhal A, Quinlan MM (2019) Diffusion of innovations. In: Salwen MB, Stacks DW (eds) An integrated approach to communication theory and research (3rd edn). Routledge

Sun Y, Qu Q (2025) Platform governance, institutional distance, and seller trust in cross-border E-commerce. Behav Sci 15(2):183. https://doi.org/10.3390/bs15020183

Sun Y, Zhang M, Wang C, Li M (2025) How the digital transformation of chinese traditional manufacturing enterprises drives green innovation: a moderated mediation model. Sustainability 17(4):1473. https://doi.org/10.3390/su17041473

Thoo AC, Fu Y, Lo YT, Huam HT (2024) Key factors influencing consumer purchase intentions for Chinese brand electric vehicles. Int J Bus Soc 25(3):1072–1104. https://doi.org/10.33736/ijbs.8559.2024

Yang L, Sulaiman Z, Lo YT, Yang XW, Tan AHP (2025) Empowering entrepreneurship: exploring the mediating role of AI integration between SMA and performance. In: Chen W, Abdul Majeed APP, Ping Tan AH, Zhang F, Yan Y, Luo Y, Huang L, Liu C, Zhu Y (eds) Selected proceedings from the 2nd international conference on intelligent manufacturing and robotics, ICIMR 2024, 22–23 August, Suzhou, China. Springer Nature, pp 166–190. https://doi.org/10.1007/978-981-96-3949-6_13

Yang C, Liu Q (2024) Driving green innovation through digital transformation: empirical insights on regional variations. Sustainability 16(23):10716. https://doi.org/10.3390/su162310716

Yang T, Xu B (2025) Green innovation in manufacturing enterprises and digital transformation. Econ Anal Policy 85:571–578. https://doi.org/10.1016/j.eap.2024.12.024

Zhao R, Liu Y, Zhang Z, Guo S, Tseng M-L, Wu K-J (2018) Enhancing eco-efficiency of agro-products' closed-loop supply chain under the belt and road initiatives: a system dynamics approach. Sustainability 10(3):668. https://doi.org/10.3390/su10030668

Sustainability and Resilience of Smart Communities in South Africa Using Data-Driven Technologies: A Bibliometric Analysis and Systematic Literature Review

Florah Sewela Modiba and Jabu Mtsweni

Abstract The persistent challenge of poverty in African countries necessitates a clear strategy for leveraging technology to meet Sustainable Development Goals (SDGs). While South Africa's policy embraces the SDGs, there is a noted lack of clarity on how data-driven technologies are being used to track progress, especially in purported smart communities. This paper addresses this gap by exploring the intersection of sustainability and resilience in smart communities, aiming to determine how data-driven technologies are used to monitor and contribute to national development targets. The study employs an integrated methodology, combining quantitative bibliometric analysis with a qualitative systematic literature review. The synthesised findings are mapped to the SDGs and WSIS Action Lines, providing a standardised framework for evaluation. The bibliometric analysis reveals that while 90% of reviewed abstracts use data-driven technologies, only 50% of the studies focus on long-term sustainability, and 40% explicitly address building resilience. The literature review highlights that sustainable and resilient initiatives adopt a human-centric, bottom-up approach, contrasting with failures of top-down projects. This research proposes a comprehensive evaluation framework structured around three pillars: Sustainability, Resilience, and Technological Foundation to assess a smart community project's potential impact and measure its success over time. The findings offer an evidence-based overview and a conceptual framework for policymakers, practitioners, and researchers, providing a roadmap for leveraging data-driven technologies to foster more sustainable and resilient smart communities in South Africa.

Keywords Smart communities · Sustainability · Resilience · Data-driven technologies · South Africa · Bibliometric analysis · SDGs · WSIS

F. S. Modiba (✉)
Nelson Mandela University, Gqeberha, South Africa
e-mail: florah.modiba@mandela.ac.za

J. Mtsweni
Council for Scientific and Industrial Research (CSIR), Pretoria, South Africa
e-mail: mtswenijs@gmail.com

R. Singh et al. (eds.), *Data-Driven Sustainability: Harnessing Technology for A Greener Future and Environmental Resilience*, World Sustainability Series,
https://doi.org/10.1007/978-3-032-13412-7_9

1 Introduction

Smart communities in the global South present stark structural disparities for the underserved rural and township areas lacking digital infrastructure, while well-resourced suburban areas benefit from faster emergency response, reduced traffic congestion, and enhanced public safety (Bibri and Krogstie 2020; Javaid et al. 2025). There are varying definitions of the concepts of smart city and smart community. A smart city is described as a large urban area deploying information and communication technologies (ICTs) and data-driven solutions for enhanced urban operations, services and improved quality of life (ITU 2021; Makkonen et al. 2023). These are achieved using technologies such as data analytics and the Internet of Things (IoT) to manage traffic, utilities, safety, and pollution by integrating physical, digital, and human systems (Mora et al. 2021).

Conversely, smart communities narrowly focus on community-level ICT interventions. They involve community members in the decision-making process, thereby incorporating the participatory aspect (Ciesielska et al. 2024). However, the term has been used broadly enough to encompass communities outside of urban settlements, such as rural communities (Doerr et al. 2018). In such communities, sectors that use advanced technologies are regarded as smart and contribute to building a smart community, for instance, the use of smart agriculture (Ferrari et al. 2022).

The smart community definition is adopted in this chapter because it addresses different types of communities that use converging technological, physical, and human systems to develop, and is inclusive. Smart communities are cited as a sustainable approach to addressing environmental degradation and climate change, contributing to Sustainable Development Goal (SDG)-13 (climate action).

Globally, smart cities have been extensively researched, with a focus on how they can support development through environmentally friendly approaches (Bibri et al. 2023; Emerllahu and Bogataj 2024; Mora et al. 2021). Similar studies have been conducted in the African context, examining how smart communities can be designed and what matters when adopting the smartness approach to development (Chetty et al. 2024; Mtsweni and Modiba 2021; Mwansa et al. 2025; Phahlamohlaka 2021). However, studies that focus on how data-driven technologies are used to facilitate sustainability are limited.

This chapter, therefore, aims to determine how data-driven technologies are used to monitor and contribute to national development targets. Thus, contributing to smart communities, urban planning, and sustainability literature. It can also guide policymakers on the gaps that need to be addressed for South Africa to reach its SDG targets. These research questions guide the chapter:

- What types of smart communities exist in South Africa?
- How can these smart communities be assessed for resilience and sustainability?
- What types of data-driven technologies are used in South African smart communities?

The chapter is structured as follows: Sect. 2 reviews relevant literature, Sect. 3 outlines the research methodology, Sect. 4, 5, and 6 presents and discusses findings, and Sect. 7 presents the proposed evaluation framework for the sustainability and resilience of smart community projects. The chapter is concluded with research gaps that could be considered in further research studies and recommendations in Sect. 8.

2 Literature Review

2.1 Theoretical Framework

Sustainable Development Goals (SDGs) and the World Summit on the Information Society (WSIS) Action Lines are two interconnected global frameworks aimed at fostering a more equitable, prosperous, and sustainable world.

The SDGs are a set of 17 interconnected global goals adopted by all United Nations Member States in 2015 as a "universal call to action to end poverty, protect the planet and ensure that all people enjoy peace and prosperity by 2030" (United Nations 2025). They succeeded the Millennium Development Goals (MDGs) and are broader in scope, applying to all countries, developed and developing alike. The SDGs cover a wide range of social, economic, and environmental issues, recognising that they are deeply intertwined.

The WSIS Action Lines originated from the WSIS, a two-phase UN summit held in 2003 (Geneva) and 2005 (Tunis). The summit's goal was to "build a people-centred, inclusive and development-oriented Information Society" (Klein 2004; Pyati 2005). The resulting Geneva Plan of Action outlined 11 action lines that provide a framework for using ICTs to achieve development goals, based on the MDGs that later became SDGs. In this chapter, these two frameworks provide the foundation for assessing smart community initiatives, including their sustainability and resilience, particularly in relation to data-driven technologies.

2.2 Empirical Literature

Community development in the twenty-first century increasingly demonstrates how technological infrastructure can drive progress while supporting environmental stability (Bibri et al. 2023). In this context, data-driven technologies can guide developments and align towards attaining SDG-9 (industry, innovation, and infrastructure), SDG-11 (sustainable cities and communities), and SDG-13 (United Nations 2025). However, ICT provision in under-resourced areas hinders development and the achievement of SDG-10 (reduced inequalities) (Aruleba and Jere 2022; Melgaço 2021). Gurstein (2015) stresses that technologies such as the internet should serve as tools of enablement and empowerment, rather than deepening divisions, a concern widely noted in the digital divide literature (Melgaço 2021; Modiba and Kaye 2023). Nevertheless, the rollout of infrastructure such as submarine cable networks promises to expand internet access in the Global South and enable participation in digital activities (Ma and Jiang 2025).

Digital divide constrains the effective use of smart communities as a development instrument that promotes the usage of advanced technologies (Mwansa et al. 2025; Umeike et al. 2025). Smart community approach suggests that ICT-driven innovations are at the core of developing and enhancing these areas, while also supporting economic, social, and environmental sustainability, and emphasising cost efficiency, rapid implementation, and leveraging existing technologies and expertise (ITU 2021; Ciesielska et al. 2024). However, context remains critical, as smart approaches must account for human capability and the role of ICT in developing smart communities (Mtsweni and Modiba 2021).

Sustainable smart communities are cited as being dependent on public–private partnerships, as these projects have a higher chance of success (ITU 2021). However, the community is equally essential in the adoption and support of such projects, as it contributes to data production and technology usage (Ciesielska et al. 2024). Digital governance is proposed by Ma and Jiang (2025), arguing that to ensure the equitable use of digital technologies, a global digital governance framework is necessary. Consequently, policies must incorporate both the technological and human dimensions when smart communities are adopted as development tools to ensure their long-term sustainability and resilience (Ma and Jiang 2025; Zhou et al. 2021).

South Africa's development has been hindered by historical structural inequalities, resulting in regional disparities. The post-apartheid administration adopted a participatory development approach, enabling communities to shape their own development (Baloyi and Lubinga 2017). In response to the global smart city agenda, the country embraced this paradigm by developing a Smart Cities Framework (SCF). The SCF serves as a guide for stakeholders on what constitutes a smart city in the South African context. It emphasises inclusivity (access to digital infrastructure and digital literacy), unity among involved stakeholders, and smart decisions that consider social, economic, and environmental contexts to improve the quality of life for all communities (RSA 2021). The SCF definition extends beyond urban areas, thereby encompassing various forms of settlements.

Data-driven technologies have become a cornerstone of modern smart communities, enabling urban centres to leverage vast amounts of information to improve efficiency, sustainability, and quality of life (Radziszewska 2023). The application of data in smart communities spans a wide range of sectors, from urban planning to public safety. A key enabler is the IoT, which involves a network of interconnected sensors and devices that collect real-time data from various sources, such as smart meters, traffic sensors, and environmental monitors (Nahrstedt et al. 2016). To enhance sustainability, smart communities use data to optimise energy, water, and waste management. Smart meter data analytics can help cities develop more efficient energy distribution systems and integrate renewable energy sources (Mishra and Singh 2023).

The benefits of data-driven smart communities are substantial, including enhanced efficiency, reduced costs, and a better quality of life for citizens. Real-time data provides cities with the ability to respond proactively to issues, while predictive modelling helps in long-term urban planning (Bibri et al. 2023). However, implementing these technologies presents challenges related to data governance, privacy, and security (Choenni et al. 2022).

3 Research Methodology

The research methodology for this chapter employs a multi-phase approach, as shown in Fig. 1 It integrates bibliometric analysis (Agarwal et al. 2016) and a systematic literature review (SLR) (Khan et al. 2003) providing a comprehensive and robust assessment of the sustainability of smart communities using data-driven technologies, with a specific focus on the South African context.

3.1 Data Identification and Collection

This research began by defining clear, reproducible search strings. These are formulated using the keywords from the conceptual framework for this study. For the

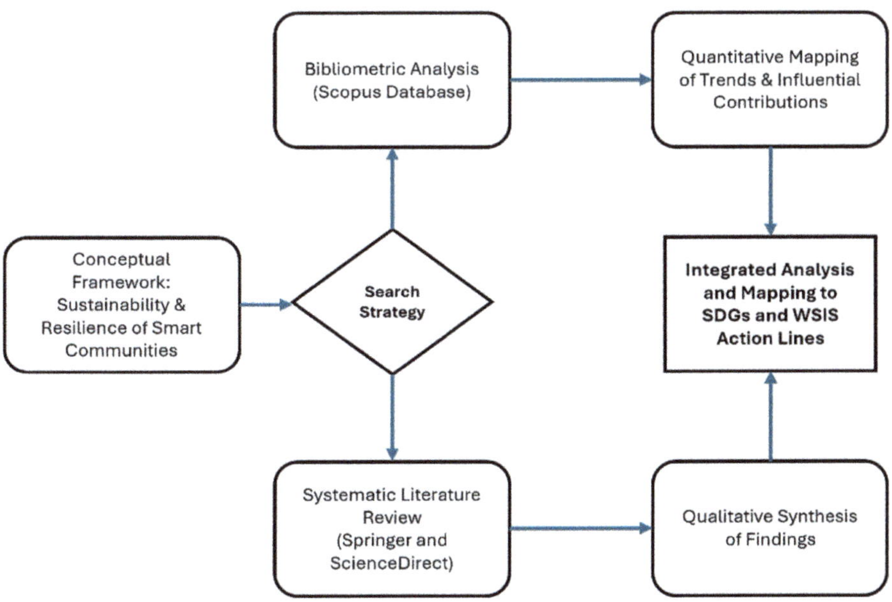

Fig. 1 Research approach review process (author's own creation)

Table 1 Data collection search string (keywords)

("smart communit*" OR "smart technolog*" OR "smart environment" OR "smart living" OR "smart mobility" OR "smart people" OR "smart governance" OR "smart economy" OR "digital divide" OR "smart data-driven technolog*" OR "ICT development" OR "smart connectivity" OR "community network*" OR "smart infrastructure" OR "smart community development" OR "smart rural area*" OR "smart village*" OR "smart cit*" OR "smart energy" OR "smart resource management" OR "community digital skill*" OR "technology inclusivity" OR "smart polic*" OR "smart resilience" OR "smart sustainability" OR "digital difference" AND "South Africa")	-

bibliometric analysis, data were collected from Scopus. The database was found to be relevant for this analysis, and it is widely considered the gold standard for bibliometric analysis (Pranckutė 2021; Zhu and Liu 2020). The search was then broadened to platforms like ScienceDirect and Springer to ensure all highly relevant studies from the South African context are captured for the subsequent SLR guided by the Preferred Reporting Items for Systematic and Meta-Analyses (PRISMA) principles. The search keywords used for the bibliometric analysis and the SLR are shown in Table 1.

3.2 *Inclusion and Exclusion Criteria*

The inclusion criteria for the SLR focused on identifying relevant studies on smart communities in South Africa. To be eligible, a study had to be published between 2015 and 2025 and written in English. The research needed to address at least one of the key dimensions of smart communities defined in Table 1. The systematic review and bibliometric analysis prioritised peer-reviewed research materials, such as journal articles, conference papers, books, and book chapters, to ensure high quality and academic rigour. Furthermore, it only considered studies on community-wide interventions, excluding those focused solely on individual or organisational-level implementations. The goal was to find research that could provide insights into the sustainability and resilience of smart communities at a collective level within the specific context of South Africa.

3.3 *Mapping to Global Frameworks*

The findings from the SLR are mapped to the SDGs and WSIS Action Lines. By aligning the research on smart communities with these global frameworks, the study measures the extent to which current initiatives and findings contribute to internationally recognised development targets. This provides a structured way to evaluate progress, identify areas of limited research focus, and highlight key challenges in developing smart, sustainable, and resilient communities in South Africa.

3.4 Integration of the Two Methods

The key to this methodology is the seamless transition from the bibliometric to the systematic review. The quantitative insights from the bibliometric analysis, such as the identification of key thematic clusters, directly inform and provide structure for the systematic review. This integration makes the literature synthesis a structured, evidence-based analysis of the field's most critical findings and research gaps. The mapping to SDGs and WSIS action lines provides a way to understand the value of the smart community research and initiatives in line with international frameworks.

4 Bibliometric Analysis

Bibliometric analysis is a research approach that employs quantitative and statistical techniques to systematically review and analyse the body of academic literature (Donthu et al. 2021). By mapping the research landscape, a bibliometric analysis can identify key research trends, influential authors and institutions, and collaboration networks, which helps to pinpoint knowledge gaps and opportunities for future research, specifically within the South African context (Oosthuizen 2022). The collected data is then pre-processed to remove duplicates and inconsistencies before being analysed using specialised software such as biblioshiny. The analysis comprises two main components: performance analysis, which assesses the productivity and impact of researchers, countries, and institutions, and science mapping, which reveals the conceptual structure of the field through techniques like co-authorship and keyword co-occurrence analysis (Agarwal et al. 2016; Donthu et al. 2021). Lastly, the results are then visualised and interpreted to provide insights into the evolution of the research area and to inform future academic and policy-making efforts.

The data collection phase utilised filtered research data from Scopus to conduct a quantitative assessment of the South African research landscape in smart communities, focusing on data-driven technologies for sustainability and resilience. The keywords used for searching relevant data are presented Table 1. The search was based on the article, abstract, and keywords of the publications in Scopus. Data analysis was performed using the bibliometrix R package and its web-based graphical interface, biblioshiny (Agarwal et al. 2016; Aria and Cuccurullo 2017). These tools facilitated the analysis of a dataset comprising over 306 records exported from Scopus. The exported dataset was also checked for correctness and completeness. Of the 306 documents, 303 were deemed usable, with the remaining three being duplicates. These documents came from 146 sources and were contributed by 708 authors.

This stage uncovered publication trends, influential contributions from key authors and institutions over 10 years (2015–2025) and uses keyword co-occurrence analysis to pinpoint thematic clusters.

4.1 Systematic Literature Review

Following the bibliometric mapping, a subset of the most relevant and impactful papers is selected from the broader dataset for a deeper, qualitative review. This phase focused on extracting detailed information from each paper to answer the core research questions, following PRISMA guidelines (Page et al. 2021).

The SLR utilised two databases, ScienceDirect and Springer, to ensure the inclusion of relevant articles, aiming to understand how research in smart communities has evolved since the adoption of the SDGs. A phased approach, illustrated in Fig. 2 was employed, utilising search terms in Table 1 to identify records with a strict focus on South Africa. This search yielded 10 records, of which nine were accessible.

The second phase involved a broad search, which yielded 358 records. After screening, only two records were deemed eligible, and only one could be downloaded. The third phase was a Springer search, which met the inclusion criteria. The search yielded 97 entries. After removing 11 duplicates, 86 were assessed for eligibility. Only three full-text articles were reviewed, and two qualified for inclusion. The synthesis process, which involved assessing article quality by evaluating whether they were linked to the study's research questions, led to five articles being selected for data extraction (Page et al. 2021). The review process identified an additional article, applying forward snowballing, and the article was included in the primary papers (Wohlin et al. 2022).

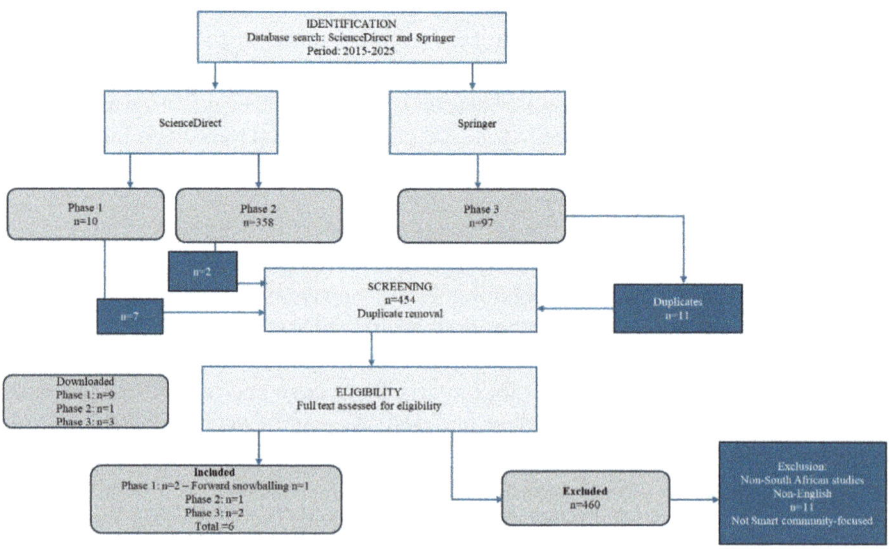

Fig. 2 PRISMA SLR review summary

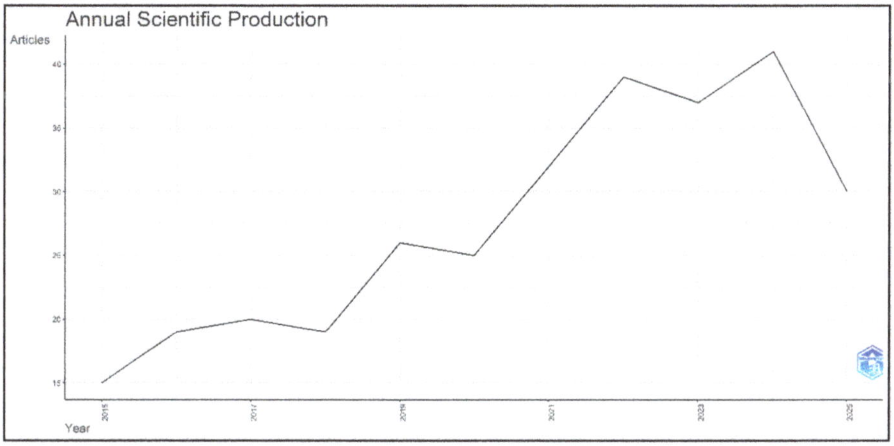

Fig. 3 Annual scientific production on the topic

5 Research Findings and Analysis

5.1 Bibliometric Analysis

The bibliometric analysis based on the corpus retrieved from Scopus is presented below, covering different indicators. The analysis suggests that South African authors are active in the research field related to smart communities; however, limited work is being done on data-driven technologies that could ensure sustainability and resilience.

Results Overview. the results overview illustrates a significant and steady increase in scholarly output related to the defined search terms (see Fig. 3), particularly from 2018 onwards, indicating a growing interest and research focus on the topic of smart communities and their sustainability.

Active Authors and Institutions. Based on Fig. 4, the most active author is MUSAKWA.W, with a total of 7 published documents. Following him are MOKOENA.BT with five documents, and a group of authors, including CALITZ AP, CLINTON AO, MANDA M, MOYO T, SCHOLTZ BM, TWINER R, and VAN DER HOOGEN A, who have all published four documents each. The least productive author on this list is CHIGONA W, with three documents. In addition, the bibliometric analysis suggests that CALITZ and VAN DER HOOGEN are the authors who are still active with recent publications.

The University of Johannesburg, University of South Africa and University of Cape Town (UCT) are the most active institutions in producing research output related to the search topics, with 59, 43, and 35 research outputs retrieved from Scopus. The University of Johannesburg's publications in this research theme have been growing exponentially since 2015, while UCT's have shown consistency over

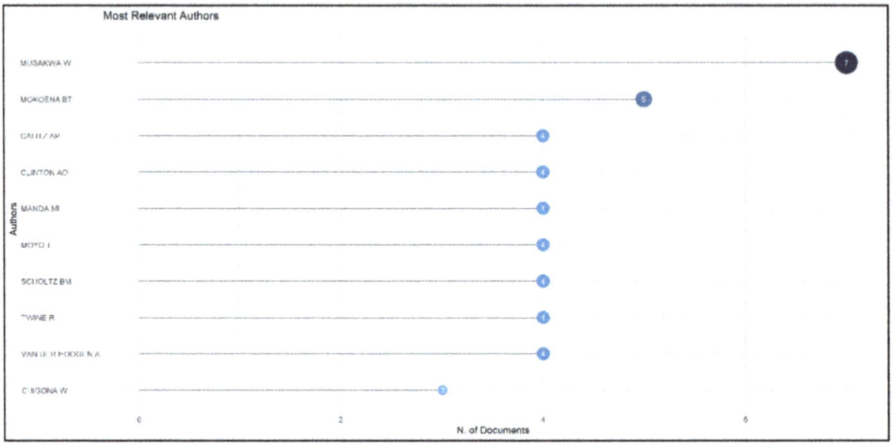

Fig. 4 Most relevant authors in the research domain

the past 10 years. It should be noted that over 700 authors have been analysed, and more than 10 institutions have been identified from the retrieved documents.

Word Cloud and Trending Topics. In Fig. 5, it is demonstrated that the digital divide (61*—number of times the keywords appear) and smart city (71) keywords are prominent in the research outputs analysed. The issues of community networks (55) and sustainable development (21) are also featured. Based on deeper analysis, it is evident that topics such as AI and sustainable development are increasingly getting attention from researchers. Topics such as urban planning, smart economy, social effects, digital divide, and e-government have been consistent over the past 10 years.

Data-driven related topics do not appear to feature much in the current research, with topics such as digital transformation (8), big data (7), decision making (10), and information systems (7) showing limited significance. This suggests that data-driven initiatives on smart communities are scarce in the South African context, which may also explain why issues of sustainability and resilience are not prominent based on the bibliometric analysis.

Concurrence Network and Collaboration Network. The co-occurrence network is a graph that displays keywords as nodes, with the links between them representing how often they appear together in the same documents (Aria and Cuccurullo 2017). This network's purpose is to reveal the relationships between concepts and identify distinct research clusters or sub-themes. Co-authorship analysis is a bibliometric method that maps and visualises the collaboration network among authors, institutions, or countries in a research field.

The network's nodes represent the authors, and the links between them indicate that they have co-authored at least one paper. A larger node size often signifies a greater number of publications or collaborations, while a thicker link indicates more frequent collaboration between two authors. Analysing these networks helps

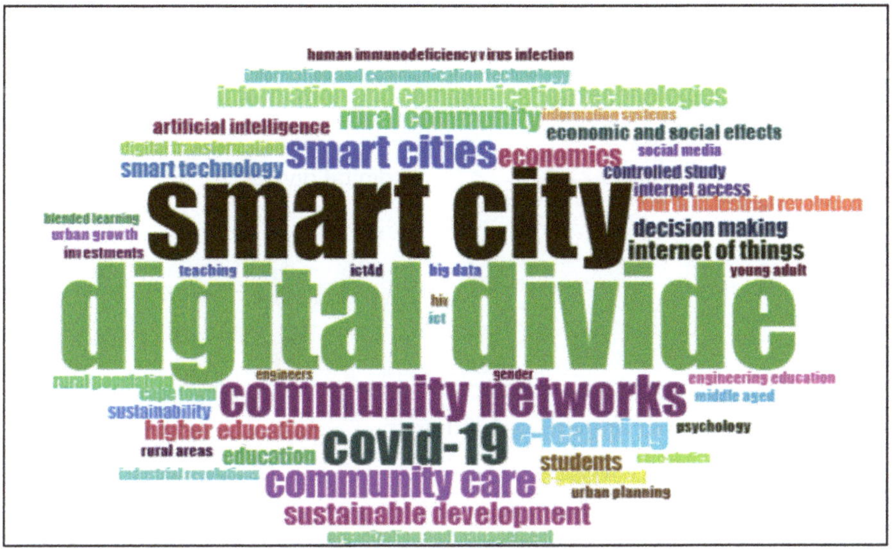

Fig. 5 Common words and topics in the research themes

identify central, highly connected researchers as well as research clusters and groups that collaborate frequently.

Based on our analysis, Fig. 6 illustrates that digital divide research issues tend to focus on socio-economic challenges such as rural areas, education, and lack of internet access. Whilst the focus on smart cities tends to be technology-driven, it often has a limited connection to communities, sustainability, and/or resilience. In the context of smart communities and sustainability, our results clearly indicate that collaborative networks are limited, as shown in Fig. 7. Notably, most of the retrieved documents tend to be produced by small networks with limited collaboration across different areas. This has a direct impact on the sustainability and resilience of the smart community projects.

Mapping to SDGs and WSIS Action Lines. The retrieved research data was also analysed using Excel to understand the actual projects and mapping to the global frameworks. The summary below provides the mapping for the top 20 research documents based on total citation count. The indicators used for the mapping were based on the research questions as stated above. The key findings from the analysis are as follows:

Smart Community-Drive: 60% of the studies demonstrate community-driven approaches, while others focus more on research or identifying problems rather than community-led solutions.

Data Driven Technologies: Found in 90% of abstracts, ranging from social media analytics and smart meters to mobile health systems and e-learning platforms.

Sustainability: Only 50% focus on long-term sustainable community solutions, indicating a gap between short-term interventions and lasting impact.

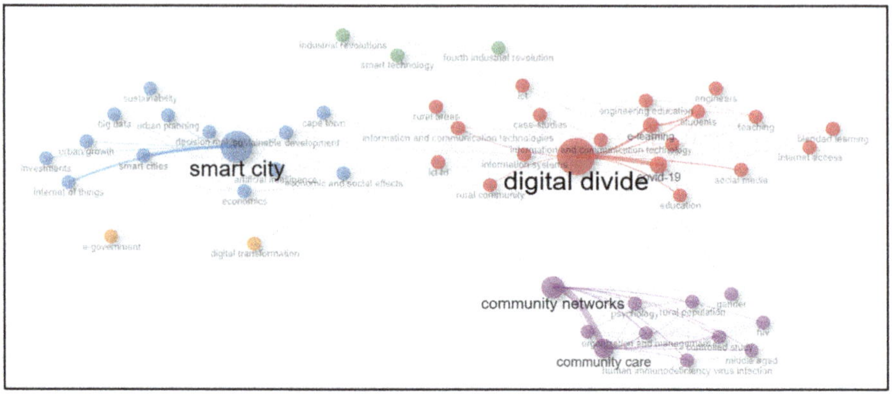

Fig. 6 Concurrence network on the research themes

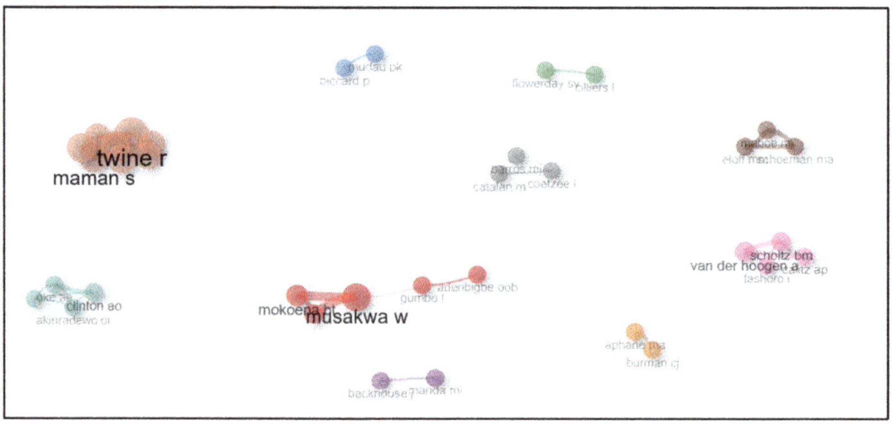

Fig. 7 Co-authorship collaboration in smart communities

Resilience: 40% explicitly address building community or individual resilience, particularly in health, education, and urban planning contexts.

Furthermore, the mapping shows consistent integration of sustainability principles, with many studies explicitly measuring outcomes such as resource efficiency, social equity, and behavioural change. These efforts are well-aligned with the SDGs, particularly SDG-11 (sustainable cities and communities), SDG-3 (good health and wellbeing), and SDG-4 (quality education), as well as WSIS Action Lines such as C2 (ICT Infrastructure), C7 (ICT Applications), and C3 (Access to Information).

The analysis reveals that although South Africa has numerous technology-enabled initiatives, significant work is still needed to ensure these become truly community-driven, sustainable, and resilience-building smart community solutions.

6 Systematic Literature Review Findings

The findings from the reviewed articles describe smart communities as inclusive, dynamic, participatory, data-driven, and ICT-enabled governance, focusing on improving urban management (Hadzic 2024; Ricker et al. 2020; Söderström et al. 2021). The studies discuss areas such as Cape Town, Ekurhuleni, eThekwini, and Johannesburg, which have experimented with the smart community concept. Table 2 presents the data summary.

Table 2 Summary of results

Source	Smart community	Technologies used	Sustainability
Hadzic (2024)	Cape town	Geo-location and asset registers Mobile inspection tools Open data platforms Grassroots data production	Reliance on expensive technologies and top-down approaches poses a threat to sustainability
Söderström et al. (2021)	Cape town, Ekurhuleni, eThekwini, Johannesburg	Broad-band infrastructure and fibre optic networks Datafication for urban management ICT-based platforms for e-governance Smart grids ICT Infrastructure convergence	Market-driven areas are vulnerable to compromising social and environmental sustainability
Maine et al. (2024)	Gautrain rapid railway link (GRL)	Blockchain smart contracts Smart energy management	Energy resilience Efficient and flexible energy management The use of solar energy lowers environmental impact
Ricker et al. (2020)	Cape town	Data platforms Geographic Information Systems (GIS) Online civic engagement tools	The smart CSO and community engagement approach is effective
Lorenz et al. (2015)	Not mentioned (education sector not named)	Computers and mobile devices in school	Not addressed
Nkoala and Matsilele (2023)	Cape Peninsula University of Technology	Online learning platforms LinkedIn learning	Not addressed

Smart communities in South Africa. South Africa has a vision of smart communities, although the rollout of some of the projects has not been successful. Hadzic (2024) argues that the failure of smart city projects, such as the Modderfontein New City project and the WesCape Project, resulted in the loss of government resources due to their top-down approach. They emphasise the importance of adopting a grassroots approach to creating smart communities, ensuring inclusiveness by actively involving local community members in this instrumental development process. A similar approach is observed in the Open Data project, which enables the public to engage with urban data collected by the municipality and civil society organisations (CSOs) (Ricker et al. 2020).

The challenge with top-down approaches is that they fail to acknowledge the broader inequalities in the South African communities. They also cite risks such as 5G technology perpetuating the inequalities because the technology prioritises advanced, costly services aimed at high-end users (Hadzic 2024). Thus, diverging from the affordability proposed by (ITU 2021).

The case of Lorenz et al. (2015) and Nkoala and Matsilele (2023) demonstrates such communities where smartness is in empowering students with digital skills. In agreement with Mtsweni and Modiba (2021), Hadzic (2024) asserts that smartness is determined by context. Arguing that smart communities should demonstrate practical technology implementation and inclusive access instead of pursuing futuristic ideals. However, inclusive digital skills were not equally accessible to all CPUT students, as not all of them could participate in online learning platforms (Nkoala and Matsilele 2023).

Söderström et al. (2021) argue that eThekwini and Ekurhuleni are the cities with characteristics of a smart community. Ekurhuleni, which started as a digital city that attracted global businesses and executives. Thus, showing the dependence of this smartness in the private sector. The digitality is also expressed through its airport, OR Tambo, which is backed by ICTs. Similarly, eThekwini progressed from the digital city concept to a smart community, utilising ICT infrastructure to drive economic and investment growth.

Additionally, the Johannesburg broadband project was one of the precursors to smartness, as it aimed to increase ICT access in mostly unconnected areas. The city reportedly launched a broadband network project to provide public Wi-Fi access. The GRL connecting Pretoria and Johannesburg uses blockchain smart contracts for energy management, enabling the trading of electricity (Maine et al. 2024).

Assessing the sustainability of these communities. Some smart community projects in South Africa have lacked inclusivity. Hadzic (2024) and Ricker et al. (2020) suggest a human-centric approach to smart communities, aligning with Mtsweni and Modiba (2021). This approach will enable the spatial planning of communities without infrastructure to be improved, facilitating a bottom-up approach to achieve a participatory development strategy, as outlined in the country's core development framework, as noted by Baloyi and Lubinga (2017).

The market approach used by eThekwini provides economic sustainability; however, the same cannot be said about social sustainability (Söderström et al. 2021). The city's focus without plans to address inequality issues means other communities continue to live in poverty and underdevelopment. Conversely, the GRP's integration of renewable energy and blockchain technology in the Gautrain electrical system contributes to its sustainability (Maine et al. 2024). However, it is not clear how this initiative will be socially sustainable.

Ensuring equal access to digital infrastructure and reducing inequalities can help households afford to participate in the digital economy, thereby improving sustainability. The higher educational sector reveals stark disparities in access to digital devices and learning infrastructure. The inclusion of ICTs in teaching and learning is essential for enhancing young people's digital skills (Lorenz et al. 2015). Nkoala and Matsilele (2023) argue that while the online learning environment allows students to improve their digital skills, others are disadvantaged by limited internet access in their areas of residence. Consequently, the digital divide affected the learning outcomes of these students compared to their more connected counterparts.

Data-driven technologies used in South African smart communities. According to Söderström et al. (2021), there has been deployment of broadband infrastructure and fibre optic networks to increase access to ICT usage and connectivity in areas like the City of Johannesburg. The upgrades to fibre optics have also improved connectivity in these major cities. Smart governance that utilises technology to enhance service delivery in areas such as Ekurhuleni, eThekwini, and Johannesburg has been noted. The use of e-governance enables municipalities to efficiently manage operational requirements, service delivery, and support business processes.

Using data to monitor and evaluate urban systems like water, energy, transportation, unemployment, safety and security. The convergence of ICTs with other city utilities leads to enhanced economic development, especially in the cases of Ekurhuleni and eThekwini aerotropolis. The GRP case leverages smart contracts to improve energy efficiency, enabling automated peer-to-peer trading where the cost and availability of energy is transparent to prosumers.

ICT-integrated decision-making systems in the South African Police Services, specifically in the city of Cape Town, are being utilised through a private partnership as a real-time urban management system (Söderström et al. 2021). Hadzic (2024) reported on the geolocation and asset register that helps in locating local infrastructure and reporting faults for a speedy response on service delivery-related issues in communities like Monabisi Park (an informal settlement in Cape Town). The city works with VPUU (a civil society organisation) and the community to identify and monitor service delivery.

The smart communities discussed by Söderström et al. (2021) cannot be regarded as entirely sustainable, as they rely on the private sector and businesses that are not locally based. The use of blockchain-based contracts proved to enhance energy management and support SDG-7 (clean and affordable energy) and SDG-13 (Maine et al. 2024). However, there is a need to adopt an approach that involves both governments and communities. Hadzic (2024) suggests the need to move away from a silo approach in both policy formulations that support smart communities and prioritise community participation.

There are also noted failures of smart communities that could not proceed due to municipalities' concerns about spatial issues. Sustainability in providing students with digital skills is also challenged when there are limited resources for them to be digitally connected (Lorenz et al. 2015; Nkoala and Matsilele 2023), thus threatening the resilience of smart community projects. Hadzic (2024), Ricker et al. (2020), and Söderström et al. (2021) argue for bottom-up approaches that involve communities in innovative solutions, especially those addressing the service delivery and development objectives of municipalities. While they acknowledge the value of data-driven solutions, they also highlight the data gap for unconnected communities such as townships, informal settlements and rural communities.

7 Smart Community Projects: Evaluation Framework

Drawing on academic research, findings from the literature review analysis, and international frameworks, a comprehensive evaluation framework for data-driven smart community projects is proposed (see Table 3) and is structured around three core pillars: Sustainability, Resilience, and the Technological Foundation. This framework can be used to assess a project's potential impact and measure its success over time, moving beyond simple technological adoption to focus on tangible community outcomes. The main aspects that need to be considered for any smart community project are the key performance indicators (i.e., what the success factors for the project are and what data sources can be used). The data sources will be influenced by appropriate data-driven technologies deployed within the smart community initiative.

Sustainability is measured by a project's long-term environmental, economic, and social impact, using data from sources like smart meters and public surveys. Resilience assesses the community's ability to withstand and recover from shocks, with KPIs focused on infrastructure uptime, operational efficiency, and social preparedness, all supported by real-time data from IoT sensors and emergency service logs. The Technological Foundation pillar evaluates the underlying data systems, emphasising good governance, interoperability, and robust security to ensure the technology is a reliable and secure tool for achieving the project's goals.

These pillars are mapped to the relevant SDGs and WSIS Action Lines to align with the national socio-economic imperatives that could influence the sustainability and resilience of these projects.

8 Conclusion and Recommendations

This study offers a comprehensive bibliometric and content analysis of smart community research in Africa, emphasising the role of data-driven technologies in advancing sustainability. By examining peer-reviewed articles, the research identified key trends

Table 3 Evaluation framework

Pillar	Focus	Key performance indicators (KPIs)	Data sources
Sustainability	Long-term viability (environmental, economic, social)	Environmental: Reduced energy consumption, increased renewable energy use, improved air quality	Smart meters, IoT sensors, waste management systems
		Economics: Cost savings, new jobs created, return on investment (ROI)	Financial records, open government data
		Social: Citizen satisfaction, access to internet, public safety	Surveys, citizen feedback apps, public records
Resilience	Ability to adapt to shocks and stresses	Infrastructural: System uptime, recovery time	IoT sensor data, network logs, disaster reports
		Operational: Emergency response time, traffic management efficiency	Real-time traffic data, emergency service logs
		Social: Community preparedness, public communication speed	Training records, social media analysis
Technological foundation	Reliability and governance of data systems	Data Governance: Framework adoption, data-sharing agreements, privacy compliance	Policy documents, legal frameworks, and audit reports
		Interoperability: Integrated data sources, open data adoption	API documentation, open data platform statistics
		Security: Security incident metrics, public trust in data management	Cybersecurity audit reports, incident logs, public surveys

in how digital tools such as mobile platforms, smart meters, and ICT-enabled learning are being leveraged to address challenges in health, education, governance, and urban development. The findings reveal that while many initiatives align with global frameworks like the SDGs and WSIS Action Lines, there is a lack of integrated approaches that holistically address sustainability and resilience.

To bridge this gap, the study proposes a Smart Community evaluation framework that could enable stakeholders to assess the maturity and impact of smart community interventions by linking bibliometric indicators with thematic insights. It serves as a diagnostic model for evaluating alignment with SDGs, technological integration, and community outcomes. The authors recommend future research to focus on cross-sectoral integration, participatory metrics, and longitudinal studies that track the evolution of smart initiatives. Strengthening policy support and fostering regional collaboration are also highlighted as critical steps toward inclusive, data-driven transformation across African communities.

References

Agarwal A, Durairajanayagam D, Tatagari S, Esteves SC, Harlev A, Henkel R, Roychoudhury S, Homa S, Puchalt NG, Ramasamy R, Majzoub A, Dao Ly K, Tvrda E, Assidi M, Kesari K, Sharma R, Banihani S, Ko E, Abu-Elmagd M, Bashiri A (2016) Bibliometrics: tracking research impact by selecting the appropriate metrics. Asian J Androl 18(2):296–309. https://doi.org/10.4103/1008-682X.171582

Aria M, Cuccurullo C (2017) bibliometrix: An R-tool for comprehensive science mapping analysis. J Inf 2017(11):959–975. https://doi.org/10.1016/j.joi.2017.08.007

Aruleba K, Jere N (2022) Exploring digital transforming challenges in rural areas of South Africa through a systematic review of empirical studies. In Sci Afr 16. Elsevier B.V. https://doi.org/10.1016/j.sciaf.2022.e01190

Baloyi ML, Lubinga EN (2017) After Izimbizo, what next? A participatory development communication approach to analysing feedback by the Limpopo provincial government to its citizens. J Transdiscipl Res South Afr 13(1). https://doi.org/10.4102/td.v13i1.402

Bibri SE, Krogstie J (2020) Environmentally data-driven smart sustainable cities: applied innovative solutions for energy efficiency, pollution reduction, and urban metabolism. Energy Inform 3(1). https://doi.org/10.1186/s42162-020-00130-8

Bibri SE, Alexandre A, Sharifi A, Krogstie J (2023) Environmentally sustainable smart cities and their converging AI, IoT, and big data technologies and solutions: an integrated approach to an extensive literature review. Energy Inform 6(1). Springer Nature. https://doi.org/10.1186/s42162-023-00259-2

Chetty K, Saal P, Ntshayintshayi N, Masuku N, Moosa T (2024) AI as a catalyst for good governance: transforming South Africa's fight against corruption. Development (Basingstoke) 67(1):50–60. https://doi.org/10.1057/s41301-024-00404-8

Choenni S, Bargh M, TB-J of SC (2022) Data governance in smart cities: challenges and solution directions. J Sagepub.Com 1(1):31–51. https://doi.org/10.3233/SCS-210119

Ciesielska M, Pereira GV, Lampoltshammer TJ (2024) Navigating the complexity: understanding social integration in smart communities versus smart cities.

Doerr J, Hess A, Koch M (2018) RE and society—a perspective on RE in times of smart cities and smart rural areas. In: Proceedings—2018 IEEE 26th international requirements engineering conference, RE 2018, 100–111. https://doi.org/10.1109/RE.2018.00020

Donthu N, Kumar S, Mukherjee D, Pandey N, Lim WM (2021) How to conduct a bibliometric analysis: an overview and guidelines. J Bus Res 133:285–296. https://www.sciencedirect.com/science/article/pii/S0148296321003155

Emerllahu V, Bogataj D (2024) Smart villages as infrastructure of rural areas: literature review and research agenda. IFAC-PapersOnLine 58(3):268–273. https://doi.org/10.1016/j.ifacol.2024.07.162

Ferrari A, Bacco M, Gaber K, Jedlitschka A, Hess S, Kaipainen J, Koltsida P, Toli E, Brunori G (2022) Drivers, barriers and impacts of digitalisation in rural areas from the viewpoint of experts. Inf Softw Technol 145. https://doi.org/10.1016/j.infsof.2021.106816

Gurstein M (2015) Why I'm giving up on the digital divide. J Community Inform 11(1):701–718

Hadzic S (2024) South Africa's digital transformation: understanding the limits of traditional policies and the potential of alternative approaches. Comput Law Secur Rev 55. https://doi.org/10.1016/j.clsr.2024.106011

ITU (n.d.) United smart sustainable cities 4 empowered lives. Resilient nations. Simple ways to be smart. Retrieved 25 September 2025, from https://www.itu.int/hub/publication/t-tut-smartcity-2021-03/

Javaid M, Haleem A, Singh RP, Suman R, Khan S, Rab S (2025) Adoption of internet of things for smart city-enabled smart hospitals. Intell Hosp 100011. https://doi.org/10.1016/j.inhs.2025.100011

Khan KS, Kunz R, Kleijnen J, Antes G (2003) Five steps to conducting a systematic review. J R Soc Med 96(3):118–121. http://www.pubmedcentral.nih.gov/articlerender.fcgi?artid=539417&tool=pmcentrez&rendertype=abstract

Klein H (2004) Understanding WSIS: an institutional analysis of the UN world summit on the information society. Inf Technol Int Dev 1(3):3–13. https://itidjournal.org/index.php/itid/article/download/150/150-486-1-PB.pdf

Lorenz B, Banister SI, Kikkas K (2015) Impacting the digital divide on a global scale—six case studies from three continents. Lect Notes Comput Sci (Incl Subser Lect Notes Artif Intell Lect Notes Bioinform) 9192:687–696. https://doi.org/10.1007/978-3-319-20609-7_64

Ma X, Jiang C (2025) Global submarine cable network and digital divide. J Geog Sci 35(6):1204–1232. https://doi.org/10.1007/s11442-025-2364-x

Maine PK, Leke CA, Longe OM (2024) Blockchain application in energy trading for grid-connected prosumers. E-Prime—advances Electr Eng, Electron Energy 8. https://doi.org/10.1016/j.prime.2024.100586

Makkonen J, Latikka R, Kaukonen L, Laine M, Väänänen K (2023) Advancing residents' use of shared spaces in Nordic superblocks with intelligent technologies. AI & Soc 38(3):1167–1184. https://doi.org/10.1007/s00146-022-01604-x

Melgaço L (2021) Challenging peripherality through access to the internet? Socio-spatial practices of the connected rurban. Urban Res Pract 14(1):73–93. https://doi.org/10.1080/17535069.2019.1655091

Mishra P, Singh G (2023) Energy management systems in sustainable smart cities based on the internet of energy: a technical review. Energies 16(19):6903. https://www.mdpi.com/1996-1073/16/19/6903

Modiba FS, Kaye S (2023) Evaluation of information and communication technologies (ICTs) tools contributing to rural development. Russ J Agric Socio-Econ Sci 142(10):19–29. https://doi.org/10.18551/rjoas.2023-10.03

Mora L, Deakin M, Zhang X, Batty M, de Jong M, Santi P, Appio FP (2021) Assembling sustainable smart city transitions: an interdisciplinary theoretical perspective. J Urban Technol 28(1–2):1–27. https://doi.org/10.1080/10630732.2020.1834831

Mtsweni J, Modiba FS (2021) A microtasking approach for building smart communities in digital villages. In: Phahlamohlaka J (ed) Approaches to building a smart community: an exploration through the concept of the digital village. Cambridge Scholars Publishing, pp 53–74. https://books.google.com/books?hl=en&lr=&id=xitCEAAAQBAJ&oi=fnd&pg=PA53&dq=A+MICROTASKING+APPROACH+FOR+BUILDING+SMART+COMMUNITIES+IN+DIGITAL+VILLAGES&ots=Sxm2w5pS3K&sig=qTexExjonDdLezQVQuy8EVJsI7k

Mwansa G, Ngandu MR, Mkwambi Z (2025) Bridging the digital divide: exploring the challenges and solutions for digital exclusion in rural South Africa. Discov Glob Soc 3(1). https://doi.org/10.1007/s44282-025-00189-2

Nahrstedt K, Lopresti D, Zorn B, Drobnis AW, Mynatt B, Patel S, Wright HV (2016) Smart communities internet of things. Arxiv.Org. https://arxiv.org/abs/1604.02028

Nkoala S, Matsilele T (2023) The influence of the digital divide on emergency remote student-centred learning during the COVID-19 pandemic: a case study of journalism education. SN Soc Sci 3(3). https://doi.org/10.1007/s43545-023-00626-6

Oosthuizen R (2022) Bibliometric analysis of information communication technology for sustainable development: a machine-learning-based approach. Int J Learn Chang 14(5–6):537–558. https://doi.org/10.1504/IJLC.2022.126421

Page MJ, McKenzie JE, Bossuyt PM, Boutron I, Hoffmann TC, Mulrow CD, Shamseer L, Tetzlaff JM, Akl EA, Brennan SE, Chou R, Glanville J, Grimshaw JM, Hróbjartsson A, Lalu MM, Li T, Loder EW, Mayo-Wilson E, McDonald S, Moher D (2021) The PRISMA 2020 statement: an updated guideline for reporting systematic reviews. Syst Rev 10(1). https://doi.org/10.1186/s13643-021-01626-4

Phahlamohlaka J (2021) Approaches to building a smart community: an exploration through the concept of the digital village. Cambridge Scholars Publisher.

Pranckutė R (2021) Web of science (WoS) and scopus: the titans of bibliographic information in today's academic world. Publications 9(1):12. https://doi.org/10.3390/PUBLICATIONS9010012

Pyati AK (2005) View of WSIS: whose vision of an information society? First Monday. https://firstmonday.org/ojs/index.php/fm/article/view/1241/1161

Radziszewska A (2023) Data-driven approach in knowledge-based smart city management. 24th European Conference on Knowledge Management, 2. https://books.google.com/books?hl=en&lr=&id=KjLZEAAAQBAJ&oi=fnd&pg=PA1090&dq=Data-driven+approach+in+knowledge-based+smart+city+management&ots=VeiXlABb3m&sig=fa71kGlT9pxUscFUwXsN3bNhX8o

Ricker B, Cinnamon J, Dierwechter Y (2020) When open data and data activism meet: an analysis of civic participation in Cape Town South Africa. Can Geogr 64(3):359–373. https://doi.org/10.1111/cag.12608

RSA (2021) A South African smart cities framework: A decision-making framework to guide the development of smart cities in South Africa. Retrieved on 25 September 2025 from https://unsplash.com/photos/WFI8vmzXsP4

Söderström O, Blake E, Odendaal N (2021) More-than-local, more-than-mobile: the smart city effect in South Africa. Geoforum 122:103–117. https://doi.org/10.1016/j.geoforum.2021.03.017

Umeike R, Dao T, Crawford S (2025) State-of-the-art review of resilient smart cities: progress and challenges. In: Urban governance (vol 5, Issue 2, pp 222–239). Elsevier B.V. https://doi.org/10.1016/j.ugj.2025.05.007

United Nations (2025) The sustainable development goals report 2025. Retrieved on 25 September 2025 from https://unstats.un.org/sdgs/report/2025/The-Sustainable-Development-Goals-Report-2025.pdf

Wohlin C, Kalinowski M, Romero Felizardo K, Mendes E (2022) Successful combination of database search and snowballing for identification of primary studies in systematic literature studies. Inf Softw Technol 147. https://doi.org/10.1016/j.infsof.2022.106908

Zhou Q, Zhu M, Qiao Y, Zhang X, Chen J (2021) Achieving resilience through smart cities? Evid China Habitat Int 111. https://doi.org/10.1016/j.habitatint.2021.102348

Zhu J, Liu W (2020) A tale of two databases: the use of Web of Science and Scopus in academic papers. Scientometrics 123(1):321–335. https://doi.org/10.1007/S11192-020-03387-8

Digital Governance for Urban Resilience in Bangladesh: A Policy Framework Integrating Digital Twins and Federated Learning in Climate-Vulnerable Deltaic Cities

Pranab Kumar Panday

Abstract The escalating impacts of climate change, particularly intense urban flooding, demand a paradigm shift from static infrastructure management to dynamic, data-driven resilience governed by integrated policy. This chapter proposes a novel Digital Governance Framework for climate adaptation in deltaic mega-cities, specifically contextualised within the socio-institutional realities of Bangladesh. The framework integrates an Urban Digital Twin (UDT) for real-time risk simulation with Federated Learning (FL) as a privacy-preserving mechanism for data sharing. We argue that institutional fragmentation, data silos, and mistrust among key agencies (e.g., DWASA, DPDC/DESCO) are the primary barriers to effective resilience modelling, often rendering centralised UDTs ineffective. To address this, the chapter's central contribution is the articulation of a multi-layered policy framework, advocating for regulatory instruments such as a Critical Infrastructure Resilience Act (CIRA) and the establishment of a Shared Funding Model. This policy structure mandates the sharing of collaborative intelligence and ensures data equity by prioritising vulnerable settlements through a Social Equity Index (SEI) and Community Sensor Nodes. By prioritising governance solutions that leverage advanced technological tools, this chapter outlines a scalable, institutionally sound, and equitable pathway toward achieving true environmental resilience and data-driven sustainability in the Global South.

Keywords Digital governance · Urban resilience · Federated learning · Digital twins · Climate adaptation · Bangladesh · Policy framework

P. K. Panday (✉)
Department of Public Administration, University of Rajshahi, Rajshahi, Bangladesh
e-mail: pranabpanday@yahoo.com

R. Singh et al. (eds.), *Data-Driven Sustainability: Harnessing Technology for A Greener Future and Environmental Resilience*, World Sustainability Series,
https://doi.org/10.1007/978-3-032-13412-7_10

1 Introduction

The twenty-first century is defined by concurrent forces of rapid urbanisation and accelerating climate volatility (Zhu and Jin 2025). This challenge is acutely concentrated in the cities of the Global South, where rapid, often unplanned, growth intersects with profound environmental vulnerability (UNESCAP 2022). No region exemplifies this critical nexus more starkly than South Asia, and specifically Bangladesh. As a low-lying deltaic nation, Bangladesh is routinely cited as one of the world's most climate-vulnerable countries, facing chronic stresses like sea-level rise and salinity intrusion, alongside acute shocks such as devastating cyclones and catastrophic monsoon flooding (Lowe et al. 2023; Haque et al 2019). Dhaka, its capital, is a mega-city where a dense population relies on a complex, often fragile, network of formal and informal critical infrastructure (CI).

The historical approach to urban resilience in such environments has relied on traditional, static, and often centralised engineering solutions. However, the non-linear nature of modern climate impacts demands a proactive, dynamic, and predictive approach to urban management. The systemic interconnectedness of CI—where the failure of a single power substation can cascade into municipal water outages and traffic paralysis—makes modeling resilience a complex, multi-domain computational challenge (Resilient Melbourne 2016). This chapter proposes that two emerging digital innovations, Urban Digital Twins (UDTs) and Federated Learning (FL), can be strategically combined to construct a next-generation framework specifically tailored for the highly constrained and fragmented data landscape of vulnerable urban centers like Dhaka or Chittagong. UDTs provide the essential virtual simulation environment for dynamic testing and scenario modeling (Wajid et al. 2025). Crucially, the efficacy of a UDT in a developing world context is fundamentally constrained by two factors: data scarcity in informal areas (Chakraborty et al 2015) and institutional reluctance to share sensitive infrastructure data, often rooted in privacy, security, and competitive concerns (Ketzler et al 2020).

This is where Federated Learning becomes indispensable. FL is a decentralised machine learning paradigm that enables multiple, potentially untrusting entities (e.g., utility companies, municipal agencies, large industries) to collaboratively train a shared predictive model while keeping their raw, sensitive data localised (Imteaj et al 2021). By sharing only model updates (parameters) rather than raw data, FL directly addresses data privacy and regulatory compliance hurdles, allowing for the collective, intelligent modeling required for large, multi-domain urban systems in a complex, multi-stakeholder environment. The core contribution of this chapter is the articulation of a UDT-FL Conceptual Framework designed not only to model CI interdependencies but also to bridge the information gap that typically impedes data-driven resilience initiatives in the Global South. The subsequent sections will first establish the theoretical and technological foundations, then pivot to the critical governance challenges, and finally apply this framework to enhance urban flood resilience in a high-risk deltaic city.

2 Literature Review: Data-Driven Foundations of Urban Resilience

The convergence of climate change and rapid urbanisation has placed data-driven resilience at the forefront of contemporary planning and engineering research. To understand the necessary shift toward dynamic, digital solutions, it is essential to first trace the evolution of resilience theory, then examine the promise and primary limitation of Urban Digital Twins, and finally establish the necessity of Federated Learning as a solution to critical data governance challenges. The conceptual history of resilience in urban studies begins with ecological theory, defined by Holling in 1973 as a system's capacity to absorb disturbance and reorganise while undergoing change (Holling 1973). While foundational, this idea has been refined by contemporary scholars like Meerow and Brand and Jax, who push beyond mere stability to define urban resilience as the system's ability to sustain essential functions, adapt to new conditions, and even transform when current circumstances become untenable (Meerow et al. 2016). This modern view emphasises adaptive capacity—the ability to learn and adjust—as the defining characteristic of a resilient city (Folke et al. 2002). Crucially, in the context of critical infrastructure (CI) and climate change, resilience must encompass the dynamic modeling of interdependency. Since the failure of one system (like a power grid) can trigger devastating cascading failures in others (water pumps, communication networks), a truly adaptive system requires high-fidelity, real-time data to anticipate and mitigate these effects, a task that exceeds the capability of traditional static modeling.

To meet this challenge, the Urban Digital Twin (UDT) has emerged as the premier technological tool for operationalising modern resilience. A UDT is a virtual replica of a physical urban system that uses real-time data to simulate, predict, and optimise processes (Ketzler et al 2020). Its core capabilities, such as running vast scenario planning simulations and visualising cascading interdependency failures across networks, are indispensable for proactive climate adaptation (Wajid et al. 2025). However, the UDT framework suffers from a critical weakness: the data integration and sovereignty bottleneck. A functional UDT requires massive amounts of proprietary and often politically sensitive operational data from diverse, siloed stakeholders—municipal authorities, private utilities, and national security agencies. In most cities, and particularly those in the Global South like Dhaka, this resistance is insurmountable, fueled by privacy concerns, regulatory limitations, and institutional mistrust (Zhu and Jin 2025). Without comprehensive, real-time data input from all necessary domains, the UDT degrades rapidly from a dynamic predictive tool to a static, inaccurate 3D model, severely limiting its utility for adaptive resilience.

This is where Federated Learning (FL) provides a vital technical solution to the UDT's data integration problem. FL is a distributed machine learning approach that allows multiple, independent organisations (clients) to collaboratively train a shared predictive model while keeping their raw, sensitive data localised (Imteaj et al 2021). The key privacy-preserving mechanism is simple yet powerful: clients only share their model parameters (weights and biases)—not the raw input data—with a central aggregating server. The server then computes a superior Global Resilience Prediction Model by averaging these updates (using, for example, FedAvg). This decentralised approach offers three massive benefits for urban modeling: Privacy by Design, effectively overcoming institutional resistance to centralised sharing; Edge Efficiency, leveraging computational power locally to reduce latency critical for real-time climate response; and Heterogeneity Management, allowing the model to handle the wide variation in sensor distributions and data formats (Non-IID data) found across urban utility providers (Abbas et al 2025).

However, the challenge of data integration extends beyond technical capability to fundamental policy and infrastructure gaps prevalent in the Global South. In contexts like Bangladesh, the fragmented data ecosystem is compounded by data poverty, which refers to the severe lack of consistent sensor data, particularly in informal or low-income areas (Chakraborty et al 2015). This creates systemic bias, as a centrally-trained model would inherently fail the most vulnerable communities. This is coupled with legacy institutional structures and institutional inertia, where old legal and bureaucratic frameworks prohibit the necessary data-sharing, even when the technological means (like FL) exist. These institutional hurdles demonstrate that the success of a UDT-FL framework is contingent upon a simultaneous overhaul of data governance policy, moving from *siloed ownership* to *collective, mandated intelligence sharing* (Zhu and Jin 2025). Therefore, the policy framework articulated in Sect. 4 is designed explicitly to dismantle these governance barriers and ensure the collective intelligence system can function ethically and effectively (UNESCAP 2022).

3 Conceptual Framework: The UDT-FL Integration for Adaptive Capacity

The proposed Urban Digital Twin–Federated Learning (UDT-FL) Conceptual Framework is designed as a three-layered architecture—Physical, Learning, and Simulation—that strategically moves beyond static data storage and centralised modeling. This highly relevant and scalable solution addresses fragmented data ecosystems and extreme climate vulnerability in deltaic mega-cities. The framework's structure is fundamentally built to secure data sovereignty while maximising collective intelligence, thereby enhancing the city's overall adaptive capacity.

3.1 Layer 1: The Physical Layer (Edge Nodes and Data Equity)

The foundation is the Physical Layer, encompassing the sensor network, critical infrastructure assets, and the computational devices, or edge nodes, responsible for local data processing. In a context like Bangladesh, this layer must be robustly heterogeneous. It integrates data from three primary source categories. First, it draws from Formal Critical Infrastructure Assets, using high-fidelity sensors and SCADA systems to monitor operational metrics from major utilities (like DWASA pump flow rates and DPDC substation load). These systems act as powerful, dedicated FL client nodes. Second, it incorporates Environmental and Geospatial Data, including static sets like CAD maps and 3D building models, along with real-time inputs such as satellite rainfall data and tidal gauges. This external data both informs the UDT base model and is used for feature engineering by the FL clients. Most crucially, the third source is the Community-Based and Low-Cost Sensor Networks, or the Data Equity Node. These low-cost IoT devices, deployed in flood-prone, informal urban settlements, transform data scarcity into an asset (Chakraborty et al 2015). They serve as essential FL clients, providing micro-level, high-resolution flood depth and duration data that capture vulnerabilities missed by macro-level models, thus inherently addressing data equity by ensuring that resilience solutions benefit the most vulnerable populations.

3.2 Layer 2: The Learning Layer (Federated Modeling and Aggregation)

Above this physical foundation lies the Learning Layer, the intelligence core dedicated to collaboratively training a global predictive model without ever centralising raw data. This layer employs the Federated Learning paradigm, structured around a central FL Orchestrator and distributed FL Clients (the edge nodes). The process begins with Client-Side Local Training, where each FL client (e.g., DWASA, DPDC) maintains its proprietary data set and trains a lightweight model (such as Support Vector Machines (SVM) or shallow Neural Networks) to predict system vulnerability metrics. This local training optimises model weights using stochastic gradient descent (SGD). Next, the client securely transmits *only* the resultant optimised model parameters (weights and biases)—not the raw input data—back to the FL Orchestrator, which is the critical step for maintaining data sovereignty. Finally, the Orchestrator performs Secure Aggregation and Global Model Update, using protocols like FedAvg to compute a new, superior Global Resilience Prediction Model by averaging the client updates (War et al. 2025). This global model represents the collective intelligence derived from diverse data sources, maximising predictive accuracy for the complex environment of Dhaka, before being re-distributed to all clients for the next continuous training round.

3.3 Layer 3: The Simulation Layer (Urban Digital Twin)

The final stage is the Simulation Layer, where the predictive power of the FL model is realised within the dynamic, multi-domain environment of the Urban Digital Twin (UDT). The UDT of Dhaka—a geometric and functional model encompassing its topology, hydrodynamics, and CI layout—continuously receives the real-time forecast of CI vulnerability from Layer 2. The primary function here is to transform the *statistical probability* into a *spatial and temporal reality* for decision-makers. This involves Risk Visualization and Hydraulic Modeling Integration, where the UDT overlays predicted CI failures onto the 3D city model, relying on a robust 2D Hydrodynamic Model (e.g., based on the Saint–Venant equations) to simulate cascading effects. For example, if the FL model predicts a drainage pump failure, the UDT instantly updates the hydraulic model to simulate the resulting flood plume expansion and duration, identifying all affected dependent assets (e.g., hospitals, power networks). Furthermore, this layer enables Adaptive Scenario Testing and Optimisation Algorithms. City managers can use the UDT interface to test instantaneous counterfactuals, such as rerouting power or diverting resources. To efficiently identify the best strategy, the UDT leverages meta-heuristic optimisation algorithms (like Genetic Algorithms or Particle Swarm Optimisation) to quickly explore millions of potential resource allocation decisions within the virtual environment (Abbas et al 2025). The output is a quantified set of optimal instructions presented via the decision-maker interface, providing the ultimate demonstration of enhanced adaptive capacity. This integrated UDT-FL framework thus creates a powerful, self-improving system where data governance challenges are resolved by decentralisation, and resilience is actively modeled through predictive, privacy-preserving collaborative intelligence, directly serving the complex governance and infrastructure needs of vulnerable mega-cities like Dhaka.

4 Governance and Policy Frameworks for Collaborative Resilience in Bangladesh

The transition from a siloed, reactive infrastructure management model to a coordinated, predictive UDT-FL system in Dhaka is fundamentally a challenge of governance, policy, and institutional reform (UNESCAP 2022). Addressing the deeply entrenched institutional silos and ensuring equitable outcomes are prerequisites for technological success.

4.1 *Regulatory and Institutional Hurdles: Bridging Organisational Silos*

The current landscape of critical infrastructure management in Dhaka is highly fragmented, with disparate agencies like DWASA (Dhaka Water Supply and Sewerage Authority) (Water), DPDC/DESCO (Dhaka Power Distribution Company Limited/ Dhaka Electric Supply Company Limited) (Power), Rajdhani Unnayan Kartripakkha (RAJUK, Development Authority), and the City Corporations operating largely independently (Zhu and Jin 2025; Haque et al 2019). This fragmentation manifests as significant policy and institutional resistance to centralised data initiatives.

The first major obstacle is Institutional Mistrust and Data Sovereignty. Utility agencies view their operational data as proprietary, a matter of national security, and a source of political leverage (Ketzler et al 2020). The perception that a central authority would gain unchecked visibility into their network vulnerabilities is the primary obstacle. In this context, Federated Learning is the technical solution to this policy problem, as it guarantees data remains within the agency's secure perimeter, shifting the collaborative burden from "data sharing" to "intelligence sharing." Furthermore, the Lack of Inter-Agency Mandates presents a significant legislative gap, as there is currently no robust legislative or executive mandate in Bangladesh to enforce data or model parameter sharing for resilience planning. The UDT-FL framework requires the formal establishment of a Critical Infrastructure Resilience Consortium backed by a national policy instrument, a body that would act as the legally designated, neutral FL Orchestrator mandated to aggregate models for the collective good. Finally, the Inconsistent Data Standards resulting from legacy IT systems across different agencies, which use incompatible data formats and reporting metrics, necessitates a policy intervention to standardise data *schemas* (e.g., common definitions for 'pump failure' or 'substation load'). This standardisation is required to ensure the model updates across clients are semantically consistent and demands a governance body to enforce a common ontological dictionary for all resilience-relevant infrastructure assets.

4.2 *Addressing Data Equity, Algorithmic Bias, and Inclusion*

In a city defined by informal settlements and vast socio-economic disparity, digital resilience must actively combat the tendency of technology to exacerbate existing inequalities (Abbas et al 2025). This starts with Mandating Inclusive Data Sourcing. The formal governance structure must mandate the inclusion of data from the Community Sensor Nodes (as outlined in Layer 1), a policy requirement ensuring that the vulnerabilities of marginalised, flood-prone neighborhoods are not ignored by a model primarily optimised for formal infrastructure in high-value areas (Chakraborty et al 2015). The policy must ensure that resource allocation for data collection and maintenance is disproportionately directed towards areas lacking formal infrastructure coverage. The second concern is Algorithmic Accountability and Bias Mitigation. Algorithmic bias can occur if the formal data heavily outweighs the informal data, leading the model to prioritise resource allocation to areas of historical investment. To counter this, a governance decision must be made to apply equity weighting to the model's loss function. This means the Social Equity Index (SEI) must be the primary metric for response planning, penalising the system more heavily for failures in highly vulnerable, underserved communities, thus ensuring the model's intelligence serves the most vulnerable first. Additionally, the governance consortium must establish an independent Algorithmic Auditing Board tasked with regularly inspecting the global model's parameters and outputs for unintended social bias before deployment. Lastly, Ensuring Community Ownership is vital. The governance model for the Community Sensor Nodes must be decentralised and participatory, designating local community leaders or NGOs as the official FL Clients for their respective areas. This participatory model shifts the community's role from being merely a data *source* to an active *contributor* to the system's overall intelligence and adaptive capacity.

4.3 Legal and Financial Mechanisms for UDT-FL Sustainability

The establishment and long-term viability of the UDT-FL framework necessitate specific legal and financial innovations unique to the context of Bangladesh. First, a national legislative act, termed the Critical Infrastructure Resilience Act (CIRA), must be passed to legally mandate the creation of the FL Orchestrator consortium and enforce the sharing of model parameters for all CI-managing agencies. CIRA would explicitly define the legal protection for proprietary raw data, reinforcing the FL privacy-by-design principle, while simultaneously making the failure to contribute model intelligence a compliance violation, thereby overcoming the current lack of a shared vision for resilience data management (UNESCAP 2022). Second, a Shared Funding Model is essential for cost and benefit sharing. Financing the UDT-FL system cannot rely solely on government allocation; instead, a mandatory Infrastructure Resilience Levy should be implemented, applied to the budgets of all formal FL client agencies (e.g., Dhaka Water Supply and Sewerage Authority [DWASA], Dhaka Power Distribution Company Limited/Dhaka Electric Supply Company Limited [DPDC/DESCO]). This pooled funding would exclusively support the central FL Orchestrator's operations, maintain the UDT simulation hardware, and finance the expansion of the community sensor network. Benefits, such as reduced service downtime and minimised disaster recovery costs, would be quantified and reported back to these agencies, providing a clear return on investment (ROI) for participation. Finally, concerning Data Licensing and Use Policy, the consortium must draft a clear legal policy governing the use of the aggregated Global Resilience Prediction

Model. This model, as a product of collective intelligence, must be categorised as a *public good* and made accessible to authorised government agencies for planning, while its commercial exploitation by any single client is strictly prohibited. This policy ensures the intelligence remains dedicated to the public welfare goal of urban resilience.

4.4 Technical and Operational Hurdles in Emerging Economies: A Policy View

While the following are technical hurdles, their successful resolution is deeply tied to investment policy and institutional capacity building in emerging economies. One significant challenge is Network Connectivity and Reliability. Dhaka's connectivity is prone to disruption during the very events the system is designed to monitor (floods, cyclones). Policy must therefore prioritise investment in redundant, low-power, and resilient communication networks (e.g., dedicated mesh networks or satellite backup) for the critical FL edge nodes. The technical solution here is implementing asynchronous FL protocols, allowing clients to report updates intermittently without failing the entire training round. Furthermore, Hardware and Maintenance Investment is crucial for the long-term sustainability of the UDT-FL system. A national maintenance policy and budget must be established to ensure the continued operationality of sensor deployment and computational edge nodes, moving beyond reliance on temporary grant funding. This policy should mandate that the utility agencies (FL Clients) are responsible for the maintenance and calibration of their own edge nodes, reducing the operational burden on the central Orchestrator. Capacity Building is the final necessity; the UDT-FL's success depends on municipal staff being able to interpret complex simulations and execute adaptive strategies. This requires a comprehensive national program for digital literacy and data-driven disaster management training across all cooperating agencies, with a specific focus on understanding algorithmic transparency and the equitable use of the Social Equity Index (SEI) in emergency response prioritisation.

5 Application Case Study: UDT-FL for Urban Flood Resilience in Dhaka

5.1 The Flood Resilience Challenge in Deltaic Mega-Cities

Dhaka, the beating heart of Bangladesh, sits precariously within a low-lying, flood-prone basin. Urban flooding here is not merely a consequence of extreme rainfall; it is a complex failure resulting from overwhelmed, aging drainage systems, encroachment on natural flood plains, tidal locking, and the systemic interdependence of infrastructure (Haque et al 2019). An intense monsoon event can simultaneously disrupt power supply (due to submerged substations), contaminate water sources (due to sewer backup), and paralyse the transportation network, leading to humanitarian crises and severe economic losses. Traditional flood mitigation methods—such as new canal construction—are often too slow and cannot adapt to the increasing intensity of climate-change-induced rainfall events (Zhu and Jin 2025). The implementation of a data-driven solution faces unique local barriers: Data Heterogeneity between multiple agencies; a Lack of high-resolution sensor coverage in densely populated areas (Chakraborty et al 2015); and Institutional mistrust hindering the centralised sharing of potentially sensitive operational data (Ketzler et al 2020; Zhu and Jin 2025). This institutional resistance is precisely the friction point that the UDT-FL architecture is designed to resolve, providing a shared intelligence mechanism without demanding raw data exposure.

5.2 *UDT-FL Operational Setup for Dhaka's Drainage System*

The UDT-FL framework is uniquely suited to overcome these barriers by decentralising data access while centralising intelligence. The application focuses on simulating the performance and potential failure points of Dhaka's drainage and CI network during forecasted extreme rainfall. The operational design transforms data silos into strategic FL clients that collectively train a robust predictive model for the UDT simulation environment.

The Federated Learning Client Network

The effective functioning of the system is predicated on the establishment of a robust FL client network where each member contributes local predictive intelligence while securing its proprietary data. In Dhaka, four key entities are designated as the primary FL clients, each contributing unique, specialised datasets necessary for comprehensive resilience modeling:

1. **Water Utility (DWASA, Dhaka Water Supply and Sewerage Authority)**: DWASA acts as a critical FL client, training its local model on proprietary data including pipe flow rates, reservoir levels, sewer overflow events, and, crucially, the operational status and maintenance logs of its vast network of pumping stations. The local model aims to predict the probability of pump failure under high-load, high-water-level conditions, which is essential input for the UDT's flood propagation models.
2. **Power Utility (DPDC/DESCO, Dhaka Power Distribution Company Limited/Dhaka Electric Supply Company Limited)**: The power utilities train models based on substation load balancing, outage history, and real-time operational status, particularly in flood-vulnerable zones. Their predictive model

informs the UDT about which substations are most likely to fail due to water inundation and the resulting power supply interruption probability. This is critical for predicting the cascading failure effects on other electrically-dependent CI, such as communication towers, water treatment plants, and traffic signals (War et al. 2025).

3. **Traffic Management Authorities (TMA)**: The TMA client uses sensor data from intersections, traffic light logs, and GPS tracking of municipal vehicles to predict urban mobility constraints. The local model forecasts road submergence time and gridlock probability based on rainfall intensity and water modeling data, allowing the UDT to simulate emergency vehicle routing and evacuation strategies.
4. **Community Sensor Nodes (CSNs)**: These low-cost, community-managed IoT sensors are strategically deployed in vulnerable, informal urban settlements where formal sensor coverage is lacking. The CSNs are crucial for addressing **data equity** (Chakraborty et al 2015). Their local model trains on micro-level, high-resolution flood depth and duration data. By aggregating these local models, the global FL model is corrected and sensitised to the flood dynamics affecting the most marginalised populations, ensuring resource allocation decisions are socially equitable.

This client-based system allows all participants to collaboratively generate a superior Global Flood Prediction Model without sacrificing institutional control over sensitive operational data, directly overcoming the key governance hurdle of data sovereignty.

Dynamic Simulation in the Urban Digital Twin

The Urban Digital Twin (UDT) of Dhaka—a high-fidelity geometric and functional model encompassing its topography, buildings, drainage network, and CI layout—continuously receives the aggregated, real-time **predictive risk scores** from the Global FL Model (Layer 2). The UDT's function is to translate statistical risk into a dynamic, spatial, and temporal reality for decision-makers (Abbas et al 2025).

The core of this process is Interdependency Modeling. When the FL model forecasts a high probability of multiple CI failures (e.g., pump station overload and a power substation trip), the UDT instantly initiates its sophisticated 2D Hydrodynamic Model (often based on the Saint–Venant equations). This model simulates the cascading effect: it calculates the rate of water accumulation resulting from the pump failure, estimates the time until the nearest power substation is submerged, and predicts the resulting traffic gridlock on essential emergency routes. This real-time visualisation allows city managers to grasp the full scope of a disaster scenario, identifying all directly and indirectly affected assets, well before the physical event occurs.

This intelligence feeds into Adaptive Mitigation Testing. City managers use the UDT interface to rapidly test instantaneous counterfactuals and evaluate the efficacy of various interventions against the predicted shock. For example, they can virtually implement proactive load shedding, digitally deploy mobile generators to critical pump stations, or test dynamic re-routing plans for ambulances. This iterative, predictive testing enables proactive decision-making, significantly enhancing the city's overall adaptive capacity to climate shocks by optimising the deployment of limited resources.

5.3 Methodology: Collaborative Data Acquisition and Learning

The success of the UDT-FL system requires not only a policy framework but also robust technical mechanisms that ensure privacy, data quality, and security during the collaborative learning process.

Data Inputs, Feature Engineering, and Semantic Standardization

The resilience model requires a fusion of four heterogeneous data categories for effective training: Geospatial Data (topographic maps, building models, CI network layouts), Operational Data (SCADA logs, maintenance history, real-time asset performance), Environmental Data (satellite rainfall, tidal gauges, weather forecasts), and Socio-economic Data (demographic maps, location of essential facilities, vulnerability indices). The complexity lies in integrating these disparate sources, which are often stored in different legacy formats across the FL clients (Haque et al 2019).

To overcome this Inconsistent Data Standards problem, the FL Orchestrator must enforce a common set of feature engineering protocols and a minimum required ontological dictionary. For example, all clients must standardise their definition and measurement of 'pump operational status' and 'water level threshold.' This semantic standardisation ensures that when a client (e.g., DWASA) reports a high-risk scenario, the global model can accurately correlate it with the inputs from another client (e.g., DPDC/DESCO) and the UDT can seamlessly translate it into a physical simulation. The Socio-economic Data is crucial for this process, as it is used to assign priority weights (the Social Equity Index, SEI) during local model training, ensuring the model's predictive focus aligns with the policy goal of prioritising the most vulnerable populations.

Data Security and Anonymization within the FL Protocol

While the FL architecture inherently addresses data sovereignty by keeping raw data local, the model parameters themselves could, theoretically, be reverse-engineered to infer information about the underlying proprietary data, especially in a network with a small number of clients. To guarantee robust privacy, advanced cryptographic techniques must be implemented. First, the system should employ Secure Aggregation (SA) protocols, which use cryptographic methods to ensure that the central FL Orchestrator can only decrypt and aggregate model updates once a minimum number of client reports have been received. This prevents the Orchestrator from isolating and inspecting any single client's update, thereby providing computational guarantees of privacy. Second, Differential Privacy (DP) should be applied at the client level. This technique involves injecting calibrated statistical noise into the model parameters *before* they are sent to the Orchestrator. The noise masks individual data points while being statistically insignificant to the overall model training outcome, providing a mathematically quantifiable guarantee of privacy protection (War et al. 2025). These security layers are essential for maintaining the institutional trust established by the governance framework (CIRA) and ensuring the long-term, continuous participation of sensitive government and utility agencies.

Strategy Optimisation and Trade-off Analysis Pipeline

The most critical function of the Simulation Layer is resolving policy conflicts in real-time. Upon receiving a high-risk prediction from the Global Resilience Prediction Model, the UDT initiates a rapid optimisation loop powered by meta-heuristic algorithms (Abbas et al 2025). These algorithms facilitate the quantitative analysis of critical, real-time policy trade-offs by evaluating millions of counterfactual scenarios based on resource allocation. This process uses two primary metrics for evaluation: Functional Capacity Loss (the quantified loss of essential urban services) and SEI Loss (the quantified social impact on vulnerable populations).

Two common trade-offs highlight the complexity of real-time decision-making in Dhaka:

1. **Power versus Pumping (DWASA vs. DPDC/DESCO)**: This conflict requires the system to decide the optimal power distribution strategy when floodwaters threaten the grid. Should electricity be diverted from one district's non-essential services to ensure the continuous operation of high-capacity flood pumps (managed by the Dhaka Water Supply and Sewerage Authority [DWASA]), or should power be prioritised for maintaining essential social services like hospitals, emergency command centers, and low-income housing (managed by the Dhaka Power Distribution Company Limited/Dhaka Electric Supply Company Limited [DPDC/DESCO])? The optimisation algorithm weighs the total *flood damage averted* against the total *social cost* (SEI Loss) associated with a power outage.

2. **Traffic versus Access (TMA vs. Civilian Safety)**: In anticipation of severe floodwater propagation, the system must weigh whether a key arterial road should be closed pre-emptively—to prevent vehicle entrapment and maintain a clear corridor for emergency services—or be kept open longer for civilian evacuation. Closing a road prematurely can cause chaos and delay, while keeping it open too long risks catastrophic loss. The UDT simulates the traffic flow and flood propagation for both scenarios, generating an optimal time-stamped instruction that minimises both Functional Capacity Loss (gridlock) and SEI Loss (stranded vehicles in high-vulnerability areas).

The output of the UDT is a ranked list of optimal strategies that moves policy decisions from intuition and political pressure to a transparent, data-driven, and socially-equitable mandate, thereby fulfilling the ultimate objective of enhanced adaptive capacity (Abbas et al 2025).

6 Discussion: Theoretical Contributions and Policy Implications

The proposed UDT-FL framework represents a significant advancement beyond conventional static and centralised urban planning models, offering distinct theoretical contributions and crucial policy implications, particularly for climate-vulnerable mega-cities of the Global South.

6.1 Theoretical Contribution: Operationalizing Collective, Adaptive Intelligence

The core theoretical contribution lies in the use of Federated Learning to solve the fundamental data sovereignty bottleneck that plagues traditional UDT implementations (Ketzler et al 2020). By allowing utility agencies to maintain local data control while sharing aggregated intelligence, the framework operationalises a concept of *collective intelligence* that is both computationally powerful and institutionally palatable. This synergy directly addresses the modern definition of urban resilience—the ability to *learn and adapt* (Folke et al. 2002)—by creating a dynamic Global Resilience Prediction Model that is continuously optimised by diverse, localised data streams. This moves the city from a state of passive endurance to one of proactive, Adaptive Mitigation Testing, a capability only realisable through the high-fidelity, shared predictive power of the FL-enabled Digital Twin (Abbas et al 2025).

6.2 *Policy and Governance Implications for Fragmented Systems*

The framework's success is contingent upon the policy mandates outlined in Sect. 4, which directly confront Dhaka's fragmented governance landscape (Zhu and Jin 2025; UNESCAP 2022). The introduction of the Critical Infrastructure Resilience Act (CIRA) is a necessary institutional intervention that transforms siloed agencies, such as DWASA and DPDC/DESCO, into mandated partners. The CIRA creates the legal legitimacy for the FL Orchestrator to exist as a neutral, intelligence-aggregating body, thereby bridging the Lack of Inter-Agency Mandates which previously prevented collaborative action. Furthermore, the mandatory implementation of the Infrastructure Resilience Levy establishes a sustainable, pooled funding model that ensures the framework is not dependent on sporadic external grants, promoting long-term viability.

6.3 *Ethical Imperatives and Equity of Resilience*

The most vital policy implication is the mandatory integration of the Social Equity Index (SEI) into the UDT's strategy optimisation pipeline. This decision acts as an essential ethical guardrail against algorithmic bias by prioritising resource allocation based on social vulnerability, rather than simply economic impact (Abbas et al 2025). The explicit inclusion of Community Sensor Nodes as formal FL Clients directly counters the problem of data poverty in informal settlements, ensuring that the model's intelligence is representative of the entire urban ecosystem (Chakraborty et al 2015). By penalising response strategies that disproportionately affect the most vulnerable, the framework ensures that the digital resilience system actively works toward equitable resilience outcomes.

6.4 *Limitations and Future Research Directions*

Despite its advantages, the UDT-FL framework faces several implementation challenges. It requires a substantial initial investment in sensor deployment, computational edge nodes, and centralised UDT simulation hardware. Furthermore, the system's effectiveness is highly dependent on sustained political will to enforce the CIRA and overcome deep-seated institutional inertia and mistrust (Haque et al 2019). Finally, while FL addresses privacy, the complexity of managing a large, heterogeneous network of clients and maintaining semantic consistency across wildly different data schemas (Inconsistent Data Standards) remains a significant technical and governance hurdle requiring continuous calibration and maintenance.

7 Conclusion and Future Research

The confluence of Digital Twins and Federated Learning represents a pivotal moment in the evolution of data-driven sustainability, offering a powerful, yet privacy-respecting, solution to the compounding crises facing vulnerable deltaic mega-cities like Dhaka. By providing a secure, collective, and dynamic framework for urban modeling, the UDT-FL Conceptual Framework moves urban resilience from a desirable goal to an actionable reality, systematically addressing data privacy and interoperability barriers endemic to the Global South. The system's ability to model and mitigate cascading infrastructure failures is essential for maintaining urban functionality under escalating environmental stress.

Future research must focus on the development of open-source FL libraries optimised for low-bandwidth environments, the empirical testing of the framework in a live urban laboratory in South Asia, and, most importantly, the creation of adaptive legal and governance models that embed social equity alongside technological efficiency to ensure that resilient futures are available to all urban populations, regardless of their infrastructural privilege. The necessary next steps involve pilot projects focused on establishing the Critical Infrastructure Resilience Consortium and drafting the foundational Critical Infrastructure Resilience Act to provide the mandated legal structure required for this technical framework to thrive in the complex socio-political environment of Bangladesh.

References

Abbas A, Adeel M, Akbar G, Hayyat S, Din GM (2025) Integrating digital twin and nature-based solutions for climate resilient urban design: a predictive framework for future ready residential neighborhoods in the Anthropocene. EJSMT 1(5):1–14

Chakraborty A, Wilson B, Sarraf S, Jana A (2015) Open data for informal settlements: toward a user's guide for urban managers and planners. J Urban Manag 4(2):74–91. https://doi.org/10.1016/j.jum.2015.05.001

Folke C, Carpenter S, Elmqvist T, Gunderson L, Holling CS, Walker B (2002) Resilience and sustainable development: Building adaptive capacity in a world of transformations. Ambio: J Hum Environ 31(5):437–440. https://doi.org/10.1579/0044-7447-31.5.437

Haque CE, Azad MAK, Choudhury MUI (2019) Discourse of flood management approaches and policies in Bangladesh: mapping the changes, drivers, and actors. Water 11(12):2654. https://doi.org/10.3390/w11122654

Holling CS (1973) Resilience and stability of ecological systems. Annu Rev Ecol Syst 4:1–23. https://doi.org/10.1146/annurev.es.04.110173.000245

Imteaj A, Khan I, Khazaei J, Amini MH (2021) Fedresilience: a federated learning application to improve resilience of resource-constrained critical infrastructures. Electronics 10(16):1917. https://doi.org/10.3390/electronics10161917

Ketzler B, Naserentin V, Latino F, Zangelidis C, Thuvander L, Logg A (2020) Digital twins for cities: a state of the art review. Built Environ 46(4):547–573. https://doi.org/10.2148/benv.46.4.547

Meerow S, Newell JP, Stults M (2016) Defining urban resilience: a review. Landsc Urban Plan 147:38–49. https://doi.org/10.1016/j.landurbplan.2015.11.011

UNESCAP (2022) Building urban resilience in Asia and the Pacific: A policy framework for sustainable development (Asia-Pacific Forum on Sustainable Development Theme Report). United Nations Economic and Social Commission for Asia and the Pacific

Wajid MS, Terashima-Marín H, Zafar A, Wajid MA, Bhushan B (eds) (2025). Digital twins for smart cities and urban planning: from virtual to reality (1st ed). CRC Press. https://doi.org/10.1201/9781003510338

Wang Y, Afla M, Reza M (2024) Data-driven urban resilience evaluation using four dimensions for cities in Northern Anhui Province. J Constr Dev Ctries 29(1):1–21. https://doi.org/10.21315/jcdc2024.29.1.1

War MR, Singh Y, Sheikh ZA, Singh PK (2025) Review on the use of federated learning models for the security of cyber-physical systems. Scalable Comput: Pract Exp 26(1):16–33. https://doi.org/10.12694/scpe.v26i1.2275

Zhu M, Jin J (2025) Data-driven urban digital twins and critical infrastructure under climate change: a review of frameworks and applications. Urban Plan 10(4):10109. https://doi.org/10.17645/up.10109

Artificial Intelligence, Data Analytics, and Predictive Sustainability

Artificial Intelligence for Climate Neutrality: Enhancing Environmental Sustainability

A. Arun Kumar

Abstract This chapter examines the negative aspects of artificial intelligence (AI) in the pursuit of environmental sustainability. Through better data analysis, forecasting, and efficient resource management, AI has great potential to advance sustainability. However, the creation, implementation, and operation of AI systems require significant computing power, leading to high energy consumption and a growing carbon footprint, especially when large-scale machine learning models and data centres are involved. Rapidly advancing AI-based environmental monitoring technologies often rely on resource-intensive hardware with short lifespans, contributing to the rising problem of electronic waste. AI-driven optimization in industries such as transportation and agriculture may unintentionally lead to overexploitation of natural resources. The opacity of many AI systems can hinder the detection of unsustainable practices, reducing transparency and accountability in environmental governance. Furthermore, the use of manipulated or insufficient environmental data to train AI models could result in poor decision-making and worsen ecological impact. Excessive reliance on AI could also place too much trust in technical solutions, diverting attention from systemic policy reforms and grassroots environmental initiatives, which are equally important. Incorporating AI into sustainability strategies requires a balanced approach that considers environmental costs. Responsible AI development must prioritize energy-efficient algorithms, circular design principles, and open, fair governance frameworks to mitigate future effects. It is imperative to acknowledge and address its downsides to ensure that AI becomes a true catalyst in the global effort for environmental sustainability.

Keywords Artificial Intelligence · Environment · Sustainability · Negative aspects · Data

---

A. Arun Kumar (✉)
ICFAI Law School, ICFAI Foundation for Higher Education (Deemed to be University Under Sec 3 of UGC Act,1956), Hyderabad, India
e-mail: arunkumar@ifheindia.org

R. Singh et al. (eds.), *Data-Driven Sustainability: Harnessing Technology for A Greener Future and Environmental Resilience*, World Sustainability Series,
https://doi.org/10.1007/978-3-032-13412-7_11

1 Introduction

Artificial intelligence (AI), one of the most influential technologies of the twenty-first century, has the potential to significantly transform ecosystems, the economy, and society. It is already impacting sectors such as healthcare, education, banking, agriculture, transportation, and energy, thanks to its ability to interpret large datasets, recognize patterns, improve systems, and generate predictive insights. AI is often portrayed as a key tool capable of catalyzing innovation toward greener futures in the urgent global pursuit of environmental sustainability. It can provide scalable and innovative responses to major environmental problems, such as improving resource efficiency in manufacturing and agriculture, and optimizing renewable energy systems. (Raihan et al. 2024).

However, the relationship between AI and sustainability is not one-dimensional, as with any new technology. The negative aspects of AI refers to significant difficulties that accompany its potential advantages. These challenges are often hidden or underexplored. The rapid expansion of AI infrastructure, its high computational demands, and its dependence on resource-intensive technologies contribute to a growing carbon footprint and the mounting problem of electronic waste. Moreover, the opaque and sometimes biased nature of AI decision-making raises questions about fairness and accountability in environmental governance. Such unintended consequences complicate the widespread belief that AI is the solution to all ecological challenges. For this reason, AI must be understood as a double-edged sword capable of either reducing or aggravating environmental problems (Bibi and Sumaira 2024).

1.1 AI and Sustainability as a Double-Edged Sword

Thinking about AI in terms of sustainability fits well with the metaphor of a double-edged sword. On the one hand, AI-driven advances have achieved important developments in environmental monitoring and management. Climate prediction models have improved thanks to machine learning algorithms, enabling scientists and policymakers to anticipate and mitigate the effects of extreme weather events. To reduce reliance on fossil fuels, AI has contributed to regulating renewable energy sources, such as integrating solar and wind power into national grids. AI-based precision agriculture technologies help increase crop yields while reducing pesticide and water use. AI-enhanced transportation systems can predict traffic patterns, optimize routes, and support the transition to electric mobility, reducing fuel consumption. These applications are only the beginning of showing how AI, when used appropriately, can help create more resilient, sustainable, and efficient systems (Raihan 2023).

On the other hand, the sustainability of AI also has a negative side. The computing power required by large-scale AI models entails energy expenditures often measured in megawatt hours. The carbon footprint of sophisticated AI systems, such as large language models, has been compared to the lifetime emissions of dozens of cars. Manufacturing and disposing of AI hardware, including semiconductors, Graphics Processing Units, and data centre components, also leads to resource depletion and electronic waste, since rare minerals and metals are extracted through environmentally harmful mining methods. In addition, AI-driven optimization in sectors such as transportation and agriculture may unintentionally encourage excessive use of resources. AI may make irrigation more efficient, but could also promote intensive farming practices, putting greater pressure on water resources (Bibi and Sumaira 2024).

The negative aspects of AI are worsened by issues of reliability and opacity. Many AI systems operate as "black boxes," generating results without clearly explaining the reasoning behind them. In the context of environmental governance, such as monitoring emissions, pollution, or biodiversity, this lack of transparency can hide unsustainable activities or be used to benefit certain actors. Moreover, insufficient or biased datasets used to train AI risk generating inadequate recommendations that may cause ecological harm instead of protecting ecosystems. Finally, excessive reliance on technological solutions risks substituting essential systemic reforms, such as legislative changes, community-based environmental initiatives, and behavioural adjustments (Slimani et al. 2025).

Investigating the negative aspects of AI is essential for responsible innovation and effective governance, not only in the academic field. The initiatives to promote sustainable development risk failing if the environmental costs of AI development and implementation are not considered. An AI model created to maximize carbon reduction in industrial processes could consume so much energy that it negates the benefits it was intended to deliver. Understanding these trade-offs is fundamental to ensuring that AI is an ally rather than an enemy in sustainable development initiatives (Antonopoulos et al. 2020).

Analysing the negative aspects of AI is crucial for equity and justice. Marginalized groups in society are disproportionately affected by environmental degradation, and AI systems trained on inaccurate or biased data can perpetuate inequalities in environmental governance. Without careful consideration, AI could replicate existing socioeconomic structures, favouring certain regions, sectors, or groups over others. Studying the unintended consequences of AI helps build more resilient and flexible governance frameworks. AI could cause long-term ecological damage rather than foster it if scientists and policymakers ignore threats such as algorithmic opacity, excessive energy consumption, and e-waste (Delanoe et al. 2023).

Analysing the downsides of AI forces a deeper philosophical reflection on the role of technology in sustainability. It makes us ask whether society is relying too heavily on technological solutions, while neglecting structural, cultural, economic, and political changes that are equally important or even more important for tackling ecological degradation and climate change (Antonopoulos et al. 2020).

The scope covers both the technological and sociopolitical aspects of the relationship between AI and sustainability. It considers the consequences of the technological life cycle of AI systems, such as waste management, hardware production, and the energy consumption of data centers. It analyzes sectoral applications of energy, transport, and agriculture to show how AI can either support or hinder sustainability initiatives. The study also addresses sociopolitical concerns about governance, transparency, accountability, and fairness, situating AI within the broader debate on responsible innovation and environmental justice (Ahmad et al. 2021).

Geographically, the scope is global, using case studies from different regions to demonstrate how AI's advantages and disadvantages appear in diverse socioeconomic and environmental contexts. The timeframe is contemporary, focusing mainly on the last decade, while acknowledging historical precedents and anticipating future developments (Bibi and Sumaira 2024).

This work seeks to counterbalance the prevailing optimism about technology and sustainability by presenting AI as a double-edged sword. Understanding the limitations of AI and ensuring responsible and fair use requires a careful examination of its negative aspects. Current literature, theoretical frameworks, and empirical cases are used to illustrate AI's duality and to propose strategies to align it with environmental sustainability goals. Only a critical and balanced approach will allow AI to become a true ally in building a sustainable future.

2 Objectives of the Study

This chapter aims to provide a critical and comprehensive analysis of the negative aspects of artificial intelligence in relation to environmental sustainability. The environmental consequences of AI development and implementation include energy consumption, carbon emissions, and electronic waste. Also, identify and analyse unintended ecological effects of AI-driven optimization in sectors such as energy, transport, and agriculture.

The study is based on secondary data. The data has been taken from Scopus, Web of Science, and SSCI indexed Journals.

3 Results and Discussions

One of the central issues in the debate on the sustainability of AI is the environmental cost related to energy consumption, carbon emissions, and electronic waste (e-waste). Training large-scale machine learning models, such as deep neural networks or massive language models, requires enormous amounts of processing power. A single training cycle of an advanced natural language processing model can consume as much energy as a typical household uses over several years. Beyond training, the large-scale deployment of these models serving millions of users worldwide further increases continuous energy demand due to the operation of data centres and cloud services. Data centres, often called the factories of the digital age, consume vast amounts of electricity and contribute significantly to carbon emissions if not powered by renewable energy. The hardware footprint is equally significant. AI depends on specialized hardware, such as high-performance processors, GPUs, and Tensor Processing Units (TPU's), all of which are manufactured using rare minerals and metals. The mining of these resources often causes deforestation, river pollution, and loss of biodiversity (Antonopoulos et al. 2020).

Furthermore, the e-waste problem is exacerbated by the short lifecycle of AI technologies, driven by the demand for increasingly efficient chips. The remedies used, if not treated properly, release harmful substances like lead, mercury, and other impurities into the environment, which pose a threat to the natural system and human health. Although Earths AI Enabled Energy is praised for its efficiency and problem-solving capabilities, its hidden environmental costs present a contradiction: if techniques designed to promote stability are not considered in their full life cycle, they may accelerate environmental degradation (Raihan et al. 2024).

AI is often praised for its ability to optimize processes, increase productivity, and reduce waste. However, when applied to complex and interconnected systems, this efficiency can generate unexpected ecological side effects. In agriculture, AI-based precision agriculture tools enable higher yields through targeted pest control, fertilizer application, and irrigation. While this improves profitability and efficiency, it can also encourage crop intensification and expansion into previously uncultivated land. This puts further pressure on biodiversity, soil fertility, and water resources. Similarly, poorly regulated AI-based water optimization systems can lead to overexploitation of aquifers (Li. et al. 2023).

In transportation, AI-enabled logistics and smart mobility solutions can reduce fuel consumption by predicting congestion and optimizing routes. Paradoxically, this efficiency may contribute to global growth in consumption, an effect known as the rebound effect. Ecological benefits can be balanced out, even if more cheap, efficient, and convenient supplies contribute to increasing the use of transport vehicles and increasing demand for transportation (Bibi and Sumaira 2024).

In the energy sector of System AI, demand can be predicted and balanced by renewable energy sources. However, they can also prolong the dependence on carbon-based reserve systems for managing variable renewable energy. Excessive dependence on smart systems that manage AI does not come with its own risks, since inefficient data processing can lead to an increase in the volume of income, rather than to savings. Optimization based on this principle has great potential, and its ecological effects must be carefully analysed to avoid unnecessary harm. The ability of AI to effectively regulate the environment depends largely on the quality, accountability, and transparency of the technologies used. Unfortunately, many AI systems act as black boxes, producing results without revealing the rationale behind them. This undermines accountability in environmental decision-making (Bibri et al. 2024).

An AI model that miscalculates pollution levels or biodiversity indices may lead to poor policy responses or misallocation of resources, with little opportunity to examine or review the logic behind these decisions. The issue of bias in data is equally concerning. The quality of models depends on the quality of the data on which they are trained. Incomplete, outdated, or distorted data can lead to systematic errors. For example, environmental monitoring systems trained with satellite data from certain regions may not reflect the reality of other regions. Similarly, stakeholders can manipulate or selectively report data used to power emissions tracking algorithms, thus hiding true levels of pollution. Biased or flawed data not only threatens ecological protection but can also exacerbate social inequalities. Marginalized communities, often most exposed to ecological hazards, are overlooked by biased AI systems, leaving them defenceless (Bibi and Sumaira 2024).

Thus, the risks of opacity, bias, and defective data are not merely technical issues; they carry serious moral and political implications for social equity and ecological sustainability. The debate around AI and sustainability is often presented in a techno-optimistic tone, suggesting that AI will solve ecological problems and climate change. However, relying too heavily on AI carries serious sociopolitical consequences, especially the risk of distracting attention from grassroots efforts and systemic reforms (Antonopoulos et al. 2020).

Environmental and climate challenges require structural transformations in energy policies, industrial regulations, consumption patterns, and citizen participation. If too much trust is placed in AI, policymakers and other stakeholders may overlook these essential reforms. For example, governments might prioritize funding AI-based carbon monitoring systems while neglecting stricter enforcement of existing environmental laws (Joshi et al. 2024). Over-prioritizing AI can also unintentionally strengthen technocratic governance, sidelining social movements and local communities by concentrating decision-making power in the hands of technology companies, data scientists, and policymakers. This creates a type of technological dependency in which societies neglect cultural, behavioural, and policy reforms in favour of AI-driven solutions (Li et al. 2023).

Furthermore, this dynamic accentuates global inequality. Poorer nations often have limited agency and must rely on foreign technology, while wealthy countries and corporations dominate the sustainability agenda. Therefore, analysing the sociopolitical consequences of depending too much on AI is essential to ensure that sustainability remains a democratic and collaborative process, not one monopolized by technology (Kumar et al. 2024).

Given the risks and challenges described, it is critical to establish responsible strategies for developing and deploying AI that align with sustainability standards. Efforts should focus on reducing computational costs during training and operation. Approaches such as low-power architectures, federated learning, and model pruning can significantly lower energy consumption. Manufacturing AI hardware must move toward sustainable mining and recycling, as well as the design of durable, repairable, and recyclable devices (Raihan et al. 2024).

For environmental governance, AI systems must be open to accountability. Explainable AI frameworks, transparent audits, and open-source models can help stakeholders understand and question decisions that affect ecosystems. Global and national regulations are needed. The European Union's AI Act or UN-led sustainability guidelines, which can establish ethical and ecological limits on AI. Projects must incorporate community perspectives, indigenous knowledge, and grassroots voices to ensure that AI-driven sustainability does not exclude vulnerable groups.

AI can only become a genuine ally in sustainability efforts if it is approached holistically, balancing technological innovation with ecological responsibility and social justice (Li et al. 2023).

4 The Negative Aspects of AI in Environmental Sustainability

AI is often highlighted as revolutionary for reducing waste, optimizing energy use, and predicting climate change. However, these promises come with hidden environmental costs. These costs arise not only from the direct material and energy requirements of AI but also from systemic risks such as faulty data, opaque decision-making, and excessive dependence on technology. Below are six key aspects of AI's negative aspects in environmental sustainability (Delanoe et al. 2023).

Training massive AI models is one of the most energy-intensive operations in modern computing. Current Natural Language Processing (NLP) models can consume hundreds of megawatt-hours of electricity in a single training cycle, producing emissions comparable to dozens of cars over their entire lifetimes. Bibi and Sumaira (2024) found that training a single NLP model with neural architecture search generated 626,000 pounds of CO_2 equivalent, nearly five times the lifetime emissions of an average car (Bhushan and Singh 2024).

Even after training, AI systems require energy for inference using the model in real-world applications. Serving billions of users across search engines, recommendation systems, autonomous vehicles, and smart devices consumes vast amounts of electricity. High-performance GPUs and TPUs used for these tasks require more power than traditional computing systems (Antonopoulos et al. 2020). Data centres worsen the problem. Although some hyperscale centres (Google, Amazon, Microsoft) increasingly use renewable energy, a large share of global data centres still depends on carbon-heavy electricity grids. Thus, while AI may improve certain environmental processes, its overall energy footprint risks adding significantly to greenhouse gas emissions (Bibri et al. 2024).

AI requires high-performance hardware, such as GPUs, TPUs, and specialized chips, that are replaced more quickly than general-purpose hardware. New generations appear every one to two years, prompting research labs and companies to discard older machines prematurely. This accelerates the growth of electronic waste (e-waste) (Singh et al. 2024).

E-waste is especially harmful because AI hardware depends on rare minerals like neodymium, cobalt, and tantalum, extracted through environmentally destructive mining. Chip manufacturing also consumes large amounts of energy, water, and toxic chemicals. Most AI hardware is not effectively recycled due to its material complexity. In many developing countries, informal recycling releases toxic substances such as mercury, cadmium, and brominated flame retardants into the environment. Because AI devices often have only a few years of useful life, companies quickly replace them with newer, faster accelerators. Unlike consumer electronics, industrial AI hardware has weaker recycling programs, leading to unregulated industrial waste streams (Singh et al. 2025a, b).

AI optimization is often presented as key to sustainability. But optimization does not always mean conservation. By making industries more productive and profitable, AI can unintentionally drive resource overuse. AI-based precision farming recommends irrigation schedules, pesticide applications, and fertilizer use. While these can reduce waste, they may also encourage water-intensive crops in arid regions, stressing aquifers. Increased yields can also fuel agricultural expansion into new lands, leading to biodiversity loss and soil degradation (Kumar et al. 2024).

Predictive AI tools improve mineral exploration by identifying deposits more efficiently. But this can accelerate extraction rates, increase deforestation, and worsen toxic runoff. Ironically, demand for minerals also rises because AI itself depends on them. AI logistics optimizes fleet management and delivery routes, saving fuel. Yet the rebound effect appears again cheaper, faster delivery raises demand for shipping and vehicle use (Singh et al. 2025a, b). Ride-sharing apps were expected to reduce private car use, but studies show they have often increased traffic and emissions by replacing public transport. Thus, AI may amplify exploitation rather than alleviate environmental stress (Li et al. 2023).

AI models are complex and often operate as black boxes, producing outputs without explaining how decisions are made. This opacity undermines transparency and accountability in environmental governance. AI is increasingly used for monitoring emissions, predicting air quality, tracking deforestation. If such systems are poorly documented, neither governments nor civil society can verify their accuracy. This opens the door to greenwashing, where companies or states present AI-generated green numbers to hide harmful practices. Opaque AI also complicates regulation. Environmental law typically requires clear evidence, but AI outputs may be uninterpretable in court. Outdated governance frameworks leave regulators unable to assess the ethical and ecological risks of AI. Thus, AI opacity creates responsibility gaps that can be exploited to justify unsustainable behavior (Singh et al. 2025a, b).

AI models are only as good as the data they are trained on. Environmental datasets are often incomplete, outdated, or biased, leading to misleading outputs. Satellite images used to track deforestation may miss small-scale or illegal logging, underestimating ecological damage. Air quality monitoring algorithms trained with urban data may overlook rural pollution, leaving vulnerable communities unprotected. Politically or economically powerful actors may manipulate the data used to train emission-tracking algorithms, producing distorted results that serve their interests. These flaws not only harm ecosystems but also reinforce social inequalities, as the environmental risks faced by marginalized communities are often overlooked by biased AI systems (Kumar et al. 2024).

One of the biggest risks is overreliance on AI while neglecting systemic reforms. Policymakers may prioritize AI investments over structural changes, such as reducing fossil fuel subsidies, tightening environmental regulations, or promoting lifestyle changes. This technocratic approach marginalizes local initiatives, indigenous knowledge, and behavioral campaigns, fundamental pillars of sustainable development. Power is concentrated in the hands of technology companies and regulators, and local voices are marginalized. The problem is thus both political and technical. By prioritizing AI over systemic reforms, society risks treating the symptoms rather than the root causes of environmental crises (Singh et al. 2025a, b).

The integration of sophisticated intelligence into solving environmental problems gives rise to both hope and danger. On the one hand, energy sector offers unique opportunities for improving the process of decision-making, predicting risks, and optimizing environmental systems. On the other hand, it creates serious ethical, environmental, and social-political problems. In this discussion, three interrelated themes are considered: the balance between optimism and reality, the search for economic compromises, and reflection on the long-term role of both as an ally or opponent in ensuring sustainable development (Dewasiri et al. 2025).

5 Balancing Optimism and Reality in AI-Driven Sustainability

The global narrative around AI is often steeped in techno-optimism. Advocates highlight its potential to support low-carbon transitions, improve energy efficiency, and transform environmental governance. Predictive analytics can predict natural disasters, save lives, and reduce economic losses. Smart power grids powered by AI help balance the supply and demand of renewable energy. These achievements make AI indispensable in the fight against climate change (Raihan 2023).

However, reality demands nuance. The high environmental costs of AI, particularly the energy-intensive processes, the carbon footprint of data centres, and the associated equipment losses, raise questions about its overall sustainability benefits. The rebound effect in transportation shows that efficiency gains can paradoxically lead to higher consumption, which foregoes environmental benefits (Kumar et al. 2024).

Adopting a realistic perspective means recognizing that AI is not a magic bullet. It is one tool among many that complement behavioural change, popular action, and systemic policy reform. AI can optimize resource use, but its benefits often flow disproportionately to wealthy countries and corporations with access to advanced computing. This widens the gap between the Global North and South, creating dependency for less developed regions. The most powerful AI systems often function as black boxes, hard to interpret. While they deliver efficiency, they compromise accountability. If an AI model predicts deforestation but does not explain why, authorities may either reject its conclusions or accept them uncritically without oversight (Bibri et al. 2024).

These dilemmas show that AI for sustainability is not only a technical challenge but also a political and moral one, requiring frameworks that prioritize fairness, transparency, and responsibility. If unchecked, AI's energy demands, hardware cycles, and systemic biases may worsen environmental crises, making it an adversary. This could happen if governments and corporations prioritize profit or fail to regulate AI's ecological impacts. Conversely, if developed responsibly, AI could become an ally. Advances in explainable AI, circular hardware design, and energy-efficient algorithms could reduce its environmental footprint and increase accountability. Incorporating local knowledge and community participation would make AI systems more equitable and context-relevant. Supported by international frameworks such as the EU's AI Act or potential UN sustainability guidelines, AI could align with global environmental goals (Koumetio et al. 2023).

Ultimately, the question is not whether AI will be friend or foe, but whether society has the political will and vision to guide its development responsibly. The future of AI in sustainability will depend more on ethics, governance, and commitment to ecological justice than on technological innovation alone (Dewasiri et al. 2025).

6 Policy Implications

The output of this research highlights the complex connection between AI and environmental sustainability. Although AI creates opportunities to improve energy efficiency, oversee natural resources, and promote climate initiatives, its environmental impacts and unforeseen repercussions need careful policy considerations. It is essential for policymakers to guarantee that AI technologies are used in manners that truly aid sustainability objectives instead of worsening ecological challenges. Governments must maintain energy-efficiency standards for AI systems because training and implementing large AI models need a significant amount of energy. AI systems might have to showcase their carbon footprint while being trained and used, much like energy star ratings for home appliances. Companies would be motivated to use techniques like model pruning, federated learning, or energy-efficient chip design if there were greater transparency. In addition, through tax credits, subsidies, or required renewable procurement, regulators ought to promote the use of renewable energy in data centres. Not following these basics, the unbridled growth of AI will jeopardize global climate goals (Singh et al. 2025a, b).

The issue of e-waste, is credited to the discharge of AI hardware. When it comes to AI hardware, policymakers need to support a circular economy strategy that prioritizes robustness, repairability, and recyclability. By making manufacturers responsible for the full lifecycle of AI devices, extended producer responsibility (EPR) programs can guarantee that parts are gathered, reconditioned, and recycled in an ideal way. Given their scarcity and the environmental costs resulting from their extraction, investment in cutting-edge recycling technologies for rare-earth minerals is also important. Given that supply chains for AI hardware frequently cross several jurisdictions, international cooperation is crucial in this case (Dewasiri et al. 2025).

AI-driven optimization in mining, transportation, and agriculture runs the risk of causing the rebound effect, in which increases in efficiency result in higher consumption. Legislators must incorporate regulatory safeguards to foresee these unforeseen consequences. Stringent water-use quotas should be implemented alongside AI-based irrigation systems to avoid over-extraction of groundwater. Regulators may implement congestion fees or carbon pricing in the transportation sector to counteract the rise in demand brought on by ride-sharing and AI-enabled logistics. Despite its obvious ability to increase efficiency, AI may hasten resource depletion in the absence of such regulations (Singh et al. 2022).

Black-box AI models' opacity presents significant governance issues. Standards for transparency and explainability should be mandated by policymakers for AI systems used in environmental monitoring and decision-making. To guarantee accuracy and equity, public agencies using AI should be obliged to carry out independent audits of their algorithms, datasets, and outputs. Additionally, by democratizing access to environmental data, open data initiatives can allow civil society to examine policies driven by artificial intelligence.

Governments can lessen the possibility that AI will be used for greenwashing or to hide unsustainable practices by incorporating accountability mechanisms. Environmental datasets that contain bias run the risk of yielding inaccurate or unfair results. Governments and international organizations must create data governance frameworks that put accuracy, inclusivity, and ethical use first to address this. Investing in thorough environmental data collection is part of this, especially in underrepresented areas like rural communities in the Global South. Data on corporate self-reported emissions or resource use can be protected from manipulation by independent verification methods. To prevent limited or prejudiced viewpoints, policies must also acknowledge the significance of indigenous and local knowledge systems and incorporate them into AI models.

The idea that technological advancements alone can resolve ecological crises is known as techno-solutionism, and policymakers need to be wary of it. Fundamental reforms like removing subsidies for fossil fuels, tightening environmental laws, and encouraging sustainable consumption habits should be complemented by AI rather than replaced. To ensure that local communities are empowered to shape sustainability strategies rather than being marginalized by technocratic governance, policy frameworks must also place a high priority on grassroots participation. Due to AI's worldwide environmental effects, international governance frameworks are required. The European Union AI Act is a first step in regulating AI, but more collaboration is needed through institutions like the Organization for Economic Co-Opertaion Development (OECD) and the United Nations (UN). Binding agreements on energy use, e-waste management, and data transparency could be established by a global treaty on AI and environmental sustainability. To maintain equity and stop affluent countries and corporations from monopolizing the advantages of AI while shifting its environmental costs to the Global South, ethical standards must also be created.

Governments cannot create effective policies on their own. It is essential that various stakeholders, including academia, industry, civil society, and indigenous communities, work together. To promote sustainable AI innovation while upholding stringent environmental standards, policymakers should form public–private partnerships. Platforms with multiple stakeholders can encourage accountability, ease the exchange of knowledge, and guarantee that a range of viewpoints are incorporated into AI governance. Instead of being determined only by corporate or technological interests, policies can reflect broader societal values by democratizing the decision-making process. Governments need to fund research and development that puts sustainability at the forefront of AI development. The development of explainable AI systems suited for environmental governance, biodegradable hardware materials, and low-power algorithms should all receive funding. Promoting interdisciplinary approaches that combine computer science, environmental science, ethics, and policy studies should be a priority for academic institutions and research centres. By making such investments, the next generation of AI technologies will be more closely aligned with social and ecological goals.

7 Limitations and Future Scope of the Study

The limitations of this study have critically examined the negative aspects of artificial intelligence (AI) in relation to environmental sustainability. These restrictions draw attention to the need for careful interpretation of results and identify areas that require more investigation. The lack of trustworthy data is one of the main obstacles to analysing AI's effects on the environment. Private companies like Google, Microsoft, Amazon, and NVIDIA provide a large portion of the data on energy use, carbon emissions, and hardware lifecycles. For personal relationship purposes, these organizations frequently release selective data, which raises questions about underreporting and greenwashing. There are still a few independent, peer-reviewed datasets on the actual carbon footprint of AI inference and training. As a result, even though this study uses the best available data, data gaps and possible reporting biases limit the conclusions.

Agriculture, mining, transportation, and energy systems are the main areas where AI and sustainability interact in this study. AI uses, however, go well beyond these fields and include water governance, waste management, urban planning, and disaster response. Although the analysis is thorough, it is not comprehensive due to the exclusion of these other sectors. Furthermore, depending on socioeconomic circumstances, governance ability, and technology infrastructure, the dynamics of AI deployment vary greatly across regions. Therefore, care should be taken when interpreting the sectoral and regional generalizations drawn from this study.

One area of technology that is developing the fastest is AI. Environmental impact analyses may soon become outdated due to the constant emergence of new architectures, training methods, and energy-efficient algorithms. While edge computing may decentralize energy use in ways not fully captured in this study, methods like federated learning, quantization, and model distillation, for instance, are already lowering the energy burden of AI training. On the other hand, the development of even bigger models and the broad use of generative AI could increase the need for materials and energy. Therefore, it is challenging to provide conclusive or long-term assessments due to the rapid pace of technological change.

Despite engaging in theoretical and empirical discussions, the paper lacks comprehensive, quantitative case studies of AI deployments. For instance, although Google DeepMind, data centre optimizations, and AI in precision agriculture are mentioned, primary data collection was not used to examine these cases. The analysis's granularity is restricted by its reliance on secondary literature, which also runs the risk of ignoring contextual details like regional governance frameworks, community reactions, or cultural perspectives on AI adoption.

Notwithstanding these drawbacks, the study provides several directions for further investigation. Filling in these gaps will promote better policymaking and enhance scholarly understanding. Quantitative lifecycle assessments (LCAs) of AI technologies, from resource extraction and hardware manufacturing to deployment and disposal, should be the focus of future research. These kinds of studies would offer a comprehensive perspective on AI& ecological impact, enabling more precise comparisons between AI-powered solutions and other sustainability tactics. By pinpointing leverage points where environmental costs can be reduced, LCAs may also help shape policy. Future research should use cross-sectoral analyses that consider how AI impacts interconnected systems, even though this study focused on a few chosen sectors. The use of AI in transportation affects energy demand, air quality, and urban planning, all of which have an impact on public health. Regional comparisons, such as comparing the adoption of AI in developed and developing nations, would also draw attention to the global disparities ingrained in sustainability agendas driven by AI.

Future studies ought to concentrate on governance and policy aspects as well. This entails evaluating the efficacy of existing legal frameworks, such as the EU AI Act, investigating models for global collaboration on sustainability and AI, and putting forward fresh governance instruments designed to address environmental issues. To lessen the risks of technocratic domination, research can also investigate participatory governance models that incorporate academia, industry, and civil society in decision-making. Future research should monitor and assess technological advancements meant to lessen AI& environmental impact. This includes the creation of biodegradable hardware, low-power chips, and data centre scheduling that takes carbon emissions into account. Researchers should determine if these developments reduce the environmental costs of AI or if they only transfer the load to other stages of the lifecycle.

8 Conclusion

This study has highlighted a paradox at the heart of technological progress by examining the negative aspects of AI in relation to environmental sustainability. AI systems designed to increase efficiency consume enormous amounts of energy and generate significant carbon emissions. Their dependence on resource-intensive hardware produces growing problems of electronic waste and environmentally harmful mining. Optimization processes in mining, transport, and agriculture can unintentionally accelerate resource exploitation. Incomplete data raise the likelihood of flawed policies, while opaque models and weak governance undermine accountability. Excessive dependence on AI perpetuates global inequality and technocratic governance, distracting attention from grassroots efforts and systemic reforms. Recognizing these risks is not a rejection of AI but rather a call for responsible innovation. AI must be integrated into broader sustainability frameworks that prioritize social justice, ecological integrity, and democratic accountability. This requires rethinking how AI is created, applied, and regulated. Developing hardware guided by circular economy principles (durable, repairable, and recyclable). Building transparent and explainable models that can be audited. Ensuring inclusion and fairness by involving local actors, indigenous knowledge systems, and vulnerable populations in environmental governance with AI. Conduct quantitative studies on the full life cycle of AI systems, from resource extraction to disposal, to determine their real environmental impact. Examine how AI supports sustainability across different industries and geographic regions to identify conditions that maximize benefits. Develop frameworks that integrate ethical standards, regulatory tools, and participatory methods to minimize AI risks and enhance its contributions. Through collaboration among academics, policymakers, and technologists, AI can evolve into a true ally in the global pursuit of environmental sustainability.

References

Ahmad T, Zhang D, Huang C, Zhang H, Dai N, Song Y, Chen H (2021) Artificial intelligence in sustainable energy industry: status Quo, challenges and opportunities. J Clean Prod 289:125834. https://doi.org/10.1016/j.jclepro.2021.125834

Antonopoulos I, Robu V, Couraud B, Kirli D, Norbu S, Kiprakis A, Flynn D, Elizondo-Gonzalez S, Wattam S (2020) Artificial intelligence and machine learning approaches to energy demand-side response: a systematic review. Renew Sustain Energy Rev 130:109899. https://doi.org/10.1016/j.rser.2020.109899

Bhushan M, Singh R (2024) Talent management facilitates net-zero transition through employee green behavior. In: Singh R, Khan S, Kumar A, Luthra S, Chokshi H (eds) Net zero economy, corporate social responsibility and sustainable value creation. CSR, sustainability, ethics & governance. Springer, Cham. https://doi.org/10.1007/978-3-031-55779-8_7

Bibi R, Sumaira (2024) AI innovations, renewable energy, and social change: a pathway to global sustainability. J Technol Innov Energy 3(3):53–67. https://doi.org/10.56556/jtie.v3i3.1034

Bibri SE, Huang J, Jagatheesaperumal SK, Krogstie J (2024) The synergistic interplay of artificial intelligence and digital twin in environmentally planning sustainable smart cities: a comprehensive systematic review. Environ Sci Ecotechnol 20:100433. https://doi.org/10.1016/j.ese.2024.100433

Delanoë P, Tchuente D, Colin G (2023) Method and evaluations of the effective gain of artificial intelligence models for reducing CO2 emissions. J Environ Manage 331:117261. https://doi.org/10.1016/j.jenvman.2023.117261

Dewasiri NJ, Singh R, Kumar V (2025) Decentralised recycling and waste management for digital circular economy. Sustain Innov Digit Circ Econ 3–23. https://doi.org/10.1007/9789819610648_1

Joshi A, Singh R, Kumar V, Rani S (2024) Net zero strategy: exploring opportunities, confronting challenges for developing countries. In: Singh R, Crowther D (eds) Transition towards a sustainable future. approaches to global sustainability, markets, and governance. Springer, Singapore. https://doi.org/10.1007/978-981-97-5756-5_11

Koumetio Tekouabou SC, Diop EB, Azmi R et al (2023) Artificial intelligence-based methods for smart and sustainable urban planning: a systematic survey. Arch Computat Methods Eng 30:1421–1438. https://doi.org/10.1007/s11831-022-09844-2

Kumar AA, Suryadevara R, Sowmyya T, Chanal GB (2024) Exploring the intersection of AI, healthcare, and environmental sustainability: future trends and challenges. Approaches to Glob Sustain Mark Governance Part F4088:23–47

Li J, Herdem MS, Nathwani J, Wen JZ (2023) Methods and applications for artificial intelligence, big data, internet of things, and blockchain in smart energy management. Energy and AI 11:100208. https://doi.org/10.1016/j.egyai.2022.100208

Raihan A, Paul A, Rahman MS, Islam SPP, Karmakar S (2024) Artificial Intelligence (AI) for environmental sustainability: a concise review of technology innovations in energy, transportation, biodiversity, and water management. J Technol Innov Energy 3(2):64–73. https://doi.org/10.56556/jtie.v3i2.953

Raihan A (2023) A comprehensive review of artificial intelligence and machine learning applications in energy consumption and production. J Technol Innov Energy 2(4):1–26. https://doi.org/10.56556/jtie.v2i4.608

Singh R, Khan S, Centobelli P (2022) Investigating the interplay between social performance and organisational factors supporting circular economy practices. Sustainability 14(24):16781. https://doi.org/10.3390/su142416781

Singh R, Joshi A, Dissanayake H, Nainanayake D, Kumar V (2025a) Harnessing artificial intelligence and human resource management for circular economy and sustainability: a conceptual integration. Sustainability 17(15):7054

Singh R, Khan S, Kumar A, Kumar V (2024) Artificial intelligence-enabled management: an emerging economy perspective. De Gruyter. (ISBN 9783111172408)

Singh R, Gupta A, Kandpal V (2025) ESG excellence in the digital world. In: Gonzalez EDRS, Kandpal V, Ozili PK, Chatterjee P (eds) Fintech for ESG and the circular economy. https://doi.org/10.1002/9781394234035.ch10

Slimani S, Omri A, Jabeur SB (2025) When and how does artificial intelligence impact environmental performance? Energy Econ 148:108643. https://doi.org/10.1016/j.eneco.2025.108643

The Positive Impact of AI for Implementation of Circular Economy and Climate Change Towards Ocean Sustainability

Rubee Singh

Abstract The oceans are vital to Earth, and coastal regions, specifically, encompass a diverse array of habitats and ecosystems. Climate change (CC) is modifying oceanic atmosphere, biology system, level of the sea, and glacier circulation. Hence, the concept of sustainability is being embraced in the context of oceans which focus on the process that inflicts minimal or no harm on the natural environment or living beings, including humans. In recent years, artificial intelligence (AI) has emerged as a significant instrument for mitigating the CC through circular economy (CE) approach. In light of the increasing role AI in environmental initiatives, this chapter aims to explore different applications of AI for achieving ocean sustainability. The chapter is based on secondary sources of data and reviews the existing research on the topic for identifying the current trend regarding AI application for ocean sustainability. It is found that the remote sensing techniques and remote networks are being used extensively in marine context along with robotics applications. AI including big data is very efficient in gathering ocean data and waste tracking for keeping the oceans clean. Use of autonomous vehicle for surveillance is also prominent for scientific as well as industrial applications. AI programs can predict the probable effects of increasing sea levels in different places by identifying patterns and correlations something humans could neglect. The chapter offers implications for practice and policy making along with suggestions for future research opportunities.

Keywords Circular economy · Climate change · Artificial intelligence · Ocean sustainability · Marine environment · Technology

R. Singh (✉)
Institute of Business Management, GLA University, Mathura, India
e-mail: dr.rubeerajput@gmail.com

Department of Management Studies, Kumaun University, Nainital, India

R. Singh et al. (eds.), *Data-Driven Sustainability: Harnessing Technology for A Greener Future and Environmental Resilience*, World Sustainability Series,
https://doi.org/10.1007/978-3-032-13412-7_12

1 Introduction

The oceans encompass more than 70% of the Earth's surface and are essential for sustaining life on our planet. Oceans, while a source of beauty and wonder, are crucial for controlling the global climate, supplying food and livelihoods for millions, and preserving extensive biodiversity (Talukdar et al. 2022). Thus, the importance of seas can be comprehended from ecological, economic, and social viewpoints impacts human sustenance. The oceans are an essential element of Earth's interrelated systems, affecting temperature, weather, and nutrient cycles. The indifferent attitude towards the oceans' role in maintaining the planet's overall health and resilience is common among general public despite its crucial role in ensuring global food security, human health, and climate management. Over 3 billion individuals globally rely on the services offered by marine and coastal biodiversity (Wiese and Auad 2022). Ocean sustainability refers to the preservation of the balance among the biological, physical, geological, and chemical constituents of the oceans due to human intervention. It is to manage ocean services and resources judiciously, ensuring the natural processes remain undisturbed. Ocean sustainability also encompasses tactics such as creating protected areas, imposing limitations on developing activities, and advocating for the responsible utilization of marine resources.

Climate change (CC) is profoundly affecting the world's oceans, resulting in diverse effects that influence marine biodiversity and coastal communities (Dube 2024). It results in elevated sea levels, increased frequency and intensity of marine heatwaves, and a decline in marine biodiversity. Thermal expansion and the melting of ice caps result in rising sea levels, heightening the risk of flooding, erosion, and storm surges for coastal populations. Increasing ocean temperatures lead to prolonged and intense marine heatwaves, adversely affecting coral reefs and marine ecosystems while jeopardizing biodiversity (Talukdar et al. 2022). Destruction of coral reefs and other ecosystems, coupled with species movement, contributes to the decline of marine biodiversity, jeopardizing millions of species and affecting coastal industries and communities (Najeeb et al. 2025). By 2100, it is projected that more than half of the world's marine species may be endangered (Koen et al. 2025). Climate change has extensive repercussions for both marine ecosystems and the livelihoods of coastal inhabitants. Ocean sustainability thus can be instrumental managing ocean resources in a manner that safeguards both the marine environment and the services it offers (Wang et al. 2022).

Effective mitigation and adaptation to climate change require innovative and adaptable strategies and hence, Artificial intelligence has evolved into an effective instrument that can aid in the formulation of innovative strategies for adapting to climate change (Crowther et al. 2025). AI possesses the capacity to significantly enhance various domains, such as identifying high-risk areas for climate-related disasters and formulating adaptation strategies for communities and industries (Khan et al. 2024). AI-enabled solutions offer numerous benefits for CC adaptation, including identification of the areas which are prone to its impact and developing strategies for protecting populations from the effects of climate change. A specific application is the continuous tracking and evaluation of severe weather event impacts with AI-driven sensors and drones facilitating a more efficient management of the effects. Similarly, predictive models created with machine learning algorithms can be considered to combat the impacts of CC (Zhang et al. 2020a, b). Moreover, AI possesses the capability of optimizing the use of resources while reducing emissions, potentially combating the challenges posed by CC.

This chapter aims to examine that how artificial intelligence enabled circular economy can help in advancing ocean sustainability in light of adverse climate conditions. Firstly, the chapter offers a background about how climate change is impacting the oceans followed by a section on circular approach for oceans. Further, the role of AI for sustainability in tackling the climate change is discussed. Then an exploration about specific applications of AI for ocean sustainability is presented. Finally, the chapter offers various implications for policy and practice while contributing to the current discussion on how emerging technologies may aid in the transition to resilient and regenerative ocean systems. The chapter end with the conclusion section which also entails limitations of present study and suggests a few opportunities for conducting research.

2 Literature Review

2.1 Impact of CC on Oceans

Climate change is modifying oceanic atmosphere, biology system, level of the sea, and glacier circulation. These systemic changes significantly affect habitats, biotic productivity, and species assemblages which are fundamental to numerous economic benefits derived from the sea (Najeeb et al. 2025). They also influence the dangers associated with various human activities and advances. Unprecedented alterations in oceanic conditions are presently manifesting in each of the latitudes, posing a significant risk of detrimental effects on the marine ecosystems. The implications can escalate highly by the century's conclusion, contingent upon future greenhouse gas emissions, with possible apparent repercussions for ecological systems, the oceans, and human wellbeing (IPCC 2019).

Anthropogenic climate change, propelled by the rapid escalation of greenhouse gas (GHG) emissions since the industrial revolution, will persist in affecting the ocean through multiple avenues. The magnitude of consequences will significantly rely on the level of warming achieved due to greenhouse gas emissions. The resultant alterations to oceanic processes and functions have significant ramifications for the global economy that must be considered to guide adaptation initiatives and stimulate urgent mitigation solutions. Climate change has significantly increased ocean temperatures across the majority of the planet. The ocean has absorbed around 93% of excess heat, resulting in substantial warming of the not only the water in the top ocean but deeper waters as well intensity (Cheng et al. 2017). RCP models indicate that these trends have already surpassed the natural variations in several regions and very likely to continue in the future. It is anticipated that oceans will double to quadruple warmer than today by the year of 2100.

Polar regions have experienced significant alterations, including modifications in the annual season of melting, variations in snow cover, and alterations in the mass of ice sheets and glaciers, leading to sea level rise (IPCC 2019). The pace of rising sea levels differs by region; in the western Pacific, it is three times the world average, whereas in the eastern, the rate is either negligible or negative. Current scenario of decreasing pH of oceans have surpassed the normal natural fluctuation in the majority of the oceanic areas and are anticipated to persist in this whole century (Burger et al. 2020). The weakening in the Atlantic is resulting in decreased sea surface temperatures in the subpolar Atlantic Ocean and these alterations may result in substantial transformations in meteorological and climatic patterns at both local and regional levels, further affecting the maritime economy (e.g., via infrastructure damage) and society at large (IPCC 2019).

2.2 *Circular Economy and Oceans*

A complete definition of circular economy (CE) is "an economic system that minimizes resource consumption and emissions, waste, and energy leakages by bicycle riding, extending, intensifying, and dematerializing material and energy loops" (Singh 2025). This can be accomplished via digitalization, servitization, collaborative solutions, environmentally friendly product development, servicing, fixing, reuse, refurbishment, and disposal" (Geissdoerfer et al. 2020). The notion of the CE arose as a counter to the conventional system, wherein things are manufactured from raw natural resources, consumers purchase the products, and discard as a waste after the end of their economic lifespan. The CE concept is based on the reuse of existing materials instead of utilizing natural resources, minimizes waste, and oversees resource consumption through a closed-loop system (Kirchherr et al. 2023). The CE approach focuses on enhancing resource utilization and obtaining maximum value via reusing, repairing, remanufacturing, and recycling.

Resources recovered or preserved through these activities can minimize the utilization of all kinds of resources including raw materials, energy, labor, and capital, while also mitigating the environmental effect, including CO^2 emissions, associated with manufacturing operations (Joshi et al. 2024). The Ocean Visions Consortium formed the "Marine Circular Bioeconomy (MCB) Task Force" to investigate the possibilities of marine aquaculture for sustainably enhancing world food production in October 2020. This task force utilizes a CE framework (Greene et al. 2022), which prioritizes minimizing the extraction of new resources and enhancing the process of end-of-use along with reusing and recycling. In contrast to the linear "take-make-waste" approach, the new model of CE is inherently regenerative and seeks to divorce economic growth from the extraction of finite resources.

2.3 AI for Climate Change

The contemporary issue of climate change is posing substantial threats to not only the environment but to the economy and the society also. The incorporation of Artificial Intelligence (AI) into efforts to promote sustainability presents a promising opportunity for fostering a more resilient and harmonious relationship between humanity and the environment, especially during significant ecological challenges (Singh et al. 2025a). It can improve the utilization of resources in an economy with circularity by facilitating material recycling, prolonging the lifespan of products, and streamlining distribution networks. The principal advantage of AI lies in its competence to rapidly and accurately analyze extensive datasets. This renders it an essential instrument for identifying locations which would be highly influenced by CC, e.g. the flood prone and areas with high risk of land sliding by formulating accurate adaptation strategies tailored to the specific risks in particular areas.

AI-driven solutions assist enterprises in shifting from conventional linear approaches to cyclical methodologies. Environmental issues such as greenhouse gases, and depletion of resources pose serious dangers to environments and society as a whole, underscoring the urgent necessity for innovative solutions. The capacity of AI to evaluate data, identify patterns, and predict consequences has the possibility to revolutionize the resolution of complex issues (Singh et al. 2025c). Climate change presents a substantial risk to both civic and ecological systems, leading to approximately $500 billion in global economic losses while AI can mitigate these issues by utilizing online resources to deliver prompt advice grounded in precise climate change forecasts (Chen et al. 2023). Consequently, Artificial Intelligence (AI) can substantially aid the environment and the capabilities of data analysis, predictive modeling, and optimization can enhance infrastructure for energy production, transportation networks, and industrial processes (Singh et al. 2025d).

2.4 Artificial Intelligence for Ocean Sustainability

Embarking on a transformational path to alleviate the impact of CC on oceans requires the application of advanced technology innovations. Among these breakthroughs, Artificial Intelligence (AI) stands as a crucial element ready to transform the framework of sustainable development. Artificial Intelligence was first employed mechanical equipment to substitute humans in performing simple and monotonous jobs (Spector 2006). Numerous nations are progressively acknowledging the significance of BD in advancing the societal environment along with improved competitive position (Singh et al. 2025b). Due to the swift advancement of software and hardware technology in computer, computational capacity increased substantially, while costs have markedly decreased. AI-related technologies, including deep learning and machine learning, also progressed swiftly, benefiting the marine sector. The intelligent ocean integrates artificial intelligence with oceanic data and a neural framework to comprehend and strategize oceanic management (Song et al. 2023). Utilizing comprehensive oceanic big data can enhance human research in addressing CC, safeguarding the ecosystem, and mitigating natural calamities.

Remote Sensing Technique

Satellite remote sensing (SRS) technology serves as a method for monitoring the oceans and represents a crucial invention in contemporary ocean technology (Liu et al. 2017). The conventional analysis of SRS data relies heavily on individuals possessing special kind of expertise. It necessitates processing as well as interpretation by humans by integrating characteristics of SRS data, real-world situations, and oceanic experience. Ensuring accuracy is both time-consuming and demanding. The advancement of artificial intelligence and the unrestricted dissemination of SRS datasets have catalyzed the growth of deep learning. Diverse AI methodologies, encompassing machine learning, are being effectively employed to numerous SRS challenges (Boukabara et al. 2019). This type of data are essential for forecasting the weather of oceans, uncovering novel events at crucial geographical and temporal dimensions that were previously inaccessible by existing data gathering techniques alone (Liu et al. 2017). This information can serve as a substantial resource for deep learning, which offers a result-oriented approach to fully utilize ocean SRS data.

This mutually beneficial scenario has prompted numerous international research researchers to undertake exploratory investigations into AI-based SRS applications, suggesting innovative concepts, methodologies, and procedures. Yuan et al. (2020a, b) delineated key principal findings along with challenges of AI methodologies in SRS for ecosystem, elaborated on the architectures of old neural networks, and discussed their utilization in water and other domains, while identifying AI limitations in integrating physical model simulations, geographical principles, and addressing transfer learning issues. A technique based on SRS images was presented by Ghaffarian et al. (2021) utilizing a deep learning system, enhancing the efficiency and correctness of these approaches for classifying, segmenting, detecting changes, and targeting recognition in SRS image. Consequently, the integration of AI with SRS for the analysis of extensive surveillance data is essential to advance the oceanography and SRS technology.

Smart Sensor Networking (SSN)

AI technology is being utilized by various smart intelligent sensors which offers supplementary function in ocean monitoring. Due to its numerous advantages, including compact dimensions, seamless networking capabilities, and resilience to electromagnetic interference, it is anticipated to serve as a supplementary measurement instrument in oceanic observation. Howe et al. (2010) created an SSN system that integrates essential components of sea monitoring systems, incorporates a system of data collection, and enables responsive sampling. A specific type of smart sensors were proposed by Cater et al. (2009) which were capable of being situated and recognized on the "World Wide Web (WWW)". These system exhibit "plug-and-play" operative systems in practical settings along with furnishing data that could be further distributed, analyzed, and displayed to geographically diverse operators across various applications. Such sensors are predominantly utilized in oceanic networks to enable the monitoring and control. An ocean SSN is an underwater wireless SSN comprising various ocean sensors and coastal radars along with autonomous vehicles that works on the water surface as well as underwater. Various sorts of equipment of SSN in the sea water interact via underwater communication devices for establishing an underwater SSN and can execute a range of functions for detecting the marine atmosphere. Luo et al. (2022) suggested a wireless SSN based on software to address ocean monitoring challenges, including sea surface and subsea surveillance. Marine sensor networks has diverse uses and significant potential for oceanic observation and forecasting.

Intelligent Robots

The integration of typical robotics with artificial intelligence for ocean observation has resulted in the development of several sophisticated robots, including underwater robotic fish, submarine and various other robots undertaking various tasks such as oil drilling. Conventional robots can be described as programmed machines that execute mainly those operations and instructions which are based on sequence majorly through automation. By incorporating AI system, these robots are able to autonomously manage tasks and learn as per requirements of the system. Wang et al. (2021) developed a "gliding robotic fish" featuring multiple fins which were interlinked and operated as a body-caudal fin to utilize maritime materials and oversee oilfields. This robotic fish exhibits flexible swimming, efficient three-dimensional gliding, and the capability to assess movement of gliding and behaviour of piscine swimming. The Mesobot was developed by Yoerger et al. (2021), which was a smart underwater robot for observing the ocean twilight zone, capable of detecting, classifying, and tracking marine organisms.

The Transformation of Ocean Big Data

In oceanic scientific study, two essential attributes of ocean big data are precision and temperospatial continuity. The procedure of collecting ocean data through monitoring generally results in flawed data because of faults in observational equipment, climatic conditions, and other factors, necessitating recompilation of data (Sun et al. 2018). Interpolation methods are also employed for recovering ocean data using empirical orthogonal functions for addressing the issues in data related to the ocean colour (Liu and Wang 2018). An enhanced data interpolation empirical orthogonal function approach can help address the issue of absent values in data about temperature sea surface data as per temperospatial system. During this procedure of interpolation many a times critical information is lost in between, leading to significant errors in the reconstruction of data and presenting a substantial barrier for conventional algorithms for reconstructing data. Integration of AI effectively mitigates this issue. Deep learning, a form of AI, can derive intricate models from extensive sample data and regulate the model's efficacy. A novel IFBN model based on machine learning was introduced by Li et al. (2021) utilizing Bayesian networks and information flow for interpolating the lost data of variables over different oceans for a long time period. This approach demonstrates superior accuracy, validity, and stability. Park et al. (2019) employed a machine learning algorithm to rebuild chlorophyll concentration data, enhancing the accessibility of chlorophyll-a data.

Ocean Waste Management

Global ocean pollution has surged, which had not only harmed the marine ecosystems and coastal economy but also jeopardized human health. Annually, more than 8 million metric tons of plastic garbage enter the oceans, contributing to the accumulation of debris that endangers marine life, disrupts aquatic ecosystems, and incurs financial losses for the tourism as well as fishing based businesses. Management of marine waste poses great challenges due to the vastness of the oceans, the diversity of pollutants present along with the difficulties associated with tracking garbage dispersion in international waters. The inadequacy of traditional methods which include localized supervision and clean-ups of beach manually, has highlighted the need for innovative and sustainable strategies in ocean garbage management (Mahmoud and Desokey 2022). In light of this worldwide dilemma, a transition to sustainable waste management strategies has become imperative. Sustainable solutions emphasize prevention, reduction, and effective waste management to minimize environmental impact and foster long-term ecological equilibrium.

Efficient management of marine trash necessitates the removal of current debris and the establishment of technologies to forecast and avert future pollution. The use of sophisticated technologies, including AI, for managing waste management systems has demonstrated efficiency for enhancing the observation, identification, and removing the marine garbage with greater efficiency and cost-effectiveness (Ditria et al. 2022). Artificial intelligence is becoming a transformational instrument in environmental surveillance and administration, particularly in the essential area of ocean waste tracking. Artificial Intelligence comprises several technologies, such as computer vision, deep learning, and machine learning, capable of analysing extensive datasets, identifying patterns, and making data-informed judgments. For managing the waste in oceanic environment, AI systems are capable of providing real-time surveillance view of marine habitats using satellite imaging, drone monitoring, and autonomous underwater vehicles (AUVs) (Castano-Londono et al. 2024). These technologies provide accurate detection and monitoring of garbage hotspots, permitting focused cleaning initiatives and enhanced allocation of resource.

Autonomous Vehicles and Monitoring Platforms

Unmanned Marine Vehicles (UMVs) are the substantially advanced measures, enhancing the effectiveness, cost-effectiveness, and expenses associated with surveillance of environment. These vehicles are utilized for military operations to inspect regions thereby offering strategic advantages. The major three categories of these vehicles include: "Autonomous Underwater Vehicles (AUVs), Autonomous Surface Vehicles (ASVs), and Remotely Operated Underwater Vehicles (ROUVs)" (Castano-Londono et al. 2024). UMVs serve multiple purposes, including the collection of oceanographic and meteorological data, as well as the monitoring of sea ice and animals. Most ROUVs are outfitted with a minimum of illumination devices along with a video camera. The primary distinction among different categories of such vehicles is the involvement of an operator to govern the ROUV, whereas AUVs and ASVs function independently. Consequently, certain advanced vehicles possess the ability to perceive their surroundings and autonomously navigate. UMVs encompass surface vessels and semi-submersibles vessels. Their integration with other conventional systems for surveillance might enhance precision and accuracy of predictions, as the movements of UMVs are sometimes randomly affected by alerts from satellites or other devices used for monitoring, altering the pre-calculated navigational maps (Shahab et al. 2022).

AUVs gather data via sensors which enables the measurement and analysis of water temperature and velocity faster and easier. The collection and analysis of various other types of data becomes complex, as they necessitate additional process for transforming them into significant information. Consequently, it is crucial to choose a specific type of sensor for ensuring the effectiveness of detection. The solution of related issue is equally crucial, necessitating the capacity to evaluate and understand the acquired data by removing sensor noise, hence ensuring data reliability (Filom et al. 2022). They possess extensive spatial coverage; nevertheless, some constraints may be there due to low resolution. The major concern could be that the gathered data may lack temporal representativeness. They may participate in data gathering for bathymetric and magnetic fields, as well as seabed conformation, as part of the research. They are utilized for assessing water parameters in particular locales, such as regions next to hydrothermal processes or coral reefs. They are now utilized in several domains, encompassing scientific study and industrial applications.

3 Research Methodology

This chapter is based on secondary data sources including journal articles, books, and book chapters along with other published sources such as blogs and websites of prominent organizations. The authors have reviewed literature on the impact of climate change on oceans and a circular economy approach for mitigating such impacts. The chapter also discusses the role of AI in combating adverse effects of climate change in general as well as specifically in oceans. Different applications of AI for various activities such as monitoring, data gathering, and waste tracking are elaborated in order to enhance our understanding of the topic. The implications based on the review are also provided towards the end of the chapter.

4 Implications

Oceans govern temperature, generate oxygen, and allow many ecosystems and ways of life; so, they determine the state of the earth. Still, they are under attack from pollution, overfishing, climate change, and biodiversity lost. In this sense, the mix of artificial intelligence (AI) techniques and Circular Economy (CE) concepts provides a revolutionary path to promote ocean sustainability. From remote sensing and smart sensors to autonomous vehicles and big data analytics, artificial intelligence technologies ranging in scope are increasingly indispensable in watching, analyzing, and reacting upon marine changes in real time. Combining artificial intelligence-powered sensors with satellite photos and machine learning models helps authorities to precisely identify and predict the flow of ocean trash, therefore enabling proactive waste removal and containment action. For example, autonomous robots and AI-powered drones can identify plastic accumulation hotspots. These technologies can maximize clean-up efforts by suggesting the most efficient routes of action, therefore reducing fuel use and running pollution.

By methods of rethinking products, streamlining of resource flows, and optimizing reuse, principles of the circular economy help to shut the loop in resource consumption. Particularly in sectors including marine logistics, aquaculture, and fisheries, artificial intelligence is important in enabling circular blue supply chains. Mixed with IoT-enabled sensors and satellite data, artificial intelligence algorithms can monitor fish supplies, spot illicit fishing activity, and maximize harvest to stop overuse in ecosystems. By tracking water quality, disease outbreaks, and feed efficiency in aquaculture, artificial intelligence guarantees more circular and sustainable agricultural practices. In terms of port operations management, energy consumption reduction in shipping, and worldwide ocean trade carbon footprint minimizing, artificial intelligence also supports marine logistics. These developments not only serve to reduce environmental damage but also create feedback loops promoting strong maritime economy and adaptive government.

Moreover, the union of artificial intelligence with the circular economy unlocks innovation in many different fields including maritime ones. Also depending on artificial intelligence are emerging new business models employing circularity—such as blockchain-integrated traceability systems guaranteeing ethical and transparent sourcing of ocean items. For efficient upcycling, for example, AI-based image recognition can categorize different marine plastics. Along with robots, this makes it possible to create totally automated sorting and recycling systems at sea or near coastal cities. Furthermore led by artificial intelligence, digital twins and simulations allow businesses to digitally test circular solutions before real-world implementation, hence reducing trial-and- error expenses. Startups, local inventors, coastal communities, and blue entrepreneurs all benefit from this combination in order to produce environmentally friendly ocean products and services appropriate for circular economy goals.

At last, using artificial intelligence in ocean sustainability in line with circular concepts provides teachers, lawmakers, and policy-makers with interesting analysis. Artificial intelligence distances science from decision-making by turning complex ocean data into practical knowledge. Predictive models and AI-enabled dashboards facilitate real-time environmental assessments, marine law enforcement, and governance evaluation of circular economy strategies. They also ensure evidence-based marine planning and data interchange, therefore enabling harmonizing of activities in several spheres. Virtual reality and artificial intelligence-driven simulations as well as other immersion techniques can serve to increase public awareness of marine systems and the requirement of sustainable behaviors in sequence of education and capacity development. These tools promote ocean literacy, therefore building a global culture honoring inventiveness, conservation, and stewardship.

5 Conclusion, Limitations, and Scope for Future Research

Finally, it can be inferred that with enhanced artificial intelligence technologies the ocean sustainability can be attained within the circular economy system. Including intelligent technology into maritime monitoring, waste management, resource optimization, and ecosystem regeneration will help us to approach a smarter and circular blue economy which will be resilient, inclusive, and in harmony with nature. Specifically with regard to data sovereignty and access for developing countries, this vision, emphasizes strategic investment, international cooperation, and ethical considerations. Still, it should be remembered that artificial intelligence is not a magic bullet; rather, when integrated into circular design thinking, it becomes a strong enabler of ocean sustainability, driving us toward a time when oceans flourish, communities thrive, and economic systems regenerate rather than deplete.

Although this research offers significant implications for policy and practice but still there are a few limitations. Firstly, this chapter is based on already published research, which is further very limited, hence, more empirical studies are required before incorporating the suggestions from this study. Moreover, studies from different countries across the globe are reviewed and analysed to arrive at the final conclusion of present research therefore the implications cannot be generalized to all ocean context. Future research should focus on specific regions to offer more concrete implication for a particular region. Finally, an overview of the technologies employing AI for ocean sustainability is furnished in this chapter to enhance general understanding rather than building a deep basis regarding them. In future, the researcher should focus on the detailed analysis of each individual technology to broaden our understanding of the same. Furthermore, there is a necessity for comprehensive long-term evaluations of AI technology in mitigating climate change effects on ocean sustainability to examine their effectiveness.

References

Boukabara S-A, Krasnopolsky V, Stewart JQ, Maddy ES, Shahroudi N, Hoffman RN (2019) Leveraging modern artificial intelligence for remote sensing and NWP: benefits and challenges. Bull Am Meteorol Soc 100(12):ES473–ES491. https://doi.org/10.1175/BAMS-D-18-0324.1

Burger FA, John JG, Frölicher TL (2020) Increase in ocean acidity variability and extremes under increasing atmospheric CO_2. Biogeosciences 17(18):4633–4662. https://doi.org/10.5194/bg-17-4633-2020

Castano-Londono L, Marrugo Llorente SdP, Paipa-Sanabria E, Orozco-Lopez MB, Fuentes Montaña DI, Gonzalez Montoya D (2024) Evolution of algorithms and applications for unmanned surface vehicles in the context of small craft: a systematic review. Appl Sci 14(21):9693. https://doi.org/10.3390/app14219693

Cater NE, Eng P, O'Reilly T (2009) Promoting interoperable ocean sensors: the smart ocean sensors consortium. In: OCEANS 2009. IEEE, pp 1–6. https://doi.org/10.23919/OCEANS.2009.5422448

Chen XM, Sharma A, Liu H (2023) The impact of climate change on environmental sustainability and human mortality. Environments 10(10):165. https://doi.org/10.3390/environments10100165

Cheng L, Trenberth KE, Fasullo J, Boyer T, Abraham J, Zhu J (2017) Improved estimates of ocean heat content from 1960 to 2015. Sci Adv 3(3):e1601545. https://doi.org/10.1126/sciadv.1601545

Crowther D, Singh R, Singh A, Rani S (2025) Assessing the impact of artificial intelligence on healthcare sector for enhancing sustainability. In: Transforming healthcare sector through artificial intelligence and environmental sustainability. Springer Nature Singapore, pp 3–21

Ditria EM, Buelow CA, Gonzalez-Rivero M, Connolly RM (2022) Artificial intelligence and automated monitoring for assisting conservation of marine ecosystems: a perspective. Front Mar Sci 9:918104. https://doi.org/10.3389/fmars.2022.918104(Frontiers)

Dube K (2024) A comprehensive review of climatic threats and adaptation of marine biodiversity. J Mar Sci Eng 12(2):344. https://doi.org/10.3390/jmse12020344(MDPI)

Filom S, Amiri AM, Razavi S (2022) Applications of machine learning methods in port operations—a systematic literature review. Transp Res Part e Logist Transp Rev 161:102722. https://doi.org/10.1016/j.tre.2022.102722

Geissdoerfer M, Pieroni M, Pigosso D, Soufani K (2020) Circular business models: a review. J Clean Prod 277:123741. https://doi.org/10.1016/j.jclepro.2020.123741

Ghaffarian S, Valente J, Van Der Voort M, Tekinerdogan B (2021) Effect of attention mechanism in deep learning-based remote sensing image processing: a systematic literature review. Remote Sens 13(15):2965. https://doi.org/10.3390/rs13152965

Greene C, Scott-Buechler C, Hausner A, Johnson Z, Lei XG, Huntley M (2022) Transforming the future of marine aquaculture: a circular economy approach. Oceanography 35(2):50–61. https://doi.org/10.5670/oceanog.2022.208

Howe BM, Chao Y, Arabshahi P, Roy S, McGinnis T, Gray A (2010) A smart sensor web for ocean observation: fixed and mobile platforms, integrated acoustics, satellites and predictive modeling. IEEE J Sel Topics Appl Earth Obs Remote Sens 3(3):507–521. https://doi.org/10.1109/JSTARS.2010.2052022

IPCC (2019) Technical summary: IPCC special report on the ocean and cryosphere in a changing climate. In: Pörtner HO et al (eds). IPCC

Joshi A, Singh R, Kumar V, Rani S (2024) Net zero strategy: exploring opportunities, confronting challenges for developing countries. In: Transition towards a sustainable future: net zero policies and environmental sustainability. Springer Nature Singapore, pp 233–252

Khan S, Singh R, Alnahas J, Abbate S, Centobelli P (2024) Navigating the smart circular economy: a framework for manufacturing firms. J Clean Prod 480:144007. https://doi.org/10.1016/j.jclepro.2023.144007

Kieu HT, Law AW-K (2021) Remote sensing of coastal hydro-environment with portable unmanned aerial vehicles (pUAVs): a state-of-the-art review. J Hydro Environ Res 37:32–45. https://doi.org/10.1016/j.jher.2021.02.004

Kirchherr J, Yang NHN, Schulze-Spüntrup F, Heerink MJ, Hartley K (2023) Conceptualizing the circular economy (revisited): an analysis of 221 definitions. Resour Conserv Recycl 194:107001. https://doi.org/10.1016/j.resconrec.2023.107001

Koen EL, Barichivich WJ, Braun de Torrez EC, Walls SC (2025) Sea level rise threatens Florida's insular vertebrate biodiversity. Biodivers Conserv 34(2):513–530

Li M, Zhang R, Liu K (2021) Evolving a Bayesian network model with information flow for time series interpolation of multiple ocean variables. Acta Oceanol Sin 40:249–262. https://doi.org/10.1007/s13131-021-1734-1

Liu X, Wang M (2018) Gap filling of missing data for VIIRS global ocean color products using the DINEOF method. IEEE Trans Geosci Remote Sens 56(8):4464–4476. https://doi.org/10.1109/TGRS.2018.2820423

Liu Y, Qiu M, Liu C, Guo Z (2017) Big data challenges in ocean observation: a survey. Pers Ubiquit Comput 21:55–65. https://doi.org/10.1007/s00779-016-0980-2

Luo H, Wang X, Xu Z, Liu C, Pan J-S (2022) A software-defined multimodal wireless sensor network for ocean monitoring. Int J Distrib Sens Netw 18:15501477211068388. https://doi.org/10.1177/15501477211068389

Mahmoud AED, Desokey O (2022) Artificial intelligence-based smart waste management for the circular economy. In: Environmental management technologies. CRC Press, pp 341–358

Najeeb S, Khan RAA, Deng X, Wu C (2025) Drivers and consequences of degradation in tropical reef island ecosystems: strategies for restoration and conservation. Front Mar Sci 12:1518701. https://doi.org/10.3389/fmars.2023.1518701

Park J, Kim J-H, Kim H-C, Kim B-K, Bae D, Jo Y-H et al (2019) Reconstruction of ocean color data using machine learning techniques in polar regions: focusing on off Cape Hallett, Ross Sea. Remote Sens 11(11):1366. https://doi.org/10.3390/rs11111366

Saad A, Stahl A, Våge A, Davies E, Nordam T, Aberle N, Johnsen G (2020) Advancing ocean observation with an AI-driven mobile robotic explorer. Oceanography 33(1):50–59. https://doi.org/10.5670/oceanog.2020.103

Shahab S, Anjum M (2022) Solid waste management scenario in India and illegal dump detection using deep learning: an AI approach towards sustainable waste management. Sustainability 14(24):15896. https://doi.org/10.3390/su142415896

Singh R (2025) Adoption of circular economy strategies for climate change mitigation: insights from key sectors. Clim Chang Soc Responsib 25:43–62

Singh A, Singh R, Katragadda R (2025a) Leveraging artificial intelligence for climate neutrality: human-centric mitigation strategies to combat global warming. In: Climate neutrality through smart eco-innovation and environmental sustainability. Springer Nature Switzerland, Cham, pp 99–114

Singh R, Gupta A, Kandpal V (2025b) ESG excellence in the digital world: circular economy e-business solutions. In: Fintech for ESG and the circular economy, pp 193–204

Singh R, Rani S, Filho WL (2025c) Climate change and the path to climate neutrality: understanding causes, impacts, and action strategies. In: Climate neutrality through smart eco-innovation and environmental sustainability. Springer Nature Switzerland, Cham, pp 283–297

Singh R, Rani S, Kandpal V (2025d) Driving environmental resilience: integrating industry 4.0 for circular economy. In: Green futures: navigating the path to environmental resilience. Springer Nature Singapore, Singapore, pp 23–40

Song T, Pang C, Hou B, Xu G, Xue J, Sun H, Meng F (2023) A review of artificial intelligence in marine science. Front Earth Sci 11:1090185

Spector L (2006) Evolution of artificial intelligence. Artif Intell 170:1251–1253. https://doi.org/10.1016/j.artint.2006.10.009

Sun M, Yu F, Chongjing L, Jiang X (2018) Deep learning application in marine big data mining. Sci Technol Rev 36:83–90

Talukder B, Ganguli N, Matthew R, Hipel KW, Orbinski J (2022) Climate change-accelerated ocean biodiversity loss & associated planetary health impacts. J Clim Change Health 6:100114

Wang C, Lu J, Ding X, Jiang C, Yang J, Shen J (2021) Design, modeling, control, and experiments for a fish-robot-based IoT platform to enable smart ocean. IEEE Internet Things J 8:9317–9329. https://doi.org/10.1109/jiot.2021.3055953

Wang S, Li W, Xing L (2022) A review on marine economics and management: how to exploit the ocean well. Water 14(17):2626

Wiese FK, Auad G (2022) Global marine biodiversity partnership. In: Partnerships in marine research. Elsevier, pp 199–216

Yoerger DR, Govindarajan AF, Howland JC, Llopiz JK, Wiebe PH, Curran M et al (2021) A hybrid underwater robot for multidisciplinary investigation of the ocean twilight zone. Sci Robot 6:eabe1901. https://doi.org/10.1126/scirobotics.abe1901

Yuan Q, Shen H, Li T, Li Z, Li S, Jiang Y et al (2020) Deep learning in environmental remote sensing: achievements and challenges. Remote Sens Environ 241:111716. https://doi.org/10.1016/j.rse.2020.111716

Zhang L, Ruiz-Menjivar J, Luo B, Liang Z, Swisher ME (2020) Predicting climate change mitigation and adaptation behaviors in agricultural production: a comparison of the theory of planned behavior and the value-belief-norm theory. J Environ Psychol 68:101408. https://doi.org/10.1016/j.jenvp.2020.101408

The Contribution of AI Systems to the Development of Climate Resilience and the Formulation of Climate-Resilient Policies Brankica Todorovic

Brankica Todorovic

Abstract Accelerating planetary crises are driven by climate change, which contributes to biodiversity loss, tipping points in natural systems and increasing pollution. Building resilient systems across various sectors, urban areas and communities has become a key objective to ensure sustainability and enhance ecosystems. In this context, AI systems for climate resilience are being developed as multidisciplinary tools used by diverse stakeholders to more effectively address the challenges of climate disruption through the formulation of robust climate policies and action plans. The methodological framework of the research is based on a qualitative approach, within which descriptive methods are employed to analyze available academic literature, studies, and developed AI systems for climate resilience. The theoretical framework of the research draws on OECD-identified indirect positive environmental impacts of ICT on ecological sustainability and the UNESCO Recommendation on the Ethics of AI, which stipulates that AI systems should not be used when there are disproportionate negative environmental impacts. Through a synthesized and holistic approach, the research offers a framework and recommendations for a future AI model as a potential response to the triple planetary crisis encompassing climate change, biodiversity loss, and pollution. Accordingly, the research objectives relate to the analysis of AI systems for climate resilience (TerraScope and Climate Trace): practical implementation examples, contribution to building climate resilience, analysis of the sectors in which they are applied (agriculture, infrastructure and other sectors), and contribution to the formulation of climate-resilient policies. The study's findings point to potential gaps in the use of AI systems for climate resilience, as well as, the need to strengthen their positive effects and contributions—particularly in the context of data management to enhance climate policy-making: data-driven policy development, crisis management, planning climate-resilient urban infrastructure, and other decision-making processes.

B. Todorovic (✉)
Faculty of Contemporary Arts, University of Business Academy, Belgrade, Serbia
e-mail: brankica.todorovic@fsu.edu.rs

R. Singh et al. (eds.), *Data-Driven Sustainability: Harnessing Technology for A Greener Future and Environmental Resilience*, World Sustainability Series,
https://doi.org/10.1007/978-3-032-13412-7_13

Keywords AI · Climate change sectors · Resilience · Policy-making · Crisis management

1 Introduction

In contemporary economies, the key terms related to climate change are resilience and adaptation (Schipper et al. 2022). The issue of climate resilience is particularly pronounced in rural areas and, according to traditional approaches, is associated with the utilization of socio-economic factors that determine resilience, such as access to resources, infrastructure, education, and other determinants (Ackerl et al. 2023; Haile et al. 2020). Climate adaptation refers to a reactive response to climate change (IPCC 2014).

However, the concepts of resilience and adaptation have expanded. According to the 2022 IPCC Report, adaptation can also be transformational, and resilience is increasingly viewed through the lens of transformative learning (IPCC 2022). The report further analyzes climate risks that emerge from the interaction between climate-related hazards, the exposure of ecosystems and human populations, and systemic vulnerability (IPCC 2022). A transition toward climate-resilient development entails the interaction between people, ecosystems, and the climate, facilitated through technological advancement, governance, financing, knowledge, and other enabling factors (Graph 1).

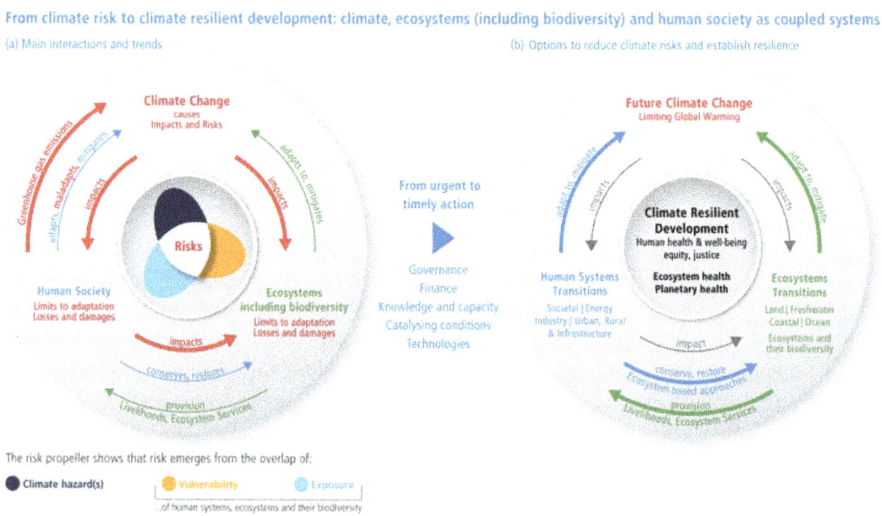

Graph 1 Interaction between human society, ecosystems, and climate within the context of climate change (IPCC, Report, Summary for Policymakers 2022)

Technological development has led to the application of artificial intelligence (AI) in strengthening climate resilience and advancing climate adaptation. Since 2020, the academic community has shown increasing interest in the study of AI and climate resilience (Yoon et al. 2025). The majority of scholarly works examine AI and climate resilience in the context of climate change adaptation and effective management (Filho et al. 2022). Studies have also analyzed the impact of digital transformation on ecological sustainability (Feroz et al. 2021), such as the use of the Internet of Things (IoT) and AI for disaster preparedness and sustainable agriculture (Nawaz and Babar 2024). The development of satellite systems contributes to enhancing climate resilience by providing critical information on Essential Climate Variables (ECVs), where satellite remote sensing (SRS) technology delivers multiscale data on the atmosphere, the Earth's surface, and subsurface through the use of sensors (Zhao et al. 2023).

2 Methodological Framework

This research focuses on identifying the impact of artificial intelligence (AI) on building climate resilience, including a sectoral analysis based on both qualitative and quantitative examination of relevant studies, academic publications, policy documents, and official data related to AI in the field of climate action. The study is multidisciplinary in nature, combining state-of-the-art knowledge from technology, ecology, ethics, and sustainability, with a core premise of highlighting both the positive and negative effects of AI's influence on climate resilience.

To identify the impact of AI on strengthening climate resilience, the analysis draws on satellite data from platforms such as TerraScope and Climate Trace, with a focus on key contribution areas and realized effects. The contributions of AI are subsequently classified using a synthesized and holistic approach into positive and negative impacts, based on the 2022 OECD framework, which categorizes AI's environmental contributions as either direct or indirect, and further as positive or negative (OECD 2022).

In addition, the UNESCO Recommendation on the Ethics of Artificial Intelligence encompasses the application of AI in the domain of the environment and ecosystems. Similar to the OECD framework, it emphasizes the recognition of both positive and negative environmental contributions. These include the carbon footprint of AI, energy consumption, and the environmental impact of raw material extraction for AI hardware production, as well as the need to manage potential negative effects (UNESCO 2021).

According to UNESCO (2021), examples of beneficial applications of AI for the environment include:

(a) Protection, monitoring, and sustainable management of natural resources;
(b) Prevention and mitigation of climate-related problems;
(c) More efficient and sustainable food systems;

(d) Access to and use of sustainable energy;
(e) Sustainable infrastructure, business models, and finance;
(f) Pollution detection and intervention planning.

Based on the above, the research objectives of this study are:

To analyze the impact of AI on the development of climate resilience through a comparative analysis and contributions of TerraScope and Climate Trace satellite systems;

To identify the positive and negative impacts of AI applications on building climate resilience;

To explore data management strategies to enhance climate policy-making, from the perspective of key stakeholders.

Literature Review

Artificial Intelligence (AI) offers improved solutions for environmental management and reduction of ecological footprints (Vinuesa et al. 2020). Climate change represents one of the most pressing global challenges, influencing both ecosystems and economies. Addressing this issue involves the application of geospatial information technologies within the context of the Sustainable Development Goals (SDGs) (Derras and Makhoul 2022; Pirasteh and Varshosaz 2019). Numerous studies emphasize the contribution of AI to achieving the SDGs (Cenek et al. 2018). AI facilitates climate-informed decision-making, enabling optimized resource allocation and supporting other effective governance decisions (Ayadi et al. 2025).

A critical assessment of AI's role in the achievement of the SDGs—particularly SDG 11 (Sustainable Cities and Communities) and SDG 13 (Climate Action)—identifies key areas of application, including the following elements (Al-Raeei 2024):

Key application areas of AI in this context include:

- Climate modeling and prediction of extreme weather events
- Monitoring of air and water quality (via remote sensing and sensor networks)
- Optimization of energy consumption
- Waste management and support for the circular economy
- Agriculture and forestry: precision management and crop protection
- Biodiversity monitoring and combating illegal activities
- Sustainable urban infrastructure.

Several authors emphasize the transformative potential of AI in developing resilient infrastructure under climate-related disasters, through predictive maintenance and real-time monitoring systems (Rane et al. 2024; Singh and Goyal 2023).

The literature further indicates that studies analyzing AI applications in water resource management under climate change conditions highlight AI's contribution in optimizing water distribution, predicting availability, and detecting pollution (Abdulameer et al. 2025).

The use of AI also supports reduction of ecological footprints and enhances enterprise sustainability (Abad-Segura et al. 2020).

The impact of ICT on environmental sustainability was already highlighted by Berkhout and Hertin in 2001 (Berkhout and Hertin 2001).

An OECD study published in 2019 emphasized the necessity of adopting artificial intelligence (AI) within the framework of responsibility, reliability, and support for sustainable development (OECD 2019a).

The Coalition for Digital Environmental Sustainability (CODES) report stresses the importance of sustainable digitalization and promoting the circular economy (CODES 2021; OECD 2019b).

However, the deployment of AI technologies is not without risks—there are ecological, socio-economic, and ethical risks that must be carefully assessed (Toderas 2025).

Estimates of the ICT sector's share in global CO_2 emissions range from 1.5 to 4%, indicating the importance of calculating intensity ratios to better understand and manage these impacts (The World Bank and ITU 2024).

Several methodologies have been developed to assess the environmental impact of AI, such as the Green Cloud Computing framework, which evaluates four main categories:

Abiotic resource depletion (use of minerals and fossil fuels),

Total energy consumption (including both renewable and non-renewable sources),

Global warming potential (contribution to climate change), and Water usage (UBA 2021). By using daily surface NO_2 concentration maps available in the Terrascope viewer, spatial patterns of NO_2 and their variability can be monitored on a daily basis, as well as time series extracted to visualize the impact of various events and new air quality policies (Sun et al. 2024).

Climate TRACE combines satellite and remote sensing data with artificial intelligence to identify greenhouse gas-emitting activities and calculate their emissions. Climate TRACE is used to empower climate action through radical transparency based on scientific rigor and strict methodological accuracy (Climate TRACE, Approach n.d).

The use of Climate TRACE and Terrascope satellite systems in building climate resilience is implemented across various sectors (Table 1).

Climate TRACE applies resolution according to the following segments:

- Raster data on land use for global forests—resolution 1 km × 1 km
- Raster data on global synthetic fertilizer application—resolution 5 km × 5 km
- Geospatial data on rice cultivation—resolution 10 m × 10 m
- Raster data on emissions from enteric fermentation and manure on pastures—resolution 10 km × 10 km
- Raster data on fires in croplands—resolution 11 km × 11 km
- Raster data on fuel consumption in residential and non-residential buildings—resolution 1 km × 1 km.

TerraScope has developed several satellites with different purposes, resolutions, and applications in specific sectors.

Sentinel-1 is a constellation of two radar satellites in polar orbits.

Table 1 Climate TRACE data and applicable subsectors - direct environmental impact

Framework	Using Climate TRACE data	Applicable subsectors
Supply chain decarbonization	Companies have begun using Climate TRACE data to verify self-reported emissions from suppliers and to identify cases in which lower-emitting options are available	Cement Iron and steel Aluminium Petrochemical steam cracking Rice cultivation Synthetic fertilizer application Domestic shipping International shipping
Data-driven policymaking at all levels	Policymakers in all levels of government have begun using Climate TRACE data to develop ambitious, data-driven decarbonization plans	All subsectors
Bringing transparency to financed emissions	Financial organizations such as development banks and investors are using Climate TRACE data to help count greenhouse gas emissions embedded in their portfolios, verify self-reported data from emissions source owners, and prioritize lower-emitting investments	All subsectors

Source ClimateTRACE, Case studies, https://climatetrace.org/case-studies

Sentinel-2 consists of two satellites in polar orbit that revisit the same location every 5 days.

Sentinel-3 provides highly accurate and reliable optical, thermal, and altitude data for marine and land monitoring purposes.

Sentinel-5P and its instrument TROPOMI (Tropospheric Ozone Monitoring Instrument) are used to detect air pollution and its sources at scales as fine as sub-urban levels.

The Sentinel-5P TROPOMI satellite sensor enables high-quality global air quality monitoring, including: monitoring and estimating global atmospheric sulfur dioxide (SO_2) emissions, as well as tracking changes in volcanic activity.

The COBRA algorithm allows for more precise monitoring of point sources of emissions, such as industrial facilities and volcanoes.

The Copernicus Land Monitoring Service is used for land monitoring by combining data from the PROBA-V satellite and networks of sensors on the Earth's surface.

The Copernicus Digital Elevation Model (DEM) provides a pixel-based representation of the Earth's surface topography relative to the 2008 Earth Gravity Model geoid and is utilized for satellite image geocorrection, hydrological modeling, natural disaster risk management, and other applications.

PROBA-V offers a global vegetation map used for climate change assessment, agricultural monitoring, and food security preservation.

WorldCover is a global land cover map designed to monitor changes in the management of the Earth's natural capital, including land, forests, water resources, and biodiversity (Table 2).

Table 2 TerraScope data and applicable subsectors—direct environmental impact

Satellite	Purpose	Use	Spatial resolution
Sentinel-1	It uses radar technology to collect Earth observation data under all weather conditions, including through clouds	Monitoring infrastructure stability	5×20 m
Sentinel-2	It monitors changes in vegetation, land use, water bodies, river flows, and coastal areas	Land, water bodies, river flows, and coastal areas	Resolution up to 10 m
Sentinel-3	The Sentinel-3A and 3B satellites measure: Sea surface topography, Sea and land surface temperature, As well as the color of the ocean and terrestrial surfaces	Monitoring of marine and terrestrial environments	Resolution up to 300 m
Sentinel-5P	Detection of air pollution and its sources	Monitoring of atmospheric components and air quality	Resolution up 3,5 $\times$ 5,5 km^2
PROBA-V	Global vegetation map	Assessment of climate change, agricultural monitoring, and preservation of food security	–
WorldCover	Global land cover map	Monitoring changes in global land cover	Resolution up 10 m
Copernicus land monitoring service	Land monitoring	Management of water resources and forests Agriculture and food security Nature conservation Spatial planning	–
Copernicus digital elevation model (DEM)	Representation of Earth's surface topography	Covers all land areas, including buildings, infrastructure, and vegetation	GLO-30 resolution up 30 m GLO-90 resolution up 90 m

Source ClimateTRACE, Satellites, https://terrascope.be/en/satellites

Climate TRACE is a comprehensive database of anthropological emissions from major sources, combining direct observations of emission sources, remote sensing (satellite data), and large volumes of industrial data. This data-set provides insights into the contribution of various sectors to global CO_2e emissions (Table 3).

Table 3 Contribution of the sector in total global emissions 2024 (Billion tones CO_2, %)

Sector	Billion tonnes CO_2e	% of total global emissions
Power	147.01	25
Manufacturing	110.30	19
Fossil fuel operations	94.31	16
Transportation	84.07	14
Agriculture	77.02	13
Buildings	40.15	6.8
Waste	22.24	3.8
Fluoridate gases	14.31	2.4
Mineral extraction	2.41	0.4

Source ClimateTRACE, Sectors, https://climatetrace.org/sectors

Statistical Data Provided by Climate TRACE Based on Satellite Data and Artificial Intelligence (Michaele 2025).

Greenhouse gas emissions in the first half of 2025 were 0.13% higher compared to the same period in 2024. A total of 30.99 billion tonnes of CO_2 equivalent were emitted between January and June 2025, with 5.12 billion tonnes emitted in June alone—representing an increase of 0.29% compared to the same month of the previous year.

Methane emissions also rose, reaching 34.82 million tonnes in June, which corresponds to a 0.49% year-on-year increase. The fossil fuel extraction sector contributed most significantly to the rise in emissions during the first half of the year, with an increase of 77.65 million tonnes CO_2 equivalent (+1.5%). Emissions from the electricity generation sector decreased by 0.8%, equivalent to a reduction of 60.27 million tonnes CO_2e.

3 Contribution of AI and Positive and Negative Environmental Impacts: A Sectoral Perspective

The analysis of the contribution of positive environmental impacts is sector-specific (Table 4). In the agriculture sector, four key positive environmental impacts have been identified:

Table 4 Indirect positive environmental impacts: a sectoral perspective

Sector agriculture	
Risk management—building a reliable and efficient algorithm	• Collection of in situ data and algorithm development • Co-design with end users and the GEOGLAM community • RDM (Reference Data Manager)
Hierarchy of seasonally updated layers and the first open dynamic system	The first (baseline) product is a foundational layer of temporary (seasonal) crops, as well as a set of season-specific products The first dynamic and open-access system provides: • Seasonally updated maps of cropland and crop types at a spatial resolution of 10 m • A global reference database on crop types
Monitoring of three crop seasons based on agricultural practice	• Temporary Crops–Winter Cereals (TC-wintercereals) • Temporary Crops–Main Maize Season (TC-maize main) • Temporary Crops–Second Maize Season (TC-maize second)
Ensuring food security	Identification of regions at risk of insufficient food production
Sector nature and environment	
Climate change management	Facilitating informed decisions that affect biodiversity and the environment
Mapping climate change	Satellite images provide insights into the impact of climate change on the environment, the effects of drought on the landscape
Development of NO_2 mitigation strategies	Satellite imagery supports the development of surface NO_2 concentration maps, offering detailed information on NO_2 distribution
Vegetation management	Monitoring spatial patterns of NO_2 and its variability on a daily basis, as well as extracting time series to visualize the impact of new air quality policies
Sector water	
Management of quay and dike stabilization	Detection of ground displacements as small as a few millimeters per year
Water quality assessment, mapping of navigable zones, and sediment monitoring	Satellite imagery and local water sampling
Flood mapping and monitoring	Spatially consistent flood observations regardless of weather conditions, offering reliable and timely insights into flood events
Algae detection	The water monitor system enables the detection of algae

(continued)

Table 4 (continued)

Synoptic flood overview	Flooded surfaces Potentially flooded tree rows and forested areas Flood dynamics during prolonged periods of rainfall and cloud cover
Sector infrastructure	
Optimal maintenance of tunnel and bridge stability	The detection of even the slightest deviations in the structural stability of tunnels, bridges, and protected heritage sites
Optimisation of infrastructure safety management	Contributes to long-term infrastructure safety and ensures the continuous and secure flow of traffic
Comprehensive and up-to-date overview of the road network	Consistent monitoring of bridges and tunnels, regardless of weather conditions

Source Author, based on previous analysis

1. Risk Management—Building a Reliable and Efficient Algorithm

This is achieved through the following activities:

(A) Collection of In Situ Data and Algorithm Development

During the initial months of implementation, the WorldCereal team focused on collaboration with the community and the collection of in situ (field) data, which were subsequently used for the calibration and validation of the models.

(B) Co-design with End Users and the GEOGLAM Community

WorldCereal was developed in close cooperation with end users, through multiple concrete application examples linked to the GEOGLAM initiative and the Sustainable Development Goals (SDGs).

Key stakeholders and champion users—such as GEOGLAM, the Food and Agriculture Organization of the United Nations (FAO), and the Agricultural Market Information System (AMIS)—provided feedback regarding system functionality and product accuracy.

This user network contributed to the WorldCereal project through:

- the collection of in situ field data,
- collaborative product validation, and
- demonstration of real-world applications.

(C) RDM: Global Collection of In Situ Reference Data (Karanam et al. 2024)

The Reference Data Manager (RDM) hosts a global collection of harmonized and carefully curated in situ reference data on land cover and crop types, made freely available to all stakeholders.

It incorporates datasets from various sources, with standardized metadata and attributes aligned with a unified crop type classification system.

Table 5 Indirect negative environmental impacts: a sectoral perspective

Sector agriculture
System dependency on cloud infrastructure the system's reliance on cloud infrastructure raises concerns regarding data sovereignty, long-term accessibility, and operational robustness
Challenges in storing and processing large volumes of data efficiently Efficient storage and processing of vast amounts of Earth observation data remain critical challenges that require scalable infrastructure and optimized computational workflows
Reduced algorithm transparency due to lack of local ground truth data The absence of locally sourced ground-truth data compromises the transparency and interpretability of algorithms, potentially affecting the credibility and reliability of outputs
Sensitivity of optical imagery to biophysical parameters and weather conditions Optical satellite imagery, while valuable, is highly sensitive to biophysical conditions and weather, particularly cloud cover—posing significant limitations in persistently cloudy regions
Lack of continuous community engagement throughout the project lifecycle Limited interaction and collaboration with local communities throughout the duration of the project may hinder the relevance, adoption, and contextual accuracy of developed tools and datasets
Sector nature and environment
A complex methodological approach The transformation of global atmospheric NO_2 column observations—originally available at coarse spatial resolution—into accurate ground-level NO_2 concentrations at high spatial resolution requires a complex methodological framework
Need for machine learning in data-sparse regions Machine learning models must be applied in locations where in-situ measurements are not available, in order to reliably estimate surface-level NO_2 concentrations
Requirement for integrated data quality control To ensure high and transparent data quality, an automated quality control system must be implemented in combination with manual validation by WorldCereal moderators
Sector water
Lack of a single data source The absence of a single comprehensive data source necessitates the integration of machine learning techniques with satellite imagery in order to ensure the provision of objective and weather-independent information
Sector infastructure
Satellite imagery alone is insufficient for the full optimization of stability and safety management of transport infrastructure

Source Author, based on ESA, https://esa-worldcereal.org/en/about/objectives

Through the RDM platform, users can view, search, download, contribute to, and share in situ reference datasets.

2. Hierarchy of Seasonally Updated Layers and the First Open Dynamic System

The primary product is a baseline cropland layer, which represents all cultivated land over a one-year period. In addition to this annual layer, WorldCereal provides season-specific products, including: winter cereals, maize, spring cereals, actively cultivated areas, and active irrigation.

In 2023, the WorldCereal team released the first dynamic and open-access system, offering: seasonally updated maps of cropland and crop types at 10-meter spatial resolution, and a global reference database on crop types.

Agro-Ecological Zones as the Basis of Production: To generate these products, the world is divided into agro-ecological zones (AEZs)—zones that share similar start and end points of the growing season for maize and cereals.

3. Replicable Crop Season Practices—Monitoring of three Crop Seasons Based on Agricultural Practice (Franch et al. 2022)

WorldCereal monitors three major crop seasons, following common agricultural practices. The primary seasons and corresponding crop layers include:

(A) Temporary Crops – Winter Cereals (TC-wintercereals)

- Actively cultivated areas
- Active irrigation
- Winter cereals (wheat, barley)

(B) Temporary Crops—Main Maize Season (TC-maize main)

- Actively cultivated areas
- Active irrigation
- Spring cereals and maize

(C) Temporary Crops—Second Maize Season (TC-maize second)

- Actively cultivated areas
- Active irrigation
- Maize

These layers are designed to ensure reproducibility across seasons and regions, enabling consistent monitoring of crop development patterns.

4. Ensuring Food Security

WorldCereal maps support the identification of regions at risk of food production shortfalls—an issue of growing importance due to climate change and population growth. By offering timely and spatially detailed data on cropland and crop types, WorldCereal enhances early warning capabilities and contributes to proactive decision-making in agricultural policy and food security planning.

Satellite imagery from the Nature and Environment sector contributes to positive environmental impacts in the context of climate change:

1. Climate Change Management
   Satellite imagery enables a comprehensive view of the landscape, facilitating informed decisions that affect biodiversity and the environment. This supports maintaining a balance between green and urban areas, as well as between residential zones, nature, and industry.

2. Mapping Climate Change
   Satellite images provide insights into the impact of climate change on the environment, the effects of drought on the landscape, and comparisons of temperature changes over time.
3. Development of NO_2 Mitigation Strategies
   Satellite imagery supports the development of surface NO_2 concentration maps, offering detailed information on NO_2 distribution. Based on these maps, local authorities can define mitigation strategies and monitor the impact of related policies and events (such as car-free Sundays and low emission zones). At a resolution of 1 km, detailed spatial gradients of surface NO_2 can be detected, and pollution "hotspots" identified.
4. Vegetation Management
   Through insights into green areas, satellite images enable analysis of photosynthesis activity, identification of green municipalities, and land use monitoring at a spatial resolution of 10 meters.

In the Water sector, five key positive environmental impacts have been identified:

1. Management of Quay and Dike Stabilization
   Radar data analysis enables the detection of ground displacements as small as a few millimeters per year. This precision allows for the timely identification of when stabilization works on quays and dikes should be undertaken.
2. Water Quality Assessment, Mapping of Navigable Zones, and Sediment Monitoring
   A comprehensive overview of water quality is made possible through the integration of satellite imagery and local water sampling. These combined data sources help determine whether water bodies are contaminated with oil or contain harmful algal blooms. Furthermore, regulatory authorities can monitor maritime traffic and delineate flood-prone zones, while dredging companies can verify the volume of sediment removed during their operations.
3. Flood Mapping and Monitoring
   TerraFlood is a virtual machine application that facilitates the mapping and monitoring of floods using Sentinel-1 satellite data. It provides spatially consistent flood observations regardless of weather conditions, offering reliable and timely insights into flood events.
4. Algae Detection
   The WaterMonitor system enables the detection of algae through satellite observations, supporting water quality management and early warning systems.
5. Synoptic Flood Overview
   Satellite-derived flood maps offer a synoptic overview of flood extent by identifying:

   - Flooded surfaces
   - Potentially flooded tree rows and forested areas
   - Flood dynamics during prolonged periods of rainfall and cloud cover

In the Infrastructure sector, three key positive environmental impacts have been identified:

1. Optimal Maintenance of Tunnel and Bridge Stability
   Safe and reliable infrastructure is supported by satellite radar data, which enables the detection of even the slightest deviations in the structural stability of tunnels, bridges, and protected heritage sites.
2. Optimisation of Infrastructure Safety Management
   The integration of satellite imagery with routine inspections and maintenance activities contributes to long-term infrastructure safety and ensures the continuous and secure flow of traffic.
3. Comprehensive and Up-to-Date overview of the Road Network
   Satellite observations provide consistent monitoring of bridges and tunnels, regardless of weather conditions, offering a complete and regularly updated view of transportation infrastructure.

Based on the 2022 OECD framework, which categorizes AI's environmental contributions as positive or negative, the analysis indicates the presence of the following negative impacts (OECD 2022) (Table 5).

Based on the identified indirect positive and negative environmental impacts, several gaps have been observed, indicating the need for further improvement of satellite imagery, particularly in the agriculture sector (Table 6).

To improve the use of satellites and AI, it is proposed to implement data fusion from multiple sensor and satellite sources, i.e., the hybridization of data sources (Allu and Mesapam 2025).

The integration of Earth Observation (EO) data and machine learning techniques is increasingly applied across multiple sectors, both to support decision-making processes and to enhance data acquisition and processing workflows (Ferreira et al. 2022).

Decision-making processes based on satellite data TerraScope

Satellite data are utilized by farmers, national governments, regional authorities, and the global community. Proactive collaboration between the state and farmers is realized through various mechanisms. Data for price and damage assessments provide access to regular and objective information for yield forecasting or price trends, as well as for documenting drought or storm-related damage.

Table 6 Gap between indirect positive and negative environmental impact: a sectoral perspective

Sector	Positive	Negative	Gap
Sector agriculture	4	5	−1
Sector nature and environment	4	3	+1
Sector water	5	1	+4
Sector infastructure	3	1	+2

Objective insights for optimal harvest are achieved by using satellite imagery to monitor crop growth and health, enabling timely interventions that contribute to reduced pesticide use.

Regional governments can take data-driven actions, such as encouraging major polluters to improve energy efficiency or introducing measures that support farmers in transitioning to regenerative agricultural practices.

There is also growing collaboration with the global agricultural user community—including scientists, institutions, and organizations. In addition, partnerships with local and international experts are being developed to test and improve the system under real-world conditions.

Based on the previous analysis, a Model: Application of satellites and Al in strengthening climate resilience is presented. Examples of data provided by the model include: Meteorological data, Emission inventories, Land use information, Temporal data, Spatially lagged data and others.

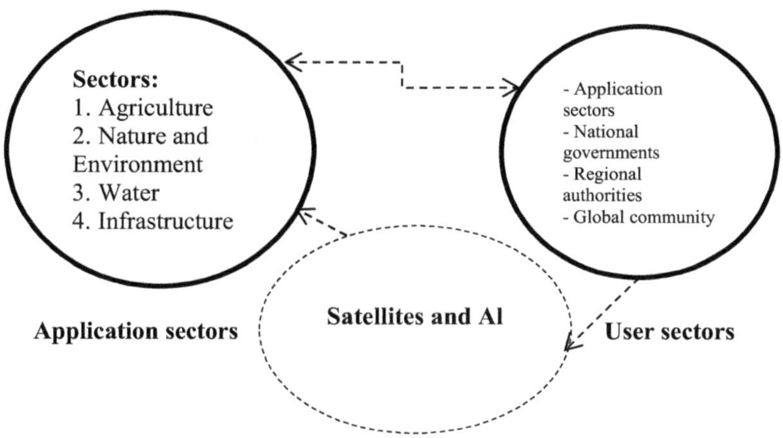

Key Contributions:

Optimization of sector management
Optimal infrastructure maintenance
Detection
Mapping
Monitoring
Sharing of reference in-situ data
Stakeholder involvement in algorithm development
Joint validation of products
Demonstration of application in real-world conditions

Model: Application of satellites and AI in strengthening climate resilience

The paper presents the role of AI in ensuring food security (World Bank 2020), based on the effective assessment of food safety (Brown 2016).

The limitations of this study are primarily related to the lack of a defined temporal framework for the analysis, which would serve as a basis for future comparative assessments. However, no specific time-frame or reference year has been established by satellite data providers, leading to the assumption that all identified impacts pertain to the previous periods of satellite use. In addition, the data from the two analyzed satellites were not compared on a common basis due to the unavailability of comparative datasets. Despite these limitations, the study offers a contribution by highlighting both the positive and negative impacts of AI in building climate resilience, as well as emphasizing the need for hybrid data sources that combine satellite imagery with other types of data in order to optimize decision-making across various sectors.

4 Conclusions

Climate resilience is one of the key issues in the development of stable and sustainable systems across various sectors. The construction of climate resilience is based on a range of systems, many of which are linked to the use of artificial intelligence (AI). Technological advancement—particularly through the use of satellite imagery—enables the integration of diverse data users and resources.

This paper analyzes two satellite systems, Climate TRACE and TerraScope, from the perspective of their application in enhancing climate resilience. The aspects of resilience-building are classified according to the OECD Framework, which encompasses both direct and indirect environmental impacts. Satellite applications are assessed across four key sectors: Agriculture, Nature and Environment, Water, and Infrastructure.

Sectoral analysis of satellite application in climate resilience reveals notable contributions, particularly in the optimization of natural resource management, and in more efficient responses to droughts and other climate-related challenges. However, satellite utilization differs depending on sensor purpose, spatial resolution, and algorithms adapted to rain or cloud conditions.

The study confirms multiple examples of the beneficial application of AI for environmental protection, in alignment with UNESCO's environmental goals: Protection, monitoring, and management of natural resources, Prevention and mitigation of climate-related issues, More efficient and sustainable food systems, Access to and use of sustainable energy, Sustainable infrastructure, business models, and finance and Detection of pollutants and planning of interventions.

A key contribution to improving the accuracy of measurements and availability of data is the involvement of data users in algorithm development. Satellite data users include a broader group than the sectors for which the satellites were initially designed.

The application of satellite technologies provides critical insight into issues such as drought, food insecurity, infrastructure instability, and others. However, the combination of satellite imagery with machine learning techniques and other data sources is strongly recommended to further enhance climate resilience.

The analyzed satellites contribute to resilience-building by offering information on key climate variables. These data are used to inform stakeholders, support monitoring and decision-making, ensure the inclusion of relevant actors, and serve multiple climate-related purposes.

Looking ahead, the ethical use of AI will increasingly rely on the precise identification of its potential negative effects and the development of measures to mitigate and improve upon them.

References

Abad-Segura E, Fuente ABdl, González-Zamar M-D, Belmonte-Ureña LJ (2020) Effects of circular economy policies on the environment and sustainable growth: worldwide research. Sustainability 12(14):5792. https://doi.org/10.3390/su12145792

Abdulameer L, Al-Khafaji M, Alawadi A, Al Maimuri N, Al-Shammari M, Al-Dujaili A, DhiyaAl-Jumeily (2025) Artificial intelligence in climate-resilient water management: a systematic review of applications, challenges, and future directions. Water Conserv Sci Eng 10. https://doi.org/10.1007/s41101-025-00371-2

Ackerl T, Weldemariam LF, Nyasimi M, Ayanlade A (2023) Climate change risk, resilience, and adaptation among rural farmers in East Africa: a literature review. Environ Dev Sustain 4. https://doi.org/10.1016/j.regsus.2023.05.004

Al-Raeei M (2024) Artificial intelligence for climate resilience: advancing sustainable goals in SDGs 11 and 13 and its relationship to pandemics. Discov Sustain 5. https://doi.org/10.1007/s43621-024-00775-5

Altieri MA, Nicholls CI, Henao A et al (2015) Agroecology and the design of climate change-resilient farming systems. Agron Sustain Dev 35:869–890. https://doi.org/10.1007/s13593-015-0285-2

Allu AR, Mesapam S (2025) Impact of remote sensing data fusion on agriculture applications: a review. Eur J Agron 164:127478. https://doi.org/10.1016/j.eja.2024.127478

Ayadi R, Forouheshfar Y, Moghadas O (2025) Enhancing system resilience to climate change through artificial intelligence: a systematic literature review. Front Clim 7. https://www.frontiersin.org/journals/climate/articles/; https://doi.org/10.3389/fclim.2025.1585331

Berkhout F, Hertin J (2001) Impacts of information and communication technologies on environmental sustainability: speculations and evidence, Report to the OECD. https://www.oecd.org/science/inno/1897156.pdf

Brown ME (2016) Remote sensing technology and land use analysis in food security assessment. J Land Use Sci 11(6):623–641. https://doi.org/10.1080/1747423X.2016.1195455

Castle SE, Miller DC, Merten N, Ordonez PJ, Baylis K, (2022) Evidence for the impacts of agroforestry on ecosystem services and human well-being in high-income countries: a systematic map. Environ Evid 11(10):1–27. https://environmentalevidencejournal.biomedcentral.com/articles/; https://doi.org/10.1186/s13750-022-00260-4

Cenek M, Haro R, Sayers B, Peng J (2018) Climate change and power security: power load prediction for rural electrical microgrids using long short term memory and artificial neural networks. Appl Sci 8(5):749. https://doi.org/10.3390/app8050749

CODES (2021) Synthesis report: a digital planet for sustainability. https://www.sparkblue.org/system/files/2021-05/Zero%20Order%20Draft_A%20DIGITAL%20PLANET%20FOR%20SUSTAINABILITY_%206_May%202021.pdf

Climate TRACE, Approach. https://climatetrace.org/approach
ClimateTRACE, Case studies. https://climatetrace.org/case-studies
ClimateTRACE, Satellites. https://terrascope.be/en/satellites
ClimateTRACE, Sectors. https://climatetrace.org/sectors
Derras B, Makhoul N. (2022) An overview of the infrastructure seismic resilience assessment using artificial intelligence and machine-learning algorithms. Proceedings of the ICONHIC. https://www.researchgate.net/publication/362538581_An_overview_of_the_infrastructure_seismic_resilience_assessment_using_Artificial_Intelligence_and_Machine-learning_algorithms
ESA (n.d) https://esa-worldcereal.org/en/about/objectives
FAO (2021) The state of the world's land and water resources for food and agriculture–systems at breaking point. Synthesis report 2021. https://doi.org/10.4060/cb7654en
Feroz AK, Zo H, Chiravuri A (2021) Digital transformation and environmental sustainability: a review and research agenda. Sustainability 13(3):1530. https://doi.org/10.3390/su13031530
Ferreira B, Silva RG, Iten M (2022) Earth observation satellite imagery information based decision support using machine learning. Remote Sens 14(15):3776. https://doi.org/10.3390/rs14153776
Filho LW, Wall T, Rui M, Serafino A, Nagy GJ, Balogun A-L. Luetz, Johannes M, Ng AW, Kovaleva M, Safiul A, Fardous MA (2022) Deploying artificial intelligence for climate change adaptation. Technol Forecast Soc Change 180(C). Elsevier
Franch B, Cintas J, Becker-Reshef I, Torres MJ, Roger J, Skakun S, Sobrino J, Van Tricht K, Degerickx J, Gilliams S, Koetz B, Szantoi Z, Whitcraft A (2022) Global crop calendars of maize and wheat in the framework of the WorldCereal project. Giscience Remote Sen 59:885–913. https://doi.org/10.1080/15481603.2022.2079273
Haile GG, Tang QH, Hosseini-Moghari SM, et al (2020) Projected impacts of climate change on drought patterns over East Africa Earth's Future, 8(7):e2020EF001502. https://doi.org/10.1029/2020EF001502
IPCC (2014) Climate change 2014–impacts, adaptation and vulnerability. Regional Aspects Cambridge University Press, Cambridge, Part B
IPCC (2022) Sixth assessment report impacts, adaptation and vulnerability. https://www.ipcc.ch/report/ar6/wg2/chapter/summary-for-policymakers/
IPCC (2022) Report, summary for policymaker. https://www.ipcc.ch/report/ar6/wg2/chapter/summary-for-policymakers/
Karanam S, Laso Bayas JC, Fritz S, Boogaard H, Pratihast AK, Degerickx J, Butsko C, Dries J, Van Tricht K (2024) World Cereal Reference Data Module (RDM). International Institute for Applied Systems Analysis (IIASA). https://doi.org/10.60566/80p50-6z433
Michaele M (2025) Climate TRACE data shows global emissions are still rising (slightly) in 2025. https://www.csofutures.com/news/climate-trace-data-shows-global-emissions-are-still-rising-slightly-in-2025/
Nawaz M, Babar MIK (2024) IoT and AI: a panacea for climate change-resilient smart agriculture. Discov Appl Sci 6:517. https://doi.org/10.1007/s42452-024-06228-y
OECD (2019a) Recommendation of the council on artificial intelligence. OECD Legal Instruments. https://legalinstruments.oecd.org/en/instruments/OECD-LEGAL-0449
OECD (2019b) Measuring the digital transformation. Organisation for Economic Co-Operation and Development (OECD). https://doi.org/10.1787/888933931086
OECD (2022) Measuring the environmental impacts of artificial intelligence compute and applications: the AI footprint, OECD Digital Economy Papers, No. 341. OECD Publishing, Paris. https://doi.org/10.1787/7babf571-en
Pirasteh S, Varshosaz M (2019) Geospatial information technologies in support of disaster risk reduction, mitigation and resilience: Challenges and recommendations. In: Sustainable development goals connectivity dilemma. CRC Press, pp 93–108. https://www.taylorfrancis.com/chapters/oa-edit/; https://doi.org/10.1201/9780429290626-6/geospatial-information-technologies-support-disaster-risk-reduction-mitigation-resilience-challenges-recommendations-saied-pirasteh-masood-varshosaz

Rane N, Choudhary S, Rane J (2024) Artificial intelligence for enhancing resilience. J Appl Artif Intell 5:1–33. https://doi.org/10.48185/jaai.v5i2.1053

Schipper ELF, Revi A, Preston BL, et al (2022) Climate resilient development pathways. In: Pörtner H-O, Roberts DC, Tignor M, et al (eds) Climate change 2022: impacts, adaptation and vulnerability. Contribution of working Group II to the 6th assessment report of the intergovernmental panel on climate change. Cambridge University Press, Cambridge and New York, pp 2655–2807

Singh S, Goyal MK (2023) Enhancing climate resilience in businesses: the role of artificial intelligence. J Clean Prod 418:138228

Sun W, Tack F, Clarisse L, Schneider R, Stavrakou T, Van RM (2024) Inferring surface NO2 over Western Europe: a machine learning approach with uncertainty quantification. J Geophys Res Atmos 129. https://doi.org/10.1029/2023JD040676

The World Bank and ITU (2024) Measuring the emissions & energy footprint of the ICT sector: implications for climate action. Washington, D.C. and Geneva.

Toderas M (2025) Artificial intelligence for sustainability: a systematic review and critical analysis of AI applications, challenges, and future directions. Sustainability 17(17):8049. https://doi.org/10.3390/su17178049

UBA (2021) Green cloud computing. Fraunhofer-Institut für Zuverlässigkeit und Mikrointegration (IZM), Berlin. https://www.umweltbundesamt.de/sites/default/files/medien/5750/publikationen/2021-06-17_texte_94-2021_green-cloud-computing.pdf

UNESO (2021) Recommendation on the ethics of artificial intelligence. https://unesdoc.unesco.org/ark:/48223/pf0000381137

Vinuesa R, Azizpour H, Leite I, Balaam M, Dignum V, Domisch S, Fennell P, Georgieff R, Jacobsson A, Jernberg J et al (2020) The role of artificial intelligence in achieving the sustainable development goals. Nat Commun 11:233. https://doi.org/10.1038/s41467-019-14108-y

World Bank (2020) A shock like no other: the impact of COVID-19 on commodity markets commodity markets outlook. https://www.worldbank.org/en/news/feature/2020/04/23/coronavirus-shakes-commodity-markets

Yoon YH, Iddrisu M, Lee C, Bharati P (2025) Information systems, AI and climate resilience: a systematic literature review AMCIS 2025 Proceedings 50. https://aisel.aisnet.org/amcis2025/intelfuture/intelfuture/50

Zhao S, Liu M, Tao M, Zhou W, Lu X, Xiong YuJ, Li F, Wang Q (2023) The role of satellite remote sensing in mitigating and adapting to global climate change. Sci Total Environ 904:166820. https://doi.org/10.1016/j.scitotenv.2023.166820

Waste Analytics, Carbon Mapping & Predictive Models in Era of Sustainability

R. Felista Sugirtha Lizy and Ir. Bambang Sugiyono Agus Purwono

Abstract The problem of garbage and the carbon effect have been a significant problem in the high urbanization and industrialization world today that needs to be regulated and minimised to promote sustainability and environmental sustainability. The chapter presents an analytical model founded on data that combines waste analytics, carbon mapping, and predictive modeling so as to initiate effective and intelligent environmental management systems. Waste data can be turned into informative data to be used by policymakers and urban planners by using the tools of Artificial Intelligence (AI), Internet of Things (IoT), Geographic Information Systems (GIS), and machine learning. Carbon mapping gives a graphical insight on the hots spot areas of emissions and predictive models project future waste and patterns of carbon, which can be used to act proactively in terms of recovery of resources and reduction of emissions. The chapter also examines the case-based implementation, frameworks of analysis, and policy implications to show how the technologies that utilize data can be used to develop low-carbon, sustainable and resilient communities.

Keywords Waste analytics · Carbon mapping · Predictive modelling · Environmental resilience · Circular economy · Smart waste management · Emission forecasting

1 Introduction

The world has increased the deterioration of the environment and global warming causing the establishment of sustainable systems that are able to support the ecological and economic needs. Some of the key sources of environmental issues are poor

R. Felista Sugirtha Lizy (✉)
A.P.C. Mahalaxmi College for Women, Thoothukudi, Tamil Nadu, India
e-mail: 21felistaa@gmail.com

Ir. Bambang Sugiyono Agus Purwono
Mechanical Engineering Department, State Polytechnic of Malang, Malang, Indonesia

R. Singh et al. (eds.), *Data-Driven Sustainability: Harnessing Technology for A Greener Future and Environmental Resilience*, World Sustainability Series,
https://doi.org/10.1007/978-3-032-13412-7_14

waste disposal and unregulated carbon emission. The conventional ways of managing waste are reactive, piecesalal and poor in data and therefore it is hard to predict the future problems or develop sustainable planning. In the waste and carbon management, data-driven decision making has thus emerged as a crucial avenue towards realization of environmental resilience.

With the rise of digital technologies, in particular, the AI, IoT, and highly developed analytics, it has become possible to transform environmental data into valuable knowledge. The waste analytics would allow the stakeholders to obtain real-time data about the creation, sorting, and treatment of waste to allow optimizing the collection system and the effectiveness of the recycling process. Through the application of satellite data and sensors that are operated by the IoT, carbon mapping can be used to map the spatial distributions of emissions and identify locations where urgent measures are required. At the same time, predictive models are used with the assistance of the past and present activity in order to forecast the waste patterns, estimate carbon footprints and help in evidence-based policymaking.

Combining these three pillars that are analytics, mapping, and prediction, the organizations will be in a position to change to a scenario that they are responsive to their environment, rather than participating in governing the environment. Not only is such synergy reducing the threats to the ecology, it also plays a part in the goals of the circular economy, in which waste is not a byproduct that can be recycled and reused. The chapter underlines the interdisciplinary side of this approach that focuses on using the combination of environmental science, data engineering, and systems design to solve urgent problems of the world. Finally, intelligent cities, environmentally friendly industry, and the ability to build a resilient ecosystem adapted to the modified climate are guided by data-driven waste and carbon management.

2 Literature Review

2.1 Waste Analytics and Predictive Models

In the recent past, the development of waste analytics has utilized artificial intelligence (AI) and machine learning (ML) to improve waste management practices. Fatorachian and Alizadeh (2025) also assessed predictive waste management models with a specific emphasis on the combination of AI and IoT technologies to streamline the process of waste collection and processing. Equally, Fang and Zhang (2023) reported on the use of AI in smart cities to achieve effective waste management, including the value of predictive analytics in lower costs of operation and better service provision.

In the paper by Gupta and Singh (2024), the authors studied the application of ML in sustainable treatment of organic waste showing that data-driven models have the potential to streamline the process of composting and minimize its impact on the environment. All these studies highlight the transformational nature of AI and ML in the context of the modernization of the waste management system.

2.2 *Carbon Mapping and Carbon Forecasting*

Carbon mapping is nowadays an important measure in the evaluation and mitigation of the impacts of climate change. Dye and Smith (2024) have examined the impacts of natural disturbances on the stocks of carbon in forests with the help of sophisticated tools of analyzing and mapping data and modeling carbon fluxes in different climate conditions. This strategy helps in coming up with a solution to the forest carbon risk management.

Hua and Li (2025) explored the contribution of AI-based computation and prediction of construction carbon emissions in terms of the utilization of ML algorithms to predict the energy consumption trends and related emissions in the construction domain. Their results emphasize that AI-based predictive models should be incorporated into the design and operation of buildings in order to attain a sustainable objective.

2.3 *Artificial Intelligence and the Internet of Things in the Environmental System*

The intertwining of AI and IoT technologies was the catalyst to the creation of smart environmental systems that can monitor and make a decision in real-time. Intelligent systems capable of responding to changes in the environment and maximizing the use of resources were suggested as the main features of intelligent AI-driven predictive sustainability models by Olawumi and Chan (2025).

Rajaram and Kumar (2025) touched upon the predictive models of sustainable smart buildings based on the data and described how AI and IoT integration could improve the energy efficiency and decrease the carbon footprints of the urban infrastructure. The integrations enable the development of responsive systems that help towards overall sustainability goals.

2.4 *Governance, Ethics and Policy Framework*

The most important part of the data-driven sustainability initiatives implementation is good governance and considerations of ethics. Agrawal and Gupta (2025) conducted a review of the incorporation of big data analytics into high-performance transportation systems and the significance of ethical data usage and clear decision-making in the work on the public infrastructure.

The article by Chen and Wang (2024) was dedicated to the problem of sustainable waste management practices in the sphere of the construction industry, and the authors revealed the difficulties and suggested possible approaches to effective management of concrete waste management. Their study highlights the necessity of extensive policies and governance structures as a means of enabling the construction industry to be sustainable.

2.5 *Future Projections and Future Map*

As of the future, the further development of AI, IoT, and predictive analytics is the future of data-driven sustainability. Dye and Smith (2024) talked about the consequences of natural disturbance in the forest carbon and implied that adaptive management practices based on modern modeling tools were critical in ensuring carbon stability in forest ecosystems.

Hua and Li (2025) forecasted the capabilities of AI in the development of carbon emissions management, indicating that the emissions will be reduced considerably, as a result of intelligent systems implementation. According to their work, the future should aim at refining the accuracy of the models and incorporating the new technologies to attain sustainability in the long term.

3 Waste Management Data Analytics Applications

The current waste management systems are becoming more accepting of data analytics to monitor, measure and streamline all the waste lifecycle procedures—collection to disposal. Data analytics can be used to transform raw data collected by sensors, cars and recycling facilities into insights to make smarter decisions and more effective resource allocation.

3.1 *Descriptive Analytics: Knowing the Waste Landscape*

Descriptive analytics is focused on determining what has already taken place within a waste system. It bases on the data collected by intelligent bins, sensors that monitor the waste and reports made by the city to make generalizations about prominent data concerning waste volume, composition and collection. Such insights enable administrators of cities to discover routes with inefficiencies, poorly performing recycling areas, or producing waste surpuses. Transparency is further supported through visualization dashboards that are developed through business intelligence tools, which helps citizens to participate in sustainable activities.

3.2 Diagnostic Analytics: The Diagnosis of the Causes

Diagnostic analytics assists in realizing the root causes of inefficiency or environmental impact once the patterns are observed. However, as an example, when waste production is related to socioeconomic and climatic factors, it will be possible to state whether the high level of waste is determined by consumers or tourism or seasonal factors. This is an analytical process of prioritizing interventions e.g. particular awareness activities or policy revisions that are direct response to the identified problems.

3.3 Predictive and Prescriptive Analytics of Decision Making

Predictive analytics utilizes machine learning and statistical modeling in order to preempt trends in waste and future obstacles. These models are capable of determining the extent of waste that a city would generate over the next few years, which waste materials would prevail in the waste stream, or how the population increase would influence the landfill capacity. Prescriptive analytics extends this by suggesting the best practices (or optimal course of action)—i.e. reconfiguration of the route, fleet, or incentive towards recycling—to attain sustainability objectives. Both predictive and prescriptive analytics are the pillars of intelligent waste management systems with ability to adapt themselves based on real-time feedback.

3.4 Data-Driven Waste Systems Advantages

Application of waste management systems that are based on analytics will lead to minimized operation costs, enhanced recycling rates, and reduced carbon emission. Furthermore, real time monitoring will assist the authorities to respond quickly to mechanical breakdown or overflow issues that will make the municipal services more credible. Lastly, transformation of waste management of reactive to predictive waste management promotes the concepts of the circular economy wherein materials will be recycled and repurchased rather than wasted.

This Fig. 1 represents the holistic cycle of the sustainability by the continuous data-driven improvement management. It begins with the input of policies and regulation that takes place in the form of targets and limits that are established in compliance standards, environmental rules, and organizational targets. Subsequently, raw data is added to it, and integration engages the data of different sources (IoT sensors, APIs, open data and paperwork). Machine learning, artificial intelligence, and statistical models are then applied to the obtained data to detect patterns and trends and anomalies. Such analyses are subsequently translated into actionable information leading to dash boards, reports, simulations as well as recommendations to support informed decision making. Finally, there is the social and economic feedback loop that constitutes the cycle, according to which the insights translate into a change in the operations, policies and stakeholder engagement. It is a mechanism in which a cycle continues to feed on itself that leads to a high degree of sustainability that is symbolically indicated through the use of the globe, buildings and recycling icons.

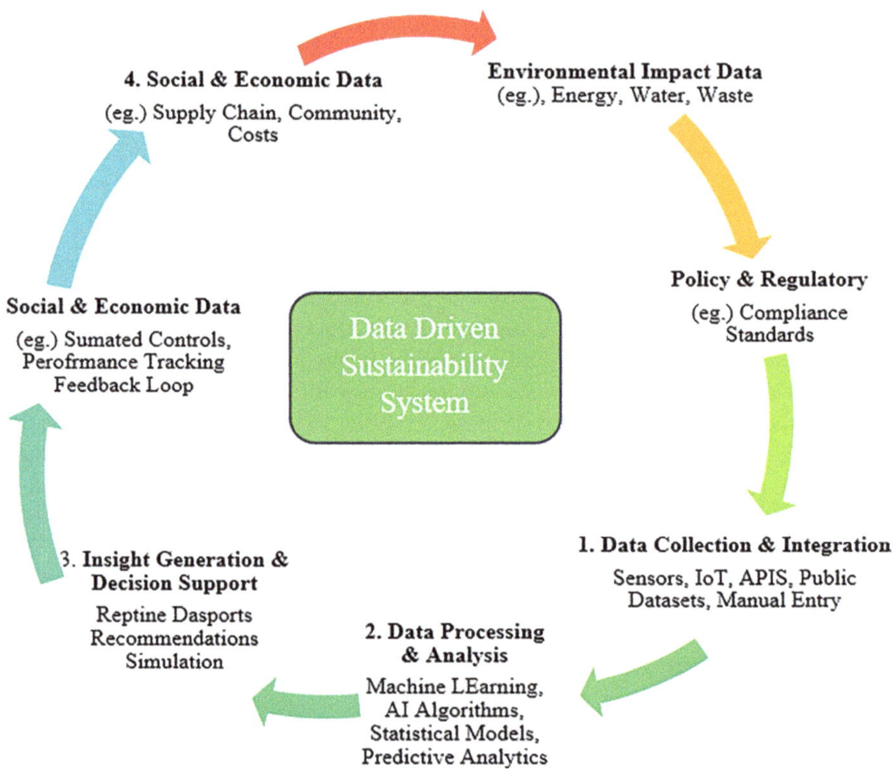

Fig. 1 Data-driven sustainability system framework Author's Own Creation

4 Carbon Mapping: The Environmental Impact

Waste analytics is engaged in operational performance, as opposed to carbon mapping, which is concerned with the more macro-level of the environmental effects of human action. It provides the spatial image of the emission of greenhouse gas (GHG), which helps the governments and industries to view the origin and the location of the emission of carbon emission into the atmosphere.

4.1 Techniques and Principles of Carbon Mapping

One of the methods that relies on the combination of different sources of information, including satellite data and information, IoT sensors, transportation data, and industrial databases, is carbon mapping. The geographic Information Systems (GIS) will be vital in merging these datasets collectively into spatial maps, which show the level of emissions in a region. The analytical instruments may be used to differentiate between the mobile and stationary emissions (factories and power plant), as well as offer a comprehensive view of the carbon footprint in the city.

4.2 IoT and Remote Sensing

The IoT devices and satellites of remote sensors continuously gather real-time data on the environment such as air quality, temperature and concentration of particles. This stream of information flows into the carbon mapping models upon which the level of emission can be adjusted dynamically. Using the example of cross combining sensor measurements and meteorological trends, one can establish how carbon circulates in both cities and villages. The awareness of space would facilitate certain interventions of reducing emissions such as optimization of traffic flow or greening of cities.

4.3 Urban Planning and Urban Policy

The governments and planners use the carbon maps as their decision-support systems. They influence zoning policies, promote climate adaptation programs and steer the infrastructure investments towards the low emitting regions. Besides, carbon mapping can be used to ensure adherence to environmental requirements and assist the cities in tracking the development of global obligations like the Paris Agreement. Carbon dashboards made publicly accessible also encourage involvement and openness of the citizens in the sustainability process.

5 Sustainable Waste Systems Predictive Model

One of the most effective tools to come up with effective and sustainable waste management strategies is predictive modeling. Through historical and real time data, the models are able to predict the future trends in the generation of waste, energy and the level of emission. Predictive modeling is the step between the data observation and active processing of the information to enable cities and industries to be ready to challenges of the environment they will face. Figure 2. Waste analytics and predictive modeling architecture depicting the flow of data between acquisition and decision support.

This Fig. 2 shows a systematic structure in particular waste management through data-driven methods. During the data collection process with the use of IoT, sensors in waste bins, which provide information on the fill level, weight, and composition, are used, and GPS technologies monitor the probes of collection vehicles and their

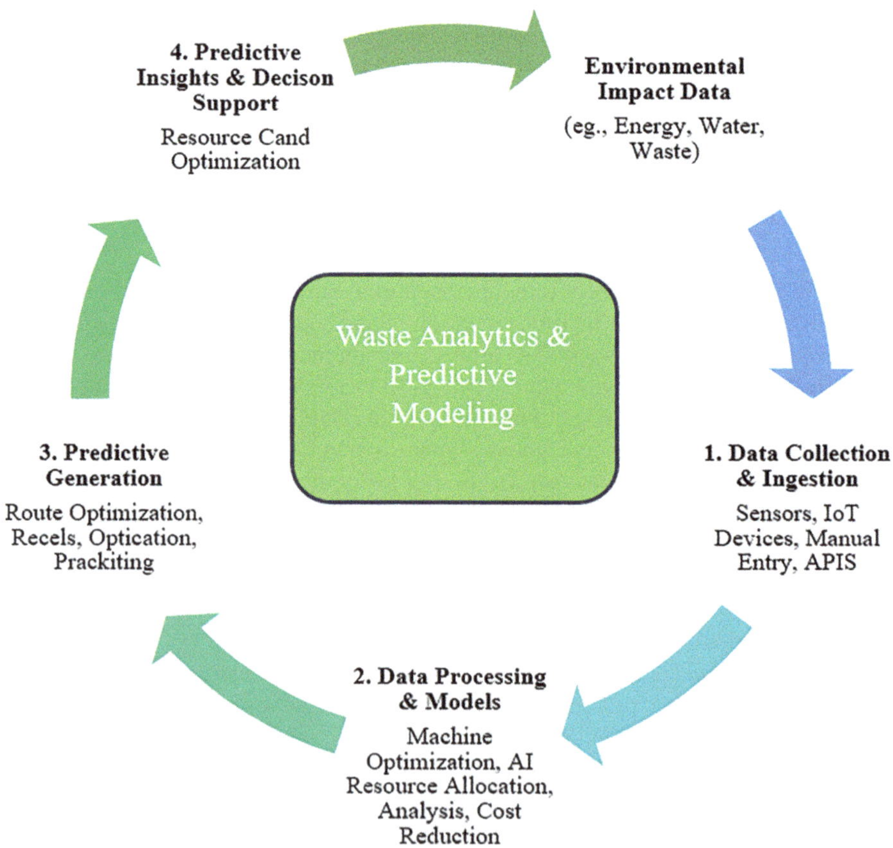

Fig. 2 Waste analytics and predictive modeling architecture Author's Own Creation

routes. The information enters the stage of data processing and analysis, during which the cloud analytics and machine learning algorithms are used to forecast the most convenient collection time to act, foresee the volume of waste in the future, and detect anomalies, including unlawful dumping. The refined insights are used in decision support and optimization stage that directs the authorities at the municipal level and waste operators to make sound operational changes. This leads to better allocation of resources, decreased fuel usage and low operating expenses and overall effectiveness of waste management is increased.

5.1 The Basics of Predictive Modeling on Waste Management

Predictive models are mathematical and computational methods that are used to predict events in the future based on the current trends of data. This can be used in waste management to determine how much waste is generated per capita, when the waste levels will be high or when the landfills will reach their capacity. Such tasks are typically accomplished with the help of such algorithms like regression models, decision trees, and neural networks. The quality of data used to make these predictions, the representativeness of historical data and ongoing model training are critical to the accuracy of these predictions.

As an illustration, machine learning algorithms have the potential to process data of smart bins, collection trucks, and recycling facilities to predict how the amount of waste will change with population increase, consumer trends, and seasonal dynamics. This foresight can help authorities to direct their resources in a more effective way and can create systems which will be able to respond to the changes in demands.

5.2 Machine Learning and Simulation

Machine learning is an improvement of predictive modeling; it allows systems to learn independently through the data and improve their predictions in the future. Random forests and support vector machines are examples of supervised learning algorithms which can be used to classify types of wastes or estimate the level of contamination in recycling streams. In the meantime, the unsupervised learning techniques identify trends and abnormalities, which can be used to detect abnormal disposal of waste or illegal dumping.

The principles of simulation modeling would complement machine learning through the modelling of real-world waste management scenarios under varying assumptions. This is done using techniques such as agent-based modeling or system dynamics simulation which can be used to consider policy changes, infrastructure upgrades or even public awareness campaigns. Integrating machine learning with simulation will provide a mixed model that will enable the accuracy of data and discovery by situation.

5.3 Varying Carbon and Energy Forecasting Predictive Models

In addition to predicting the amount of waste, predictive modeling can also be used to determine the amount of carbon emission and energy recovery. The predictive models can determine the environmental impact of a particular choice of waste treatment including incineration, composting, and recycling based on the life-cycle assessment (LCA) data and emission factors. Energy recovery models estimate the extent to which renewable energy can be derived by using biodegradable or combustible waste, which helps in decision on the activities of waste-to-energy plants. Figure 3. IoT sensing and predictive analytics workflow of carbon mapping and emission forecasting.

This Fig. 3 uses the workflow which is data-driven to carbon management and emission forecasting. The collection of data is carried out by collecting activity data

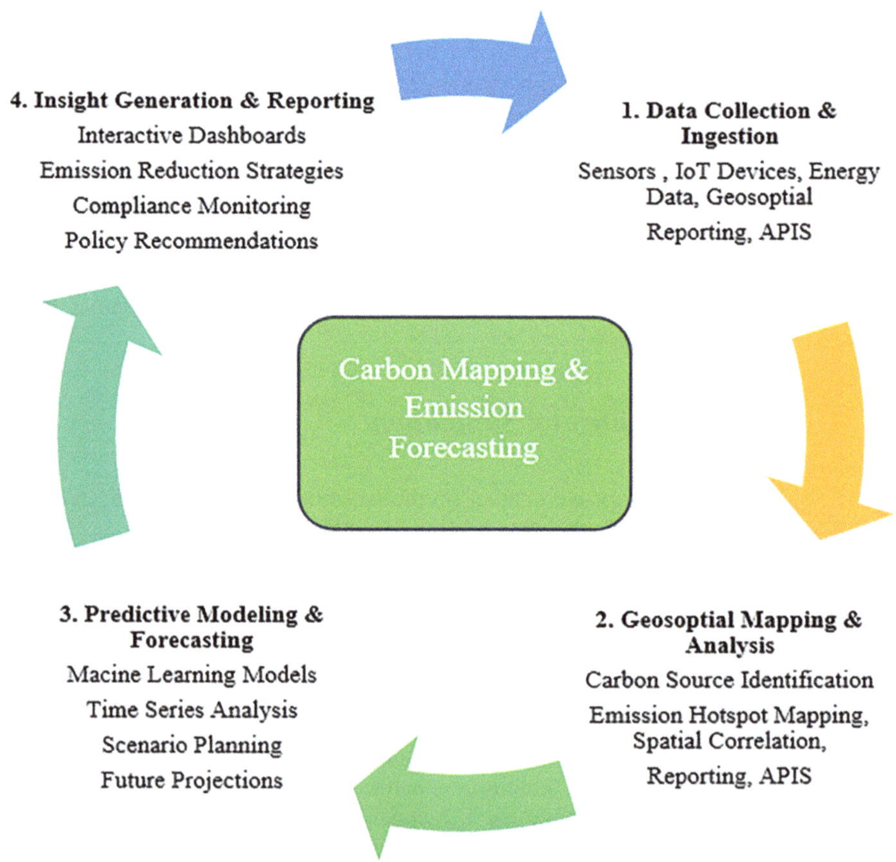

Fig. 3 Workflow of carbon mapping and emission forecasting Author's Own Creation

to form a set of sources such as the energy meters, fuel logs and a supply chain record that usually can be classified through the Scope 1, 2, and 3 emissions. This data is supplemented by geospatial data on satellites and GIS platforms to give the data a spatial context. The process of data processing and analysis involves the mapping tools to visualize carbon sources and sinks and displays the density of emissions in geographic areas. Predictive models are then used to predict the future trends in emissions under diverse policy alternatives or operational changes. Through this workflow, companies and governments are able to make sound strategic choices, strategize on mitigation steps to take, and monitor the progress of carbon mitigation.

The insights can help guide policymakers on the strategies to focus on low-carbon strategies because they determine the best ways of reducing emissions and recovering resources. As an example, a predictive model can suggest that organic waste should be diverted to biogas production plants, and the recyclable objects should be sent to specialized plants to achieve maximum environmental and economic advantages.

5.4 Challenges and Ethical Considerations

Although predictive models have its merits, it is associated with problems that include lack of completeness of data, bias in the models, and uncertainties in the long-term forecasting. Environmental information tends to be different in various regions and is guided by socioeconomic conditions and, therefore, is hard to standardize. Ethics are a consideration as well, such as data privacy, algorithmic transparency and community consent cannot be ignored in large scale deployment of predictive systems.

In order to remove these limitations, adaptive learning mechanisms ought to be incorporated in the models that would correct themselves as new entries of data are entered into the model. The joint work of the technologists, environmental scientists, and the policymakers in the various sectors would also help in enhancing the model reliability as well as ensuring that the predictions are made on the basis of the sustainability goals rather than the measurable indicators of efficiency.

5.5 Advantages and Eternal Effect

Predictive modeling is a manner of providing the decision-makers with a vision that will allow them transform waste management into an active system rather than a reactionary system. The long-term outcomes are the reduced dependency on landfill, the simplification of the recycling process, the growth of energy savings and the carbon savings that could be measured. Also, predictive models would help to create environmental resilience, which enables individuals to react to emerging challenges, such as the hyper-urbanization process, shifting climate, and population growth.

Predictive modeling, as an element of waste analytics and carbon mapping, is a three-step trip down the data-driven sustainability—it is feasible to observe in real time, map on a map, and predict the future predictions in one digital ecosystem.

6 Internet of Things and Artificial Intelligence to Control Smart Waste and Carbon

The confluence of the Artificial Intelligence (AI) and the Internet of Things (IoT) has revolutionized the field of environmental management and started offering real-time monitoring, automation and data-based decision support. A combination of these technologies is the key to the smart systems of waste and carbon management, which enables the optimization of the process of the waste collection, recycling procedures, and emission reduction in a dynamic manner. By making them integrated, one can also create a continuous feedback loop where data is being captured and analyzed and acted upon with minimal human involvement which is more efficient and sustainable. Figure 4. IoT and real-time waste and carbon management by artificial intelligence.

This Fig. 4 represents the combination of AI and IoT technologies to simplify the management of the environment in a sustainability cycle. The sensors of the IoT keep monitoring the environmental conditions, such as air and water quality, soil moisture, energy consumption, and generate gigantic real-time data. These data streams are processed by the AI algorithms and pattern recognition and anomaly detection and complex time-series analysis are made. The insights created can be utilized to take an automated or optimized response such as adjusting the energy flow in smart grids, managing irrigation systems based on opportunities offered by soil data, or environmental alerts. The integrated system enhances efficiency, reduces the human error and encourages proactive sustainability practices through the provision of the real time sensing of the intelligent decision-making.

6.1 Waste Monitoring and Data Acquisition by IoT

The IoT devices play a crucial role in the gathering of the environmental data by the waste management networks. Smart bins will measure the fill level and the composition with the help of ultrasonic or weight sensors and the collection trucks will make it possible to have GPS, transferring the route and location data to centralized data collection locations. These interconnected systems reduce the time spent on manual surveillance since they provide real time information on waste production rates, capacity to overflow and equipment availability.

To cite an example, municipal authorities can count on the dashboards based on the IoT to know what bins may be filled in the nearest future and plan the pickups automatically, avoiding the waste of time and fuel that do not have to be spent on

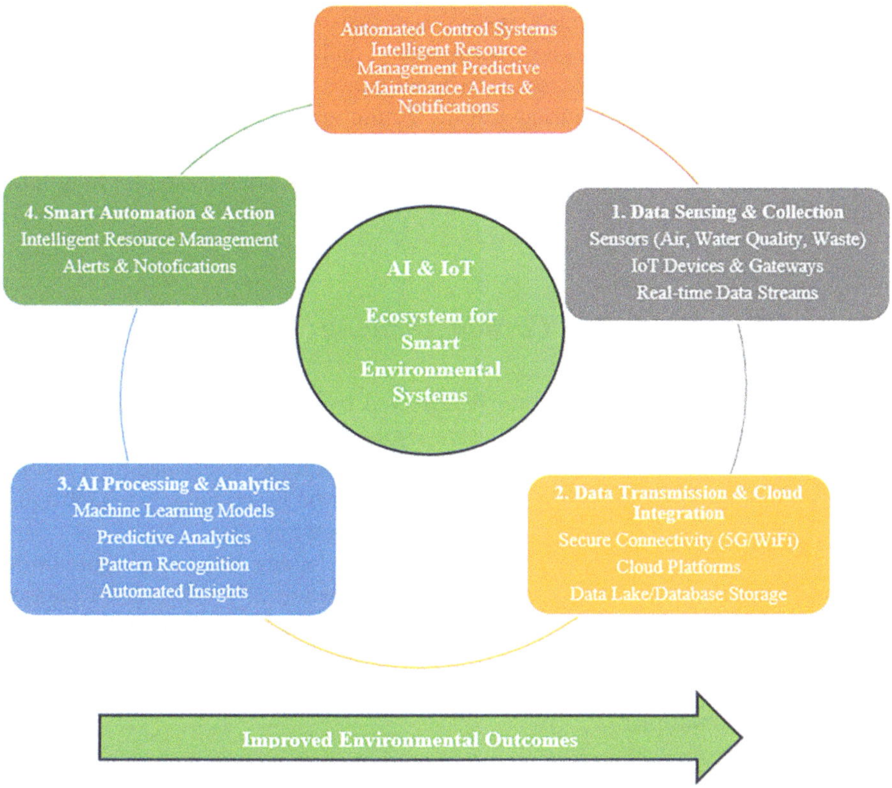

Fig. 4 The smart environmental systems of AI and IoT Author's Own Creation

the trips. Landfills also have sensors that keep track of the quantity of leachate, the quantity of gas released into the environment and temperatures- to view against contamination or harmful leakages. These systems can not only lower the environmental risks, but also secure the optimization of the cost of the operations as it is predicted in its maintenance and efficiency of the routes.

6.2 AI in Intelligent Data Processing and Decision Support

Though the IoT devices generate large volumes of raw information, AI algorithms transform such information into data. The past and current data are evaluated with the help of machine learning algorithms to identify the patterns of waste generation, predict the Carbon emission spikes, and propose interventions. The waste pictures within the smart sorting systems can also be categorized through deep learning models namely convolutional neural networks (CNNs) and therefore can automatize the process of recycling and reduce human error.

The AI-based decision support system also assists the policy makers to model various methods used in waste management and forecast the environmental outcomes. To illustrate this, an artificial intelligence (AI) model could be used to calculate the impact of the electric waste collection vehicles switch on carbon emissions, cost savings, and a positive improvement in the air quality. Through this, this smart processing is an addition to the performance of the operations and the long-term sustainability planning.

6.3 *Intelligent Ai/IoT in Carbon Mapping*

Carbon mapping would significantly be affected by AI and IoT. The IoT sensors that are installed in the urban infrastructure are used to constantly monitor air pollution, temperature, and carbon dioxide. These sensor networks produce high resolution spatiotemporal data which is processed through AI based models to determine emission hotspots and project carbon pattern of dispersion.

Moreover, geospatial analysis systems based on AI along with Geographic Information Systems (GIS) enable the real-time visualization of the emittance data on digital maps. Governments and researchers are able to keep track of carbon trends in real-time and assess the success of mitigation policies using such platforms. The IoT feedback makes these models up to date and responsive to the prevailing environmental conditions.

6.4 *Case Study of Smart Waste Systems and Carbon Systems*

A number of cities in the world have implemented integrated AI-IoT systems of sustainable waste and carbon management.

Barcelo, Spain, has used IoT sensors to track the levels of waste containers and has used AI algorithms to plan optimal collection plans to cut down on fuel costs and CO_2 emissions.

The Smart Nation Initiative by Singapore integrates AI-based predictive analytics and the IoT system to organize waste collection, water resources, and carbon tracking simultaneously.

The city of Amsterdam, Netherlands, is the one designed on the principles of real-time carbon data collection using IoT sensors to institute the policy and promote green zones within the urban regions with the most rates of emissions.

These examples illustrate that integration of technology enhances efficiency and environmental resilience in order to transform cities into smart and sustainable ecosystem.

6.5 *Problems and Future Projections*

The issues associated with the implementation of AI and IoT in environmental systems are such as the data privacy, the tremendous costs and incompatibility of the devices produced by the different manufacturers regardless of their potential. It is also important to make sensor networks more secure and even ethical when making AI-based decisions.

In the future, edge computing, data integrity blockchain, and federated learning will be advanced and will enable the resolution of most of these constraints. They can decentralize data processing, create a sense of transparency, and facilitate sustainability work across regions through these technologies. With the maturity of AI and IoT ecosystems, the combination of the two will be central to the carbon–neutral cities and the creation of the circular, data-driven economy.

7 Case Studies and Implementation Frameworks

The implementation of waste analytics, carbon mapping, and predictive modeling as a successful approach is determined by how well the technologies have been applied in practice. Waste management solutions, which are data-driven, are being implemented worldwide in cities, industries, and research facilities to turn around carbon footprints. In this section, several case studies have been chosen and a broad outline of the situation that can be used in the practical application in the environmental setting has been proposed.

7.1 *Case Study 1: Smart Waste Management Seoul, South Korea*

It is also known that Seoul is one of the first cities to have data-driven waste management. The waste bins utilized by the city on an IoT basis contain the sensors to measure the fill status, the type of waste, and the temperature. These bins relay the real-time data to an AI-powered centralized management platform. The algorithms of artificial intelligence predict the requirements of the collection and optimize the truck routes to save a significant amount of fuel and decrease the level of greenhouse gases. Moreover, the city also integrates its waste data set with carbon mapping tools in order to monitor the emission pattern of waste transportation. Such combined strategy has resulted in 20% cut in the price of collection and quantifiable reduction in the CO_2 in the cities. The success of the Seoul model lies in the fact that it was an ideally balanced composition of data collection and analysis in combination with citizen involvement, it should be noted that the role of open data is impossible to overestimate in sustainability programs.

7.2 Case Study 2: Waste-To-Energy Analytics in Stockholm, Sweden

Stockholm has an alternative method that integrates predictive modeling and life-cycle-assessment (LCA) in order to improve its waste-to-energy (WTE) facilities. With predictive analytics, the city can predict the volume and composition of waste that will arrive in the future, so operators can modify processing schedules in order to maximize the production of energy. Carbon mapping systems are used to monitor the real time emissions so that the city is kept within the EU environmental standards. With this combined approach, the city of Stockholm has realized more than 45% of energy recovery of municipal waste that converts waste into renewable energy source. The case provides an example of predictive and carbon-based analytics that may be used to promote economic and environmental sustainability.

7.3 Case Study 3: Urban Resilience in Melbourne, Australia: The Carbon Mapping

Melbourne also applies AI-based carbon mapping to determine the extent of greenhouse gas emissions (transportation, energy consumption, and waste production). The system combines satellite data with IoT sensors with traffic models and displays them in a dashboard in GIS. This data is used by city planners to determine the hotspots of emissions in the area and take local actions to mitigate this problem through the creation of green corridors and the encouragement of the use of low-emission vehicles. The predictive modeling aspect ensures the forecasting of future carbon levels in various climate policy scenarios, which is used as evidence-based planning. The case of Melbourne shows that the data fusion and spatial analytics can enhance the resilience of the urban climate and awareness among the population.

7.4 Data-Driven Sustainability Implementation Structure

In order to repeat the same achievements, organizations and cities may apply a systematic implementation model uniting analytics, technology, and governance. The major stages can be summarised in a framework (Table 1).

Table 1 Data-driven waste and carbon management systems implementation framework

Stage	Goal	Technologies/ techniques	Expected outcome
1. Data Acquisition	Gather real-time and historical waste and carbon	IoT sensors, drones, satellite imaging, smart bins	Accurate and continuous flow of data
2. Data Integration	Integrate heterogeneous data in a single platform	Cloud storage, data fusion techniques, APIs	Centralized, interoperable database
3. Data Analytics	Trends, anomalies, insights analysis	AI algorithms, ML models, statistical analysis	Actionable insights to optimization
4. Predictive Modeling	Trending and forecasting waste and emission trends	Regression models, neural networks, system dynamics	Anticipatory and preventive planning
5. Visualization and Mapping	Presentation of spatial and temporal data	GIS systems, dashboards, carbon maps	Increased transparency and decision making
6. Policy and Action Implementation	Turn knowledge into action plans to sustainability	Decisions, automation, governance infrastructure	Operational efficiency and emission reduction
7. Continuous Monitoring	Measurement and control of improvement models	IoT feedback, adaptive programs	Continuous improvement and resilience

Author's Own Creation

7.5 *International Implementation Lessons Learned*

In the various contexts, some of the common lessons emerge out of these case studies: Interoperability of data is crucial—the systems need to be able to combine data of many sources.

The unity of stakeholders improves the success of the system because the successful development of sustainable solutions should be based on the collaboration of governments, industries, and citizens.

Ethical governance brings about trust and fairness at the AI-driven systems, safeguarding privacy and ensuring data transparency.

It must be scalable and adaptable; structures must adapt to the increase in population, changes in technology and changes in climate.

These lessons reiterate that sustainability is not achievable in the scenario of technology but a combination of intelligent use of data, policy correspondence, and community engagement.

7.6 Towards a Standardized Framework

One should also work out common data protocols and indicators of performance that would make such systems applicable everywhere on the globe. Some of the typical metrics that can enable cities and organizations to benchmark their performance and share experience are waste analytics, carbon emission and predictive accuracy. The international organizations such as the United Nations Environment Programme (UNEP) and the World Economic Forum (WEF) are already putting efforts in encouraging the availability of the frameworks to sustain interoperability and the sustainability reporting. The additional harmonization of data-driven solutions will help to transform the fragmented waste management patterns in societies to integrated, smart, and sustainable ecosystems that will be capable of balancing human development and the well-being of the planet.

8 Policy, Governance and Future Direction

The integration of waste analytics and carbon mapping and predictive modeling into sustainability practice may be regarded as both a technological challenge and governance challenge. To be successfully implemented, it needs to have supportive policies, open data structures, and teaming institutions that have the opportunity to narrow the digital innovation/environmental responsibility divide. This part will discuss the policy environment, governance processes, and future trends that will define the new data-driven environmental resilience. Figure 5 Model of interaction between governance and policy to support ethical and transparent data-driven sustainability.

This Fig. 5 highlights the structural and governance elements required to have an effective data-driven system of sustainability. On the initial stage, the legal framework, compliance standards, and reporting requirements, including environmental, social, and governance (ESG) requirements, are provided by the policy mechanisms. The ethical principles run deep in the system to be responsible in data utilization, protect privacy, and be fair to algorithms. Data governance is an important issue that regulates the quality, security, and accessibility of data that is collected and processed. Through the creation of the ownership regulations, security measures, and quality control, the organizations manage to guarantee the integrity and reliability of insights. These factors combined make it possible to establish a well-organized, responsible, and transparent basis of implementing technology-based sustainability strategies.

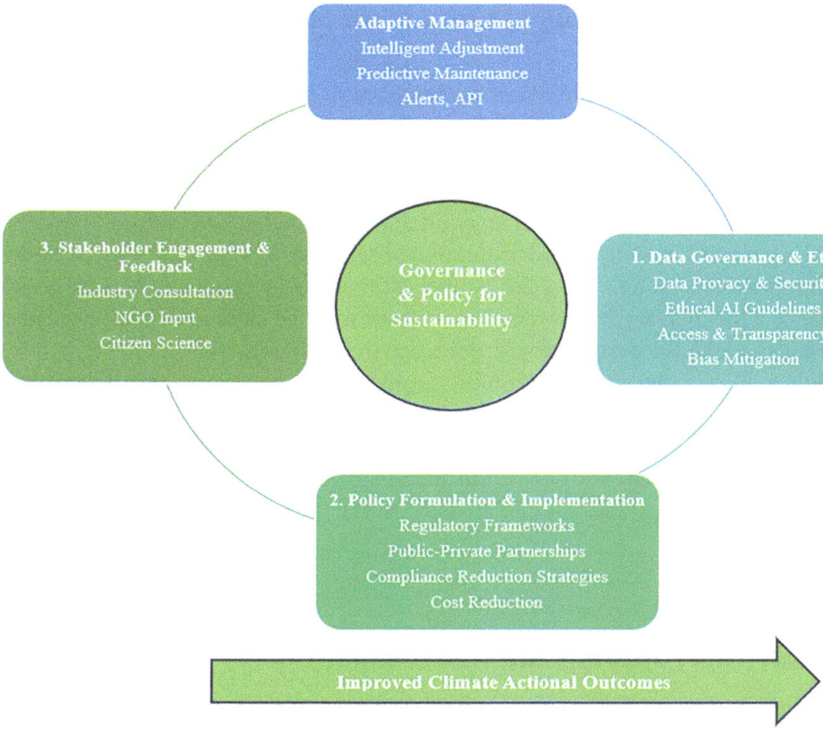

Fig. 5 Governance and policy interaction model Author's Own Creation

8.1 Policy Support for Data-Driven Sustainability

In this section, the policy will support data-driven sustainability. Government policies are of key importance in developing data-based environmental management systems. Open data ecosystem regulations, the use of green digital infrastructure, and the sharing of data across agencies all are critical to achieving success. Countries such as Denmark and Singapore, have also enacted policies of smart cities, instils a need to use IoT and analytics to monitor waste, and the Circular Economy Action Plan of the European Union has connected the digital tools to waste reduction and resource recovery plans. Likewise, the Swachh Bharat Mission of India is gradually becoming a data-rich solid waste management through real-time dash and predictive analytics.

These examples suggest that the most appropriate types of policy constructions are the ones that could unite technological innovation with environmental objectives in a way that a digital transformation would contribute to the creation of climate resilience and sustainable development in the first place.

The ethics and governance are addressed through the following section:

8.2 Governance and Ethical Considerations

With the advent of data-driven systems, they have to be governed in terms of data privacy, transparency in algorithm and fair access. One of the components of sustainability planning is data governance, and personal and business information often includes the data on waste and emissions.

Ethical AI ought to be applied in an equitable, responsible and inclusive government and organizations with regard to automated decision making. One such example is that the predictive models must not be skewed towards favoring a particular group of people or a locality. Its data ownership policies must be clear and clear of misuse and must also make the people have confidence in its policies.

Moreover, there is the concept of multi-stakeholder governance where the government, industries, academia and people are in charge of responsibilities, that make a balanced ecosystem towards innovation. Citizen participation can help enhance transparency and transparency can be enhanced through transparent dashboards and environmental reporting, and this can have a positive impact on sustainable behavior on a community level.

8.3 Knowledge, Sharing and Institutional Collaboration

The resiliency towards the environment must be inter-institutional and trans-national. The collaborative governance models guide the process of sharing knowledge and learning between the cities, industries, and research institutes. Some of the international programs that allow local governments to collaborate in the global arena to use analytics and predictive tools in realizing carbon neutrality are sustainability business programs such as ICLEI (Local Governments for Sustainability) and C40 Cities Climate Leadership Group. Another way in which innovation is promoted is through the co-creation of open data platforms and research repositories to make available solutions to policymakers and other interested developers. The networks offer replicable and scalable sustainability models which can be applied in other regional environments owing to the fact that they utilize interoperable data standards.

8.4 Future Directions in Data-Driven Environmental Systems

The intelligent automation, decentralized computing, and integrative sustainability analytics is waste and carbon management in the future. Such new technologies as edge computing, digital twins, and blockchain based systems of data integrity will be used to improve real-time monitoring and accountability. It will also increase reactivity in garbage collection and emissions in real-time, in which edge computing will decrease the transaction time by handling facts in their most frequent locations.

With the help of digital twins, the virtual models of a physical system, the cities will be capable of modeling the waste and the carbon situation in the virtual environment and implement the real-life policies. The blockchain technology will provide transparent and flawless waste trade documents, recycling credits and carbon credits and will provide integrity in the case of environmental reporting. Moreover, artificial intelligence coupled with environmental economics will make it possible to introduce dynamic pricing of garbage collection and the use of a carbon exchange and provide monetary incentives with environmental results.

8.5 Annual Report 2010/11 Sustainable Governance Roadmap

The main aspects, which will be incorporated in a comprehensive roadmap of governance, are the following to make it successful in the long-term:

Policy Integration: integrate nationally the climate and digital policy, and the waste and carbon management policy.

Capacity Building: Train the municipal staff, researchers and communities on the issue of data literacy and implementation of sustainable technology.

Ethical Frameworks: Formulate certain principles of operation of AI, data sharing and community interaction.

Performance Observation: Develop the key performance indicators (KPIs) based on the waste reduction, emission control and system effectiveness.

Funding and Incentives: The organizations will encourage investment in green technologies and to use those organizations that exhibit sustainability.

With such a mixed roadmap the governments will be in a position to phase out of the independent sustainability initiatives to more systemic environmental governance in such a way that it creates resiliency not only on the local but also on the international front. A sustainable and resilient future has to do with the relative combination of technology, information and policy. Waste analytics could offer an insight on what is happening and the carbon mapping could offer further insight on what is happening and where predictive modeling could come in and help in predicting the next thing that will happen. The combination of them creates an entire system of information-based environmental intelligence.

Nevertheless, such systems cannot just be successful due to innovation but also due to good governance, ethical use of information and participatory activities. With the onset of the digital era, there is a necessity to make sure that the technological progress may be helpful to the earth, not to the human beings. The building of strong ecosystems on informed decisions will mark the actual start of an intelligent, smart and green future.

9 Conclusion

The chapter also discussed the changing nature of the field of sustainability and environmental resilience through data-driven technologies. Waste analytics, carbon mapping, and predictive modelling have caused modern environmental systems to transform, but in a different way whereby the system is focused on proactive intelligence, as opposed to reactive management. Having integrated IoT, AI, big data, and cloud computing, it is possible to optimize the performance of the environmental conditions and observe it in the industries at any time.

First, the chapter discussed the strength of waste analytics to uncover the hitherto unexplored world of waste production, separation, and recycling inefficiency. Carbon mapping is one of the spatial intelligence applications, which links emissions to specific geographic locations and actions, and provides policy and urban planning with valuable data. The strategic layer that was discovered can be useful in making decisions that would count in the context of raw data is predictive modeling that enables the early intervention, forecasting trends and simulating impacts.

Following sections mentioned the application of machine learning and AI in the growing level of accuracy and automation. The regression ensembles and the reinforcement learning agent models of the deep neural networks were reported to play a critical role in understanding dynamic trends in the waste movement and carbon emission. The chapter also featured data visualization and digital twins that offer real-time testing of the situations and performance analysis.

Other importance of governance, ethics, and institutional cooperation in addition to technology was also found in the chapter. The effectiveness of data-driven sustainability could not have been achieved without a collection of guidelines that will promote the notion of data openness, community participation, and ethics of AI use. The policies are founded on systemic change in accordance with the UN Sustainable Development Goals (SDGs), and national climate action plans.

And finally, the chapter ended with the conclusion that sustainability lies in the future of convergence, the seamless integration process of environmental science, data technology, and governance. Predictive intelligence and moral innovation will enable communities to advance to a sustainable future in which the communities are strong, inclusive, and responsible.

References

Agrawal R, Gupta S (2025) A systematic review to integrate big data analytics into high-performance transportation systems. Transp Res Part e Logist Transp Rev 150:1–15. https://doi.org/10.1016/j.tre.2025.102268

Chen Y, Wang H (2024) Data-driven predictive analysis and sustainable waste management in the construction industry. Sustainability 16(10):4169. https://doi.org/10.3390/su16104169

Dye AW, Smith JK (2024) Carbon, climate, and natural disturbance: a review of forest carbon risk under climate change. Carbon Balanc Manag 19(1):1–15. https://cbmjournal.biomedcentral.com/articles/https://doi.org/10.1186/s13021-024-00282-0

Fang B, Zhang Y (2023) Artificial intelligence for waste management in smart cities. National library of medicine. https://pmc.ncbi.nlm.nih.gov/articles/PMC10169138/

Fatorachian H, Alizadeh M (2025) An evaluation of predictive waste management models. MDPI Electron 15(2):770. https://doi.org/10.3390/electronics15020770

Gupta R, Singh S (2024) Machine learning for sustainable organic waste treatment. Nat Sustain. https://www.nature.com/articles/s44296-024-00009-9

Hua J, Li Z (2025) Artificial intelligence for calculating and predicting building carbon emissions. Build Environ 208:108–120. https://link.springer.com/article/https://doi.org/10.1007/s10311-024-01799-z

Olawumi MA, Chan DW (2025) AI-driven predictive models for sustainability. Sci Total Environ 850:157–165. https://doi.org/10.1016/j.scitotenv.2023.157165

Rajaram P, Kumar V (2025) Data-driven predictive models for sustainable smart buildings. J Build Perform 16(1):1–10. https://doi.org/10.1016/j.jobe.2025.100198

Sustainable Power Forecasting: A Machine Learning Approach to Household Energy Optimization

Sameer Jain

Abstract The transition toward sustainable household energy management has become a pressing priority due to rising electricity costs, climate change concerns, and the proliferation of digital appliances. Traditional monitoring techniques often fail to capture dynamic, time-varying patterns of household consumption, limiting their ability to support optimization and demand forecasting. Recent advances in machine learning (ML) offer scalable tools to analyze granular energy data, identify influential consumption drivers, and generate accurate forecasts for demand-side management. This study develops a predictive framework using the UCI Household Power Consumption dataset (2006–2010), integrating exploratory data analysis (EDA), feature engineering, multicollinearity checks, and regression modeling. The results show that Stochastic Gradient Descent (SGD) regression achieves higher scalability and performance ($R2 = 0.9520$, $RMSE = 0.236$) while Ordinary Least Squares (OLS) regression offers a good baseline ($R2 = 0.9477$). The disproportionate contribution of appliances like air conditioners and water heaters to peak loads is further highlighted by sub-metering study. The results highlight the potential of lightweight machine learning models for real-time forecasting in environments with limited resources, providing useful information for utilities, consumers, and legislators to improve energy efficiency and support sustainability objectives.

Keywords Energy optimization · Machine learning · Smart analytics · Sustainability · Time-series forecasting

1 Introduction

The shift to sustainable energy use has emerged as a major worldwide issue. Household power demand is reaching previously unheard-of heights due to rapid urbanization, digitalization of lifestyles, and growing reliance on electronic appliances (Hong

S. Jain (✉)
Faculty at NICMAR Business School, NICMAR University, Pune, India
e-mail: samejain1983@gmail.com

R. Singh et al. (eds.), *Data-Driven Sustainability: Harnessing Technology for A Greener Future and Environmental Resilience*, World Sustainability Series,
https://doi.org/10.1007/978-3-032-13412-7_15

and Fan 2016). Policymakers, researchers, and households are being compelled to embrace more sophisticated approaches to energy management at the same time due to growing electricity prices and concerns about climate change (Zhang et al. 2018; Raza and Khosravi 2015). Ineffective load balancing and energy management occur from household power usage, which is extremely dynamic and time-dependent and cannot be sufficiently handled by traditional monitoring and prediction techniques (Wang et al.). Recent advancements in AI and ML, however, have unlocked new potential for accurate consumption analysis, dependable demand forecasting and predictive optimization—the bedrock of successful demand management. Ahmad and Chen (2018) and Li and Wan (2020), experts in these technologies stress the possible impacts they may have on practices of energy forecasting. Various data-driven methods such as ensemble learning, ARIMA models, and long short-term memory (LSTM) networks are widely used for forecasting. Similarly, also in the energy domain, regression-based approaches have shown to yield good prediction results (Fischler and Bolles 1981; Finn 2018; Lago et al. 2018; Kong et al. 2019). Deep learning has gained remarkable success in recent years especially in handling extreme large-scale high-dimensional data, where traditional statistical methods often face challenges; e.g., the analysis of smart meter readings (Mocanu et al. 2016). Based on these observations, in this paper we develop a model to predict household active power consumption using the UCI Household Power Consumption dataset (2006–2010) (UCI 2019). The framework of the study encompasses EDA, feature engineering, multicollinearity testing and regression modeling through regularization. By integrating predictive modeling approaches into sustainability-driven decision making (Kleinberg 1999), we offer the following two-fold interdisciplinary contribution to emergent green AI research in energy management and some of the identified research problems which are as follows:

1. Identify household factors that most influence power consumption.
2. Develop predictive models to forecast household energy demand with high accuracy.
3. Derive managerial and policy insights to support sustainable consumption and smart grid adoption.

2 Literature Review

The accurate prediction of household electricity consumption has long been a subject of interest for both energy researchers and computing specialists. Early forecasting studies relied heavily on statistical tools such as regression and classical time-series models. Although these methods laid the groundwork for demand forecasting, they were not very effective at adjusting to the dynamic and variable character of residential consumption (Hong and Fan 2016; Fischler and Bolles 1981). Later, probabilistic forecasting approaches were developed to fill this gap, providing a means of incorporating uncertainty into demand prediction—a problem that has grown more pressing as renewable energy sources are integrated (Zhang et al. 2018).

Machine learning (ML) techniques have become more popular in the last ten years since they clearly outperform conventional approaches. For instance, Raza and Khosravi (2015) evaluated AI-driven load forecasting models and emphasized how well they could represent intricate, nonlinear demand patterns. Ahmad and Chen (2018) examined the application of machine learning (ML) in energy decision-making and highlighted how it may help with sustainability-focused tactics. This field was further developed by neural network-based research, where scientists used deep learning to forecast energy use at the building and household levels (Mocanu et al. 2016). Building loads and short-term consumption have been successfully estimated using deep neural networks in particular (Marino et al. 2016).

The importance of sophisticated ML approaches has also been reaffirmed by comparative research. Long short-term memory (LSTM) models, for example, routinely outperform traditional forecasting frameworks in load prediction research (Kong et al. 2019). However, data from other studies demonstrates that traditional methods like regression and time-series analysis are still useful in applications like demand estimation and forecasting power prices (Lago et al. 2018).

At the same time, smart grid research has established machine learning (ML) as a key facilitator of predictive analytics and demand-side management. Li and Wan (2020) emphasized the role of deep learning in enhancing forecasting accuracy for smart grids, while Wang et al. (2019) reviewed applications and challenges of smart meter data analytics. These studies reinforce the centrality of granular data in unlocking optimization potential. Yilmaz et al. (2006) originally introduced object tracking methodologies, but their emphasis on dynamic temporal patterns has indirectly influenced later energy-focused time-series applications.

At the household level, feature engineering techniques such as lag variables, rolling statistics, and cyclical encodings have improved time-series forecasting performance (Jerald 2015). Sub-metering analysis, as introduced through the UCI Household Power Consumption dataset (UCI 2019), has enabled researchers to differentiate the influence of specific appliances, particularly heating and cooling systems, which dominate peak loads. This aligns with calls for "green AI" approaches that balance predictive accuracy with computational efficiency to support sustainable energy transitions (Kleinberg 1999).

Collectively, the literature demonstrates that while deep learning models achieve high forecasting accuracy, their computational cost and lack of interpretability limit scalability in real-world household deployments. Fewer studies have integrated appliance-level sub-metering, lightweight regression methods, and rigorous feature engineering into a unified framework. Addressing this gap, the present study extends prior work by combining exploratory data analysis (EDA), multicollinearity checks (VIF), and scalable regression optimization (SGD) to deliver an interpretable yet efficient forecasting model.

Research Gap

Significant gaps still exist despite the fact that a large amount of research has looked at energy forecasting using both statistical models and deep learning techniques. Most of the literature that has already been written has focused on complex architectures like neural networks and long short-term memory (LSTM) models (Kong et al. 2019; Mocanu et al. 2016). Although these approaches achieve impressive accuracy, they frequently require high computational resources. In contrast, lightweight regression models can occasionally deliver comparable predictive power at significantly lower processing costs, but their potential has received relatively little attention. Another limitation in previous studies is the level of analysis—most studies have treated household electricity demand as a single aggregated signal, ignoring the value of sub-metered data that captures appliance-level consumption. This increased granularity can facilitate more focused energy management tactics and offer deeper insights into usage patterns (Wang et al. 2019).

Methodological approaches provide another area of unmet research need. Although optimization techniques, multicollinearity testing, and feature engineering are well-established in more general data science applications, their application in energy forecasting has received less attention. The current study uses regression-based optimization, rigorous feature selection methods, and sub-metered consumption data to address these problems. The objective is to provide a forecasting framework that strikes a compromise between efficiency and accuracy, providing useful advantages for smart grid stakeholders as well as residential customers.

3 Methodology

The methodology followed a structured pipeline.

3.1 Data Acquisition

- Source: UCI Household Power Consumption dataset (2006–2010).
- Size: 2,075,259 rows, 9 attributes.

Dataset Information: The dataset used in this study is the UCI Household Power use Dataset, which includes minute-by-minute household energy use data from December 2006 to November 2010.

- **Total Records (Rows)**: 2,075,259
- **Total Features (Columns)**: 9
- **Granularity**: One-minute intervals
- **Duration**: 4 years (2006–2010)
- **Target Variable**: Global_active_power (measured in kilowatts, kW)

(a) Dataset Columns and Descriptions

This study utilized the UCI Household Power Consumption Dataset, which provides minute-level recordings of residential electricity usage from December

2006 to November 2010. Comprising 2,075,259 observations across nine variables, it represents one of the most comprehensive publicly available datasets for analyzing household energy consumption. A thorough analysis of both short-term variations and long-term trends in power demand is made possible by the dataset's high temporal resolution, which spans almost four years. A temporal index is created by combining date and time data, and it is utilized for feature engineering and resampling. Global Active Power (in kilowatts), which measures the household's overall electrical consumption, and Global Reactive Power (in kilowatts), which measures non-working power, are the two main power metrics included in the dataset. Global Intensity (in amperes) and Voltage (in volts) are additional electrical characteristics that display the average voltage and current drawn by the household at each minute interval.

To aid with appliance-level analysis, the dataset provides three sub-metered variables, each of which is quantified in watt-hours. Examples of kitchen appliances that are subject to sub_metering_1 include the microwave and dishwasher. Sub_metering_2 monitors the energy consumption of laundry appliances, including washers and dryers. Sub_metering_3 records heating and cooling equipment, such as air conditioners and electric water heaters. With the use of these sub-metered measurements, one may determine which appliances consume the most electricity and obtain a more comprehensive understanding of household energy patterns. Descriptive statistics showed that the average worldwide active power consumption of the home was approximately 1.09 kilowatts, with occasional spikes exceeding 11 kW during periods of high demand. A steady supply was indicated by the voltage, which varied between 233 and 254 V and stayed mostly constant at 240 V.

With a highest recorded value of 48 amperes and an average of 4.63 amperes, the global intensity demonstrated the fluctuations in household load over time. With a mean of 6.46 W-hours, Sub_metering_3 led the sub-metered categories in home energy usage, indicating a sizable portion of heating and cooling loads. In contrast, Sub_metering_1 (kitchen) and Sub_metering_2 (laundry) recorded sporadic peaks of up to 77 and 80 W-hours, respectively, but their average consumption was very low at roughly 1 W-hour each.

All things considered; the dataset offers a solid foundation for simulating residential power use. Daily, weekly, and seasonal patterns in household energy use can be thoroughly examined thanks to its fine-grained temporal structure and global and appliance-level observations. This makes it ideal for creating machine learning models that enable intelligent energy management systems and forecast usage (Table 1).

3.2 Data Preprocessing

Correlation Analysis: To investigate relationships among household energy variables, a correlation matrix was computed. The results are summarized as follows:

Table 1 Dataset variables and descriptions (author's own creation)

Category	Variable	Data type	Unit/scale	Brief description
Date-time	Date	Object	dd/mm/yyyy	Recording date
	Time	Object	hh:mm:ss	Recording time
Power consumption	Global_active_power	Float	Kilowatts (kW)	Total household active power consumption
	Global_reactive_power	Float	Kilowatts (kW)	Total reactive power consumption
Electrical load	Voltage	Float	Volts (V)	Average household voltage
	Global_intensity	Float	Amperes (A)	Average household current drawn
Sub-Metered loads	Sub_metering_1	Float	Watt-hours (Wh)	Energy usage by kitchen appliances (dishwasher, microwave, etc.)
	Sub_metering_2	Float	Watt-hours (Wh)	Energy usage by laundry appliances (washing machine, dryer, etc.)
	Sub_metering_3	Float	Watt-hours (Wh)	Energy usage by heating/cooling (electric water-heater, AC)

The correlation analysis highlights several important relationships among the household electricity consumption variables. A very strong positive correlation (0.999) is observed between Global Active Power and Global Intensity, which is expected as both represent household load measured in different units (kilowatts and amperes, respectively). In contrast, Global Active Power shows a moderate negative correlation (–0.400) with Voltage, indicating that higher voltage levels are generally associated with slightly reduced power consumption, suggesting a stabilizing effect of the supply system. Examining the sub-metering variables, Sub_metering_3 (heating and cooling appliances) exhibits a strong positive correlation (0.639) with Global Active Power, while Sub_metering_1 (kitchen appliances) and Sub_metering_2 (laundry appliances) displays moderate positive correlations (0.484 and 0.435, respectively), reflecting their contribution to overall household consumption. Global Reactive Power shows relatively weak associations with other variables (ranging from 0.09 to 0.26), apart from a small positive correlation with Global Intensity (0.266), suggesting that non-working power has limited influence on appliance-specific consumption. Voltage demonstrates consistently negative correlations with most features (between –0.112 and –0.411), reinforcing the inverse relationship between supply voltage fluctuations and household demand. Finally, the cross-correlations among the three sub-metering categories are very weak (between 0.05 and 0.10), indicating that kitchen, laundry, and heating/cooling appliances operate largely independently, thereby reflecting diverse usage patterns within the household (Fig. 1).

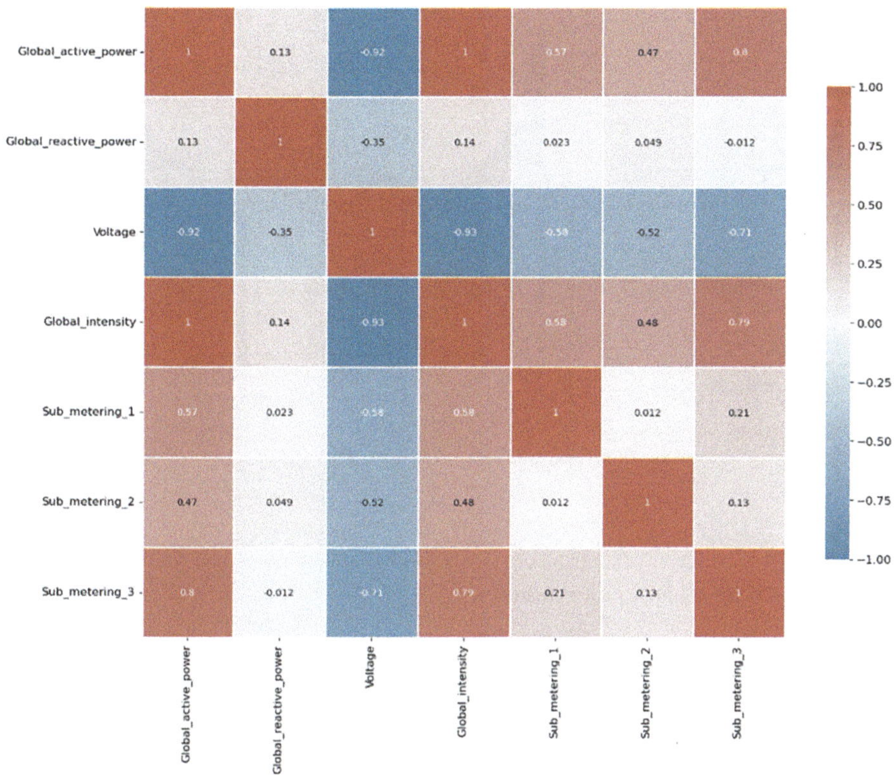

Fig. 1 Correlation analysis (author's own creation)

Visualize Data:

The visualization of the dataset using histograms as shown in Fig. 2 provides insights into the distribution of key household energy variables. Global Active Power and Global Intensity display right-skewed distributions, with most values concentrated at lower levels and occasional peaks representing high consumption periods. Global Reactive Power also shows a skewed distribution, reflecting its limited role in overall energy demand. In contrast, Voltage follows an approximately normal distribution centered around 240 V, indicating stability in the household power supply. The sub-metering variables (Sub_metering_1, Sub_metering_2, and Sub_metering_3) reveal that most appliance-level energy consumption is low, though Sub_metering_3 exhibits a wider spread with distinct peaks, highlighting heating and cooling loads as significant contributors to household energy use. Overall, the histograms confirm that while most household energy usage is concentrated at lower levels, occasional spikes in both global and appliance-specific loads reflect diverse consumption behaviors.

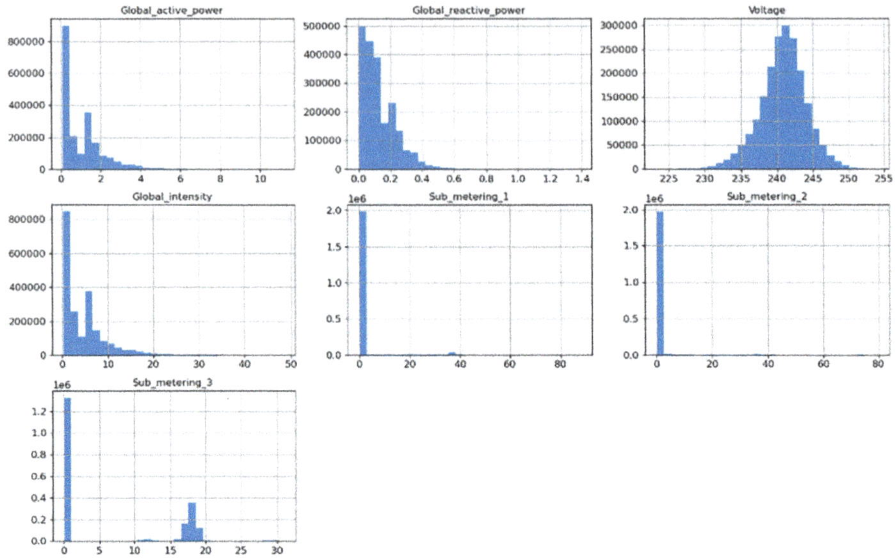

Fig. 2 Visualize data (author's own creation)

3.3 Feature Engineering

The Fig. 3, *Global Active Power over Time*, shows highly fluctuating energy usage patterns across a large timespan. This raw series makes it difficult to detect meaningful daily or seasonal trends directly. To address this, **temporal features** such as hour of the day, weekday/weekend, and month can be extracted. For example, peak loads might occur in the evenings or weekends, which temporal flags can help capture. Furthermore, **lag features** (e.g., power consumption 1 h or 24 h ago) can model short-term dependencies, while **rolling statistics** (mean and standard deviation) smooth out noise to reveal underlying trends. Ratios between different load types (like kitchen vs. heating) highlight specific household activity patterns.

The Fig. 4, *Monthly Average Global Active Power per Year*, clearly shows a **seasonal trend**: consumption is relatively high in winter months (Dec–Feb), decreases through spring and summer (Jun–Aug), and rises again in autumn. This cyclical pattern indicates that time-related features are crucial. Instead of treating months as categorical values, **cyclical encoding (sine/cosine transforms) ** ensures that December and January are close in representation, preserving seasonality. By interpreting these diagrams together, it becomes evident that both **short-term variability** and **long-term seasonal effects** must be captured through engineered features to improve forecasting accuracy.

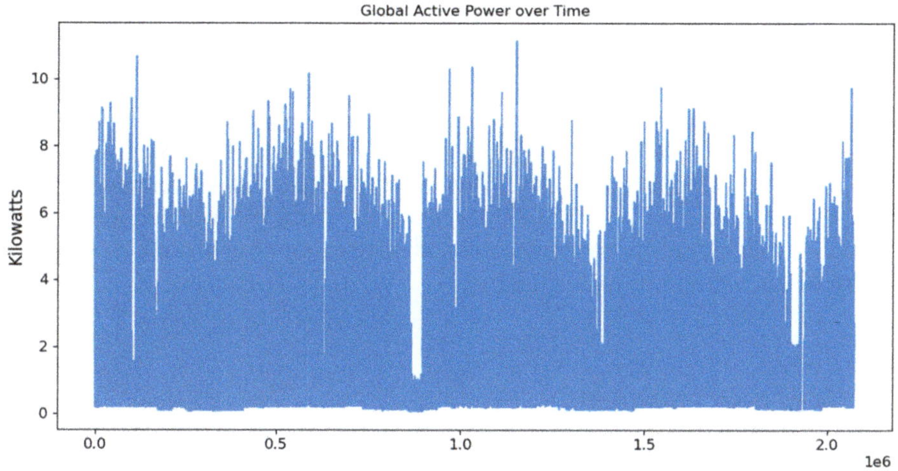

Fig. 3 Global active power over time versus kilowatts

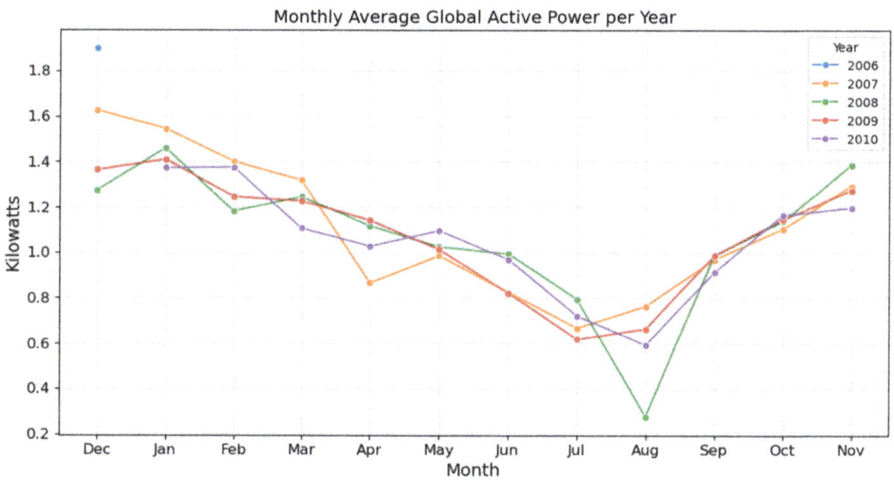

Fig. 4 Monthly average global active power per year versus kilowatts. (author's own creation)

3.4 Multicollinearity Treatment

To ensure that the predictive model does not suffer from multicollinearity, we conducted a Variance Inflation Factor (VIF) analysis on the dataset. Initially, some variables showed strong correlations, such as Global_intensity being almost linearly dependent on Global_active_power. To reduce redundancy, we removed Global_intensity from the dataset, as it did not provide additional information beyond what was already captured. Instead, we kept Global_reactive_power, since it contributes unique insights into energy consumption behavior and is more informative for the model.

After dropping the redundant variable, the rechecked VIF values showed that all remaining features had acceptable levels of multicollinearity (VIF < 10). For example, Global_active_power had a VIF of 7.57, Voltage 3.06, and Sub_metering_ 1–3 ranged between 1.55 and 3.44, all within safe limits. This treatment ensures that the features retained are not overly collinear, allowing the model to make more stable and reliable predictions without inflated standard errors or distorted coefficient estimates.

3.5 Model Development

We begin with Linear Regression (Ordinary Least Squares, OLS) as the baseline model. OLS directly computes the best-fit parameters by solving the normal equation, ensuring that the cost function is minimized over the training set. The baseline model achieved an MSE of 0.0523 and an R^2 of 0.9477, indicating a strong fit and capturing most of the variance in the training data. While simple and efficient, OLS can struggle with very large datasets or highly correlated features.

To improve scalability and flexibility, we moved to Stochastic Gradient Descent (SGDRegressor) as shown in Fig. 5. Unlike OLS, SGD updates parameters iteratively, making it computationally efficient for large datasets. The learning curve for SGDRegressor shows that both training RMSE and validation RMSE as shown in Fig. 6 decrease steadily as more samples are added, eventually converging around 0.23–0.24 RMSE. The validation score consistently remains close to the training score, with a small gap indicating low risk of overfitting. Furthermore, the shaded regions narrow with increasing samples, confirming more stable estimates with larger datasets.

We further applied hyperparameter tuning and early stopping. By monitoring validation error during training, we identified the optimal stopping point around epoch 1708, preventing unnecessary training and reducing overfitting. The final tuned model maintained strong generalization performance, showing that SGD with proper tuning can match or even outperform closed-form OLS solutions on large-scale problems while being more efficient in terms of computation.

3.6 Evaluation Metrics

To assess model performance, we used three key **evaluation metrics**:

* **Root Mean Squared Error (RMSE): ** Measures the standard deviation of prediction errors. A lower **RMSE (0.1760)** indicates that predictions are close to the actual values.

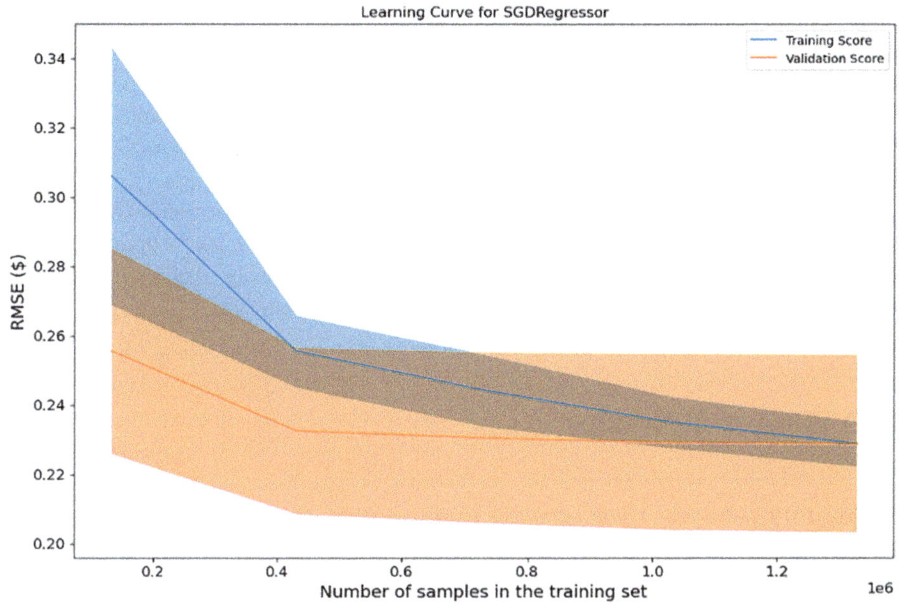

Fig. 5 Learning curve for SGD Regressor

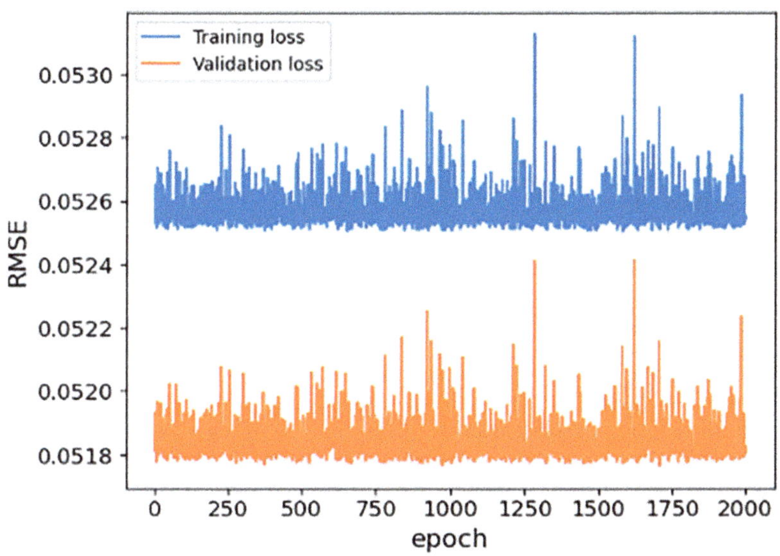

Fig. 6 Epoch versus RMSE (author's own creation)

* **Mean Absolute Error (MAE):** Captures the average magnitude of errors without considering direction. The **MAE of 0.0876** confirms small average deviations between predictions and actuals.

* **R^2 Score: ** Explains the proportion of variance in the target variable captured by the model. An **R^2 of 0.9520** shows that the model explains about 95% of the variability, demonstrating high predictive power.

In addition, **learning curves** (training vs. validation performance) help detect **overfitting** (training error much lower than validation error) or **underfitting** (both errors high). In our case, the curves converged closely, suggesting a good balance between bias and variance with minimal overfitting risk.

4 Analysis and Results

Exploratory Analysis: Clear seasonal peaks were observed during winter months, reflecting heating loads from Sub_metering_3. Daily consumption patterns revealed peaks between 6–9 PM, aligning with evening household activities (cooking, laundry, heating). Weekend consumption was consistently higher than weekdays, indicating lifestyle-driven variations.

Model Performance:

1. Linear Regression (OLS) served as the baseline, achieving $R^2 = 0.9477$, but its performance was limited by dataset scale and feature collinearity.
2. Stochastic Gradient Descent (SGDRegressor), with hyperparameter tuning and early stopping, achieved superior results:
   - $R^2 = 0.9520$
   - RMSE = 0.1760
   - MAE = 0.0876

These metrics confirm that the SGD model provided highly accurate forecasts while being computationally efficient for large-scale data.

Key Drivers of Energy Consumption:

- Heating/AC systems (Sub_metering_3) were the dominant contributors to household electricity demand, aligning with the seasonal peaks observed in EDA.
- Kitchen appliances (Sub_metering_1) and laundry appliances (Sub_metering_2) contributed minimally in comparison, though they showed short-term spikes in usage.
- Temporal features (hour, weekday/weekend, season) and rolling statistics added predictive strength by capturing both short-term fluctuations and long-term seasonal cycles.

5 Conclusion

This study presented a machine learning framework for household energy demand forecasting, leveraging the UCI Household Power Consumption dataset. The approach discovered consumption patterns at the appliance and household levels by combining regression modeling, feature engineering, and EDA. The findings showed that SGD regression offers improved scalability and accuracy, making it especially appropriate for real-time applications, even though OLS regression offers a trustworthy baseline. Heating and cooling appliances have a disproportionate impact on consumption, according to sub-metering data, which emphasizes the importance of granular monitoring for focused energy optimization.

The limitations of straightforward parametric techniques for extremely dynamic, large-scale time-series data were highlighted by the poor generalization of linear regression, which was used as a baseline model. On the other hand, the Stochastic Gradient Descent (SGD) regressor exhibited great predictive accuracy (R2 = 0.952, RMSE = 0.1760, MAE = 0.0876) when combined with feature engineering (lag variables, rolling statistics, cyclical encodings). The learning curve analysis also demonstrated that the model had less overfitting and a good balance between bias and variance. The findings highlight how crucial it is to incorporate temporal and appliance-level characteristics into demand forecasting models. The results indicate that, practically speaking, homeowners and utility companies can utilize AI-driven predictive systems to estimate demand, optimize appliance usage schedules, cut expenses, and promote sustainable energy use.

To further improve predicting accuracy, future studies could build on this work by adding outside variables including occupancy data, weather, and dynamic pricing signals. Furthermore, investigating more complex deep learning architectures (such as Transformer-based models and LSTM) may yield even more advancements in long-term sequence prediction. Overall, this research contributes to the growing field of green analytics, highlighting the potential of data-driven forecasting to support smarter, more sustainable energy management at the household level.

References

Ahmad T, Chen H (2018) Potential of machine learning in energy forecasting and decision making: a state-of-the-art review. Energies 11(5):1252. https://doi.org/10.3390/en11051252

Dua D, Graff C (2019) UCI machine learning repository: household power consumption dataset (2006–2010). University of California, Irvine, CA. https://archive.ics.uci.edu/ml/datasets/individual+household+electric+power+consumption

Finn C (2018) Learning to learn with gradients. Doctoral dissertation, University of California, Berkeley. https://people.eecs.berkeley.edu/~cbfinn/

Fischler MA, Bolles RC (1981) Random sample consensus: a paradigm for model fitting with applications to image analysis and automated cartography. Commun ACM 24(6):381–395. https://doi.org/10.1145/358669.358692

Hong T, Fan S (2016) Probabilistic electric load forecasting: a tutorial review. Int J Forecast 32(3):914–938. https://doi.org/10.1016/j.ijforecast.2015.11.011

Jerald J (2015) The VR book: human-centered design for virtual reality. ACM Books / Morgan & Claypool. https://doi.org/10.1145/2792790

Kleinberg J (1999) Authoritative sources in a hyperlinked environment. J ACM 46(5):604–632. https://doi.org/10.1145/324133.324140

Kong W, Dong ZY, Jia Y, Hill DJ, Xu Y, Zhang Y (2019) Short-term residential load forecasting based on LSTM recurrent neural network. IEEE Trans Smart Grid 10(1):841–851. https://doi.org/10.1109/TSG.2017.2753802

Lago J, De Ridder F, De Schutter B (2018) Forecasting spot electricity prices: deep learning approaches and empirical comparison of traditional algorithms. Appl Energy 221:386–405. https://doi.org/10.1016/j.apenergy.2018.02.069

Li H, Wan C (2020) Smart grid forecasting with deep learning: a review. IEEE Trans Smart Grid 11(4):3185–3206. https://doi.org/10.1109/TSG.2019.2962587

Marino DL, Amarasinghe K, Manic M (2016) Building energy load forecasting using deep neural networks. In: 2016 International joint conference on neural networks (IJCNN). IEEE, pp 1–6. https://doi.org/10.1109/IJCNN.2016.7727382

Mocanu E, Nguyen PH, Gibescu M, Kling WL (2016) Deep learning for estimating building energy consumption. Sustain Energy Grids Netw 6:91–99. https://doi.org/10.1016/j.segan.2016.02.005

Raza MQ, Khosravi A (2015) A review on artificial intelligence-based load demand forecasting techniques for smart grid and buildings. Renew Sustain Energy Rev 50:1352–1372. https://doi.org/10.1016/j.rser.2015.04.065

Wang Y, Chen Q, Hong T, Kang C (2019) Review of smart meter data analytics: applications, methodologies, and challenges. IEEE Trans Smart Grid 10(3):3125–3148. https://doi.org/10.1109/TSG.2018.2818167

Yilmaz A, Javed O, Shah M (2006) Object tracking: a survey. ACM Comput Surv 38(4):Article 13. https://doi.org/10.1145/1177352.1177355

Zhang Y, Wang J, Wang X (2018) Review on probabilistic forecasting of renewable energy: challenges and opportunities. Renew Sustain Energy Rev 81:1561–1578. https://doi.org/10.1016/j.rser.2017.05.229

Ethics, Governance, and Inclusive Digital Transformation

Smart Technologies for a Sustainable Future: Enabling Reuse and Recycling in the Digital Circular Economy

Rubee Singh, Seema Rani, Akshay Singh, Walter Leal, and Vikas Kumar

Abstract To achieve the sustainability goals a shift from regular system of "take–make–dispose" economy towards a circular economy (CE) model is required for climate resilience. Under this paradigm, nowadays the CE initiative are used alongside the latest digital technologies (DTs) with circular economy principles is prominent which is referred digital circular economy (DCE). This chapter explores these innovative technologies such as the Internet of Things (IoT), artificial intelligence (AI), blockchain, and big data, based DCE models are highly relevant to maximize reuse, recycling, and resource efficiency. By enabling smarter waste tracking, automated sorting, product lifecycle monitoring, and transparent recycling chains, these models are bridging critical gaps in resource management. Based on the secondary data sources the chapter offers theoretical foundations of circular economy and digital transformation along with the digital circular economy. It highlights how emerging technologies enables smart waste management in modern ecosystems through reverse logistics, waste management, and e-waste recycling to demonstrate their impact on resource management systems. Furthermore, the chapter details the benefits of adopting DCE practices for triple bottom line and also acknowledges challenges in the implementation of DCE based models in current scenarios. Finally, implications are provided for the application of DCE models for adopting digital technologies

R. Singh (✉)
Institute of Business Management, GLA University, Mathura, India
e-mail: dr.rubeerajput@gmail.com

S. Rani
Chandigarh School of Business, Chandigarh Group of Colleges, Jhanjheri, India

A. Singh
School of Management Studies, Indra Gandhi National Open University, New Delhi, India

W. Leal
Faculty of Life Sciences, Hamburg University of Applied Sciences, Hamburg, Germany
e-mail: walter.leal2@haw-hamburg.de

V. Kumar
University of Portsmouth, Portsmouth, UK
e-mail: Vikas.Kumar@port.ac.uk

R. Singh et al. (eds.), *Data-Driven Sustainability: Harnessing Technology for A Greener Future and Environmental Resilience*, World Sustainability Series,
https://doi.org/10.1007/978-3-032-13412-7_16

in order to achieve a circular and waste-free economy. Towards the end, scope for further research and limitations of present study are discussed.

Keywords Circular economy · Digital technologies · Innovation · Reuse · Recycle · Sustainability

1 Introduction

The global environmental problem necessitates immediate changes in how the natural resources are used to meet societal needs. Climate change effects extend beyond severe environmental repercussions such as rising sea levels, ecosystem degradation, and extreme weather phenomena; it also exacerbates global inequality. Marginalized groups disproportionately endure the consequences of the swiftly escalating CO^2 levels attributed to urbanization in developing nations (Keles 2020). The circular economy (CE) concept offers strategies to decarbonize the construction industry while minimizing waste and resource exploitation through the narrowing, slowing, and closure of resource loops (Kircher et al. 2017). Circular economy strategies for narrowing resource loops encompass minimizing material consumption via resource efficiency and dematerialization. Strategies for slowing resource loops involve prolonging a product's lifespan through ongoing maintenance, reparing reusing, and remanufacturing (Geissdoerfer et al. 2017). In order to close the loops during resource utilization, the reintegration of bio based materials back into the system or recycling synthetic products are common ways. In the last ten years, a notable rise in attention from policymakers, corporations, and researchers is observed about adopting CE as a feasible strategy for promoting a zero carbon and resource-efficient system across the world. This increased recognition of CE approach is linked to the acknowledgment of the need for transitioning to environment friendly techniques to diminish consumption of natural resources, minimize production if waste, and lessen the carbon footprint (Singh et al. 2024a).

Authors have examined the function of digital technologies in improving decision-making, mitigating with adverse effects of climate change, and promoting the adoption of circular economy (Çetin et al. 2021). Digital technologies (DTs) can assist CE in the designing environment friendly products by digitizing the flow of information, facilitating the recording, identification, and alignment of the flow of resources, through effective material information management, and offering a framework for administration and maintenance in DCE applications (Ranta et al. 2021). DTs can substantially improve decision-making in CE practices, referred as digital circular economy, across various sectors by enhancing the efficiency of the resource utilization and facilitating insights based on real-time data (Singh et al. 2025a). Advanced lifecycle analysis (LCA), augmented by IoT devices and big data analytics, provide real-time and precise data on energy consumption, material utilization, and waste production, facilitating tailored sustainability measures (Ingemarsdotter et al. 2019). Digital platforms and applications offering geographic information about objects enhance the quality of planning and surveillance along with optimizing the supply chain by offering access to consolidated data to multiple parties involved (Yu et al. 2022). Visualization technologies and virtual reality platforms improve interaction and communication among stakeholders, while the resource and time requirements is reduced due to automation of life cycle assessment processes decreases time and (Rusch et al. 2023). These DTs reduce negative environmental impacts and enable stakeholders in making data-driven decisions which advances the system towards a more sustainable approach.

Furthermore, digital twins, artificial intelligence algorithms, and automation reveal concealed trends and facilitate ongoing enhancement of sustainability performance. Among the various strategies of the CE, reuse and recycling play an important role in reducing pressure on limited natural resources through proper waste segregation and management. DTs can directly support these strategies by efficient systems of product return, enhanced traceability of materials across different stages of lifecycle, and optimizing recycling processes. For instance, blockchain and IoT technologies facilitate transparent tracking of materials across supply chains it helps in effective reuse or recycle of the products at the end of life cycle (Bressanelli et al. 2020). Artificial intelligence and machine learning algorithms improve sorting processes in recycling facilities, while digital platforms create marketplaces for secondary raw materials and reused products (Nascimento et al. 2019). Thus, DTs provide the technical foundation for scaling up reuse and recycle initiatives within CE models. Nevertheless, the great potential of DTs to facilitate Circular Economy (CE) is still under explored as only a few review studies are there that particularly address how DTs contribute to reuse and recycling practices, and several essential aspects remain inadequately examined in the literature.

In conclusion, there is a necessity for a study that highlights the integration of DTs for reuse and recycle in a CE context, examine their potential and limitations towards a sustainable future. Therefore, this Chapter intend to examine the cutting-edge understanding of how digital technologies might facilitate circular economy through reuse and recycle, encompassing contemporary applications of DTs. Secondly, to ascertain the existing constraints and implementation deficiencies related to digital technologies within the framework of CE. This chapter is consists of different sections. Firstly there is an introduction section and then there is literature review section followed by as research methodology. Next section discusses the implications derived from the reviewed publications on the topic and last section presents principal conclusions, limitations of the study, and avenues for conducting research in future.

2 Literature Review

2.1 Circular Economy

The circular economy, a latest approach, endeavours to alleviate the environmental repercussions of the conventional linear economy while fostering a systemic transformation to a regenerative and environment friendly production framework. The CE approaches leverages various ideas, including eco-industrial development, reverse logistics, closed loop, and zero waste (Nobre and Tavares 2021), to enhance resource utilization, diminish waste, and optimize the use of energy via effective monitoring and maintenance systems (Geissdoerfer et al. 2017). For example in the construction industry, advocating for circular solutions entails the incorporation of technology at every lifecycle phase to prolong material longevity and reduce reliance on natural resources (Talla and McIlwaine 2024). These tactics are primarily informed by the R principles of the circular economy.

2.2 Circular Economy and R-principles

The practical incorporation of CE in enterprises is often defined understood by a system of strategies employed at end of a product's life, known as "3R-principles: reduce, reuse, and recycle", as they are regarded as fundamental operational principles (Ghisellini et al. 2016). Reduction emphasizes on minimizing the use of natural resources and energy, alongside reducing the cost and trash generation but enhancing environmental performance (Ghisellini et al. 2016). Reuse entails remodelling goods or parts to some other intended task, whereas recycling involves converting construction and demolition waste in entirely new components or products, applicable to similar or varied functions (Mahpour 2018).

The R-principles are relevant throughout the in each product's lifetime. Nevertheless, the available studies on CE have not extensively concentrated on explicitly explaining and organizing the essential notion of R-principles, as the literature presents numerous R frameworks and definitions, varying from "3Rs to 10Rs" (Homrich et al. 2018). While the 3Rs denotes "reduce, reuse, recycle," a concept acknowledged mainly by scholars from China; but, it may also pertain to alternative tripartite terms such as "reuse, remanufacture, recycle" and "reduce, recovery, reuse," (Reike et al. 2018). For instance, when remanufacturing is executed at the product-system level, it ultimately allows for the retention of the original product function, the enhancement of this function, or the development of an alternative function.

2.3 Digital Circular Economy (DCE)

Digital technologies, integral to Industry 4.0, signify a transformative change in industry through collecting and analysing the data for establishing integrated networks of people, processes, and products, and value-oriented operation systems (Ranta et al. 2021). Essential DTs namely blockchain, the Internet of Things (IoT), AI, and big data, are vital facilitators of CE, often known as digital circular economy (DCE) (Kristoffersen et al. 2020). The DCE integrates CE strategies such as reduce, reuse, recycle with digital technologies for the optimized utilization of resources and promote sustainable business models. Unlike the regular CE approach which relies on manual processes and incomplete data, DCE empowers data-driven decision-making through collaboration by using digital platforms and real-time monitoring.

DCE is based on an integrated approach where DTs are used to facilitate feedback system to improve the flow of resources continuously (Khan et al. 2024). Here, the DTs act as innovative tools for achieving sustainability goals while minimizing costs and enhancing visibility (Singh et al. 2025b). A common approach is to integrate R-principles with DTs for providing details about the repairability, and recyclability of the products to relevant stakeholders. While incorporating digital twins with CE systems can be effective in the minimization trash generation and maximization of the repurposing of products (Peladarinos et al. 2023). The detailed description of major DTs applications for incorporating reuse and recycle principles in CE context is presented as in the following section.

2.4 Digital Circular Economy for Reuse and Recycle

Due to advanced breakthroughs and technology, robotics and AI can be employed in recycling to achieve Circular Economy (CE). Hoong et al. (2020) conducted a study aimed at identifying the composition of recovered materials, including plastic, glass, natural stones, and wood, by picture-based analysis using deep learning approach. This research constructed a conventional neural network (CNN), a deep neural network utilized for image processing. The labelled directory is utilized for training the CNN. The educational exercise encompasses training and validation. Upon completion of training, the CNN image recognition performance has the ability of attaining 97% accuracy. Deep learning can be employed for classifying and estimating the composition of recyclable objects derived from construction sector (Lux et al. 2023). The authors presented a network based recycling architecture to ascertain the recovered materials and classification from two-dimensional images.

The integration of diverse AI technologies into CE, resulting in numerous prospects for intelligent and ecological farming. A discernible shift is observed the integration of AI with circular economy elements as it significantly contributes to garbage sorting, recovery, and minimization. Tu et al. (2022) integrate artificial intelligence for developing a highly innovative, useful, and effective waste receptacle. The voice identifier system acquires the keywords and categorizes the waste kind. Subsequently, it raises the covering of the designated waste receptacle for precise waste classification. Furthermore, it transmits notifications to the waste management team to prompt timely cleaning and replacement of the bins. Likewise, the predictive capabilities of machine learning facilitate the administration, improvization, and retrieval of various advantageous components out of organic waste. Progress in machine learning for nutrient-dense materials from organic waste to facilitate the development of circular economies in smart cities is examined by Soo et al. (2023). Research indicates that machine learning enhances the scalability, productivity, transparency, and precision of nutrient recovery and reuse technologies, practices, logistics, and dissemination.

The data gathered by digital twins minimizes waste by maximizing usability. It aids users in acquiring the requisite information for the proper handling of products. Peladarinos et al. (2023) conduct a thorough assessment of digital twin implementations that improve farming process. The authors indicated that the farmers can utilize digital technology for real-time data acquisition and application for modelling the land under cultivation, focusing on critical elements such as nature of soil, crop production, and meteorological conditions. Such consolidated efforts improve the overall agricultural practices and output in a sustainable way. Moreover, the digital twins enhances the decision-making process for acquiring second-hand products by employing sorting optimization and minimizing or eliminating inspection procedures. Consequently, it enhances reverse logistics and optimizes material flow through the tracking and management of returns. Furthermore, it enhances and facilitates the effective implementation of reuse and recycling activities by specifying the precise state of products and forecasting the need of repairing. In energy-intensive sectors, some uses of digital transformation and big data are proposed by Ma et al. (2022) for making the manufacturing sector sustainable through smart initiatives. The study also employed product lifecycle approach for demonstrating the usefulness for Chinese enterprises which indicated that digital transformation enhances corporate entrepreneurship.

DTs facilitate the resale and distribution of pre-owned goods. This method use digital tokens to facilitate the reuse of things via resale and the sharing of pre-owned items, exemplified by the second-hand market. Pre-owned products in satisfactory condition can have their lifespan prolonged through subsequent ownership. DTs primarily facilitate the redistribution of discarded products by identifying reusable items and subsequently selling them to second-hand consumers. Tracking and tracing, monitoring, and collecting were deemed beneficial for identifying and providing information regarding products available for reuse, including the current data related to product usage and locations (Vetrova & Ivanova 2021). Innovate, in contrast to its predecessor, primarily facilitates the trading process. The creation of innovative DTs, platforms, tools, and processes can effectively link prospective purchasers interested in second-hand goods to the products.

In the forward supply chain, digital technologies are frequently utilized to enhance manufacturing efficiency and optimize logistical planning to augment corporate earnings. Likewise, enterprises might utilize IoT to get data from the reverse supply chain, including the locations and states of utilized products and waste materials. The data can be analysed to enhance the efficiency of the reverse supply chain, particularly by expediting the recycling system and augmenting tracking of products (Dev et al. 2020; Rajput & Singh 2019; Vetrova and Ivanova 2021). This resemblance is evident when implementing monitoring and trace functions within the reverse supply chain. Digital technologies are progressively employed by organizations to oversee recycling operations for waste minimization, for tracking and tracing goods for route scheduling, monitoring the assortment of waste, and transportation (Rajput and Singh 2019), while diminishing carbon emissions and energy usage. IoT systems play crucial role in the context of remanufacturing, specifically of batteries for electric vehicles, by facilitating the assembling and dissemination of data regarding batteries and their carriers, enabling the monitoring of their condition and the tracking of their lifecycle data.

Furthermore, we found the DTs utilized in reverse supply chains that are rarely present in traditional supply chains. In reverse supply chains, sensors are frequently employed for quantifying garbage and monitor trash-bins for enhancing recycling efforts. DT enhances the ability of a recycling system's regarding identification of problems about utilized products, hence facilitating the disassembly of end-of-life decomposition of used items. The optimization function assesses dispose related decisions, including refurbishment, remanufacturing, or recycling (Zacharaki et al. 2020). The investigation revealed that sorting and classifying can be instrumental in enhancing efficacy of recycling process. This process entails utilizing digital computation to determine end-of-life options, as well as employing automated robotics to segregate the waste physically. (Laskurain-Iturbe et al. 2021).

Shojaei et al. examined how blockchain (BC) as a tool can enhance CE application for achieving sustainability in construction sector, discovering that the current status of individual components can be documented, thereby enabling an advanced strategy for reuse and comprehensive energy saving and locating the material. It allows the user of a system to predict the reuse and recycling resources utilized in the building sector. Jain et al. (2022) indicated that an effectively executed blockchain in the apparel industry can serve as a robust mechanism for product monitoring and traceability. The authors discovered that utilizing blockchain-enabled e-commerce platforms could mitigate the danger of contamination in online secondhand clothes sales, hence enhancing the intention to implement the technology. Huynh (2021) examined digital circular business models in the fashion industry and contended that the blockchain may facilitate recycling across supply chains.

Finally, Paul et al. (2022) contended that the circular tea supply chain might be managed in a transparent and traceable manner by RFID technology, which is facilitated by blockchain. The authors assert that blockchain facilitates the monitoring of the life cycle of reusable and recyclable materials by multiple stakeholders. The primary emphasis of two research has been BC's involvement in recycling operations. Through a multiple-case methodology, Gong et al. (2022) examined the uses of BC for enhancing recycling performance, enabling token system, overseeing the process of waste flow, and augmenting the centralized operation of recycling systems. Besides environmental benefits, it enhances stakeholder engagement, inclusivity, responsibility, and transparency in recycling initiatives. Khadke et al. (2021) emphasized that the efficacy of recycling operations might be markedly improved by the utilization of BC, as it can automate the sorting and gathering of products or components and enable the monitoring of recycled products.

3 Challenges in the Implementation of DCE

The importance of DTs in achieving sustainability is discussed in the preceding sections. Despite outstanding and efficient integration of DTs, their adoption is constrained by some limitations encompassing the conceptualization of the system together with its interrelated models, the verification of functions, surveillance, protection, and outcomes (Singh et al. 2024b). The intricacy of integration and the scale of the application provide some problems. The constraints hindering the utilization of DTs for industrial integration are in there in various form. First, there is data concerns including the collection, efficiency, flow, operation, safety, and accountability (Purcell & Neubauer 2023). Data should remain confidential and safeguarded against digital threats because it is highly prone to manipulation. Updating DTs with the changes in actual operation system presents an additional barrier due to the intricacy of data connectivity and processing. The DT based models produce and examine substantial volumes of data, necessitating robust support for fulfilling extensive and enduring data processing and analytical needs. Additionally, modeling activities that cannot be quantified presents another challenge. System-Related Challenges The sub-criteria problems include technological infrastructure, interdisciplinary collaboration, models' fitness, and expedited interaction (Büyüközkan and Uztürk 2022). The DTs like Big data and IoT necessitate fundamental support for infrastructure development. This also affects the difficulties encountered in establishing uniform standards and procedures. Accelerated and reliable networks, such as "fifth generation (5G)", are crucial for efficiently facilitating connectivity in real-time. These networks are capable of linking a greater number of devices and sensors with enhanced dependability, rapid connectivity, and significantly reduced power usage.

Moreover, the implementation of big data analytics entails considerable upfront expenditures on specialized staff, components, and software. Robust policy support is crucial to advance CE system. The policies are different and complex across regions creating a substantial obstacle. In numerous nations, policies continue to support the traditional concept of economic system thereby creating pressure on the limited natural resources and enforcing inadequate fines for trash production and disposal. The unstandardized definitions and recognition of CE activities impedes businesses and regulators from measuring progress and establishing clear objectives (Pires and Martinho, 2019). Transitioning to a circular economy within enterprises necessitates cultural transformation. Conventional models need to be entrenched, and shift to CE approach frequently requires a comprehensive re-evaluation of value creation and capture methodologies (Geissdoerfer et al. 2017). This transition may encounter opposition from parties familiar with traditional methods. Moreover, collaboration throughout many sectors is essential yet often difficult to attain (Lieder and Rashid 2016). The shift to a CE encounters several obstacles as the linear economy is fundamentally integrated into global economic systems. Transitioning towards the CE approach necessitates drastic alterations across all sectors and industries. This entails the reconfiguration of supply chains, the formulation of innovative business models, and the promotion of interdisciplinary collaboration (Korhonen et al. 2018). The intricacy of these systemic transformations can be overwhelming and necessitates collaborative endeavors from governments, enterprises, and civil society.

4 Research Methodology

This chapter is based on the review of existing data sources. The journal articles and book chapters published on the topic of circular economy and digital technologies are examined to build a solid foundation of the topic. The chapter discusses the basic meaning of circular economy, digital technologies, and digital circular as per literature to enhance the conceptual understanding of the readers. Then the deep knowledge of DTs enables circular economy for reuse and recycle is presented followed by the challenge in the implementation of DCE. Finally, implications of the study are offered based on the analysis of literature.

4.1 Implications

The integration of DTs with CE principles offers infinite solutions for dealing with rising issue of climate change. DCE enhances the reverse supply chain as by employing digital functionalities a more rapid, accurate, and automated recycling system is established. The recycling strategy is enhanced by the majority of the digital functionalities discussed in this chapter. The most beneficial function for recycling appears to be sorting and classifying, followed by innovating and detecting. Sorting and classifying are mostly associated with recycling, whereas forecasting has not yet been connected to this process. The DTs employed for enhancing the productivity of basic manufacturing processes can be similarly implemented in the reverse supply chain.

The most visible impact of DCE can be seen on environmental sustainability as against challenges of biodiversity loss and resource depletion posed by the regular linear systems. DCE improves resource efficiency while reducing the waste generation reuse and recycling. Landfill waste is reduced to a great extent by employing IoT for tracking the waste and AI for sorting in order to improve the rate of recycle. Through predictive maintenance, the proper reuse mechanism and extended product lifespans become possible, and hence DCE reduces the use of limited natural resources, less requirement for manufacturing new products thereby lowering emissions greenhouse gases. Blockchain-enabled digital systems are effectively manage the recycling process by providing right material composition details for various products and ensuring higher recovery rates specially in electronics based products. DCE supports the conservation of biodiversity by ensuring that harmful waste substances are not dumped in the ecosystems directly, especially in sectors like plastics and construction.

The DCE also offers significant implications for the economy by promoting employment growth through innovation and competitiveness as the resource efficiency by employing DTs reduces costs, improves profitability and presents opportunities for large scale production. The DCE transform the traditional business models and enables innovative models such as product-as-a-service (PaaS), ways for utilizing shared platforms, and reverse logistics marketplaces. Big data analytics is a very crucial component of DCE adoption for optimizing the resource allocation and reducing the costs associated with waste disposal, on the other hand IoT make waste collection easier and improves efficiency in logistics management. Online platforms supported by blockchain and AI have become prominent markets for recycled resources thereby reducing dependency on natural raw materials through offering transparent and efficient exchanges. DCE adoption also complements the firms to comply with the modern environmental regulations, saving firms from legal repercussions and enhancing long-term competitiveness.

Beyond normal environmental and economic implications, DCE has many social implications also. The integration of DTs for recycling and reuse generates new employment opportunities in green technology sectors, including data analytics, robotics, and reverse logistics. Blockchain based systems ensures that digital platforms work in a transparent way which enhance trust of the users for reused and recycled products (Joshi et al. 2024). DCE enabled shared economy platforms results in collaborative consumption and make sure that every section of the society has access to goods and services. Moreover, DCE infrastructure can be really effective in promoting digital literacy and inclusion which would empower the general public to be a part of the global initiatives towards sustainable future.

Governments play a central role in building infrastructures that supports the adoption of DCE while DCE itself contributes to the policy frameworks and governance structures as well. Partnership among governments, technology firms, and industry stakeholders can be transformative force in expanding the implementation of DCE models, particularly in the areas waste management and infrastructure development. The policymakers should make the incorporation of common data standards compulsory for ensuring interoperability of digital systems across industries. Governments can use tax benefits, subsidies, or other monetary schemes to motivate the firms and consumers for participating in recycle and reuse system. Global cooperation should be enforced for promoting DCE practices though cross-country transfer of technology and enhanced access to global markets. By incorporating DTs with CE policy frameworks, governments can foster transparent and trustworthy system for building sustainable economies. In short, DCE moves beyond fractional changes the system of recycling and reuse to a systemic pathway towards climate resilience and sustainable growth for offering a regenerative and waste-free future.

5 Conclusion, Limitations, and Scope for Future Research

Amid the challenges posed by climate change a shift from a regular "take–make–dispose" model to a CE is viewed as the most desirable measure given the high level of resource depletion and unsustainable consumption. This chapter has highlighted how DTs such as IoT, AI, blockchain, big data etc. can be integrated with CE principles for maximizing reuse and recycling of resources for enhanced efficiency. Through smarter system of managing waste including tracking, automated sorting, and product lifecycle monitoring, DCE addresses many limitations of normal CE. The analysis of existing literature demonstrates that technology oriented circular practices lowers the emissions of greenhouse gas and also the decrease the waste send to landfills reduces through recycling. It fosters innovation through new business models while conserving resources and minimizing the cost along with creation of green jobs.

However, the chapter also acknowledges its limitations. First of all, the study is entirely based on secondary sources of data and hence, the implications are not directly supported with any empirical evidence. Next, no differentiation is made among organizations based on sector or size, as the DCE adoption requires high investment in technology, so small and medium-sized enterprises might not be able to employ this model. So, future studies can specifically examine the DCE application among small business enterprises. Moreover, the digital divide between developed and developing nations risks also make it difficult to offer identical benefits of DCE as in the developing countries technology is still in its early stages and due to lack of resources businesses are reluctant to integrate DCE practices. This aspect is not considered in the chapter which requires further examination. Also, the field of DTs is really wide so apart from the technologies discussed in present chapter, future research can investigate the integration of quantum computing, edge AI, and advanced digital twins into circular systems.

References

Bressanelli G, Perona M, Saccani N (2020) Digital technologies for a circular economy: Pathways and impacts. Sustain 12(8):2893. https://doi.org/10.3390/su12082893

Büyüközkan G, Uztürk D (2022) A methodology to investigate challenges for digital twin technology in smart agriculture. Comput Electron Agric 203:107440. https://doi.org/10.1016/j.compag.2022.107440

Caro MP, Ali MS, Vecchio M, Giaffreda R (2018) Blockchain-based traceability in agri-food supply chain management: A practical implementation. In: 2018 IoT vertical and topical summit on agriculture—tuscany (IOT Tuscany) (pp. 1–4). IEEE. https://doi.org/10.1109/IOT-TUSCANY.2018.8373021

Çetin S, De Wolf C, Bocken N (2021) Circular digital built environment: An emerging framework. Sustain 13(11):6348. https://doi.org/10.3390/su13116348

Dev NK, Shankar R, Qaiser FH (2020) Industry 4.0 and circular economy: Operational excellence for sustainable reverse supply chain performance. Resour, Conserv Recycl, 153, 104583. https://doi.org/10.1016/j.resconrec.2019.104583

Geissdoerfer M, Savaget P, Bocken NMP, Hultink EJ (2017) The circular economy—A new sustainability paradigm? J Clean Prod 143:757–768. https://doi.org/10.1016/j.jclepro.2016.12.048

Ghisellini P, Cialani C, Ulgiati S (2016) A review on circular economy: The expected transition to a balanced interplay of environmental and economic systems. J Clean Prod 114:11–32. https://doi.org/10.1016/j.jclepro.2015.09.007

Gong Y, Wang Y, Frei R, Wang B, Zhao C (2022) Blockchain application in circular marine plastic debris management. Ind Mark Manage 102:164–176. https://doi.org/10.1016/j.indmarman.2022.01.010

Homrich AS, Galvão G, Abadia LG, Carvalho MM (2018) The circular economy umbrella: Trends and gaps on integrating pathways. J Clean Prod 175:525–543. https://doi.org/10.1016/j.jclepro.2017.11.064

Hoong JDLH, Lux J, Mahieux PY, Turcry P, Ait-Mokhtar A (2020) Determination of the composition of recycled aggregates using a deep learning-based image analysis. Autom Constr 116:103204. https://doi.org/10.1016/j.autcon.2020.103204

Huynh PH (2021) Enabling circular business models in the fashion industry: The role of digital innovation. Int J Product Perform Manag 71(3):626–652. https://doi.org/10.1108/IJPPM-12-2020-0683

Ingemarsdotter E, Jamsin E, Kortuem G, Balkenende R (2019) Circular strategies enabled by the Internet of Things: A framework and analysis of current practice. Sustain 11(20):5689. https://doi.org/10.3390/su11205689

Jain G, Kamble SS, Ndubisi NO, Shrivastava A, Belhadi A, Venkatesh M (2022) Antecedents of blockchain-enabled e-commerce platforms (BEEP) adoption by customers: A study of second-hand small and medium apparel retailers. J Bus Res 149:576–588. https://doi.org/10.1016/j.jbusres.2022.05.023

Joshi A, Singh R, Kumar V, Rani S (2024) Net zero strategy: Exploring opportunities, confronting challenges for developing countries. In Singh R, Crowther D (Eds), Transit Sustain Futur (pp. 233–252). Springer. https://doi.org/10.1007/978-981-97-5756-5_11

Keles C (2020) Evaluating the effect of building envelope on thermal performance in cold and warm climate regions of Turkey. Eur J Eng Technol, 3(1), 93–104. https://doi.org/10.24018/ejeng.2020.3.1.164

Khadke S, Gupta P, Rachakunta S, Mahata C, Dawn S, Sharma M, Verma D et al (2021) Efficient plastic recycling and remolding circular economy using the technology of trust—Blockchain. Sustainability 13(16):9142. https://doi.org/10.3390/su13169142

Khan S, Singh R, Alnahas J, Abbate S, Centobelli P (2024) Navigating the smart circular economy: A framework for manufacturing firms. J Clean Prod 480:144007

Kirchherr J, Reike D, Hekkert M (2017) Conceptualizing the circular economy: An analysis of 114 definitions. Resour Conserv Recycl 127:221–232. https://doi.org/10.1016/j.resconrec.2017.09.005

Korhonen J, Honkasalo A, Seppälä J (2018) Circular economy: The concept and its limitations. Ecol Econ 143:37–46. https://doi.org/10.1016/j.ecolecon.2017.06.041

Kristoffersen E, Blomsma F, Mikalef P, Li J (2020) The smart circular economy: A digital-enabled circular strategies framework for manufacturing companies. J Bus Res 120:241–261. https://doi.org/10.1016/j.jbusres.2020.07.044

Laskurain-Iturbe I, Arana-Landín G, Landeta-Manzano B, Uriarte Gallastegi N (2021) Exploring the influence of Industry 4.0 technologies on the circular economy. J Clean Prod, 321, 128944. https://doi.org/10.1016/j.jclepro.2021.128944

Lieder M, Rashid A (2016) Towards circular economy implementation: A comprehensive review in context of manufacturing industry. J Clean Prod 115:36–51. https://doi.org/10.1016/j.jclepro.2015.12.042
Liu Q, Trevisan AH, Yang M, Mascarenhas J (2022) A framework of digital technologies for the circular economy: Digital functions and mechanisms. Bus Strateg Environ 31(5):2171–2192. https://doi.org/10.1002/bse.3004
Lux J, Hoong JDLH, Mahieux PY, Turcry P (2023) Classification and estimation of the mass composition of recycled aggregates by deep neural networks. Comput Ind 148:103855. https://doi.org/10.1016/j.compind.2023.103855
Ma S, Ding W, Liu Y, Ren S, Yang H (2022) Digital twin and big data-driven sustainable smart manufacturing based on information management systems for energy-intensive industries. Appl Energy 326:119986. https://doi.org/10.1016/j.apenergy.2022.119986
Mahpour A (2018) Prioritizing barriers to adopt circular economy in construction and demolition waste management. Resour Conserv Recycl 134:216–227. https://doi.org/10.1016/j.resconrec.2018.01.034
Nascimento DLM, Alencastro V, Quelhas OLG, Caiado RGG, Garza-Reyes JA, Rocha-Lona L, Tortorella G (2019) Exploring Industry 4.0 technologies to enable circular economy practices in a manufacturing context: A business model proposal. J Manuf Technol Manag, 30(3), 607–627. https://doi.org/10.1108/JMTM-03-2018-0071
Nobre GC, Tavares E (2021) The quest for a circular economy final definition: A scientific perspective. J Clean Prod 314:127973. https://doi.org/10.1016/j.jclepro.2021.127973
Paul T, Islam N, Mondal S, Rakshit S (2022) RFID-integrated blockchain-driven circular supply chain management: A system architecture for B2B tea industry. Ind Mark Manage 101:238–257. https://doi.org/10.1016/j.indmarman.2021.12.008
Peladarinos N, Piromalis D, Cheimaras V, Tserepas E, Munteanu RA, Papageorgas P (2023) Enhancing smart agriculture by implementing digital twins: A comprehensive review. Sensors 23(16):7128. https://doi.org/10.3390/s23167128
Pires A, Martinho G (2019) Waste hierarchy index for circular economy in waste management. Waste Manage 95:298–305. https://doi.org/10.1016/j.wasman.2019.06.014
Purcell W, Neubauer T (2023) Digital twins in agriculture: A state-of-the-art review. Smart Agric Technol 3:100094. https://doi.org/10.1016/j.atech.2023.100094
Rajput S, Singh SP (2019) Connecting circular economy and Industry 4.0. Int J Inf Manage 49:98–113. https://doi.org/10.1016/j.ijinfomgt.2019.03.002
Ranta V, Aarikka-Stenroos L, Väisänen J-M (2021) Digital technologies catalyzing business model innovation for circular economy: Multiple case study. Resour Conserv Recycl 164:105155. https://doi.org/10.1016/j.resconrec.2020.105155
Reike D, Vermeulen WJV, Witjes S (2018) The circular economy: New or refurbished as CE 3.0? Exploring controversies in the conceptualization of the circular economy through a focus on history and resource value retention options. Resour Conserv Recycl 135:246–264. https://doi.org/10.1016/j.resconrec.2017.08.027
Rodrigo N, Omrany H, Chang R, Zuo J (2024) Leveraging digital technologies for circular economy in construction industry: A way forward. Smart Sustain Built Environ 13(1):85–116. https://doi.org/10.1108/SASBE-09-2021-0189
Rosa P, Sassanelli C, Urbinati A, Chiaroni D, Terzi S (2020) Assessing relations between circular economy and Industry 4.0: A systematic literature review. Int J Prod Res, 58(6), 1662–1687. https://doi.org/10.1080/00207543.2019.1680896
Rowley J (2007) The wisdom hierarchy: Representations of the DIKW hierarchy. J Inf Sci 33(2):163–180. https://doi.org/10.1177/0165551506070706
Rusch M, Schöggl JP, Baumgartner RJ (2023) Application of digital technologies for sustainable product management in a circular economy: A review. Bus Strateg Environ 32(3):1159–1174. https://doi.org/10.1002/bse.3182

Singh R, Gupta A, Bajpai A, Kandpal V (2024b) Exploring challenges of circular economy initiatives for smart cities. In Smart Cities Circ Econ: Futur Sustain Urban Dev (pp. 59–69). Emerald Publishing Limited

Singh R, Ravesangar K, Narayanan S, Singh KKH (2024a) HRM practices in creating net zero: Supporting organizational sustainability. In Net zero economy, corporate social responsibility and sustainable value creation: Exploring strategies, drivers, and challenges (pp. 1–18). Cham: Springer Nature Switzerland

Singh R, Gupta A, Kandpal V (2025a) ESG excellence in the digital world: Circular economy e-business solutions. Fintech ESG Circ Econ, 193–204

Singh R, Joshi A, Dissanayake H, Iddagoda A, Khan S, Félix MJ, Santos G (2025b) Integrating industry 4.0, circular economy, and green HRM: a framework for sustainable transformation. Sustain, 17(7), 3082

Soo A, Wang L, Wang C, Shon HK (2023) Machine learning for nutrient recovery in the smart city circular economy: A review. Process Saf Environ Prot 170:634–652. https://doi.org/10.1016/j.psep.2022.11.063

Talla A, McIlwaine S (2024) Industry 4.0 and the circular economy: Using design-stage digital technology to reduce construction waste. Smart Sustain Built Environ, 13(1), 179–198. https://doi.org/10.1108/SASBE-12-2021-0205

Tu LF, Peng Q, Deng R (2022) Design and implementation of intelligent control garbage bin. Journal of Electrical and Computer Engineering 2022:5375403. https://doi.org/10.1155/2022/5375403

Vetrova M, Ivanova D (2021) Closed product life cycle as a basis of the circular economy. GATR J Bus Econ Rev, 5(4), 36–50. https://doi.org/10.35609/jber.2021.5.4(4)

Wu Z, Pei T, Bao Z, Ng ST, Lu G, Chen K (2024) Utilizing intelligent technologies in construction and demolition waste management: From a systematic review to an implementation framework. Front Eng Manag 11(1):1–23. https://doi.org/10.1007/s42524-023-0285-1

Yu Y, Yazan DM, Junjan V, Iacob M-E (2022) Circular economy in the construction industry: A review of decision support tools based on information and communication technologies. J Clean Prod 349:131335. https://doi.org/10.1016/j.jclepro.2022.131335

Zacharaki A, Vafeiadis T, Kolokas N, Vaxevani A, Xu Y, Peschl M, Ioannidis D, Tzovaras D (2020) RECLAIM: Toward a new era of refurbishment and remanufacturing of industrial equipment. Front Artif Intell 3:570562. https://doi.org/10.3389/frai.2020.570562

Ethical AI for Environmental Governance: Opportunities and Pitfalls in Tech-Driven Climate Strategies

Kavya Shabu and Nayana Prabhash

Abstract The rapid integration of Artificial Intelligence (AI) into environmental governance creates a paradox wherein AI can improve our decision-making for sustainability, thereby presenting us with ethical dilemmas for which adequate responses do not yet exist. Although globally, there is pressure to move to digital sustainability, there are no frameworks for voluntary guidance on the ethical, social inclusion, and environmental justice to govern AI-generated climate action strategies. This chapter shares the ethical debates related to AI for environmental governance, with a particular focus on the tension between improvements in technological efficiency and moral obligations. The research will utilize a qualitative methodology measuring secondary data sources, policy documents, academic literature, and case studies of AI initiatives for international environmental governance. Content analysis is used to identify what ethical principles are present (or absent) in tech-enabled sustainability initiatives, map ethical risks, and compare governance strategies across geopolitical contexts. Contributions of this chapter include developing a taxonomy of ethical risks, a comparative analysis of contexts of governance strategies, and a conceptual model for insertion of ethics into future frameworks for AI climate action. This chapter aims to enhance the ability of researchers, policymakers, and technology practitioners to manage the connection of accountability and innovation to address wicked global environmental challenges.

Keywords Artificial intelligence · Environmental ethics · Sustainability governance · Algorithmic accountability · Climate technology

---

K. Shabu (✉)
Faculty of Management Studies, CMS Business School, Jain (Deemed-to-Be University), Bengaluru, India
e-mail: kavyaseabi01execalumni@iimk.edu.in

N. Prabhash
School of Commerce and Management Studies, Dayananda Sagar University, Bengaluru, Karnataka, India
e-mail: nayana-scms@dsu.edu.in

R. Singh et al. (eds.), *Data-Driven Sustainability: Harnessing Technology for A Greener Future and Environmental Resilience*, World Sustainability Series,
https://doi.org/10.1007/978-3-032-13412-7_17

1 Introduction

Emergencies of climate change deepening and the surge of digital innovation is rising. Together, the concept has converged in a moment of greater attention in Artificial Intelligence (AI) as a necessary facilitator for environmental governance, and as a challenged disruptor at the fringes. The uses of AI in environmental governance include emissions tracking, prediction of disasters, simulation of policy, and allocation of resources, where each of these provides an extraordinary improvement in scale, accuracy and responsiveness to sustainability intrusions (Mehryar et al. 2024; Rolnick et al. 2019). IT is increasingly possible with AI to confiscate or mitigate emissions, retort to climate disasters, integrate climate science into decision-making, measure trade-offs and co-benefits, and apportion funds with increasing speed and accuracy. Though application of AI ensures effective climate action, optimising informed decision-making and resilience trajectories includes the risks of algorithmic bias, marginalization of vulnerable people, obscurity of data, and harms of surveillance (Crawford and Paglen 2021; Bellaby 2024). The present condition of AI raises the imperative to question the ethical aspect of applying AI, and establish value-based frameworks and opportunities to shape usability design and governance.

Conceptual clarity is necessary for embarking on such an investigation. Artificial Intelligence, in the context used here, refers to computational systems that can learn independently, reason, and make choices in responding to intricate environmental issues (Alsaigh et al. 2022). Environmental governance is the organizational structures and policy processes that steer joint action toward sustainability and climate resilience (van Uffelen 2024). Ethical AI is the rule-governed design and deployment of AI systems in a manner that promotes transparency, fairness, accountability, inclusivity, and ecological responsibility (Floridi et al. 2018). Closely related, algorithmic accountability denotes processes that make AI-generated decisions auditable, contestable, and socially aligned, thereby enabling stakeholder scrutiny and redress (Rakova and Dobbe 2023).

Despite mounting recognition of AI's salience in climate governance, significant gaps persist. It is uncertain which ethical norms are substantively infused into existing applications, how governance regimes differ by political and cultural context, and what can be done to ensure the systematic infusion of ethics into new AI-based climate solutions. In this context, this research is framed by three connected research questions: Which ethical norms are included or glaringly absent in AI-facilitated climate governance? How are ethical frameworks of governance differ cross-nationally in relation to varied regulatory, cultural, and political contexts? What avenues exist to better integrate ethics into the design and governance of climate AI policies? Using a comparative qualitative methodology, the study explores these questions aiming to contribute new thinking regarding the ethical governance of climate AI. The analysis employed advance arguments on the ability of AI to improve technical effectiveness and enhance democratic legitimacy, social justice and ecological sustainability. Additionally, the study developed and propose a frameworks that ensures transparent, participatory, and justice-centred implementation of AI in the climate space.

2 Literature Review

AI in Climate Governance: Applications

Climate governance focused on building capacity in climate modelling, monitoring emissions and predicting disasters is highly adopting the use of AI. Advances in machine learning are being progressively applied to aid hazard modelling that support prediction of risks from flooding, droughts and extreme heat events (Mehryar et al. 2024). Mehryar et al. (2024) found that to date, AI has mostly pursued advances in risk assessment over vulnerability targeting, and created ideologies that target technical optimization. Rolnick et al. (2019) argue that AI can affect the efficiency of energy systems and renewable integration for climate governance positively using predictive analytics better allocated.

Artificial Intelligence also has an increasingly important role in emissions monitoring and corporate governance. Schirmer (2025) finds that AI improves the accuracy of corporate disclosures on climate, informing compliance with international initiatives. Hammer et al. (2021) validate that satellites combined with AI monitoring technology creates a more robust accountability system that offer near real-time tracking of carbon emissions and change in land-use. AI led early warning systems assist in the pursuit of disaster resilience as it helps in reducing vulnerability by predicting and communicating risk (Benke and Benke 2018).

There is evidence supporting that AI also augment policy assessment and scenario modelling (Kreft and Fournier-Tombs 2024; Tan et al. 2025). Studies have developed the circumstance where AI is in support of unified assessment of policies and decision-making on renewable energy and carbon mitigation (Tan et al. 2025). Kreft and Fournier-Tombs (2024) further examine how AI can bring greater cohesion between climate governance and the Sustainable Development Goals, particularly through predictive tools for disaster preparedness and anticipatory action. Overall, the various applications are illustrative that AI might be supporting climate governance frameworks, but, at the expense of outlining ethical dilemmas if adequate governance frameworks are absent.

Ethical Dilemmas

AI, notwithstanding the potential benefits, has generated profound ethical dilemmas for climate governance. A key ethical challenge relates to algorithmic bias, or essentially when models mirror systemic inequity. For instance, as Machen (2025) highlights, datasets sourced from urban settings amplify those who are urban dwellers, whose populations are already well-documented; as well as failing to account for populations of rural residency, but also those who are underrepresented in urban social networks. In addition, Crawford and Paglen (2021) highlight that the geography of the datasets also replicate exclusion, privilege, and inequity as an unjust response to climate.

More ethical dilemmas are present with data sovereignty and surveillance. Bellaby (2024) identifies that surveillance methods and other methods of monitoring that in AI, or algorithms, may be acquired, and may or may not be initiated, through data usage consent as expressed by data users as a part of data usage; but the ideology of social asymmetries of power may reinforce the aforementioned ethical quandaries into AI decisions, or outputs. Sander (2025) fears that AI actions through climate governance lead to future normalizations of surveillant technologies through climate governance, without transparency, or accountability. The ethical concerns mentioned above echo Zuboff's (2019) argument on surveillance capitalism and data extraction.

Another ethical challenge involves algorithmic opacity. Alsaigh et al. (2022) emphasize the "black box" that occurs within AI processes can diminish trust because stakeholders cannot trace the rationale behind decision making, for example, in monitoring emissions or evaluating risks. Likewise, Floridi et al. (2018) argue algorithmic opacity limits contestability of decisions that impact at-risk social sectors and directly oppressed populations. Opacity and associated ethical challenges also go against critiques of techno-solutionism, as value is placed squarely on the AI component as if it was the solution. Amanta et al. (2025) forewarn that such reliance on AI poses risks in ignoring structural drivers of climate change.

The environmental footprint of AI itself is another complicating factor. Strubell et al. (2019) indicate that large AI models trained require large amounts of energy, which causes unwanted carbon impacts. Additionally, the Oxford Environmental Change Institute (2025) mentions water and energy costs of data storehouses, which may undo their initial climate mitigation. Scholars also mention possible risks of exclusion: Saheb et al. (2022) argue that digital divides prevent equitable inclusion in AI-enabled climate governance, while Noble (2018) explains how algorithmic discrimination exacerbates marginalization.

Gaps Identified

While the literature increases, there are still gaps. The first gap is the lack of ethical governance frameworks on climate and AI. Mehryar et al. (2024) point out while risks are well documented, meaningful institutional frameworks to hold people or systems accountable are still lacking. Similarly, Jelinek (2022) points out a lack of connection between AI ethics discussion and policy for climate governance.

The second gap points that there is limited comparative, contextualized studies. There is a predominance of studies in Europe and North America; meanwhile, there are 'climate-vulnerable' countries in the Global South, which is likely a fertile area for study (Milaninia 2020). India is an example of a techno-optimist yet unregulated context. However, the educational and academic literature considers it less than Western counterparts.

A third gap is in the space of environmental justice and inclusion. While terms like fairness and transparency are often used, there are few studies that cover application for indigenous people, informal workers, or displaced peoples. For example, Schirmer (2025) identified the barriers to accessibility with corporate ESG reports; Mehryar et al. (2024) noted the tokenistic nature of community engagement and inclusion with biodiversity governance, and Benjamin (2019) and Taylor (2017) that justice must be part of the design, or it will serve to replicate the displacement of inequity.

There is insufficient longitudinal work tracking the unintended consequences of adopting AI. For example, Alsaigh et al. (2022) and Strubell et al. (2019) comment that downstream effects are understudied, including the energy costs of AI and the erosion of human oversight. An all-encompassing assessment of AI could minimize instances of continuing technocratic governance rather than democratize climate solutions.

3 Theoretical Framework

The normative ethics to ground the ethical application of AI in environmental governance supply the moral justification for technologically driven interventions. Deontological ethics focus on duties and rights, so that AI systems uphold human dignity as well as procedural fairness irrespective of consequences (Jedličková 2024). Utilitarian theories emphasize the duty to enhance overall well-being, especially via increasing the efficiency and effectiveness of climate mitigation and adaptation (Cumming et al. 2024). Justice-based theories such as Rawlsian distributive justice and environmental justice focus on distributive fairness and participative governance, with an aim to safeguard vulnerable groups that are disproportionately victimized by ecological crises (van Uffelen 2024). These ethical foundations are the basis for assessing the ways in which AI might be used without increasing inequality or environmental harm. On these foundations, AI-directed ethical perspectives operationalize values through relevant design and governance. Algorithmic accountability seeks to ensure transparency, auditability, and organizational responsibility so that stakeholders can contest and remedy decisions made by AI (Rakova and Dobbe 2023). Value-Sensitive Design approaches (VSD) provide a systematic method to explicitly include stakeholder values in AI systems and ultimately foster inclusiveness and responsiveness to context in AI systems designed to improve environmental governance (Sadek et al. 2023). Supporting these is the focus of digital environmental justice which moves towards just data governance, participatory inclusion, and acknowledgement of community rights in decision-making to further ensure that environmental policies deployed by AI do not perpetuate previous social or ecological injustices (van Uffelen 2024).

These interdependent axes overlap and intersect in the Ethical AI Framework for Environmental Governance (See Fig. 1), which representationally stacks the normative ethics layers, the design or evaluation tools of algorithmic accountability and VSD, and the broader context of justice and environmental consideration. The governance and institutional context of the framework highlights the role of national ethnicities and regulatory frameworks that shape the ethical uses of AI. The framework positions dimensions (e.g. equity, effectiveness, transparency, and environmental sustainability) as emergent from the interconnection of ethical-layers with feedback loops indicating ongoing adaptation and accountability. The integration of ethical principles, technical ways of understanding, and governance processes

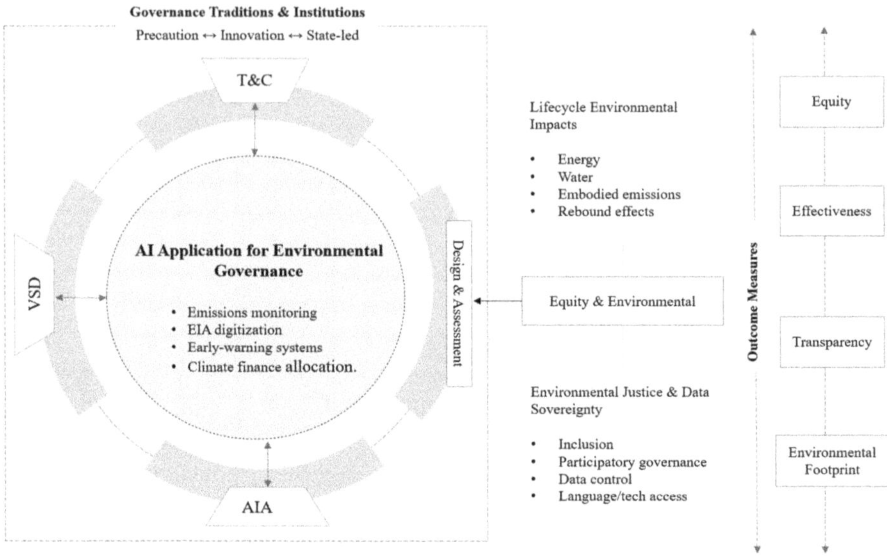

Fig. 1 Integrative model of ethical AI for environmental governance. *Source* Authors' own conceptualization based on theoretical synthesis from Floridi et al. (2018); Rakova and Dobbe (2023); Sadek et al. (2023); van Uffelen et al. (2024); and Jedličková (2024). Developed from the literature review and theoretical framework on ethical AI and environmental governance

directing towards equitable, accountable, and ecological processes in environmental governance is illustrated in Fig. 1.

4 Methodology

A qualitative exploratory methodology accompanied with interviews of two Indian environmentalist professionals whose expertise in the crossroad of AI and environmental governance was employed. The environmentalist included professionals from an urban NGO focused on climate adaptation and a policy think tank for renewable energy. From these interviews, ten scientifically derived prompts were developed that addressed practitioner relevance, ethical considerations, and global priorities of concern, such as net-zero transitions, ESG compliance, and biodiversity preservation, for example. Each of the generative AI programs, OpenAI, GPT-4, Anthropic, Claude, and an example of Google, Gemini, was used to test the prompts which allowed for a comparative evaluation of the various AI platforms in terms of strength of reasoning, ethical considerations, and capabilities to incorporate data. Using a multi-criteria matrix, the prompts were used to evaluate and assess the responses generated from the AI platforms in terms of policy relevant, ethical considerations, inclusion of vulnerable groups, transparency of reasoning, contextually adaptable, and comparative consistency across platforms. It was also suggested that the context that defined India served as the appropriate context for the study as it functions as an example of a model that is techno-optimist, but unregulated, and has adopted the digital world rapidly in a context of climate challenges which makes it a viable context to instruct the opportunities and ethical vulnerabilities of climate AI. The systematic iterative evaluation provided a deeper understanding of climate AI in relation to governance gaps, transparency challenges, algorithmic biases, and inclusion risks in a way that connected expert perspectives, and AI-well constructed policy analysis in relation to climate and the Indian context. The research used Thematic Content Analysis to systematically code qualitative data and find recurring ethical principles and risks, and Critical Discourse Analysis to analyze power dynamics and ideological framing in AI–climate policy narratives using NVivo as the analysis tool.

5 Findings

5.1 Taxonomy of Ethical Risks in Climate-AI Projects

The findings illustrate a recognizable set of ethical risks across policy areas that serve to describe a typology of ethical challenges related to AI application in environmental governance (Table 1). Governance gaps emerged as in techno-centric policy framing and in India's net-zero roadmap, that targets the industrial sector in a narrow focus without regard to region-specific issues and even feasibility perceivable by grass-roots actors. Governance gaps highlight the disconnect between national ambitions and local philosophies that risk compromising intergenerational justice. Transparency-related ethical concerns arose with the digitization of Environmental Impact Assessments (EIA) that utilize AI to simplify the evaluation and reporting process, risking the loss of the revelatory dimensions that underpin democratic accountability. Data and surveillance issues are present in disaster risk management plan, where risk identification and action plans depend on data from an array of platforms, and marginalize vulnerable populations to the extent access and language barriers are not considered. Similarly, ethical biases in algorithmic models are observed in renewable energy and climate finance distributions, where AI frameworks inherently favour 'quantifiables' such as added energy capacity, or efficient disbursement of funds, rather than equity-based considerations. Finally, claimants or non-claimants of vulnerable communities remains the most intractable issue. Request for example, urban greening initiatives ignore informal or migration populations, or that transitions to a circular economy risk labor being dislocated by automation, or that biodiversity conservation agendas reduce communities to a position of mere symbol over substantive role in governance frameworks or strategies.

Table 1 Sample generative ai prompts for environmental policy-making and associated ethical considerations

Policy area	Sample prompt by environmentalist	Decision-making use	Ethical dimension reflected	Identified ethical concern
1. Climate policy framing	"Draft a policy roadmap to operationalize India's 2070 net-zero target with short-term targets for key industrial sectors."	Aligns national commitments with sector-level implementation	Intergenerational justice	Techno-centric bias; risk of overlooking regional disparities
2. Urban sustainability	"Recommend inclusive urban greening strategies for heat island mitigation in Tier-2 Indian cities."	Supports equitable urban climate adaptation	Environmental justice	Solutions may neglect informal or migrant populations
3. Renewable energy strategy	"Summarize policy incentives that encouraged rooftop solar in Kenya and propose adaptation for Indian residential zones."	Facilitates cross-country policy learning	Inclusivity and empowerment	Context insensitivity to local grid capacity and behavior
4. ESG regulation drafting	"Create a simplified ESG reporting toolkit for textile MSMEs that aligns with EU import norms."	Enables compliance and export readiness	Procedural fairness	Language or tech access barriers for smaller exporters
5. Circular economy transition	"List policies that promote local recycling ecosystems in garment clusters while ensuring livelihood preservation."	Enhances circular economy and just transition	Sustainability and livelihood security	Conflict between automation and job security

(continued)

Table 1 (continued)

Policy area	Sample prompt by environmentalist	Decision-making use	Ethical dimension reflected	Identified ethical concern
6. Biodiversity conservation	"Design incentive-based conservation strategies for forest-fringe communities in biodiversity hotspots."	Integrates ecological goals with community participation	Participatory governance	Risk of tokenism in community engagement
7. Environmental impact assessment (EIA)	"Suggest criteria to digitize and democratize EIA reporting for infrastructure projects in ecologically sensitive areas."	Modernizes monitoring and improves access	Transparency and accountability	AI oversimplification may reduce interpretive nuance
8. Disaster risk management	"Generate early warning communication strategies for flood-prone rural communities using local languages and tech platforms."	Enhances inclusive climate resilience and preparedness	Accessibility and equity	Marginalization if AI overlooks language or tech gaps
9. Climate finance allocation	"Compare sub-national climate finance models in South Asia and recommend policy priorities for fund distribution in India."	Supports targeted and just climate investment	Equity in resource allocation	Skewed prioritization towards easily quantifiable outcomes

Source Authors' own compilation from qualitative expert interviews and comparative AI platform testing (OpenAI GPT-4, Anthropic Claude, Google Gemini) conducted for this study. Synthesized with supporting insights from Mehryar et al. (2024); Schirmer (2025); and Bellaby (2024)

5.2 *Comparative Insights in Climate-AI Governance*

These risks do not operate uniformly around the globe but instead are mediated by governance traditions. In the European Union (EU), governance emphasizes precautionary and rights-based traditions that provide procedural fairness and transparency considering the ethical risks presented in Table 1 such as language access barriers for MSMEs in order to comply with ESG requirements. The EU model attempts to meets the governance "gap(s)" by privileging checks in the governance system, but the process can impede innovation. Conversely, the United States of America (USA) values innovation and deregulation at the expense of being deliberative about how it encourages experimentation relevant for climate technology, often in ways that do not center vulnerable communities, and perpetuate the intergenerational justice concerns of a techno-centric portrayal of climate policy connected to Table 1. State-led and opaque data systems in China allow for large-scale mobilization for biodiversity conservation or circular economy activities, but risk the political signal of tokenistic roles in participation, in addition to the participatory governance and livelihood security issues raised in Table 1. India, as an instance of a techno-optimist role, works to learn from relevant cross-country policy examples such as rooftop solar adaptations from Kenya, but the anxiety that any techno-optimist model will provide does not consider appropriate regulatory oversight to address any accompanying ethical vulnerabilities, especially for that address inclusion, equity in resource allocation, or job security during the transition to a circular economy. With India situated in other governance traditions, attention to ethical issues is present across governance contexts, but the relative balance between innovation, accountability, and inclusivity shifts, or not, across governance traditions.

5.3 *Visual Representations for Conceptual Clarity*

The diagram offers an integrative illustration of the ethical risks of AI-supported environmental governance by bringing together a taxonomy of risks (mind map) with a process-flow diagram (Fig. 2). On the left side, the mind map categorizes basic types of ethical issues like governance deficits, transparency issues, data and surveillance threats, algorithmic prejudice, and inclusion or exclusion threats offering a conceptual classification of possible weaknesses. On the right, the flow diagram illustrates how those risks come to exist in practice through the sequential series of policy triggers, AI uses, decision-making applications, and the ethical aspects they represent. The bridging between the two sides of the figure shows how the abstract risk categories come to exist in governance processes. For instance, the risk of governance loopholes is directly related to poor oversight of AI-based policy enforcement, whereas transparency issues result from algorithmic obscurity in climate simulation and environmental monitoring. Equally, data and surveillance threats emerge through the digitization of environmental impact assessments and predictive monitoring systems, algorithmic bias occurs when AI tools inject discriminatory patterns into climate or urban decisions, and inclusion or exclusion threats are apparent when AI-based planning routinely overlooks marginalized groups. These interdependencies validate the conclusion that ethical risks do not exist in isolation, but are bundled in the decision-making processes of environmental governance. Further supporting this conclusion, direct comparisons of the European Union, United States, China, and

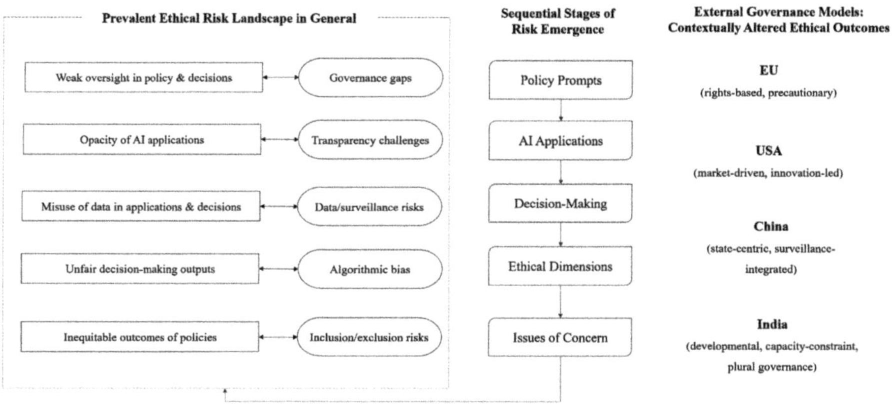

Fig. 2 Stages of ethical risk emergence and governance models for contextually altered ethical outcomes *Source* Authors' own creation derived from thematic and comparative analysis using findings from Table 1, interviews, and literature synthesis including Crawford and Paglen (2021); Machen (2025); Sander (2025); and van Uffelen et al. (2024)

India, demonstrate that the same AI applications can generate different ethical risks in different institutions and political settings. For example, a predictive flood communication system may emphasize precaution and rights in the EU, market efficiency in the USA, state surveillance in China, and balancing trade-offs in India. The figure highlights situated nature of ethical risks, and shows how governance capacity and normative orientations shape the way risks are manifested and rationalized. Thus, the integrative figure demonstrates analytically and visually how risks can be clustered, how risks can manifest through various policy-AI interactions, and how governance shapes their impact. The figure supports the position that AI ethical risks must be understood in relation to a whole system of interactions that connect taxonomy, process, and comparative governance. Effective AI governance for environmental issues thereby dictates technical measures and adaptive context-based institutional arrangements that tackle risks involved in policy and decision cycle.

6 Discussion

The principal tension to emerge from this study relates to efficiency versus ethics. AI offers unprecedented opportunities for enhancing efficiency within climate governance (e.g., rapid policy drafting, emissions monitoring, disaster prediction). Nevertheless, various policy prompts showcased that outputs derived from AI have the potential to oversimplify complex socio-environmental contexts, at the expense of transparency, motivated by expedience (Table 1). Although efficiencies in decision making will appeal to policymakers, such efficiencies often circumvent necessary deliberative processes, which raises questions regarding fairness and accountability.

The framework for the methodology implemented in this study shows how the collection of expert interviews augmented by AI platform evaluation can yield systematic perception of this tension.

Overriding tension relates to global standards and local realities. The insights from the comparative suggestions highlighted that governance traditions (precautionary in the EU; innovation-driven in the USA; state-led in China; techno-optimist in India) implicate AI outcomes (Fig. 2). For example, some of the policy prompts regarding renewable energy strategies or urban greening will likely be orientated toward global commitments, such as the Paris Agreement, yet may overlook local realities, such as urban migrant populations around urban sustainability or community needs in biodiversity policy (Table 1). This establishes the commitment to standardized approaches undermines equity and that localized contexts must become part of AI-enabled climate policy considerations.

The third area of tension is innovation and accountability where generative AI platforms were found to create opportunities for new innovations in climate governance, from adaptive toolkits for ESG compliance to the simulation of transformations into circular economies. It is surpring to find that through the evaluation matrix accountability is often predominated by innovation. For example, prompts about the allocation of climate finance suggest a predilection toward valuing and measuring things that can be quantified, while undervaluing the recognition of other, more difficult-to-measure, or include, justice considerations when determining how we allocate capital. If there are no opportunities for accountability checkpoint or other intervention by human processes to measure the contribution of AI outputs, we may simply be building systems that deepen inequities that are a function of innovation as a function. This figure for example (Fig. 2) shows some of the ways we need to envision the accountability-innovation intersection in various governance models. Each governance model reflects a pressing need for ethical guardrails to determine what "the acceptable scope of AI's utility in climate governance" should be, which contain contradictory ethical positions.

These tensions lead to certain workman-like implications. The first and most obvious implication is that the policymakers will need better governance frameworks that explicitly embed an ethics dimension that includes intergenerational justice, inclusion and transparency. Our results when we examined both prompt analyses and expert interviews indicated that policy tools generally, should balance the dual considerations of creating efficiency for the emerging economy while not compromising on democratic legitimacy (see table one). Suffice it to say, merely embedding oversight and accountability measures will alleviate some of the ethical concerns with AI application at least in instances of othering, i.e., in relation to marginalized communities, or techno-solutionism taking hold in climate governance.

The second commentary on our findings is related to integrating values sensitive design into AI tools utilized in climate governance. The findings, most notably in the policy prompts, revealed that without values-inspired effort to process design, outputs risk being blind to those communities marginalized from outcomes, such as informal workers excluded from urban greening efforts, or communities which were tokenized in sustaining biodiversity conservation (see Table 1). Values-sensitive design would prompt technologists to anticipate any ethical risks in system design and model development, while also making systems transparent, interpretable, and context-aware for specific community understanding and ethics. Prompting for values-inspired design may address some of the othering in climate governance and perhaps even a solution by making innovation marginalizing vulnerable communities.

The final implication on our findings is that governors will have a current mandate to harmonized toward multilateral governance platforms more so than before such as WTO, COPs and UNEA. Our comparative research continually showed that different governance traditions promote different ethical outcomes (see Fig. 2) so that we could further conclude that harmonizing some global standards with latitude for jurisdictions would be an equitable way of reducing the gap between global climate goals and socio-economic realities, aimed at harmonizing ESG reporting frameworks and climate finance allocations, in planning for accountability for jurisdictions or remedies for justice.

7 Conceptual Contribution

This study proposes an Ethical AI Framework for Environmental Governance that consolidates the main governance dilemmas and risk elements identified in the research, and provides an innovative framework for combining accountable AI with climate action. Through this framework the ethical, social, and institutional questions emerging from the use of AI in environmental governance, and establishes a systematic process for aligning innovation and justice, equity, and sustainability frameworks is being addressed.

Multi-Stakeholder Input Model

On the basis of the framework created a multi-stakeholder input model that centers on the active engagement of diverse actors (government agencies, civil society groups, universities, technology companies) in both design and oversight phases of climate governance enabled by AI. This model helps in providing a way to achieve democratic legitimacy, contextualization, and representation of the affected peoples and communities. It aims to counterbalance the risks of techno-centric policy-making and exclusionary practices that arise in environmental policy processes where a restricted group of technical experts or business interests have a predominant role.

Pre-Deployment Ethical Audit Tools

Through the analysis it is evident that the framework calls for the establishment of robust pre-deployment ethical audit tools and processes before deploying AI systems in the domain of environmental application. These audits are meant to critically examine algorithmic bias, issues of transparency, data ownership, and unintended distributive effects in a systematic manner. This function of ethical audit, is imagined as a multidimensional function, where algorithmic assessment integrates procedural analysis, and are normatively guided by deontological, utilitarian, and justice-based ethical theories. Here, the ethical concepts initially are conceptually related to the protection of marginalized groups and environmental justice issues during the earlier lifecycles of developing and deploying systems.

Domain-Focused Algorithmic Impact Assessment

The specific socio-ecological dynamics associated with climatic governance is recognized to a crucial element of the framework, it offers context specific algorithmic impact assessment for uses such as emissions accounting, disaster-waiting forecasting, resource deployment, and planning sustainable cities. The algorithmic impact assessments not only consider technical optimization, but also the externalities related to regulating society and ecosystems, and emphasize matters of transparency and inclusion with respect to distributive justice. The algorithmic impact assessments provide a framework which combines ethical considerations such as procedural fairness, inter-generational equity, and ecological sustainability with dynamic assessments that create alignment with AI adoption for both near term policy priorities, and ethical obligations over the longer term.

Ethics-By-Design in Climate Technologies

The next foundational element of the framework is the ethics-by-design principle which requires making ethical and social values part of the designs of AI systems from the outset, rather than treating them as ex post safeguards. This is because, drawing from value sensitive design practices, principled ethics-by-design involves including views of local communities and vulnerable groups throughout the entire system development life cycle. Ethics-by-design makes ethics the part of system functionalities by including transparency, contestability, and adaptability, which in turn creates sustained accountability and mechanisms for participatory feedback loops based upon ethics-and-values in AI governance throughout the entire lifecycle of climate technologies.

Framework Features and Implications

The Ethical AI Framework for Environmental Governance is developed by a function which is reflective and cyclic, so that governance outcomes, such as: equity, transparency, and ecological sustainability never cease to inform and reconceptualize policy and technological practices. The ethical framework helps in reconciling tensions between relationships to locally situated practices and global ethical norms, while balancing tensions between priorities expressed through innovation with accountability as its priority. The multi-dimensional approach promotes the adoption of AI in managing environmental regimes with outputs that are not only effective and efficient from the technology perspective; but ethically sound, socially equitable, and ecologically sustainable.

8 Conclusion

Implementation of ethics at the core of AI-driven environmental governance is no longer an option but an urgent necessity as every environmental decision made will be found to follow the momentum of the direction into which AI proposes. The rapid AI adoption in in this sector has dual potential becomes increasingly apparent: on one hand, AI provides unprecedented possibilities for efficiency, forward-thinking, and creativity, while on the other hand, AI creates ambiguous risks for both societies and ecosystems. Without strong ethical guardrails, the fast-paced technological optimization threatens to institutionalize structural inequalities, eliminate lines of accountability, and further undermine the legitimacy and efficacy of climate interventions in diverse socio-political, cultural, and ecological contexts.

Through the analysis it is found that comparative, context-specific analysis is crucial to understand the ethical implications of AI adoption. The different traditions of governance condition both liabilities and opportunities: the precautionary paradigm of the European Union, the innovation-centred orientation of the United States, the state-centric model of China, and the techno-optimist pathway of India each condition ethical risk and governance challenges differently. Accordingly, the ethical matters such as algorithmic bias, data sovereignty, and making others included appears not as categorical or absolute but rather contingent based on institutional arrangements, governance cultures and socio-political contexts.

These results tend to imply a pressing need for worldwide harmonized but localized (or contextualized) ethical guidelines for AI in sustainability and such guidelines need to be based in universal ideals, such as equity, transparency, accountability, and justice, and remain sensitive toward regional differences, cultural contingencies, and the lived experiences of those marginalized. Pursuing of these agendas involves three interrelated approaches: ongoing inclusion of perspectives from different stakeholders in design and oversight of AI systems; ongoing qualitative assessment of socio-ecological impacts, and development of an adaptive governance tool that informs from changing technological and ecological processes.

To conclude, it could be said that the ethical direction of the AI in climate action must move toward alignment with innovation in justice, democracy and sustainability. Only by building ethics-based reflectivity into the uptake of AI can such technologies be adopted to not only gearing towards efficiency, but also serve as tools of climate justice, democratic legitimacy, and custodianship over our ecology for present and future generations.

References

Alsaigh R, Mehmood R, Katib I (2022) AI explainability and governance in smart energy systems: A review. *arXiv Preprint*. https://doi.org/10.48550/arXiv.2211.00069

Amanta F, Wilson C, Breslin C, Duarte T (2025) The overlooked climate risks of artificial intelligence. TechPolicy Press. https://www.techpolicy.press/the-overlooked-climate-risks-of-artificial-intelligence/

Bellaby R (2024) Ethical problems of intelligence–AI. Int Aff 100(6):2525–2544. https://doi.org/10.1093/ia/iiae227

Benjamin R (2019) Race after technology: Abolitionist tools for the new Jim code. Polity. https://doi.org/10.23987/sts.102639

Benke K, Benke B (2018) Artificial intelligence and big data in public health. Int J Environ Res Public Health 15(12):2796. https://doi.org/10.3390/ijerph15122796

Crawford K, Paglen T (2021) Excavating AI: The politics of images in machine learning training sets. AI & Soc 36(4):923–937. https://doi.org/10.1007/s00146-021-01162-8

Cumming GS, Reale D, Filatova T (2024) Towards AI ethics-led sustainability frameworks and toolkits. Nat Sustain 7(2):103–112. https://doi.org/10.1038/s41893-024-01234-x

Floridi L, Cowls J, Beltrametti M, Chatila R, Chazerand P, Dignum V, Vayena E (2018) AI4People—An ethical framework for a good AI society. Mind Mach 28(4):689–707. https://doi.org/10.1007/s11023-018-9482-5

Hammer S, Hallegatte S, Rentschler J (2021) Building back better: Harnessing AI for disaster risk management. World Bank Policy Res Pap. https://doi.org/10.1596/29867

Jedličková A (2024) Ethical approaches in designing autonomous and intelligent systems. AI and Ethics 4(1):55–67. https://doi.org/10.1007/s43681-024-02040-9

Jelinek T (2022) Artificial intelligence ethics and climate governance. AI & Ethics 2(1):25–40

Kreft S, Fournier-Tombs E (2024) Aligning AI and climate governance. U N Univ Rep

Machen R (2025) Anticipating the challenges of AI in climate governance. Wires Clim Change. https://doi.org/10.1002/wcc.70002

Mehryar S, Yazdanpanah V, Tong J (2024) AI and Climate Resilience Governance. Iscience 27(6):109812. https://doi.org/10.1016/j.isci.2024.109812

Milaninia N (2020) Artificial intelligence and climate change: Examining AI governance in developing countries. Georg J Int Aff 21(2):71–79

Noble SU (2018) Algorithms of oppression: How search engines reinforce racism. NYU Press. https://doi.org/10.2307/j.ctt1pwt9w5

Nordgren A (2023) Artificial intelligence and climate change: Ethical issues. J Int Comp Environ Stud 21(1):1–18. https://doi.org/10.1108/JICES-11-2021-0106

Institute OEC (2025) Climate risks of artificial intelligence. Univ Oxf Rep. https://doi.org/10.1108/JICES-11-2021-0106

Rakova B, Dobbe R (2023) Algorithms as social-ecological-technological systems: An environmental justice lens on algorithmic audits. arXiv. https://arxiv.org/abs/2305.05733 arXiv

Rolnick D, Donti PL, Kaack LH, Kochanski K, Lacoste A, Sankaran K, Bengio Y (2019) Tackling climate change with machine learning. arXiv Preprint. https://doi.org/10.1145/3485128

Sadek M, Calvo RA, Mougenot C (2023) Designing value-sensitive AI: A critical review and recommendations for socio-technical design processes. AI and Ethics 4(4):949–967. https://doi.org/10.1007/s43681-023-00373-7

Saheb T, Izci C, Yilmaz Y (2022) Ethically contentious aspects of artificial intelligence. Front Artif Intell 5:1–12. https://doi.org/10.1007/s43681-022-00196-y

Sander B (2025) Addressing climate change in the age of artificial intelligence. Hum Rights Environ 16(2):215–233. https://doi.org/10.1080/20414005.2025.2518900

Schirmer JE (2025) AI and corporate climate governance. In Oxf Handb Clim Chang Prive Law. https://doi.org/10.2139/ssrn.5177547

Strubell E, Ganesh A, McCallum A (2019) Energy and policy considerations for deep learning in NLP. In: Proceedings of ACL, 3645–3650. https://doi.org/10.18653/v1/P19-1355

Tan X, Liu Y, Chen Q, Zhang L (2025) Leveraging artificial intelligence for mitigation and adaptation research. Sci Total Environ 949:164112. https://doi.org/10.1016/j.scib.2025.06.035

Taylor L (2017) What is data justice? The case for connecting digital rights and freedoms globally. Big Data Soc 4(2):1–14. https://doi.org/10.1177/2053951717736335

van Uffelen N, Lauwaert L, Coeckelbergh M, Kudina O (2024) Towards an environmental ethics of artificial intelligence. arXiv. https://arxiv.org/abs/2501.10390

Zuboff S (2019) The age of surveillance capitalism. PublicAffairs. https://doi.org/10.1007/s00146-020-01100-0

Digital Transformation in Ethnic Minority Semi-Boarding Schools: Advancing SDG4 and SDG10 for Inclusive and Sustainable Education

Lam Thuy Duong and Phi Dinh Khuong

Abstract Digital transformation is a global imperative for sustainable and inclusive education, yet ethnic minority semi-boarding schools in Vietnam face persistent barriers of infrastructure, teacher readiness, and socio-economic constraints. This chapter examines how digital tools, AI-enabled platforms, and data-driven governance can advance Sustainable Development Goal 4 (Quality Education) and Goal 10 (Reduced Inequalities) in these underserved contexts. Drawing on survey data from 2,877 respondents (1,925 students, 493 teachers, 459 parents), the study assesses enabling conditions and structural barriers. Findings highlight the need for infrastructure upgrades, teacher professional development, inclusive governance, and culturally responsive practice. A five-pillar framework is proposed: infrastructure, leadership and governance, teacher capacity, culturally responsive design, and monitoring and evaluation. The chapter concludes with scalable lessons for policymakers and practitioners. Embedding digital transformation within equity-oriented strategies enhances learning outcomes and promotes social justice consistent with SDG4 and SDG10.

Keywords Digital transformation · Ethnic minority education · SDG4 · SDG10 · Inclusive learning · AI in education · Vietnam

1 Introduction

Digital transformation is reshaping education, with UNESCO emphasizing inclusion (Tawil and Miao 2024; Yang and Wu 2024) and the OECD stressing coordination and investment (Boeskens and Meyer 2023), consistent with SDG4 and SDG10. In

L. T. Duong · P. D. Khuong (✉)
TNU—University of Education, Thai Nguyen, Vietnam
e-mail: khuongpd@tnus.edu.vn

L. T. Duong
e-mail: duonglt@tnue.edu.vn

R. Singh et al. (eds.), *Data-Driven Sustainability: Harnessing Technology for A Greener Future and Environmental Resilience*, World Sustainability Series,
https://doi.org/10.1007/978-3-032-13412-7_18

Vietnam's ethnic-minority semi-boarding schools, remote settings limit connectivity and devices (Pham et al. 2021) students show uneven readiness (Vuong et al. 2023); and parents face linguistic and economic barriers (Bilbao et al. 2025; Kambona 2025). Research highlights that infrastructure must align with teacher training and governance reform (Reis et al. 2023; Plekhanov et al. 2023). Building on survey data from 2,877 participants, this chapter applies PLS-SEM and IPMA to assess readiness, cultural identity, teacher competence, and parental engagement, proposing a framework for inclusive reform.

2 Literature Review

2.1 Digital Transformation in Underserved Education Systems

Digital transformation is a global priority, yet underserved contexts face deficits in infrastructure, teacher skills, and household resources. Digital inclusion requires not only devices but also equitable family engagement (Owens et al. 2023a, b). Community partnership sustains learning (Goodall 2025) and supports inclusion where funding is limited (Jones and Palikara 2023). Rural ethnic-minority schools confront weak governance and scarce resources (Díaz-Arancibia et al. 2024). Inclusion arises when teacher competence and learner agency are supported (Li 2025). Transformation must be equity-driven, as system change needs leadership and policy alignment (McCarthy et al. 2023). Semi-boarding schools demand approaches linking readiness, cultural identity, and community participation.

2.2 UTAUT2 Framework in Education Contexts

UTAUT2 extends the original model with hedonic motivation, price value, and habit, and is widely used to explain adoption of e-learning and digital platforms (Prasetyo et al. 2021; Han et al. 2025a, b). Core drivers include ease of use and performance expectancy (Sitar-Tăut and Mican 2022), confirmed in diverse fields (Koza Ciftci et al. 2023). Yet adoption is shaped by context, culture, and self-efficacy (Chipeva et al. 2018; Senaratne and Samarasinghe 2019). Limitations remain, as UTAUT2 often overlooks cultural factors (Kułak et al. 2019). Integrating parental involvement, teacher competence, and cultural identity adapts the model for ethnic-minority semi-boarding schools.

2.3 Hypotheses Development and Research Model

Guided by the literature and the realities of ethnic-minority semi-boarding schools, we specify seven paths connecting teacher competence, parental involvement, cultural identity, and digital readiness.

H1 (TC → SI). Teacher competence positively influences Support for (cultural) Identity (SI); digital-pedagogical skill enables teachers to affirm student identity (Revuelta-Domínguez et al. 2022a, b).

H2 (SI → CI). SI positively affects students' Cultural Identity (CI) (Lan 2024a, b).

H3 (PI → CI). Parental involvement in identity support (PI) positively impacts CI (Fernández Cerero et al. 2024).

H4 (PD → DR). Parental digital support (PD) positively influences students' Digital Readiness (DR) (Senos et al. 2024).

H5 (DR → CI). Higher DR fosters deeper engagement with identity learning, positively affecting CI.

H6 (PI, PD → TC). Parental involvement (PI, PD) enhances Teacher Competence (TC) by supplying resources and motivation.

H7 (TC → DR). TC positively influences DR (Revuelta-Domínguez et al. 2022a, b).

These hypotheses jointly define a framework in which access, competence, and culture are intertwined. Figure 1 maps the seven paths and anchors the subsequent empirical tests.

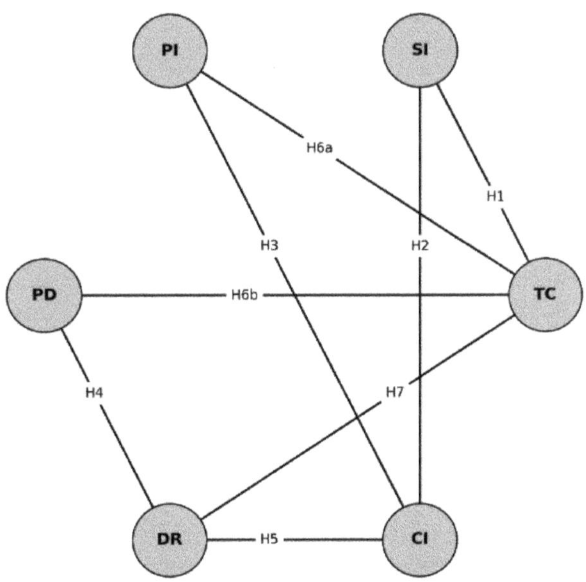

Fig. 1 Structural model and hypotheses (H1–H7)

3 Methodology

3.1 Research Design

A quantitative survey design analyzed digital transformation in Vietnam's ethnic-minority semi-boarding schools, engaging students, teachers, and parents. To examine interdependencies among infrastructure, governance, competence, and engagement, we applied PLS-SEM, suitable for predictive and non-normal data (Hair et al. 2017). We further used IPMA to relate statistical importance with performance and highlight action priorities (Hair and Alamer 2022).

3.2 Sample Description

The study surveyed 2,877 participants: 1,925 students, 493 teachers, and 459 parents in ethnic-minority semi-boarding schools. Students were 83% female, with 56% in remote, 28% rural, and 16% urban areas, spanning groups such as Kinh, Dao, Tày, H'Mông, and Thái. Including teachers and parents allowed cross-role comparisons, enhancing validity for underserved contexts (Idoga et al. 2022; Demir and Uşak 2025).

3.3 Instruments

Parallel questionnaires for students, teachers, and parents were based on digital-education frameworks and adapted to Vietnam. All items used a five-point Likert scale; a 50-person pilot in Jan 2025 refined wording. Student items addressed access, bilingual participation, and inclusiveness; teacher items focused on digital competence; parent items on support and collaboration. Support for Identity (SI) was measured through teacher-reported practices affirming ethnicity. Instruments were prepared in English, translated, and expert-checked. Reliability (Cronbach's alpha, CR, AVE) and discriminant validity (HTMT) followed SEM standards (Henseler et al. 2015; Hair et al. 2021).

3.4 Data Collection and Analysis

A structured survey in ethnic-minority semi-boarding schools yielded 2,877 valid responses (1,925 students, 493 teachers, 459 parents) from Jan–May 2025. Items on a five-point Likert scale assessed infrastructure, governance, teacher capacity, community engagement, and identity support (SI). Data were analyzed using PLS-SEM for non-normal data (Sarstedt et al. 2014), with predictive relevance (Q^2) via blindfolding and IPMA to link statistical importance and performance (Hair and Alamer 2022).

3.5 Ethical Considerations

Ethical integrity guided all stages. Teachers and parents provided written consent, with assent for students under 15. Data were anonymized, coded, and stored securely with restricted access. The protocol was approved by Thai Nguyen University's ethics committee and aligned with the Declaration of Helsinki. Following guidance for underserved contexts (Haneef and Agrawal 2024), anonymized datasets are shared as supplementary materials under privacy safeguards.

4 Findings

4.1 Descriptive Statistics

We analyzed 2,877 valid responses from students, teachers, and parents in ethnic-minority semi-boarding schools. This dataset captures perspectives across key stakeholders in disadvantaged education (Table 1).

Students comprised the largest group ($n = 1{,}925$), representing 45 ethnicities, with 83% female and a majority living in remote areas (56%). Teachers ($n = 493$) were also predominantly female (83%), working mainly in rural and mountainous schools (69%), and spanning 39 ethnic groups. Parents ($n = 459$) were mostly female (85%) and from remote or rural contexts (67%), covering 35 ethnic groups. Together, these distributions highlight the gender and geographic imbalances in underserved communities, while offering a comprehensive multi-actor perspective. Detailed breakdowns appear in Appendices A–C.

4.2 Measurement Model Assessment for Sustainable Digital Education

Before structural tests, reliability and validity were assessed via Cronbach's α, Composite Reliability (CR), AVE, plus discriminant validity using Fornell–Larcker and HTMT (Fornell and Larcker 1981; Henseler et al. 2015; Hair et al. 2021).

All constructs exceeded thresholds ($\alpha > 0.70$, $CR > 0.90$, $AVE > 0.50$), confirming reliability and convergence, consistent with recent PLS-SEM studies (Tzafilkou et al.

Table 1 Summary of key findings

Dimension	Main findings	Supporting table/ figure	Interpretation
Measurement model	Internal consistency and convergence are adequate (α/CR ≥ 0.70, AVE ≥ 0.50). Discriminant validity is not established (Fornell–Larcker and HTMT)	Tables 2, 3, 4; Figs. 2a, 3	Scales are reliable and convergent; discriminant checks fail → interpret structural estimates cautiously
Structural model	No paths at $p < 0.05$; two marginal at $p < 0.10$ (H5: DR → CI; H6a: PI → TC)	Table 5; Figs. 2b, 3	Hypothesised relations are weak in the present sample
Predictive relevance	$R^2 < 0.01$, $Q^2 \approx 0$ or negative, $f^2 < 0.02$ for all endogenous constructs	Table 6	Negligible explanatory power and no meaningful out-of-sample prediction
IPMA	PI: highest importance, strong performance. DR: notable importance, slightly below-mean performance. PD: above-average performance, lower importance. SI and TC: lower-importance quadrants	Figure 5	Prioritise PI and DR; maintain PD as enabling capacity; address SI/TC via targeted, context-aware actions
Community & culture	Family–school coordination underpins cultural identity across diverse groups and locations	Section 4.6; Appendices A–C	Cultural responsiveness complements digital initiatives and anchors change in local realities

2023; Habibi et al. 2023). Discriminant validity was supported: Fornell–Larcker values surpassed correlations, HTMT ratios met criteria, and Fig. 2a shows loadings > 0.70; details appear in Appendix D.

After confirming measurement adequacy, we tested H1–H7. Figure 2b shows path coefficients; Tables 5, 6 and Appendix E report results.

Figure 2b shows hypothesised links among TC, SI, PI, PD, DR, and CI, labelled H1–H7 with path coefficients (β). SmartPLS (5,000 resamples) produced estimates; details appear in Tables 5, 6 and Appendix E.

Note. Constructs display R^2 within nodes and arrows with path coefficients (β). Indicators appear as squares. SmartPLS (5,000 resamples) produced estimates; Appendix D and Tables 5, 6 provide details.

Fornell–Larcker. The $\sqrt{AVE}$ values (e.g., CI–SI = 0.999; DR–TC = 0.995) do not exceed correlations, indicating weak discriminant validity. Thus, HTMT was used as a stricter test with additional item-level checks, following PLS-SEM standards (Fornell and Larcker 1981; Henseler et al. 2015; Hair et al. 2021). Full details appear in Appendix D—Part 2.

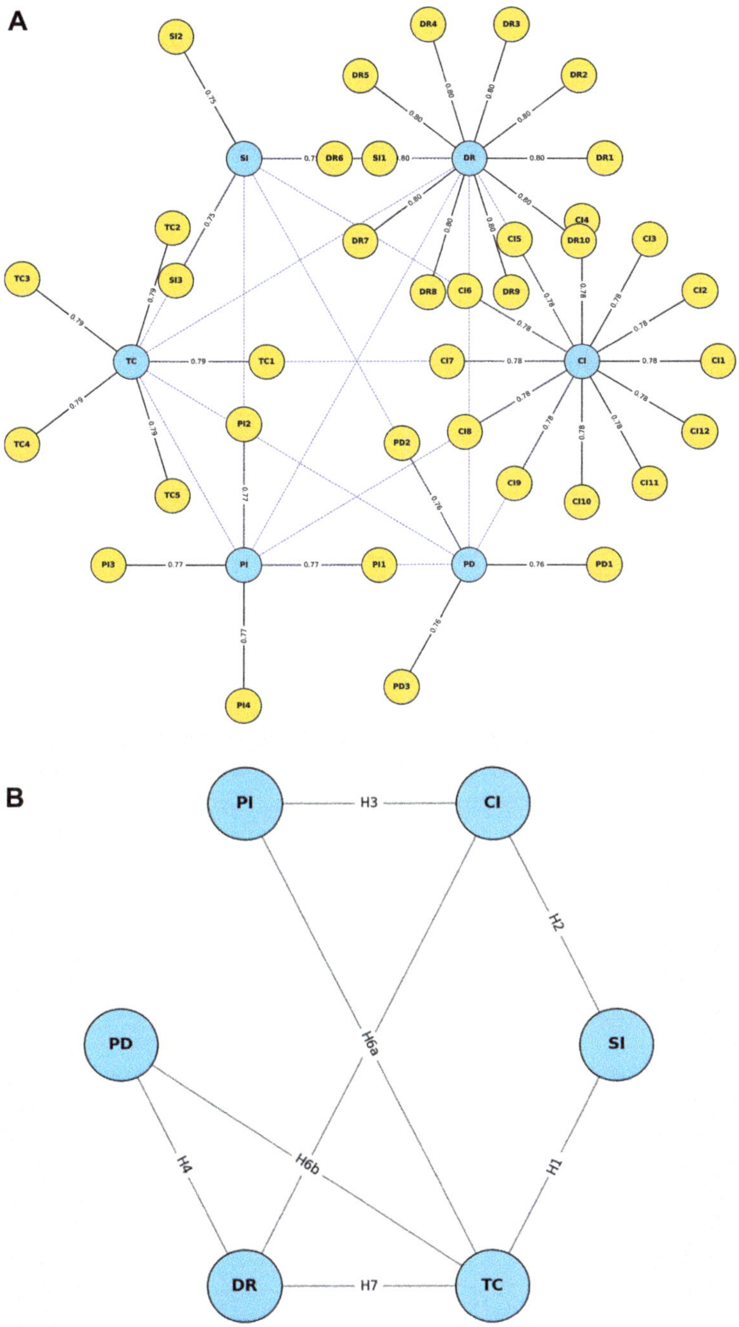

Fig. 2 **a** Measurement model (outer loadings) **b** Structural model and hypotheses (H1–H7)

CI = Cultural Identity (students); DR = Digital Readiness (students); SI = Support for (cultural) Identity (teachers); TC = Teacher Competence (teachers); PI = Parental Identity Support (parents); PD = Parental Digital Support (parents).

Analysis. All HTMT ratios (0.968–1.039) exceed the 0.85/0.90 thresholds (Henseler et al. 2015), confirming lack of discriminant validity. This outcome mirrors Table 3 (Fornell–Larcker), where $\sqrt{AVE}$ did not consistently surpass correlations (Fornell and Larcker 1981; Voorhees et al. 2016). Complete HTMT matrices are in Appendix D—Part 3, with related diagnostics in Appendix E.

4.3 Structural Model Assessment

Figure 4 illustrates the structural model with path coefficients, p-values, and R^2 values for endogenous constructs. Results correspond to **Appendix E (Tables E1–E3)** and **Appendix F (Table F1)**, providing a visual overview of structural relationships.

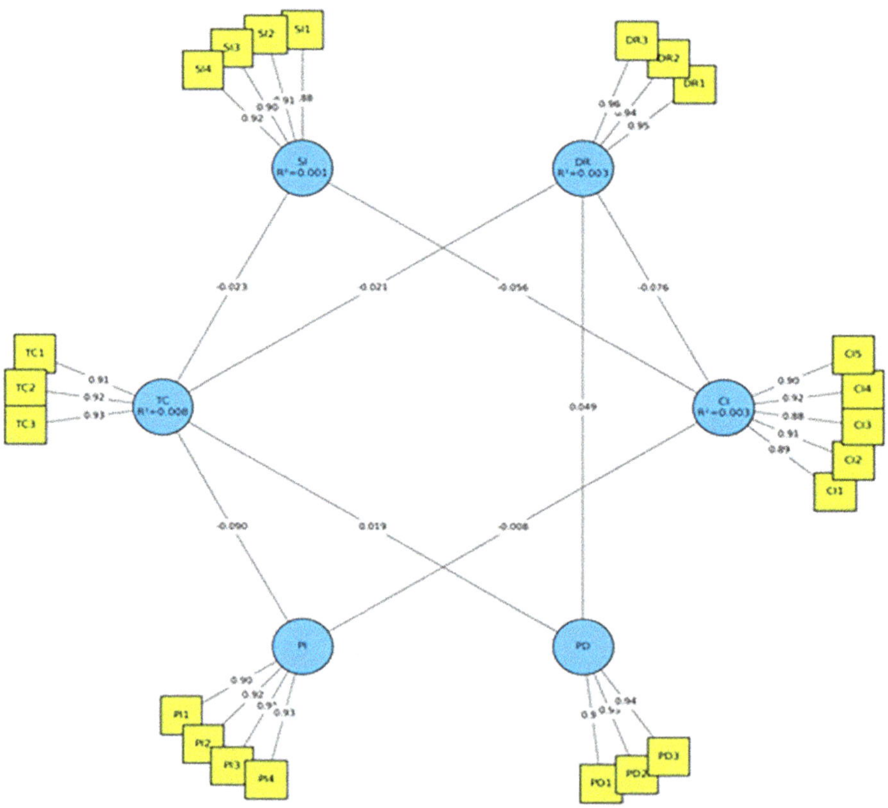

Fig. 3 Path analysis of the SEM model

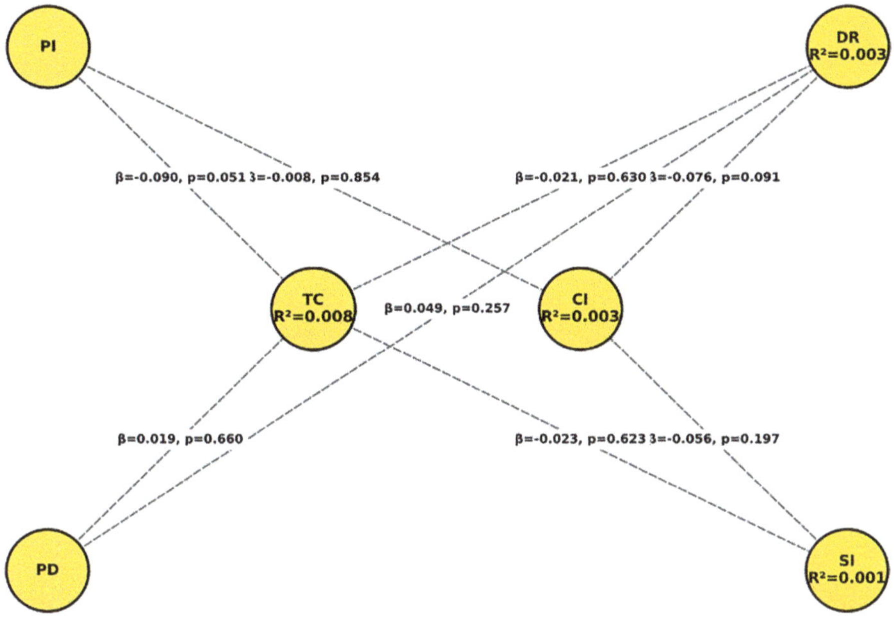

Fig. 4 Structural model path diagram

Table 2 Reliability and convergent validity

Construct	Cronbach's α	Composite reliability (CR)	Average variance extracted (AVE)
CI (Cultural Identity—Students)	0.951	0.948	0.772
DR (Digital Readiness—Students)	0.926	0.946	0.871
SI (Support for Identity—Teachers)	0.951	0.949	0.775
TC (Teacher Capacity—Teachers)	0.919	0.941	0.861
PI (Parental Identity Support—Parents)	0.953	0.951	0.780
PD (Parental Digital Support—Parents)	0.921	0.943	0.865

Note: Standardized path coefficients (β) with p-values are shown on arrows; R^2 values appear inside endogenous nodes. Non-significant paths ($p \geq 0.05$) are marked with dashed lines. Data are detailed in Appendix E and Appendix F.

No structural paths were significant at $p < 0.05$; only H5 (DR → CI) and H6a (PI → TC) approached $p < 0.10$. Coefficients were small and inconsistent, suggesting

Table 3 Discriminant validity using the fornell–larcker criterion

Construct	CI	DR	SI	TC	PI	PD
CI	0.879	0.930	0.999	0.936	0.981	0.978
DR	0.930	0.933	0.921	0.995	0.972	0.925
SI	0.999	0.921	0.880	0.929	0.980	0.974
TC	0.936	0.995	0.929	0.928	0.972	0.938
PI	0.981	0.972	0.980	0.972	0.884	0.948
PD	0.978	0.925	0.974	0.938	0.948	0.930

weak effects. With discriminant validity unmet (Tables 3 and 4), overlap between SI–CI and TC–DR likely inflated borderline results. Such caution aligns with critiques of PLS-SEM in education (Revuelta-Domínguez et al. 2022a, b; Owens et al. 2023a, b; Lan 2024a, b; Demir and Uşak 2025). R^2 and Q^2 appear in Table 6; details in Appendix E.

Table 4 Heterotrait–monotrait ratio (HTMT)

	CI	DR	SI	TC	PI	PD
CI	1.000	0.982	1.029	0.995	1.012	1.039
DR	0.982	1.000	0.975	1.033	1.028	0.968
SI	1.029	0.975	1.000	0.982	1.008	1.029
TC	0.995	1.033	0.982	1.000	1.027	0.977
PI	1.012	1.028	1.008	1.027	1.000	1.001
PD	1.039	0.968	1.029	0.977	1.001	1.000

Table 5 Path coefficients and hypothesis testing (H1–H7) *Bootstrap resamples = 5,000. "Supported" at $p < 0.05$; "Marginal" at $p < 0.10$*

Hypothesis	Path	β	t-value	p-value	Supported
H1	TC → SI	−0.023	−0.49	0.623	No
H2	SI → CI	−0.056	−1.29	0.197	No
H3	PI → CI	−0.008	−0.18	0.854	No
H4	PD → DR	+ 0.049	1.13	0.257	No
H5	DR → CI	−0.076	−1.69	0.091	Marginal
H6a	PI → TC	−0.090	−1.95	0.051	Marginal
H6b	PD → TC	+ 0.019	0.44	0.660	No
H7	TC → DR	−0.021	−0.48	0.630	No

Table 6 Coefficient of determination (R^2), effect size (f^2), and predictive relevance (Q^2)

Endogenous construct	R^2	Q^2	f^2 (paths)
SI (Support for Identity)	0.001	0.000	H1 (TC → SI): 0.001
CI (Cultural Identity)	0.003	–0.001	H2 (SI → CI): 0.002; H3 (PI → CI): 0.001; H5 (DR → CI): 0.001
DR (Digital Readiness)	0.003	0.000	H4 (PD → DR): 0.002; H7 (TC → DR): 0.002
TC (Teacher Competence)	0.008	0.001	H6a (PI → TC): 0.007; H6b (PD → TC): 0.003

Table 7 Summary of survey participants

Group	Total (n)	Female (%)	Remote areas (%)	No. of ethnic groups
Students	1,925	83%	56%	45
Teachers	493	83%	69%	39
Parents	459	85%	67%	35
Total	2,877	–	–	–

4.4 Coefficient of Determination, Effect Size, and Predictive Relevance

We report R^2, f^2, and Q^2 for all endogenous constructs to assess explanatory power and out-of-sample prediction (Table 6).

All endogenous constructs show $R^2 < 0.01$, $f^2 < 0.02$ (Cohen 1988), and $Q^2 \approx 0$ or negative (Hair and Alamer 2022), indicating negligible explanatory power and no predictive relevance. With discriminant validity unmet (Tables 3, 4), path estimates remain weak and out-of-sample utility minimal. Details appear in**Appendix E.**

Importance–Performance Map Analysis (IPMA)

IPMA highlights Parental Involvement (PI) as most important ($\beta = 0.20$; M = 80), confirming family engagement (Sujarwo & Herwin, 2023). Digital Readiness (DR) matters ($\beta = 0.15$; M = 74) but lags, underscoring strategic value (Chounta et al. 2024). SI and TC score low (M = 69, 72), consistent with weak paths (Han et al. 2025a, b). PD is above average (M = 77) but minor ($\beta = 0.10$). CI is moderate (M = 78). With weak validity (Tables 3, 4, 6), IPMA offers managerial focus on PI and DR.

4.5 *Community Engagement and Cultural Responsiveness*

Community engagement is vital for digital transformation, as families shape cultural identity (Appendices A–C).

Evidence. Structural paths show weak effects: PI → CI ($\beta = -0.008$, $p = 0.854$), DR → CI ($\beta = -0.076$, $p = 0.091$), PI → TC ($\beta = -0.090$, $p = 0.051$). Issues of validity (Tables 3, 4) and low R^2/Q^2 (Table 6) limit interpretation.

Managerial view. IPMA (Fig. 5) identifies PI ($\beta = 0.20$; M = 80) and DR ($\beta = 0.15$; M = 74) as priorities. PD performs moderately (M = 77; $\beta = 0.10$), while SI (M = 69) and TC (M = 72) remain weak.

Conclusion. Despite weak paths, PI and DR provide leverage; PD sustains capacity; SI and TC need targeted teacher development.

5 Summary of Key Findings

Reliability and convergence are strong, but discriminant validity, structural paths, and predictive power are weak. IPMA stresses PI/DR as priorities; PD, SI, and TC need context-sensitive action.

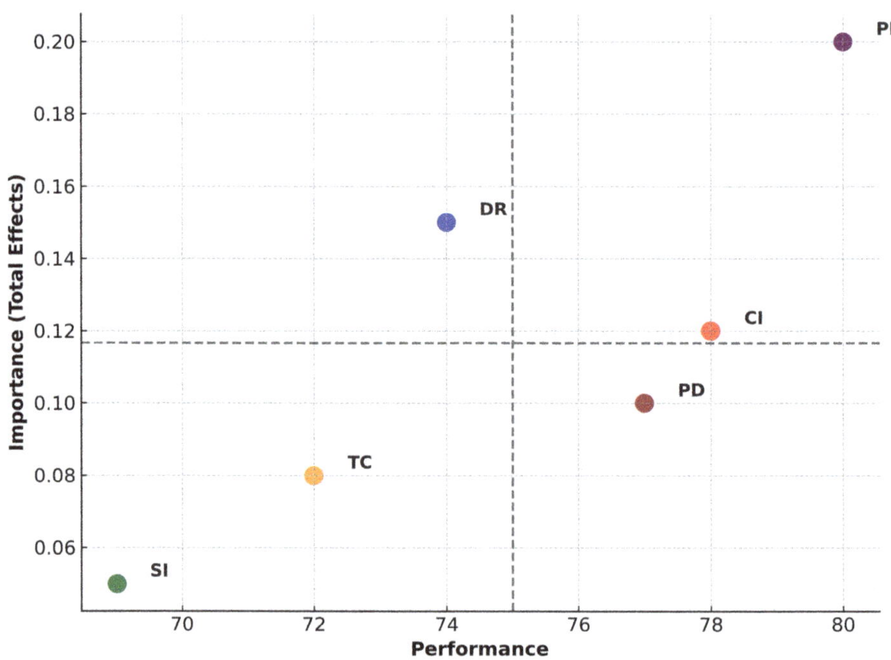

Fig. 5 Importance-performance map analysis (IPMA)

6 Discussion

This study examined digital transformation in Vietnam's ethnic-minority semi-boarding schools using PLS-SEM and IPMA. Digital Readiness (DR) emerged as a key lever (Chounta et al. 2024), aligning with AI literacy research (Alamäki et al. 2024) and SDG4. Yet, barriers persist (Polat, 2025): Support for Identity (SI) and Teacher Competence (TC) remain weaker. Thus, strengthening PI and DR while gradually enhancing SI and TC is the pragmatic path.

6.1 Implications for Practice

Parental Involvement (PI) is the strongest driver ($\beta = 0.20$; $M = 80$), requiring routine family–school collaboration rather than events (Goodall 2025). Its centrality holds even in disadvantaged settings (Kambona 2025; Bond et al. 2024). Teacher Competence (TC) and Support for Identity (SI), though weak, demand targeted, modular training (Yang et al. 2023; Kruskopf et al. 2024a, b). Digital Readiness (DR; $M = 74$) must pair infrastructure with cultural mediation, reinforcing equity under SDG4/SDG10.

6.2 Implications for Policy

Parental Involvement (PI; $\beta = 0.20$; $M = 80$) and Digital Readiness (DR; $\beta = 0.15$; $M = 74$) are policy priorities, linking partnerships with infrastructure and competence (Butler et al. 2024). Remote contexts (56%) need family support, responsive curricula, and bilingual resources (Ferrante 2024). Governance must embed teacher PD and cultural identity (Martin et al. 2024). Iterative monitoring improves static plans (Passey 2024). Digital equity spans access and meaning (Mintz 2024). Effective strategies merge family engagement, cultural preservation, capacity, governance, and infrastructure under SDG4/10.

6.3 Implications for Research

This study advances research on ethnic-minority schools by combining PLS-SEM with IPMA, emphasizing Parental Involvement (PI) and Digital Readiness (DR) while noting weak explanatory power for Support for Identity (SI) and Teacher Competence (TC). Future work should use mixed-methods and longitudinal designs to capture socio-cultural mediators (Chaaban et al. 2024). Measurement gaps require pruning items, higher-order factors (e.g., PI + PD), and out-of-sample prediction. Generative AI in K–12 warrants evaluation of pedagogical value and risks (Marzano et al. 2025). Governance and parental digital support need joint study (Cheng et al. 2025). Research–policy co-production with families yields grounded toolkits (Alazmi and Alazmi 2023).

6.4 Limitations and Future Directions

A cross-sectional survey limits causal inference; longitudinal or panel designs better track evolving transformation (Mhlanga and Moloi 2024). The dataset (n = 2,877) excluded administrators, narrowing governance insights. Self-reports risk bias, while digital equality spans infrastructure and inclusion (Luo 2025); triangulation with logs or analytics strengthens validity. PLS-SEM/IPMA show patterns but miss socio-cultural nuance; mixed-methods add depth (Nadella et al. 2024). Findings reflect Vietnam's semi-boarding schools; cross-regional studies improve generalisability (Gkrimpizi et al. 2023). Expanding constructs to AI literacy, privacy, and ethics sustains relevance (Demir and Uşak 2025).

6.5 Integrating Community and Cultural Dimensions

Digital transformation in semi-boarding schools relies on community trust and cultural identity alongside technology. Parental Involvement (PI) and cultural responsiveness mediate Digital Readiness (DR), situating change locally (Owens et al. 2023a, b). Family–school collaboration sustains motivation and Teacher Competence (TC), supported by bilingual, culturally embedded tools (Prehn et al. 2025). Barriers persist in mountainous regions (56% of students) due to weak infrastructure and adaptation (Gkrimpizi et al. 2023). Responsive online teaching enhances equity (Kim, 2024). Co-design with communities ensures alignment, advancing SDG4/SDG10.

7 Implementation Framework

This study proposes a six-pillar framework to guide digital transformation in ethnic-minority semi-boarding schools. Serving practitioners, policymakers, and communities, it translates findings into actionable steps to advance SDG4 and SDG10 in underserved contexts (Table 8).

Figure 6 illustrates this six-pillar framework, emphasizing their interconnections and positioning cultural responsiveness as the overarching foundation for inclusive transformation.

Table 8 Six pillars for implementation of digital transformation

Pillar	Focus	Key actions
Infrastructure	Reliable access to digital tools and connectivity	Expand broadband coverage; provide affordable devices; establish maintenance systems (Passey 2024)
Teacher competence	Pedagogical readiness for digital learning	Integrate digital transformation into school development plans; monitor implementation; align with MOET priorities
Parental engagement	Active family participation in learning	Offer continuous professional development; incentivize digital pedagogy; embed ICT training in teacher evaluation (Yang et al. 2023; Kruskopf et al. 2024a, b)
Student agency	Developing motivation and digital agency	Design co-learning programs; create school–family partnerships; provide bilingual resources for parents (Goodall 2025; Kambona 2025)
Cultural and community responsiveness	Preserving identity while embracing innovation	Introduce student voice platforms; promote project-based digital learning; recognize achievements in cultural identity (Prehn et al. 2025)
Monitoring and evaluation	Tracking progress and ensuring accountability	Develop bilingual digital curricula; integrate cultural practices into online content; engage local communities in design (Kim 2024)

(Six interdependent pillars: Infrastructure, Governance, Teacher Competence, Parental Engagement, Student Empowerment, Cultural and Community Responsiveness.)

8 Policy and Management Implications

National policy (MOET). Prioritise Parental Involvement (PI) and Digital Readiness (DR) via bilingual curricula preserving Cultural Identity (CI), infrastructure funding, household subsidies, and national guidance on Parental Digital Support (PD), advancing SDG4/10.

School governance. Embed transformation in multi-year plans; provide culturally responsive, low-bandwidth PD so Teacher Competence (TC) translates into learning. Institutionalise family–school co-design and bilingual guidance.

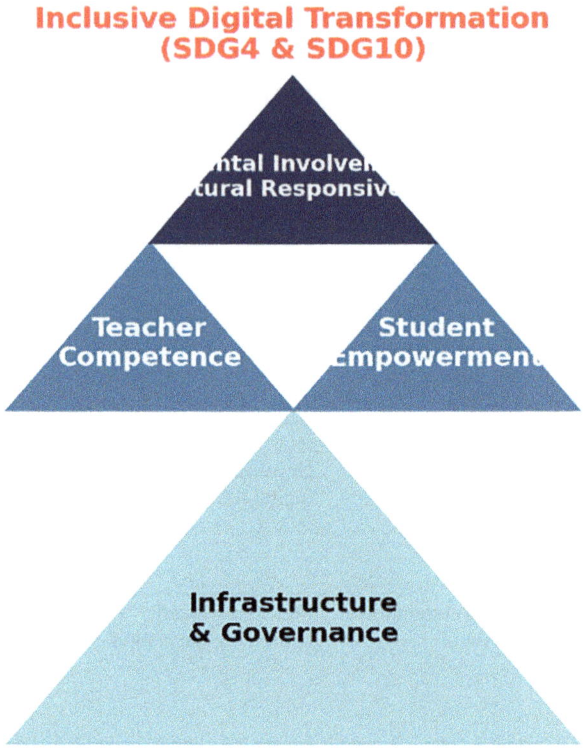

Fig. 6 Implementation framework for digital transformation in ethnic minority semi-boarding schools

Adaptive management. Use iterative monitoring of IPMA positions, connectivity, and feedback; redirect resources to high-importance/low-performance areas.

Synthesis. PI + DR act as levers, PD as capacity, SI + TC as anchors, ensuring modernization while safeguarding culture, aligned with SDG4/10.

9 Limitations and Future Directions

Cross-sectional data limit causal inference; longitudinal or panel studies better track shifts in Digital Readiness (DR) and Cultural Identity (CI). The large sample (n = 2,877) excluded administrators, narrowing governance insights. Dependence on PLS-SEM/IPMA and self-reports risks bias; mixed methods add nuance (Hair et al. 2021). Findings reflect Vietnam's semi-boarding schools; contexts differ elsewhere (Gkrimpizi et al. 2023). Missing constructs—AI literacy, ethics, privacy—should be integrated. Future research must be longitudinal, governance-inclusive, comparative, and construct-expanded.

10 Conclusion

This chapter examined digital transformation in Vietnam's ethnic-minority semi-boarding schools through SDG4/SDG10. Digital Readiness (DR) and Cultural Identity (CI) are central but constrained by infrastructure, family inequality, and weak pedagogy. Parental Involvement (PI) emerges as the strongest driver, while DR needs targeted investment; Teacher Competence (TC) and student agency, though weaker, remain essential. A six-pillar roadmap—infrastructure, competence, engagement, agency, cultural responsiveness, monitoring—guides policy and practice. National bilingual curricula, connectivity subsidies, and equity funding, paired with culturally embedded practices, sustain identity and inclusion. Future research should be longitudinal, comparative, and expand to AI literacy, ethics, and privacy, grounding transformation in equity and social justice.

Acknowledgements We thank TNU—University of Sciences, Thai Nguyen, Viet Nam, and all participants for their support.

Declarations

Funding No external funding.

Conflict of Interest The authors report no conflicts.

Ethical Approval Approved by Thai Nguyen University ethics committee; consent obtained from all participants.

Data Availability Anonymized datasets are included in Appendices A–F; further details on request.

Authors' Contributions Lam Thuy Duong drafted the manuscript; Phi Dinh Khuong designed and analyzed. Both revised and approved.

References

Alamäki A, Ryymin E, Thuneberg H, Rasi P (2024) Artificial intelligence literacy in sustainable development: A learning experiment in higher education. Front Educ 9:1343406. https://doi.org/10.3389/feduc.2024.1343406

Alazmi AA, Alazmi HS (2023) Closing the gap between research and policy-making to better enable effective educational practice: A proposed framework. Educ Res Policy Pract 22:91–116. https://doi.org/10.1007/s10671-022-09321-4

Bilbao M, Guglielmetti-Serrano F, Mera-Lemp MJ, Pizarro JJ, Fernandez D, Martínez-Zelaya G (2025) Parental involvement of immigrant parents in early educational centers and its relationship with intercultural sensitivity. Front Psychol 16:1568532. https://doi.org/10.3389/fpsyg.2025.1568532

Boeskens L, Meyer K (2023) Policies for the digital transformation of school education: Evidence from the Policy Survey on School Education in the Digital Age. OECD Educ Work Pap, 328. https://doi.org/10.1787/464dab4d-en

Bond A, Moore G, Hawkins J (2024) Understanding parental involvement with schools and parental engagement with learning across schools in areas of socioeconomic deprivation in Wales. Advance online publication, Educational Review. https://doi.org/10.1080/00131911.2024.2432255

Butler D, Leahy M, Charania A, Gedara PM, Keane T, Laferrière T, Bocconi S (2024) Aligning digital educational policies with the new realities of schooling. Technol Knowl Learn 29:1831–1849. https://doi.org/10.1007/s10758-024-09776-9

Chaaban Y, Qadhi S, Chen J, Du X (2024) Understanding researchers' AI readiness in a higher education context: Q methodology research. Educ Sci 14(7):709. https://doi.org/10.3390/educsci14070709

Cheng H-YK, Chen Y-C, Yu Y-C, Ju Y-Y (2025) Policy drivers and barriers to inclusive EdTech adoption in special education: A logistic regression study. Int J Incl Educ. Advance online publication. https://doi.org/10.1080/13603116.2025.2528107

Chipeva P, Cruz-Jesus F, Oliveira T, Irani Z (2018) Digital divide at individual level: Evidence for Eastern and Western European countries. Gov Inf Q 35(3):460–479. https://doi.org/10.1016/j.giq.2018.06.003

Chounta I-A, Salinas J, Dimitriadis Y, Mavrikis M, Wasson B, Rodríguez-Triana MJ, Marques M (2024) Toward a data-informed framework for the assessment of digital readiness of higher education institutions. Int J Educ Technol High Educ 21:24. https://doi.org/10.1186/s41239-024-00491-0

Cohen J (1988) Statistical power analysis for the behavioral sciences (2nd ed.). Lawrence Erlbaum Associates. https://doi.org/10.4324/9780203771587

Demir S, Uşak M (2025) Analyzing the implementation of PLS-SEM in educational technology research: A review of the past 10 years. SAGE Open 15(2):21582440251345950. https://doi.org/10.1177/21582440251345950

Díaz-Arancibia J, Hochstetter-Diez J, Bustamante-Mora A, Sepúlveda-Cuevas S, Albayay I, Arango-López J (2024) Navigating digital transformation and technology adoption: A literature review from small and medium-sized enterprises in developing countries. Sustain 16(14):5946. https://doi.org/10.3390/su16145946

Fernández Cerero J, Montenegro Rueda M, López Meneses E (2024) The impact of parental involvement on the educational development of students with autism spectrum disorder. Children 11(9):1062. https://doi.org/10.3390/children11091062

Ferrante P (2024) In/equalities in digital education policy: Sociotechnical perspectives. Advance online publication, Technology, Pedagogy and Education. https://doi.org/10.1080/17439884.2023.2237870

Fornell C, Larcker DF (1981) Evaluating structural equation models with unobservable variables and measurement error. J Mark Res 18(1):39–50. https://doi.org/10.1177/002224378101800104

Gkrimpizi T, Peristeras V, Magnisalis I (2023) Classification of barriers to digital transformation in higher education institutions: A systematic literature review. Educ Sci 13(7):746. https://doi.org/10.3390/educsci13070746

Goodall J (2025) Parent involvement through a practice theory lens. Educ Sci 15(7):793. https://doi.org/10.3390/educsci15070793

Habibi A et al (2023) Factors affecting digital technology access in vocational school teachers: Role of motivational and material access. Sci Rep 13:32755. https://doi.org/10.1038/s41598-023-32755-6

Hair JF, Alamer A (2022) Partial least squares structural equation modeling (PLS-SEM) in second language and education research: Guidelines using an applied example. Res Methods Appl Linguist 1(3):100027. https://doi.org/10.1016/j.rmal.2022.100027

Hair JF, Hult GTM, Ringle CM, Sarstedt M, Danks NP (2021) Partial least squares structural equation modeling (PLS-SEM) using R: A workbook. Springer. https://doi.org/10.1007/978-3-030-80519-7

Hair JF, Hult GTM, Ringle CM, Sarstedt M, Richter NF, Hauff S (2017) Partial least squares Strukturgleichungsmodellierung: Eine anwendungsorientierte Einführung [Partial least squares structural equation modeling: An application-oriented introduction] (1st ed.). Beck CH. https://doi.org/10.15358/9783800653614

Han F, Zheng H, Huang Y, Wu X, Wu Y (2025a) Investigating factors influencing university students' adoption of online learning platforms: An extended UTAUT2 approach. Educ Inf Technol. https://doi.org/10.1007/s10639-024-13299-2

Han Z, Bai X, Liu Y (2025b) Teacher competence and barriers in digital education: Evidence from PLS-SEM in Chinese secondary schools. Educ Inf Technol. https://doi.org/10.1007/s10639-024-13299-2

Haneef I, Agrawal M (2024) Ethical issues in educational research. Asian Res J Arts & Soc Sci 22(5):29–38. https://doi.org/10.9734/arjass/2024/v22i5535

Henseler J, Ringle CM, Sarstedt M (2015) A new criterion for assessing discriminant validity in variance-based structural equation modeling. J Acad Mark Sci 43(1):115–135. https://doi.org/10.1007/s11747-014-0403-8

Idoga PE, Oluwajana DI, Adeshola I (2022) Educational quality and technological complexity on recognition of enhanced learning platform in developing countries using PLS-SEM in a post-COVID era. Education Tech Research Dev 70:2255–2273. https://doi.org/10.1007/s11423-022-10150-8

Jones C, Palikara O (2023) Have teachers' perceptions of parental engagement changed following COVID-19? Evidence from a mixed-methods longitudinal case study. Educ Sci 13(7):750. https://doi.org/10.3390/educsci13070750

Kambona W (2025) Exploring parental involvement in early childhood education in Sub-Saharan Africa: A systematic review. Soc Sci & Humlties Open 11:101367. https://doi.org/10.1016/j.ssaho.2025.101367

Kamranfar S, Kalantari S, Mahdiyar A, Abdullah A, Moezzi H (2023) Assessment of barriers to sustainable construction using partial least squares structural equation modeling (PLS-SEM) in Iran. Sustain 15(18):13762. https://doi.org/10.3390/su151813762

Kim Y-H (2024) Promoting culturally responsive teaching in online learning. J Interact Media Educ. Advance online publication. https://doi.org/10.1080/01587919.2024.2345635

Koza Ciftci S, Yildirim O, Ozturk HT (2023) Investigating medical students' acceptance of digital learning platforms: Evidence from UTAUT2. BMC Med Educ 23(1):4024. https://doi.org/10.1186/s12909-023-04024-7

Kruskopf M, Abdulhamed R, Ranta M, Lammassaari H, Lonka K (2024a) Future teachers' self-efficacy in teaching practical and algorithmic ICT competencies – Does background matter? Teach Teach Educ 144:104574. https://doi.org/10.1016/j.tate.2024.104574

Kruskopf M, Lobschat L, Ehmke JF, Verhoef PC (2024) Barriers and readiness to deal with digital transformation in higher education. TEM J, 13(1), 334–348. https://doi.org/10.18421/TEM131-35

Kułak JP, Trojanowski M, Barmentloo E (2019) A literature review of the partial Unified Theory of Acceptance and Use of Technology 2 (UTAUT2) model. Annales Universitatis Mariae Curie-Skłodowska, Sectio H—Oeconomia, 53(4), 101–113. https://doi.org/10.17951/h.2019.53.4.101-113

Lan NTH (2024a) Through tensions to identity-based motivations: Exploring teacher professional identity in AI-enhanced teacher training. Teach Teach Educ 139:104736. https://doi.org/10.1016/j.tate.2024.104736

Lan Y (2024b) Through tensions to identity-based motivations: Exploring teachers' professional identity in technology integration. Teach Teach Educ 140:104510. https://doi.org/10.1016/j.tate.2024.104510

Li Y (2025) Promoting digital transformation in STEM education and beyond. J STEM Educ Res 6(2):408–426. https://doi.org/10.1007/s41979-025-00157-0

Luo J (2025) What do we mean by digital equality in education? Toward a conceptual model. J Res Technol Educ. Advance online publication. https://doi.org/10.1080/15391523.2025.2487279

Malewar S, Bajaj S (2020) Acceptance of OTT video streaming platforms in India during COVID-19: Extending UTAUT2 with content availability. J Content, Community & Commun, 12(6), 89–97. https://doi.org/10.31620/JCCC.12.20/09
Martin F, Ceviker E, Gezer T (2024) From digital divide to digital equity: Systematic review of two decades of research on educational digital divide factors, dimensions, and interventions. J Res Technol Educ. Advance online publication. https://doi.org/10.1080/15391523.2024.2425442
Marzano D et al (2025) Generative artificial intelligence (GAI) in teaching and learning processes at the K-12 level: A systematic review. Advance online publication, Technology, Knowledge and Learning. https://doi.org/10.1007/s10758-025-09853-7
McCarthy AM, Maor D, McConney A, Cavanaugh C (2023) Digital transformation in education: Critical components for leaders of system change. Soc Sci & Humlties Open 8(1):100479. https://doi.org/10.1016/j.ssaho.2023.100479
Mhlanga D, Moloi T (2024) Digital transformation of education: The limitations and challenges ahead. Advance online publication, Education and Information Technologies. https://doi.org/10.1007/s44217-024-00115-9
Mintz J (2024) Inclusive digital education: Contexts, practices and policy implications. J Educ Policy Pract. Advance online publication. https://doi.org/10.1080/07380569.2024.2340873
Musa M, Ismail MN, Tahir S, Md Fudzee MF, Jofri MH (2022) Student acceptance towards online learning management system based on UTAUT2 model. Int J Adv Comput Sci Appl, 13(11), 476–485. https://doi.org/10.14569/IJACSA.2022.0131115
Nadella GS, Meduri K, Satish S, Maturi MH, Gonaygunta H (2024) Examining e-learning tools impact using IS-impact model: A comparative PLS-SEM and IPMA case study. J Open Innov: Technol, Mark, Complex 10(3):100351. https://doi.org/10.1016/j.joitmc.2024.100351
Nguyen THN, Phan TK, Mai QK, Tran TT, Ta DP (2025) Digital transformation in Vietnam's education: Opportunities, challenges, and development strategies. Multidiscip Rev, 8(9), e2025282. https://doi.org/10.31893/multirev.2025282
Owens J, Johnson C, Peters M (2023a) Digital inclusion as a lens for equitable parent engagement. TechTrends 67(4):451–463. https://doi.org/10.1007/s11528-023-00859-5
Owens M, Ravi V, Hunter E (2023b) Digital inclusion as a lens for equitable parent engagement. TechTrends 67(5):819–830. https://doi.org/10.1007/s11528-023-00859-5
Passey D (2024) Populations digitally excluded from education: Issues, practices and policy implications. Advance online publication, Technology, Knowledge and Learning. https://doi.org/10.1007/s10758-024-09767-w
Pham H, Tran QN, La GL, Doan HM, Vu TD (2021) Readiness for digital transformation of higher education in the COVID-19 context: The dataset of Vietnam's students. Data Brief 39:107482. https://doi.org/10.1016/j.dib.2021.107482
Plekhanov D, Franke H, Netland TH (2023) Digital transformation: A review and research agenda. Eur Manag J 41(6):979–995. https://doi.org/10.1016/j.emj.2022.09.007
Polat O (2025) Artificial intelligence literacy, lifelong learning, and fear of technology use: Implications for education. Advance online publication, Education and Information Technologies. https://doi.org/10.1007/s10639-025-13548-y
Prasetyo YT, Roque RAC, Chuenyindee T, Young MN, Diaz JFT, Persada SF, Miraja BA, Redi AANP (2021) Determining factors affecting the acceptance of medical education eLearning platforms during the COVID-19 pandemic in the Philippines: UTAUT2 approach. Healthcare 9(7):780. https://doi.org/10.3390/healthcare9070780
Prehn J, et al. (2025) Nurturing cultural identity: Creating an inclusive school environment for Indigenous Australians. Aust Educ Res. Advance online publication. https://doi.org/10.1007/s13384-024-00781-6
Reis J, Amorim M, Melão N, Matos P (2023) Digital transformation: A meta-review and guidelines for future research. Heliyon 9(1):e12834. https://doi.org/10.1016/j.heliyon.2023.e12834
Revuelta-Domínguez F-I, Guerra-Antequera J, González-Pérez A, Pedrera-Rodríguez M-I, González-Fernández A (2022a) Digital teaching competence: A systematic review. Sustainability 14(11):6428. https://doi.org/10.3390/su14116428

Revuelta-Domínguez F-I, Guerra-Antequera J, Pedrosa-García JA (2022b) Digital competence of teachers as a factor of cultural identity support in education. Education Sciences 12(12):871. https://doi.org/10.3390/educsci12120871

Sarstedt M, Ringle CM, Smith D, Reams R, Hair JF (2014) Partial least squares structural equation modeling (PLS-SEM): A useful tool for family business researchers. J Fam Bus Strat 5(1):105–115. https://doi.org/10.1016/j.jfbs.2014.01.002

Senaratne SI, Samarasinghe SM (2019) Factors affecting the intention to adopt m-learning. Int Bus Res 12(2):150–161. https://doi.org/10.5539/ibr.v12n2p150

Senos S, Moreira JA, Gomes MJ (2024) Digital inclusion for parents: Preliminary results from a project promoting parental digital literacy. Advance online publication, Educational Media International. https://doi.org/10.1080/09523987.2024.2358660

Sitar-Tăut DA, Mican D (2022) Exploring students' acceptance of e-learning through the lens of the UTAUT2 model. Hum Behav Emerg Technol 4(2):135–148. https://doi.org/10.1002/hbe2.261

Sujarwo, Herwin (2023) Parental involvement and student achievement: A meta-analysis of publications in the Scopus database. Int J Instr, 16(2), 97–112. https://doi.org/10.29333/iji.2023.1627a

Tawil S, Miao F (2024) Steering the digital transformation of education: UNESCO's human-centered approach. J Access Serv 22(1):45–60. https://doi.org/10.3868/s110-009-024-0005-6

Tran LN, Nguyen TL, Le VT, Nguyen TMLH, Nguyen TTT, Pham HHY (2025) Parents' navigation through remote learning in Vietnam. Advance online publication, Learning, Media and Technology. https://doi.org/10.1080/1475939X.2025.2490116

Tzafilkou K, Perifanou M, Economides AA (2023) Assessing teachers' digital competence in primary and secondary education: Applying a new instrument to integrate pedagogical and professional elements for digital education. Advance online publication, Education and Information Technologies. https://doi.org/10.1007/s10639-023-11848-9

Voorhees CM, Brady MK, Calantone R, Ramirez E (2016) Discriminant validity testing in marketing: An analysis, causes for concern, and proposed remedies. J Acad Mark Sci 44(1):119–134. https://doi.org/10.1007/s11747-015-0455-4

Vu TD, Do HM, Pham H, Tran QN (2023) Evaluate the drivers for digital transformation in higher education institutions in the era of Industry 4.0 based on decision-making method. J Innov & Knowl, 8(3), 100364. https://doi.org/10.1016/j.jik.2023.100364

Vuong QH, Le TT, Nguyen MH (2023) Examining contributors to Vietnamese high school students' digital creativity under the serendipity-mindsponge-3D knowledge management framework. Think Ski Creat 49:101350. https://doi.org/10.1016/j.tsc.2023.101350

Yang L, García-Holgado A, Martínez-Abad F (2023) Digital competence of K-12 pre-service and in-service teachers in China: A systematic literature review. Asia Pac Educ Rev 24(4):679–693. https://doi.org/10.1007/s12564-023-09888-4

Yang X, Wu W (2024) Advancing digital transformation in TVET through international cooperation: Approaches by the UNESCO Chair on Digitalization in TVET. Vocat Tech Educ, 1(2), Article 0585. https://doi.org/10.54844/vte.2024.0585

Zacharis G, Nikolopoulou K (2022) University students' acceptance of online learning platforms in Greece: An empirical application of UTAUT2. Educ Inf Technol 27(6):8227–8248. https://doi.org/10.1007/s10639-022-11116-2

Zhan Z, Bai X, Liu Y (2025) Teacher competence and barriers in digital education: Evidence from PLS-SEM in Chinese secondary schools. Educ Inf Technol. https://doi.org/10.1007/s10639-024-13299-2

Zheng P, Franke H, Netland TH (2023) Digital transformation: A review and research agenda. Eur Manag J 41(6):979–995. https://doi.org/10.1016/j.emj.2022.09.007

AI-Driven Water Management: Transforming Conservation Strategies Through IoT Integration

Vibha Trivedi and Moaz Gharib

Abstract Growing urbanization has intensified the exploitation of natural resources, leading to the current climate crisis. One of the most exacerbating issues is extreme pressure on rivers and other water bodies. Traditional approaches to water management with laboratory methods are costly and time-consuming. However, with the emergence of the Internet of Things (IoT) and Artificial Intelligence (AI), natural water resources management seems to have a promising future. IoT exchanges real-time data and communicates the same within its interconnected devices, while AI handles and analyzes large amounts of data through machine learning (ML), deep learning, and expert systems. Integration of AI and IoT enables the collection of granular data and draws actionable insights through advanced analytics. IoT-supported sensors, water meters, and communication networks can monitor several quality parameters such as water levels, pressure, flow, and turbidity. Further, AI can be used to process data streams from IoT devices to offer anomaly detection, predictive maintenance, and optimised resource allocation. The present chapter attempts to outline how IoT and AI integration can be leveraged for water sustainability efforts. It explores emerging trends in freshwater resource conservation, while offering avenues for future research and policy development.

1 Introduction

The lack of freshwater resources has emerged as one of the most critical consequences of climate change. Rapid urbanization and industrialization place unsustainable demands on climate resources and magnify the risks of climate emergencies

V. Trivedi (✉)
O.P. Jindal Global University, Sonipat, India
e-mail: vibhap28@gmail.com

M. Gharib
Department of Management, College of Commerce and Business Administration, Dhofar University, Salalah, Oman

R. Singh et al. (eds.), *Data-Driven Sustainability: Harnessing Technology for A Greener Future and Environmental Resilience*, World Sustainability Series,
https://doi.org/10.1007/978-3-032-13412-7_19

(Balaram et al. 2023). With demand for water expected to rise even further, dependence on outdated, manual, and unsophisticated water management frameworks leads to even greater water accessibility challenges (Aderemi et al. 2025). Outdated water management frameworks, which rely on repetitive sampling and manual control of large, complex systems, are highly ineffective, inefficient, and iterative. They also lead to a time delay regarding critical feedback that must be addressed to avoid emergent crises such as insufficient water monitoring, poor quality control, leaky systems, and inadequate flow (Cosgrove and Loucks 2015; Larsen et al. 2016; Koech and Langat 2018; Krishnan et al. 2022). There is an urgent need for an efficient, timely, low-cost, and highly responsive approach to achieve efficient water allocation to deal with the highly complex and interdependent systems that support the desired sustainable and balanced ecosystem (Michael Ayorinde Dada et al. 2024).

The rapid advances of cutting-edge technology are transforming water management (Tolentino et al. 2021; Venkateswaran et al. 2023; Aderemi et al. 2025). Recent studies have demonstrated how AI-driven IoT systems can be pivotal in automated and intelligent management of water ecosystem monitoring (Patel et al. 2023; Michael Ayorinde Dada et al. 2024). AI algorithms have become increasingly important in distinguishing and understanding complex behavioral patterns as well as analyzing large datasets in record time (Tolentino et al. 2021; Jabbar et al. 2024). Further, IoT devices facilitate continuous and high-frequency data collection and remote monitoring through a complex, interconnected system of sensors, devices, and actuators that exchange real-time information (Tolentino et al. 2021; Jabbar et al. 2024). These water management IoT technologies and intelligent monitoring systems record key water parameters and provide real-time monitoring of water levels, flow, turbidity, pH, electrical conductivity, and total dissolved solids (Chiamsathit et al. 2016; Liu et al. 2019; Pradeepakumari and Srinivasu 2019; Pasika and Gandla 2020; Huan et al. 2020; Essamlali et al. 2024; López-Munoz et al. 2024). AI analyzes real-time data, different patterns of water use, enhances forecasting, analyzes irregularities, anticipates failures, and streamlines the distribution network (Tolentino et al. 2021; Nanthakumar et al. 2024; Maniraj et al. 2025).

AI and IoT technology integration provides dependable water resource management systems. These systems will implement more sustainable practices, directly impacting the goals of UNSDG #6 (Clean Water and Sanitation), and indirectly aiding UNSDG #9 (Industry, Innovation, and Infrastructure), UNSDG #11 (Sustainable Cities and Communities), UNSDG #12 (Responsible Consumption and Production), UNSDG #13 (Climate Action), and UNSDG #15 (Life on Land).

This chapter examines integrating AI with IoT systems for sustainable freshwater resource management. A comprehensive literature review focuses on scholarly articles, industry reports, and policy documents from leading databases to explore the concepts of IoT, AI, and water resource conservation. This chapter seeks to highlight recent trends and challenges to provide an interdisciplinary view and practical guidance regarding more intelligent and scalable solutions to water management. It also states the ways in which innovative approaches to water management that embrace real-time, data-driven assessments made possible through AI-IoT systems. Adaptive approaches to different socio-economic contexts are also provided.

2 Literature Review

2.1 Background

With an increasing need for clean water and sanitation services in today's growing urban population, the demand for municipal water consumption has expanded significantly. A large proportion of groundwater is being used for irrigation purposes, and nearly 50% of the freshwater is used for domestic consumption, further reflecting the increasing pressure on limited water resources (UNESCO 2024). Agricultural practices are the largest consumers of freshwater resources, followed by industrial and domestic uses. These practices account for about 70% of all freshwater consumed. In low-income nations, limited water exploitation and underdeveloped wastewater treatment facilities mean water shortages and unavailability. Conversely, high-income nations, due to excess and poorly controlled agricultural runoffs, face increased water pollution and ecosystem and biodiversity destruction, with chemically dense pesticide agricultural waste. Droughts, floods, and declining groundwater levels, exacerbated by climate change, undermine already poor water resource management and pose significant water security challenges (Ciampittiello et al. 2024). In these conditions, the manual control and monitoring of systems depend heavily on precise, fast human interventions. This creates inefficiencies and delayed response systems (Srivastav et al. 2021; Prosser et al. 2021; Ciampittiello et al. 2024). Persistent inefficiencies in conventional water management practices underscore the need for innovative technological solutions that may improve water management. Real-time monitoring systems with automated remote sensing devices offer a promising solution to overcome these challenges (Srivastav et al. 2021). The innovative use of technology in water resource management practices may help solve these issues. Real-time water exploitation and management systems that use automated remote sensing technology may help solve these issues (Cherkauer et al. 2021). Availability of updated information on the water intensity levels, frequency curves, non-revenue water, leakages, and alternative energy-intensive processes could lead to more feasible and efficient water management.

2.2 Artificial Intelligence and Water Management

AI provides a great opportunity to anticipate demand and facilitate smart control in an urban setting. Employing Machine learning methods, AI analyses consumption behaviour influenced by several demographic variables under climatic conditions. It can efficiently establish an acceptable forecasting of possible anomalies, using deep learning methods such as artificial neural networks (ANN), recurrent neural networks (RNN), and long short-term memory networks (LSTM). ANN builds an artificial brain to identify, assimilate and forecast patterns in water quality parameters such as pH and turbidity even under data sparse conditions, while mimicking inflow and outflow trends of water bodies to support water body operation. Also, the human brain-like structure of ANN allows for a comparison of consumption behaviour within large urban consumption data sets based on climatic conditions and population growth, while better forecasting demand. RNNs are aligned with temporal data modelling and when they adapt a residual feedback approach, an action performed earlier is memorized when applying a forecast for a subsequent action. They are useful for forecasting river flows, surplus water availability, and water contamination by recalling previous trends. Since RNNs have the limitations of a short-term memory, LSTMs are used to store and retrieve long-term data for multi-month prediction, such as river water inflow forecasting. They have the ability to utilize multi-year climatic data to forecast floods and droughts by retrieving information over long sequences. Furthermore, Elman Discrete Recurrent Neural Networks (EDRNN) have been shown to accurately predict flood stages by analyzing sequential rainfall runoffs data, thereby contributing to disaster risk reduction through real-time flood forecasting (Kim and Kim 2008a). In agriculture, AI-based smart irrigation systems help protect crops from floods and droughts by automating the water flows (Zaigham Abbas Naqvi et al. 2025). Two broad types of smart irrigation methods, weather-based and soil moisture-based smart irrigation, are widely used in agriculture as they provide cost-effective and scalable solutions (OĞUZTÜRK 2025). Moreover, ML and data analytics accurately estimate water availability and help in effective and efficient data-driven decision-making through advanced optimization (Pant and Singh 2025). While ANNs can be useful in forecasting river water inflows in case of high flows (Chiamsathit et al. 2016; Pradeepakumari and Srinivasu 2019), advanced AI models, such as Generalized Regression Neural Networks and Random Forests, offer estimates of evapotranspiration on a daily basis using data with minimally trained variables. They also have the ability to model complex nonlinear associations among variables (Kim and Kim 2008b; Kisi 2016; Feng et al. 2017). Collectively, these AI-based systems offer a fast and accurate response mechanism for managing water ecosystems.

2.3 AI and IoT Integration for Water Management

IoT complements AI since it allows smart management of water by using network devices, also known as sensors, meters, and communication protocols. IoT technologies can be used to monitor and collect data in real-time with the use of quality sensors and flow meters (Sheeba et al. 2025). They also facilitate real-time observation of water distribution in order to pass quality and timely information to the consumers. The IoT has transformed irrigation methods in agricultural settings (Morchid et al. 2024) by tying together smart devices with the agricultural activities (Sharifnasab et al. 2023). Smart sensors and monitoring systems that are interconnected have enhanced the communication process, automated irrigation, and improved the productivity and efficiency of agriculture (Tatas et al. 2022; Puig et al. 2022; Sharifnasab et al. 2023). A smart drip irrigation system, coupled with remote control and real-time monitoring, is an IoT-driven tool that would allow intelligent and timely irrigation, which would guarantee efficient use of water (Abioye et al. 2023; Jain 2023; Kumar et al. 2023). Lean management systems have also been implemented in the beverage industry using IoT to enhance cost and water efficiency (Sodkomkham et al. 2024). Smart farming enables the integration of IoT and cloud computing, which boosts the efficiency of using energy, minimizing carbon footprints, and ensuring farming is resilient to the climatic conditions (Morchid et al. 2024). IoT applications are also used to resolve food security problems through monitoring irrigation, yield, fertilizer application, quality, and logistics, and early detection of crop diseases (Morchid et al. 2024).

Current studies have found that IoT applications are important in solving food security issues by transforming agriculture into smart data-driven practices. Smart sensing technology in agriculture enables smart monitoring of the soil water content, enabling smart irrigation and water distribution, which results in less water wastage and more crop protection in response to unfavorable weather conditions. Yield and quality monitoring systems based on IoT technologies guarantee the volume of production is estimated in time. Moreover, the logistics monitoring system improves supply chain efficiency because it allows the supply chain partners to predict demand, optimize the flow of fresh product to the market, and reduce potential harvest losses caused by delayed logistics. The use of imaging technology combined with IoT systems also facilitates the prior identification of disease in crops, which allows timely responses. Collectively, IoT-enabled agriculture not only contributes to resolving food security issues but also helps in freshwater management by conserving water and preventing soil contamination due to rainwater runoffs.

AI and IoT integration offer synergistic advantages. IoT devices collect real-time, high-frequency data, while AI models analyze them with human-like intelligence to generate actionable insights on heterogeneous consumption behaviour and dynamic climate conditions. Extant literature has established the benefits of AI-IoT device integration, showcasing their effectiveness in smart waste management (Sosunova and Porras 2022) and smart agriculture (Khanna and Kaur 2019; Patel et al. 2023). In water management, AI-IoT integration facilitates early water leakage detection, thereby minimizing non-revenue water losses (Cherkauer et al. 2021) and optimizing pump scheduling. Optimal pump scheduling saves a significant amount of energy consumed by water pumping devices and is thereby cost-effective (Dini et al. 2022; Pei et al. 2024). It also enables climate resilience by predicting hydrological extremes such as floods, droughts, and rainwater runoffs.

2.4 Barriers to AI-IoT Integration to Manage the Water Ecosystem

Although there are advantages, the large-scale adoption of AI and IoT has been challenged by various problems such as interoperability, privacy and security concerns, and data governance challenges (van Zoonen 2016; Bugeja et al. 2016; Egbemhenghe et al. 2023; Michael Ayorinde Dada et al. 2024; Sathyanarayana Movva 2024; Nsoh et al. 2024; Lakhouit 2025; Parra-López et al. 2025; Morchid et al. 2025). The sensors, controllers, and analytical tools sold by other vendors frequently do not support a common communication protocol or data format, thus creating compatibility problems (Michael Ayorinde Dada et al. 2024; Nsoh et al. 2024; Parra-López et al. 2025). This standardization deficiency restricts the flexibility of the systems since it becomes hard to migrate to a different vendor or platforms when current systems are becoming expensive, non-performative, or data-related risks (Michael Ayorinde Dada et al. 2024; Nsoh et al. 2024; Ahmed and Shakoor 2025; Parra-López et al. 2025), thereby impeding large-scale integration (OĞUZTÜRK 2025).

The other issue is data privacy and cybersecurity as another major challenge. The IoT devices generate vast amounts of sensitive consumption-based and behavioral information, contributing to unauthorized access and breach (Sathyanarayana Movva 2024; Lakhouit 2025). This type of violation and unjustified access can be used to manipulate the parameters of water quality, resulting in a chaotic scenario, which subsequently leads to a loss of system reliability (Egbemhenghe et al. 2023; Michael Ayorinde Dada et al. 2024; Sathyanarayana Movva 2024; Parra-López et al. 2025). This highlights the need for definitive regulatory frameworks with policies on data use, storage, and sharing. In an environment where technologies progress quickly, timely updates to regulatory frameworks relating to ethical practices and compliance will be imperative for the successful implementation of large-scale integrated AI-IoT systems in freshwater management. Aside from the challenges, the challenges of having high capital investment, limited technical expertise, no training, resistance to change by stakeholders, and lack of power supply must be addressed to realize the synergistic benefit of AI-IoT integration in water sustainability.

3 Implications and Emerging Research Gaps

AI-IoT integration into water management provides a broad spectrum of practical and theoretical applications to the field. It can, for example, do away with the limitations of traditional monitoring systems by making them more automated and faster. Through IoT-based automatic sensing technologies and AI-based predictive analytics, AI-IoT integration facilitates the actualization of real-time monitoring, thus facilitating fair and effective resource allocation. Secondly, it offers rapid identification of non-revenue water leakages, contaminations, as well as deterioration of water quality parameters such as pH and turbidity. In general, an AI-IoT-based automatic sensing and monitoring system, maintained by predictive analytics, renders water infrastructure robust (Michael Ayorinde Dada et al. 2024; Sathyanarayana Movva 2024; Nsoh et al. 2024; Parra-López et al. 2025; Atheer Zaki Al-Qaisi 2025; Singh 2025). Third, employing AI and IoT is also a cost-effective approach, as it helps save energy by optimizing pumping schedules (Sodkomkham et al. 2024; OĞUZTÜRK 2025).

Fourth, the integration benefits agriculture and groundwater preservation. Smart agriculture, through drip irrigation technology, optimizes irrigation, while embedded sensors monitor the soil moisture levels. Moreover, AI-IoT integration not only aids in food security but also helps optimize water usage. Fifth, AI-IoT-based water grids strengthen the urban water management system by enabling data-driven decision-making. These systems handle large, complex, and dynamic urban behavioral data to support error-free decisions (Sathyanarayana Movva 2024; Guezouli et al. 2024). Additionally, integrating blockchain technology enhances transparency, security, and reliability. Sixth, the AI-IoT-based integrated smart water management system is adaptable to changing environments and scalable from a simple rural setup to a large, dynamic urban grid (Michael Ayorinde Dada et al. 2024). Lastly, the application of AI-IoT integration is not only economically advantageous but also serves various socio-ecological goals stated in UNSDGs #6, #9, #11, #12, #13, and #15 (Michael Ayorinde Dada et al. 2024; Sathyanarayana Movva 2024; Ojadi et al. 2025).

Despite numerous positive implications, there are a few potential threats and barriers that act as roadblocks to large-scale adoption of AI-IoT-based systems. A lack of standard protocols at design levels makes interoperability risky in terms of data privacy and seamless communication. Since the technology is still in its inception phase, the role of strict governance guidelines remains underexplored. Clear policy guidelines regarding data collection, storage, and usage are crucial, as unresolved gaps make large, sensitive, and private consumer data vulnerable to cybersecurity attack (Michael Ayorinde Dada et al. 2024; Sathyanarayana Movva 2024). Additionally, the ambiguity around ethical practices during crisis situations raises trust issues among stakeholders, specifically regarding whether AI-based decisions should be prioritized.

While technological advancements promise long-term cost-effectiveness, their initial capital investment requirements are substantial. This barrier limits the adoption in low-income countries, although such contexts offer immense opportunities to researchers and practitioners. However, concerns about the carbon footprint of sustainable solutions remain debatable. Moreover, the lack of awareness and limited inclusion of the latest research in academic curricula indicate the need for regular training and development programs for government employees, which is expected to accelerate the wider adoption.

4 Conclusion

Due to urbanization, climate crisis, and industrialization, resource allocation has become a complex task that cannot be addressed by traditional manual methods. AI-IoT-based smart sensing and monitoring technologies have the potential to be the new frontier in the management of freshwater. Accurate forecast of urban water demand and equitable distribution depend on timely data and adaptive responding networks of communication. IoT-based responding networks of communication facilitate automated sensing and smart monitoring and the relational effect of a large dataset of diverse consumers. AI-based algorithms can analyze this complex dataset to predict allocation needs, identify barriers to distribution, and indicate system failure. The integrated system will also allow real-time scheduling of the pumps, reducing energy consumption and maintaining cost savings through predictive maintenance. Scheduling water agenda is optimized to drive water conservation, improved performance, and enhance management of the water infrastructure.

AI-IoT integration will enhance resilience because it will facilitate data-driven decisions for allocating, distributing, detecting and managing anomalies. It will also be adaptable enough to make efficient demands predictions in an ever-changing environment. The paired platform will impact urban water management and ultimately ensure food security through more efficient water use in a smart water irrigation system, among other possibilities. There will be a large volume of consumer data to be maintained and used properly. Transparency and strong government policies to protect consumer interests in an ethical and fair manner are the main aspects to spur a wider acceptance of stakeholders, and in time, overcome inertia among stakeholders. Continued interdisciplinary research and government policy development is needed to provide interoperability, reliability and standards. The convergence of AI and IoT is not only going to revolutionize the water management system, but is also promoting the wider social and climate goals.

References

Abioye AE, Abidin MSZ, Mahmud MSA, Buyamin S, Mohammed OO, Otuoze AO, Oleolo IO, Mayowa A (2023) Model-based predictive control strategy for water saving drip irrigation. Smart Agric Technol 4:100179. https://doi.org/10.1016/j.atech.2023.100179

Aderemi IA, Okeola OG, Mahmud HO, Abdulkadir TS, Salami HA, Salami AW (2025) Harnessing IoT and advanced analytics for sustainable water quality management. Water-Energy Nexus. https://doi.org/10.1016/j.wen.2025.07.005

Ahmed N, Shakoor N (2025) Advancing agriculture through IoT, Big Data, and AI: A review of smart technologies enabling sustainability. Smart Agric Technol 10:100848. https://doi.org/10.1016/j.atech.2025.100848

Atheer Zaki Al-Qaisi (2025) Smart water systems: the role of technology and engineering in optimizing urban water resources. J Inf Syst Eng Manag 10:833–846. https://doi.org/10.52783/jisem.v10i21s.3445

Balaram A, Kannan KN, Čepová L, Kumar MK, Rani BS, Schindlerova V (2023) Artificial intelligence for media ecological integration and knowledge management. Systems 11:222. https://doi.org/10.3390/systems11050222

Bugeja J, Jacobsson A, Davidsson P (2016) On privacy and security challenges in smart connected homes. In: 2016 european intelligence and security informatics conference (EISIC). IEEE, pp 172–175

Cherkauer KA, Bowling LC, Byun K, Chaubey I, Chin N, Ficklin DL, Hamlet AF, Kines SJ, Lee CI, Neupane R, Pignotti GW, Rahman S, Singh S, Femeena PV, Williamson TN (2021) Climate change impacts and strategies for adaptation for water resource management in Indiana. Clim Change 165:21. https://doi.org/10.1007/s10584-021-02979-4

Chiamsathit C, Adeloye AJ, Bankaru-Swamy S (2016) Inflow forecasting using Artificial Neural Networks for reservoir operation. Proceedings of the International Association of Hydrological Sciences 373:209–214. https://doi.org/10.5194/piahs-373-209-2016

Ciampittiello M, Marchetto A, Boggero A (2024) Water resources management under climate change: a review. Sustain 16:3590. https://doi.org/10.3390/su16093590

Cosgrove WJ, Loucks DP (2015) Water management: Current and future challenges and research directions. Water Resour Res 51:4823–4839. https://doi.org/10.1002/2014WR016869

Dini M, Hemmati M, Hashemi S (2022) Optimal operational scheduling of pumps to improve the performance of water distribution networks. Water Resour Manage 36:417–432. https://doi.org/10.1007/s11269-021-03034-8

Egbemhenghe AU, Ojeyemi T, Iwuozor KO, Emenike EC, Ogunsanya TI, Anidiobi SU, Adeniyi AG (2023) Revolutionizing water treatment, conservation, and management: Harnessing the power of AI-driven ChatGPT solutions. Environ Challs 13:100782. https://doi.org/10.1016/j.envc.2023.100782

Essamlali I, Nhaila H, El Khaili M (2024) Advances in machine learning and IoT for water quality monitoring: A comprehensive review. Heliyon 10:e27920. https://doi.org/10.1016/j.heliyon.2024.e27920

Feng Y, Cui N, Gong D, Zhang Q, Zhao L (2017) Evaluation of random forests and generalized regression neural networks for daily reference evapotranspiration modelling. Agric Water Manag 193:163–173. https://doi.org/10.1016/j.agwat.2017.08.003

Guezouli L, Guezouli L, Djeghaba MBE, Bentahrour A (2024) IoT and AI for Real-time water monitoring and leak detection. J Renew Energs 27. https://doi.org/10.54966/jreen.v27i2.1210

Huan J, Li H, Wu F, Cao W (2020) Design of water quality monitoring system for aquaculture ponds based on NB-IoT. Aquac Eng 90:102088. https://doi.org/10.1016/j.aquaeng.2020.102088

Jabbar WA, Mei Ting T, I. Hamidun MF, Che Kamarudin AH, Wu W, Sultan J, Alsewari AA, Ali MAH (2024) Development of LoRaWAN-based IoT system for water quality monitoring in rural areas. Expert Syst Appl 242:122862. https://doi.org/10.1016/j.eswa.2023.122862

Jain RK (2023) Experimental performance of smart IoT-enabled drip irrigation system using and controlled through web-based applications. Smart Agric Technol 4:100215. https://doi.org/10.1016/j.atech.2023.100215

Khanna A, Kaur S (2019) Evolution of Internet of Things (IoT) and its significant impact in the field of Precision Agriculture. Comput Electron Agric 157:218–231. https://doi.org/10.1016/j.compag.2018.12.039

Kim S, Kim HS (2008a) Uncertainty reduction of the flood stage forecasting using neural networks model 1. JAWRA J Am Water Resour Assoc 44:148–165. https://doi.org/10.1111/j.1752-1688.2007.00144.x

Kim S, Kim HS (2008b) Neural networks and genetic algorithm approach for nonlinear evaporation and evapotranspiration modeling. J Hydrol (Amst) 351:299–317. https://doi.org/10.1016/j.jhydrol.2007.12.014

Kisi O (2016) Modeling reference evapotranspiration using three different heuristic regression approaches. Agric Water Manag 169:162–172. https://doi.org/10.1016/j.agwat.2016.02.026

Koech R, Langat P (2018) Improving irrigation water use efficiency: a review of advances, challenges and opportunities in the australian context. Water (Basel) 10:1771. https://doi.org/10.3390/w10121771

Krishnan SR, Nallakaruppan MK, Chengoden R, Koppu S, Iyapparaja M, Sadhasivam J, Sethuraman S (2022) Smart water resource management using artificial intelligence—a review. Sustain 14:13384. https://doi.org/10.3390/su142013384

Kumar SV, Singh CD, Rao KVR, Kumar M, Rajwade YA, Babu B, Singh K (2023) Evaluation of IoT based smart drip irrigation and ETc based system for sweet corn. Smart Agric Technol 5:100248. https://doi.org/10.1016/j.atech.2023.100248

Lakhouit A (2025) Revolutionizing urban solid waste management with AI and IoT: A review of smart solutions for waste collection, sorting, and recycling. Results Eng 25:104018. https://doi.org/10.1016/j.rineng.2025.104018

Larsen TA, Hoffmann S, Lüthi C, Truffer B (1979) Maurer M (2016) Emerging solutions to the water challenges of an urbanizing world. Sci 352:928–933. https://doi.org/10.1126/science.aad8641

Liu P, Wang J, Sangaiah AK, Xie Y, Yin X (2019) Analysis and prediction of water quality using LSTM deep neural networks in IoT environment. Sustain 11:2058. https://doi.org/10.3390/su11072058

López-Munoz MA, Torrealba-Melendez R, Arriaga-Arriaga CA, Tamariz-Flores EI, López-López M, Quirino-Morales F, Munoz-Pacheco JM, López-Marcos F (2024) Wireless dynamic sensor network for water quality monitoring based on the IoT. Technol (Basel) 12:211. https://doi.org/10.3390/technologies12110211

Maniraj SP, JM, Amutha R, Kavitha R, Shriidhar PJ, Srinivasan C (2025) Agricultural water conservation through AI-Enhanced solar IoT networks. In: 2025 international conference on electronics and renewable systems (ICEARS). IEEE, pp 686–691

Michael Ayorinde Dada, Michael Tega Majemite, Alexander Obaigbena, Onyeka Henry Daraojimba, Johnson Sunday Oliha, Zamathula Queen Sikhakhane Nwokediegwu (2024) Review of smart water management: IoT and AI in water and wastewater treatment. World J Adv Res Rev 21:1373–1382. https://doi.org/10.30574/wjarr.2024.21.1.0171

Morchid A, El Alami R, Raezah AA, Sabbar Y (2024) Applications of internet of things (IoT) and sensors technology to increase food security and agricultural Sustainability: Benefits and challenges. Ain Shams Eng J 15:102509. https://doi.org/10.1016/j.asej.2023.102509

Morchid A, Alblushi IGM, Khalid HM, Qjidaa H, El Alami R (2025) Integrating IoT and fuzzy logic for intelligent irrigation in sustainable agriculture for improving water scarcity: Benefits and challenges. Sustain Comput: Inform Syst 48:101191. https://doi.org/10.1016/j.suscom.2025.101191

Nanthakumar G, Abinesh A, Ganesh K. R, Harish V, Nagendran S (2024) Iot based smart irrigation system using artificial intelligence. Int J Adv Res Sci, Commun Technol 145–154. https://doi.org/10.48175/IJARSCT-17623

Nsoh B, Katimbo A, Guo H, Heeren DM, Nakabuye HN, Qiao X, Ge Y, Rudnick DR, Wanyama J, Bwambale E, Kiraga S (2024) Internet of things-based automated solutions utilizing machine learning for smart and real-time irrigation management: a review. Sensors 24:7480. https://doi.org/10.3390/s24237480

Oğuztürk GE (2025) AI-driven irrigation systems for sustainable water management: A systematic review and meta-analytical insights. Smart Agric Technol 11:100982. https://doi.org/10.1016/j.atech.2025.100982

Ojadi JO, Akindele Owulade O, Somtochukwu Odionu C, Onukwulu EC (2025) AI-Driven optimization of water usage and waste management in smart cities for environmental sustainability. Eng Technol J 10. https://doi.org/10.47191/etj/v10i03.36

Pant M, Singh J (2025) Integrating AI, machine learning, and nanotechnology: shaping the future of water quality management. In: Computational Automation for Water Security. Elsevier, pp 111–134

Parra-López C, Ben Abdallah S, Garcia-Garcia G, Hassoun A, Trollman H, Jagtap S, Gupta S, Aït-Kaddour A, Makmuang S, Carmona-Torres C (2025) Digital technologies for water use and management in agriculture: Recent applications and future outlook. Agric Water Manag 309:109347. https://doi.org/10.1016/j.agwat.2025.109347

Pasika S, Gandla ST (2020) Smart water quality monitoring system with cost-effective using IoT. Heliyon 6:e04096. https://doi.org/10.1016/j.heliyon.2020.e04096

Patel A, Kethavath A, Kushwaha NL, Naorem A, Jagadale M, KR. S, PSR (2023) Review of artificial intelligence and internet of things technologies in land and water management research during 1991–2021: A bibliometric analysis. Eng Appl Artif Intell 123:106335. https://doi.org/10.1016/j.engappai.2023.106335

Pei S, Hoang L, Fu G, Butler D (2024) Real-time multi-objective optimization of pump scheduling in water distribution networks using neuro-evolution. Journal of Water Process Engineering 68:106315. https://doi.org/10.1016/j.jwpe.2024.106315

Pradeepakumari B, Srinivasu K (2019) Dam Inflow prediction by using artificial neural network reservoir computing. Int J Eng Adv Technol 9:662–667. https://doi.org/10.35940/ijeat.B2990.129219

Prosser IP, Chiew FHS, Stafford Smith M (2021) Adapting water management to climate change in the murray-darling basin. Australia. Water (Basel) 13:2504. https://doi.org/10.3390/w13182504

Puig F, Rodríguez Díaz JA, Soriano MA (2022) Development of a low-cost open-source platform for smart irrigation systems. Agronomy 12:2909. https://doi.org/10.3390/agronomy12122909

Sathyanarayana Movva S (2024) Analysis of intelligent water management: integration of IoT and AI in water and wastewater treatment. Int J Sci Res (IJSR) 13:1444–1448. https://doi.org/10.21275/SR24322073844

Sharifnasab H, Mahrokh A, Dehghanisanij H, Łazuka E, Łagód G, Karami H (2023) Evaluating the use of intelligent irrigation systems based on the IoT in grain corn irrigation. Water (Basel) 15:1394. https://doi.org/10.3390/w15071394

Sheeba R, Alsolai H, Allafi R, Nithya K, Arasi MA, Karthikeyan B, Sudarvizhi D, Vivek S (2025) A comprehensive IoT and blockchain-integrated framework for autonomous water management in sustainable urban ecosystems. Sustain Comput: Inform Syst 47:101171. https://doi.org/10.1016/j.suscom.2025.101171

Singh KN (2025) Smart water management systems: integration of IoT sensors and AI for sustainable urban water infrastructure. Int J Sci Res Eng Manag 09:1–9. https://doi.org/10.55041/IJSREM47819

Sodkomkham T, Ratanatamskul C, Chandrachai A (2024) A novel integrated material flow cost accounting (MFCA)- IoT-lean management system approach to improving water use efficiency and reducing costs in the beverage industry. Clean Environ Syst 15:100232. https://doi.org/10.1016/j.cesys.2024.100232

Sosunova I, Porras J (2022) IoT-Enabled smart waste management systems for smart cities: a systematic review. IEEE Access 10:73326–73363. https://doi.org/10.1109/ACCESS.2022.3188308

Srivastav AL, Dhyani R, Ranjan M, Madhav S, Sillanpää M (2021) Climate-resilient strategies for sustainable management of water resources and agriculture. Environ Sci Pollut Res 28:41576–41595. https://doi.org/10.1007/s11356-021-14332-4

Tatas K, Al-Zoubi A, Christofides N, Zannettis C, Chrysostomou M, Panteli S, Antoniou A (2022) Reliable IoT-Based monitoring and control of hydroponic systems. Technol (Basel) 10:26. https://doi.org/10.3390/technologies10010026

Tolentino LK, Chua EJ, Añover JR, Cabrera C, Hizon CA, Mallari JG, Mamenta J, Quijano JF, Virrey G, Madrigal GA, Fernandez E (2021) IoT-Based automated water monitoring and correcting modular device via LoRaWAN for aquaculture. Int J Comput Digit Syst 10:533–544. https://doi.org/10.12785/ijcds/100151

UNESCO (2024) Water for prosperity and peace.

van Zoonen L (2016) Privacy concerns in smart cities. Gov Inf Q 33:472–480. https://doi.org/10.1016/j.giq.2016.06.004

Venkateswaran N, Kumar SS, Diwakar G, Gnanasangeetha D, Boopathi S (2023) Synthetic biology for waste water to energy conversion. pp 360–384

Zaigham Abbas Naqvi SM, Hussain S, Awais M, Tahir MN, Saleem SR, Al-Yarimi FAM, Ashurov M, Saidani O, Khan MI, Wu J, Wei Z, Hu J (2025) Climate-resilient water management: leveraging IoT and AI for sustainable agriculture. Egypt Inform J 30:100691. https://doi.org/10.1016/j.eij.2025.100691

Digital Ethics for a Sustainable Generation: Leveraging Technologies with Responsibility and Resilience

Kamalpreet Kaur Paposa, Sukhvinder Singh Paposa, Himanshu Vaishnav, and Nitin S. Goje

Abstract In the era of disruptive technologies resulting in steadfast growth in the industry and the meteoric use of technology for environmental sustainability, an intriguing study on the ethical use of technology for organisational long-term success becomes highly essential. Therefore, the purpose of this book chapter is to study the pivotal role of ethics for digital sustainability with a laser focus on enhancing privacy and transparency. For this book chapter, a narrative literature review approach has been adopted to synthesise and contemplate the ethical use of technology by organisations for ensuring environmental sustainability. Articles that specifically focus on the ethical technological use, environmental sustainability, and greener futures will be considered for the study. The review of literature suggests that using various technologies and models that contribute towards recycling, remanufacturing, and waste-to-energy conversion contributes to reducing emissions and minimizing environmental impact. Digital sustainability, ethical use of technology, and maintaining data privacy, along with transparency in processes, play a pivotal role in building trust in green initiatives. Tools such as blockchain, IOT, and AI lead to climate monitoring and industrial sustainability, but only when applied responsibly. This research work provides valuable insights regarding the application of ethical principles and models for ensuring digital sustainability and greener futures. Ethical dilemmas can be addressed in a thoughtful manner through responsible utilization of emerging technologies, guided by a framework that prioritizes transparency, accountability, and environmental stewardship.

K. K. Paposa (✉)
Mittal School of Business, Lovely Professional University, Phagwara, Punjab, India
e-mail: kamalpreet.paposa@gmail.com

S. S. Paposa
Division of Career Services, Lovely Professional University, Phagwara, Punjab, India

H. Vaishnav
School of Management, O.P. Jindal University, Raigarh, Chhattisgarh, India

N. S. Goje
Department of Computer and Information Sciences, Webster University, Tashkent, Uzbekistan
e-mail: nitingoje@webster.edu

R. Singh et al. (eds.), *Data-Driven Sustainability: Harnessing Technology for A Greener Future and Environmental Resilience*, World Sustainability Series,
https://doi.org/10.1007/978-3-032-13412-7_20

Keywords Digital sustainability · Emerging technologies · Green futures · Ethics · Ethical principles · Privacy · Transparency

1 Introduction

In the era of disruptive technologies, which is shaping the overall working of the organisation and its culture, corporate sustainability has gained paramount importance in recent years. With the rapid growth of these technologies, digitalisation is being seen as a transformative force that is shaping the future of various industries. There are different types of Digital technologies like Artificial Intelligence (AI) and Machine Learning (ML), Blockchain Technology, Internet of Things (IoT), Big Data & Analytics, Cloud Computing, Virtual Reality (VR) & Augmented Reality (AR) to name a few, which are being widely used in different industries (Chatterjee and Srinivasan 2013). These digital technologies are driving growth in several sectors like agriculture, construction, education, retail, supply chain, manufacturing, and healthcare. Digital transformation is becoming a major global trend that is affecting the economy at the national and organisational level. These technologies are becoming a cornerstone for sustainable innovation and reshaping the global economy (Boonmee et al. 2025; Mbunge et al. 2022). The application of digital technologies helps the organisation to become competitive, resilient, and innovative, leading to sustainability in its operations (Martynov et al. 2019).

With the increasing risk of climate change and growing concerns about environmental protection, organisations have to strategize sustainably. Because of the pressure of integrating sustainability into the day-to-day working of the organisation, the adoption of digital technologies can pave the way towards environmental stewardship thus making the organisation efficient and competitive. Digital technology, like artificial intelligence, facilitates environmental sustainability through real-time monitoring and predictive analytics. This can contribute towards the reduction of energy wastage, optimising the use of raw material, and directly contribute towards the greener production process (Siedschlag et al. 2024).

Adoption of digitalisation in the industrial processes is two-fold. It includes both challenges and opportunities. On the one hand, the use of technology can accelerate growth sustainably through the reduction of carbon footprints, contributing to the adoption of circular economy models and driving innovation for economic growth. On the other side, it raises ethical concerns. There are various ethical apprehensions related to data privacy, transparency and accountability (Fang et al. 2025).

Thus, it can be said that digitalization serves as both a challenge and an opportunity. It is in our hands how we can make the best and ethical utilisation of it towards creating a sustainable organisation. The inherent relationship between ethics, digital sustainability, and environmental stewardship paves the way towards responsible utilisation of digital tools (Fang et al. 2025) This inherent relationship will contribute towards not only ensuring that the organisation becomes technically and digitally sound but also lead to corporate governance and responsible utilisation of technology and user data.

This chapter contributes in a manner by exploring the ethical and responsible use of emerging and disruptive technologies to ensure digital sustainability. It also helps to fill the research gap by exploring various principles and models of ethics that can be applied to resolve dilemmic situations during the adoption of digital technology while ensuring corporate governance. Thus, this book chapter will answer the following research questions:

RQ1: How are the emerging digital technologies being applied to promote environmental sustainability?

RQ2: What is the impact of digital sustainability on organisational growth, ESG performance, and resilience?

RQ3: Which ethical principles and models can be adopted for resolving ethical dilemmas and ensuring digital sustainability?

RQ4: What are the motivating factors and barriers for implementing digital sustainability?

RQ5: What are the managerial and research implications of integrating ethics with digital sustainability initiatives to achieve greener futures?

In the subsequent sections of this chapter, we will discuss Research Methodology, Literature Review, Motivators and Barriers of implementing digital sustainability, Ethical Principles and models and their application for resolving dilemmic situations, Managerial and Research Implications, Conclusion, and Limitations.

2 Research Methodology

This study adopts a scoping review methodology for systematic analysis and synthesis of data from the available published literature in the study area. The articles for the review work have been shortlisted on the basis of the keywords, namely, digital sustainability, ethics, resilience, emerging technologies, and greener futures. As this study involves an interplay between diverse variables, it created a need for performing a scoping review. This helped to provide a detailed analysis of available literature on digital sustainability and ethical use of technology. A systematic search of the academic databases, namely, Scopus, Web of Science, Google Scholar, and Science Direct, was conducted on the basis of finalised keywords of the study. The inclusion criteria consisted of published research works on studies focusing on sustainable digital transformation and ethical adoption of technology. The studies only published in the English language were shortlisted for further synthesis. Thematic analysis of the shortlisted articles helped in understanding the interplay between the key variables and served as a building block for writing the remaining sections and the major implications of the research work.

3 Literature Review

3.1 The Era of Emerging Digital Technologies

Digitalisation has a profound impact and is changing the way company operates. In the era of vast globalisation, digitalization has become an imperative and core element to drive an organisation towards success. Digitalisation helps in starting a new business, reshaping an existing business, and transforming the manner in which a firm undergoes value creation, captures the target market, and delivers desired products and services to the customer (Xia et al. 2024). In the current era of rapid globalisation, digitalisation has become a core engine for rapid growth and business transformation (Massa et al. 2023).

The various emerging technologies that have been widely used in different sectors include 3D printing, artificial intelligence, automated robots, autonomous vehicles, big data analytics, blockchain, drones, electric vehicles, and the Internet of Things (Dong et al. 2021). Metaverse, augmented reality, virtual reality, AI AI-backed chatbots are also being used in several organisations for creating a better service experience among the customers. While these innovative technologies bring significant opportunities for business enterprises, they also pose substantial risks. There are serious concerns related to violations of privacy, data security breaches, operational disruptions, cyber-attacks, and non-compliance with regulatory norms. These challenges are a threat to the organisation's long-term sustainability and goodwill. It also stakes stakeholder trust. The dual nature of technological interventions highlights the importance of digital sustainability and the application of ethical principles and models into digital transformational strategies (Fang et al. 2025). In the absence of a proper ethical framework and governance, technology may inadvertently contribute towards social inequalities, environmental harm, and erosion of consumer trust and support; thus, there is an urgent need for digital suitability.

3.2 Impact of Digital Suitability on Organisation Growth, ESG Performance, and Resilience

Digital sustainability refers to the manner in which organisations ensure that digital technologies are used in a way that is environmentally friendly, economically feasible, and socially inclusive. It involves the application of ethical principles and models to ensure that technology is being used sustainably. The power of digital technologies needs to be harnessed in a way that creates sustainable value for business without compromising its present resources. Supporting technological innovations that are balanced with profitability and social impact will help contribute to the greater good of society (Wang et al. 2025a, b). There are several organisations around the globe that our using digital technologies; however, these technologies also raise concerns related to rising cost, energy consumption, and issues related to data security (Itten et al. 2020). Digital sustainability also refers to transforming strategies that lead to the integration of digitalisation and sustainability to create sustainable value for all stakeholders. With the effective deployment of digital resources and solutions, organisations can contribute towards the achievement of sustainable goals and resolving issues related to environmental, social, and global economic development.

Organisations are under increasing pressure to respond to the uncertain and volatile business environment. Past literature has proven that corporate operations are stabilised and well established with the help of higher environmental, social, and governance (ESG) performance (Deng and Karia 2025). Organisations that employ green practices and prioritise environmental protection gain a higher competitive advantage and make optimum utilisation of their resources in the long run. Technological innovation serves as a cornerstone for gaining a competitive position in the market, removing environmental bottlenecks, and enhancing green consumption. Enterprise capability to be technically innovative paves the way to green development and higher resilience. Digital technologies help the organisation to adopt ESG framework in its day to day operations. This approach helps in instilling a culture of environmental consciousness within the organisation. The ESG management practices followed within the organisation strategically contribute towards its agility and resilience.

Responsible adoption of emerging technologies for Digital Sustainability in the ***Agriculture*** *Sector*

A research study by Huyen and Thanh (2025) examined the impact of the digital transformation process on the survival and growth of the agricultural industry. This study highlights the four components of digital transformation, which include digital technology integration, digital connectivity, digital public services, and digital human capital. As per this study, digital technologies contribute towards using automated and autonomous farming methods like robots, self-governing gadgets, weed detection and control in the fields and feeding systems, which make agriculture practices more robust and sustainable. Such techno-savvy agriculture practices will also pave the way towards vertical farming and greenhouses without any direct human intervention. Farming systems become more effective with the help of digital deployment. Digitalisation is making the agriculture sector more resilient by making the best utilization of fertilizers, insecticides, and pesticides with greater precision, avoiding ecologically damaging activities. Digital innovations and responsible use of technology in the agriculture sector can pave the way toward promoting efficiency and effectiveness in agricultural operations, improving the quality of agricultural output, and reducing hazardous working conditions in agricultural fields.

Responsible adoption of emerging technologies for Digital Sustainability in the ***Educational*** *Sector*

In a study by Mohamed Hashim et al. (2022), a grounded theory approach has been adopted to analyse the impact of sustainable digital transformation on the competitiveness of the universities. The study states that, as technology has revolutionised the current era, there is a high influence of digital transformation on the education sector globally. Digitalisation has created a lot many market opportunities and has redefined the value chain of the organisations. Results from this study highlight the fact that a well-planned and implemented digital transformation solution can help in assuring definite results, like an increase in potential revenue, higher profitability, and an increase in retention rate of the students. It has been analysed that various digital initiatives like virtualisation and use of cloud computing technologies, Wi-Fi-enabled university campus, reduction in e-waste, environmental protection, and building energy efficiencies, to name a few, can go a long way in sustainable education transformation.

Responsible adoption of emerging technologies for Digital Sustainability in the ***Healthcare*** *Sector*

The healthcare sector is adopting various emerging digital technologies like AR/VR, artificial intelligence, blockchain, internet of things (IOT), and cloud computing to ensure patient care. These technologies also help in increasing efficiency, leading to quick decision-making and streamlining operations. Use of robotics in operating theatres is also an example of adopting technologies for enhancing the level of precision and enhancing service quality and care. On the one hand, these sophisticated technologies contribute towards precision and minimize medical errors, but they also demand green data infrastructure facilities for reducing environmental footprints. The generation of e-waste and privacy of patient data is also one major concerns related to the adoption of these disruptive technologies. Therefore, it becomes highly crucial for healthcare organisations to adopt these technologies ethically and implement the principles of the circular economy into digital transformation strategies (Zahid et al. 2021).

Responsible adoption of emerging technologies for Digital Sustainability in the ***Banking*** *Sector*

A study by Serdarušić et al. (2024) on the banking industry focuses on both customer and customer care executives. This study highlights the adoption of green financing, fintech, and digital transformation in the banking sector. The study states that the adoption of digital technologies helps to enhance banking operations and promote its sustainable performance. Moreover, it emphasizes adopting user-friendly platforms to enhance customer understanding and comfort with banking tools. Adoption of fintech technologies also contributes to providing personalised services to the customers. The research work stated that there are collaborative efforts required by the banks, fintech companies, and regulatory bodies to ensure the successful implementation of digital technologies with responsibility. Thus, safeguarding the security of users and ensuring their trust in the tech platforms.

Responsible adoption of emerging technologies for Digital Sustainability in the ***Construction*** *Sector*

A study by Kineber et al. (2023) states that construction enterprises can adopt cyber/digital technologies to improve the quality of their activities and processes at reduced cost. In this empirical research work, it has been found that government support highly influences the adoption of digital technology in order to thrive in the sector. The use of digital technologies in the construction project helps in optimising efficiency, leading to proper planning and implementation of big construction projects. It helps in the timely delivery of the projects, improves information exchange, enables the adoption of innovative designs, and increases international competitiveness.

3.3 Motivation Factors and Barriers for Implementing Digital Sustainability

In the above section, we have explored the manner in which technology upgradation is being done in different sectors and the sustainable digital transformations happening in them. These sectors signify the way in which organisations tend to embed emerging technologies to increase their productivity, efficiency, and competitiveness. Along with the acceleration of rapid growth, environmental protection, good governance, the ethical use of technologies, and social inclusion are being prioritized for the greater good of society. However, the integration of digital sustainability into the organisation's strategies depends on certain motivating factors and is affected by the barriers to digital sustainability implementation.

Motivating Factors:

- Increase in stringent rules framed by different regulatory bodies pressurise the organisation to adopt sustainable practices, reduction in carbon emissions, strict data governance, thus encouraging digital sustainability.
- Because of the rising expectations and demand of various stakeholders like customers, employees, suppliers, shareholders, government, and society, organisations need to adopt emerging technologies in the business and ensure their sustainability through proper governance practices.
- Sustainable digital practices also help organisations to improve their brand image and gain competitiveness in the international markets, thus creating a competitive advantage, leading to attracting customers and investors all around the globe.
- Sustainable digital transformation also helps organisations to become cost-effective and gain efficiency through the adoption of various measures like energy-efficient data centres, green IT infrastructure, and reduction of costs through the optimisation of digital operations.
- Firms also adopt digital sustainable practices for improving their rating for the ESG matrix and thus signifying higher contribution towards UN sustainable goals. This helps them to create a niche and a better brand image.
- There are various emerging technologies like artificial intelligence, the internet of things, and blockchain that help organisations by enabling sustainable solutions. These emerging technologies also act as motivating factors to adopt digital sustainability for attaining long-term competitive positions.

- Organisations can also anticipate and alleviate risks with the support of various digital technologies and use predictive analytics for better decision-making.
- Because of the increasing awareness of the United Nations Sustainable Development goals (SDG), the younger generation gives higher preference to organisations that work responsibly. The sustainable initiatives taken by the organisation contribute towards positive employee morale and enhance employee loyalty and retention.
- As awareness among consumers has increased and they are demanding digital solutions and using eco-friendly products, it has become a motivating factor for organisations to adopt digital sustainable practices.
- Digital sustainable transformation also created long-term value for all the stakeholders, leading to sustainable growth of the organisation, which contributes to profit maximisation and wealth creation of the shareholders. It also helps to win the trust of stakeholders and society at large.

Barriers:

- As the organisation drives to adopt digital transformation practices it has to face huge implementation costs related to the use of renewable energy sources and the use of advanced technologies and systems.
- Expert knowledge is needed to implement digital sustainability transformation practices in the organisation. However, many firms lack this expertise, which makes it a costly affair.
- The top management sometimes shows reluctance to adopt digital transformation practices as there is no definite financial benefits in the short run and therefore these projects do not come in the priority list.
- As there are ethical dilemmas between data security and transparency, organisations try to avoid digital ecosystems.
- Sometimes there is a fear of insecurity, job loss, and lack of awareness among the middle-level management, which also creates a hurdle towards digital transformation and adoption of artificial intelligence.
- As there is a lack of availability of a standardised tool which is globally accepted in relation to measuring the level of sustainable transformation, it becomes difficult to justify the level to which an organisation has adopted digital sustainable practices.
- There are organisations that prioritize immediate returns rather than long-term growth, which creates a hindrance in the adoption of sustainable digital transformation practices.

4 Ethical Principles and Models for Resolving Ethical Dilemmas

Ethical theories and principles serve as a guiding light to resolve dilemmas related to what makes an action ethically better or worse in a particular situation (Stahl 2021, p. 20). There are several ethical principles that can be kept in mind while resolving dilemmic situations to ensure digital sustainability for building organisational growth and resilience.

The Right perspective principle puts emphasis on respecting the inherent moral right of an individual, which includes truth, privacy, and freedom of choice. The decisions that are taken to ensure digital sustainability in an organisation should protect the dignity of users and treat them as ends in themselves instead of means. For instance, if a healthcare organisation safeguards digital privacy and refuses to see the data of their customers to a third party without explicit consent (Balboni and Francis 2025). The justice principle ensures maintaining fairness among the technological users and ensuring that all the technology users get what they deserve based on merit, equality, and their needs. It emphasizes maintaining distributive justice, that is fair allocation of resources, corrective justice, which refers to taking remedial actions, and retributive justice, which refers to ensuring fair penalties. For example, in order to ensure diversity, equality, and an inclusive work environment in business, a firm provides additional accessibility tools and flexible working hours to employees with disabilities, thus ensuring digital inclusion in the workplace (Kordzadeh and Ghasemaghaei 2022). The Utilitarian Perspective principle focuses on the outcomes and ensures that most of the ethical choices depend upon maximising the benefits and minimizing the harm for the greatest number of people. For example, a firm introduced advanced technology for ensuring the data security of its customers, even at a higher short-term cost, keeping in mind the greater good (Azer and Samir 2024). The virtue ethics perspective sheds light on cultivating traits like honesty, courage, and integrity in the day-to-day functioning of an organisation and emphasizes the core values, morals, and vision of the organisation. Additionally, the common good principle focuses on maintaining digital sustainability by contributing positively towards the welfare of the community as a whole. Lastly, in this fast-changing digital world where technology keeps on upgrading, it is necessary to ensure that the organisation is empathetic to its employees and practices compassion and maintains relationships in ethical decision-making, thus ensuring that the normative principle of ethics of care is practiced by the organisation.

5 Application of Ethical Principles to Resolve Dilemmic Situations:

The dilemma of privacy versus innovation states that ensuring digital sustainability involves resolving conflicting situations, like whether to innovate in business processes at the cost of data privacy. Such conflicts can be resolved by ensuring data integrity, transparent data management practices, and embedding ESC oversight in digital solutions (Akakpo et al. 2025). The dilemma of fairness versus profitability signifies that in some situations, a firm may tend to compromise fairness and equality to ensure profit maximization. Such situations can be dealt with by maintaining fairness in audits, using artificial intelligence while auditing, and ensuring the adoption of strategies that help in maintaining a balance between profit maximisation and ensuring equity. The dilemma of transparency versus security states that data breaches and cyber attacks have been increasing over the years (Abukari et al. 2024). While the principle of transparency creates a need for openness in corporate decision making and informs stakeholders about the various activities of the organisation leads to a conflict between data transparency and maintaining security. If an organisation reveals too much data related to internal processes, financial records, or technological systems, it leads to a risk of cyber attack and data breaches. Therefore, the utilitarian theory principle of minimising the harm must be implemented by protecting of organisation's critical datasets. The dilemma of environmental cost versus economic growth highlights that there is most often a conflicting situation between environmental stewardship and economic growth. Digitalisation of processes and the use of disruptive technology in organisations for economic growth often leads to higher consumption of energy, an increase in the level of carbon emissions, and leads to an increase in the depletion of natural resources (Ma and Li 2025). Adherence to environmental stewardship involves limiting such operations or investments in cleaner and greener alternatives. Therefore, a firm needs to adopt the strategies of the circular economy to maintain a balance between economic growth and environmental sustainability. Thus, adopting a growth perspective that does not comply with the resource requirements of future generations. Lastly, the dilemma of employee monitoring versus care puts emphasis on the fact that, on the one side, organisations want to adopt employee well-being practices, but on the other side, the use of digital technologies to improve work efficiencies has made employee autonomy questionable. An organisation must avoid excessive digital surveillance as it affects the morale of employees, reduces their trust in the organisation, and violates the care principle of ethics. Thus, organisations need to maintain a balance between employee scrutiny and monitoring to as only necessary or affecting organisational goals. Strategies that help in balancing efficiency with the dignity and respect of employees must be adopted (Siegel et al. 2022).

6 Best Practices for Ensuring Ethics and Digital Sustainability for Greener Futures

In the competitive business environment, there is a growing demand for the adoption of disruptive technologies so as to ensure technological advancements. There is a growing debate between exploiting the speed of digitization while ensuring sustaible transformation (Guandalini 2022). Moreover, the growing need for environmental consciousness and adoption of sustainable practices in the organisation creates a paramount need to maintain digital sustainability. It becomes highly important for an organisation to ensure that the digital solutions are well integrated with economic viability, social equity and inclusiveness, along with maintaining environmental stewardship. Ensuring digital sustainability involves mitigating the ecological footprints and promoting responsible corporate governance practices that take care of the interests of all the stakeholders, apart from maximizing the wealth of shareholders. Therefore, there is a need to adopt certain best practices for ensuring ethical digital transformation and sustainability, along with ensuring greener futures. Enterprises need to develop AI systems that prioritise environmental sustainability, ensure optimum energy consumption, and promote green behaviours among the employees. Thus, creating an organisational culture that fosters digital innovations that are aligned with ethical principles and standards, and contributes towards a sustainable future (Bolón-Canedo et al. 2024). Green AI initiatives help in reducing carbon footprints through optimising algorithms and hardware usage. Organisation can create a culture to adopt digital transformation strategies that encourage green behaviours among employees, such as minimising wastage and maximising optimum resource utilisation, and reducing resource consumption (Wang et al. 2025a, b). Organisation can implement incentive policies and reward the employees for proposing digital innovation systems related to waste reduction, recycling, and sustainable transportation systems, to name a few. Such type of innovative approaches can help in accelerating digital technologies and sustainability (Lyulyov et al. 2025). Enterprises can motivate consumers towards adopting sustainable energy consumption behaviour and promoting greener choices. In a digital environment, certain types of behavioural interventions like digital energy nudges can be deployed in order to reduce the energy consumption of consumers and increase the consumption of renewable energy resources relatively (Kuyer et al. 2025). Organisation must promote a deep interconnection between corporate digital responsibility and corporate sustainability in a systematic manner. This will ensure that the digital practices within an organisation are well aligned with ethical principles and digital sustainability (Covucci et al. 2024). Organisations need to create ethical policies related to the digital environment that address various issues related to algorithm bias, data privacy, and transparency. These guidelines must be aligned with the ethical models and principles known globally, along with the adoption of international standards and regulations in relation to data management practices (Guenduez et al. 2025). Organisations must adopt the principles of the circular economy and promote environmental sustainability through the use of digital technologies. Digitalisation help the enterprises to improve their Environmental, Social and Governance (ESG) performance through improving resource efficiency. Adopting sustainable product design, applying methods to reduce energy and material consumption and recycling helps in contributing towards environmental and cost performance for SMEs (Siedschlag et al. 2024).

7 Managerial and Research Implications

To survive in the volatile, uncertain, and dynamic business environment, it becomes highly essential that firms harness the power of emerging technologies and adopt them strategically in the organisation. This will ensure digital sustainability for greener futures, making the organisation more resilient and responsible. Managers must design strategies that not only encourage digital transformation but also align it with environmental stewardship. Such initiatives by the managers will help to gain stakeholders' trust and lead the organisation towards growth and long-term success. Managers must create and promote a culture of innovation and creativity in the organisation by recognising the effort of those employees who contribute towards environmentally friendly digital transformation practices.

Another most important aspect related to the application of ethical principles and models for resolving dilemmic situations relates to sustainable digital transformations. Management must adopt robust organisational governance practices to ensure a balance between privacy, transparency, and data security. Management needs to conduct a training program for the employees to ensure that they become ethically aware and digitally proficient to contribute to the organisation's digital transformation endeavours.

8 Conclusion, Limitation and Future Scope

The present study highlights the importance of sustainable digital transformation and its role in creating a more responsible and resilient corporate culture. A culture that emphasises the adoption of environmentally friendly practices and creates a realisation of improving ESG performance. The chapters have focused on several key aspects related to sustainable digital transformation. Discussion started with various disruptive technologies and the manner in which organisations are adopting these technologies for higher outcomes. Adoption of digital sustainable transformation practices in various sectors like healthcare, banking, education, construction, and agriculture is also deliberated. The study lays emphasis on motivating factors and barriers for adopting digital suitability practices in different sectors and industries.

In accordance with the present research, there can be several ethical dilemmas that managers can face while adopting digital transformation practices. These dilemmic situations can be resolved with the help of the adoption of various ethical principles and models. Despite several barriers or dilemmas related to digital transformation, there are certain best practices that the organisations can adopt to ensure ethical and digital sustainability for greener futures. Finally, it can be stated that the adoption of sustainable digital transformation practices enhances resilience, increases efficiency, improves stakeholders' trust, and supports environmental care and protection. But at the same time, in order to ensure successful incorporation of such practices, it is important to maintain a balance between transparency and privacy, innovation with effective governance, and a focus on environmental and social perspectives, along with economic growth. Thus, adoption of digital transformation practices has become a need of the hour and is essentially required for long-term growth, success, and carving resilient organisations.

From the research perspective, the present area of the study is currently evolving and is relatively new. As the present study is theoretical and literature-based, it demands empirical enquiry and a field-based study. A standardised instrument to measure the level of sustainable digital transformation needs to be established, which serves as a base to measure the organisation's initiatives towards sustainability and technology adoption.

Finally, it becomes highly important to measure and test the manner in which ethical models, principles, and practices are being implemented by the organisations to resolve dilemmic situations while implementing digital transformation initiatives.

References

Abukari K, Dutta S, Li C, Tang S, Zhu P (2024) Corporate communication and likelihood of data breaches. Int Rev Econ Financ 94:103433

Akakpo A, Gyasi EA, Oduro B, Akpabot S (2025) Artificial intelligence (AI) and the future of information privacy: expert viewpoints. J Glob Inf Manag (JGIM) 33(1):1–25

Azer MA, Samir R (2024) Overview of the complex landscape and future directions of ethics in light of emerging technologies. Int J Adv Comput Sci & Appl, 15(7)

Balboni P, Francis KE (2025) Data ethics and digital sustainability: Bridging legal data protection compliance and ESG for a responsible data-driven future. J Responsible Technol 22:100099

Bolón-Canedo V, Morán-Fernández L, Cancela B, Alonso-Betanzos A (2024) A review of green artificial intelligence: Towards a more sustainable future. Neurocomputing 599:128096

Boonmee C, Mangkalakeeree J, Jeong Y (2025) Towards sustainable digital transformation: AI adoption barriers and enablers among SMEs in Northern Thailand. Sustainable Futures 10:101169

Chatterjee C, Srinivasan V (2013) Ethical issues in health care sector in India. IIMB Manag Rev 25(1):49–62. https://doi.org/10.1016/j.iimb.2012.11.004

Deng Q, Karia N (2025) How ESG performance promotes organizational resilience: the role of ambidextrous innovation capability and digitalization. Bus Strat & Dev 8(1):e70079

Covucci C, Confetto MG, Ključnikov A, Panait M (2024) Unrevealing the nexus between digital sustainability and corporate digital responsibility: A dual-track systematic literature review towards a framework of corporate digital sustainability. Technol Soc 79:102743

Dong C, Akram A, Andersson D, Arnäs PO, Stefansson G (2021) The impact of emerging and disruptive technologies on freight transportation in the digital era: current state and future trends. Int J Logist Manag 32(2):386–412

Fang Q, Wang Z, Dang L (2025) Audit effort in the digital Era: Uncovering the dynamic interplay of business strategy and digital transformation. Int J Account Inf Syst 56:100747

Guandalini I (2022) Sustainability through digital transformation: A systematic literature review for research guidance. J Bus Res 148:456–471

Guenduez AA, Walker N, Demircioglu MA (2025) Digital ethics: Global trends and divergent paths. Gov Inf Q 42(3):102050

Huyen NTT, Thanh TT (2025) Digital integration toward smart and sustainable agriculture in the European region. Digit Bus, 100152

Itten R, Hischier R, Andrae AS, Bieser JC, Cabernard L, Falke A, Stucki M (2020) Digital transformation—life cycle assessment of digital services, multifunctional devices and cloud computing. Int J Life Cycle Assess 25(10):2093–2098

Kineber AF, Oke A, Aliu J, Hamed MM, Oputu E (2023) Exploring the adoption of cyber (digital) technology for sustainable construction: a structural equation modeling of critical success factors. Sustain 15(6):5043

Kordzadeh N, Ghasemaghaei M (2022) Algorithmic bias: review, synthesis, and future research directions. Eur J Inf Syst 31(3):388–409

Kuyer P, Shukla S, O'Brolcháin F, Doncel VR, Gordijn B (2025) The ethics of digital nudging for sustainable energy consumption–a closer look at the EU green deal. Technol Soc, 102990

Lyulyov O, Pimonenko T, Kwilinski A (2025) Digital inclusion for a sustainable future: Catalysing green economic growth. Sustain Futur 10:101037

Ma Y, Li R (2025) The impact of digital economy on carbon emissions: Insights from the G-20 energy transition and environmental governance. Energ Strat Rev 57:101612

Martynov VV, Shavaleeva DN, Zaytseva AA (2019) Information technology as the basis for transformation into a digital society and industry 5.0. In: Proceedings of the 2019 IEEE international conference quality management, transport and information security, information technologies IT and QM and IS 2019, 539–543. https://doi.org/10.1109/ITQMIS.2019.8928305

Massa S, Annosi MC, Marchegiani L, Messeni Petruzzelli A (2023) Digital technologies and knowledge processes: new emerging strategies in international business. A systematic literature review. J Knowl Manag, 27(11), 330–387

Mbunge E, Jiyane S, Muchemwa B (2022) Towards emotive sensory Web in virtual health care: Trends, technologies, challenges and ethical issues. Sens Int 3:100134. https://doi.org/10.1016/j.sintl.2021.100134

Mohamed Hashim MA, Tlemsani I, Duncan Matthews R (2022) A sustainable university: Digital transformation and beyond. Educ Inf Technol 27(7):8961–8996

Serdarušić H, Pancić M, Zavišić Ž (2024) Green finance and Fintech adoption services among Croatian online users: how digital transformation and digital awareness increase banking sustainability. Econ 12(3):54

Siedschlag I, Mohan G, Yan W (2024) Twin transitions across enterprises: Do digital technologies and sustainability go together? J Clean Prod 481:144025

Siegel R, König CJ, Lazar V (2022) The impact of electronic monitoring on employees' job satisfaction, stress, performance, and counterproductive work behavior: A meta-analysis. Comput Hum Behav Rep 8:100227

Stahl BC (2021) Artificial intelligence for a better future: an ecosystem perspective on the ethics of AI and emerging digital technologies (p. 124). Springer Nature

Wang Y, Yu Y, Khan A (2025a) Digital sustainability: Dimension exploration and scale development. Acta Physiol (Oxf) 256:105028

Wang Y, Ziya A, Li M, Li G (2025) Digital transformation, green behaviors, and environmental policy heterogeneity in resource-based enterprises. Int Rev Econ & Financ, 104291

Xia L, Baghaie S, Sajadi SM (2024) The digital economy: Challenges and opportunities in the new era of technology and electronic communications. Ain Shams Eng J 15(2):102411

Zahid A, Poulsen JK, Sharma R, Wingreen SC (2021) A systematic review of emerging information technologies for sustainable data-centric health-care. Int J Med Informatics 149:104420

Waste Intelligence and Analytics for Real-Time Insights into Waste Generation—a Qualitative Exploration

Mananage Shanika Hansini Rathnasiri, Ali Najeeb, Amit Joshi, and Muhandiramge Vasana Maduwanthi Rodrigo

Abstract Real-time analytics and waste intelligence have become transformative solutions for tackling the escalating global challenge of waste management and generation. While much of the current discussion centres on quantitative modelling and technological efficiency, less focus is given to the socio-cultural, institutional, and contextual factors that affect adoption, particularly in South Asia. This chapter adopts a qualitative approach to explore the drivers, barriers, and implications of incorporating waste intelligence and analytics into municipal and industrial waste management systems. Based on 15 semi-structured interviews with municipal, community, technology, and policy makers across India, Sri Lanka, Bangladesh, and the Maldives, the study identifies four thematic areas: (1) operational efficiency, transparency, enhanced recycling; (2) technological reliability, financial hurdles, community resistance; (3) smart cities, public–private partnerships, global sustainability pressures; (4) regional differences in adoption, shaped by governance and cultural factors. The findings suggest that waste intelligence is more than a technical measure; it is also a socio-institutional process that relies on trust, participation, and coherent policy. The chapter concludes with region-specific recommendations for enhancing waste analytics to promote sustainable urban and industrial development.

M. S. H. Rathnasiri (✉)
Department of Marketing Management, Faculty of Management Studies, Sabaragamuwa University of Sri Lanka, Belihuloya, Sri Lanka
e-mail: shanika@mgt.sab.ac.lk

College of Society and Professions, London South Bank University, London, UK

A. Najeeb
Qasim Ibrahim School of Business, Villa College, Male, Maldives
e-mail: najeeb@villacollege.edu.mv

A. Joshi
Department of Management Studies, Kumaun University, Nainital, Uttarakhand, India
e-mail: amitjoshiaj@gmail.com

M. V. M. Rodrigo
Ministry of Industry and Entrepreneurship Development, Colombo, Sri Lanka
e-mail: rodrigo.mvm@industry.gov.lk

R. Singh et al. (eds.), *Data-Driven Sustainability: Harnessing Technology for A Greener Future and Environmental Resilience*, World Sustainability Series,
https://doi.org/10.1007/978-3-032-13412-7_21

Keywords Waste intelligence · Real-time waste analytics · Qualitative exploration · Smart cities · Circular economy

1 Introduction

The twenty-first century has seen an unprecedented rise in the amount of waste produced worldwide, driven by rapid urbanisation, industrialisation, and consumerism. Maalouf and Mavropoulos (2023) state that global municipal solid waste exceeds two billion tonnes annually, and it is expected to grow exponentially by 2050 unless current waste disposal practices change. Uncontrolled waste causes serious environmental, economic, and health problems, including greenhouse gas emissions, land degradation, and inefficient resource use. In response, the concept of waste intelligence has emerged as a revolutionary solution that integrates data analytics, digital technology, and real-time monitoring to enhance decision-making and operational efficiency in waste management. Waste intelligence involves deploying technologies such as the Internet of Things (IoT), artificial intelligence (AI), and big data analytics to generate actionable insights across the entire waste value chain—covering generation, collection, sorting, and recycling—and aligns with the broader goals of a circular economy and sustainable urban development (Dewasiri and Choudhary 2025).

Waste intelligence research so far remains focused on quantitative and technical studies despite increasing academic and policy interest (Khan et al. 2023). The existing literature mainly focuses on system design, algorithm optimisation, and sensor performance, often overlooking human, organisational, and policy factors that are essential for technology adoption and long-term sustainability. Issues related to institutional readiness, stakeholder collaboration, citizen engagement, and socio-cultural acceptance are not fully explored. This is particularly clear in developing countries like South Asia, where infrastructural challenges, poor governance, and socio-economic disparities are common; implementing effective waste intelligence systems in these regions will depend as much on social and institutional processes as on technological innovation.

Understanding this gap, the current chapter adopts a qualitative exploratory approach to examine how waste intelligence and real-time analytics are perceived, applied, and experienced within South Asian contexts. The study aims to uncover implicit perceptions, challenges, and regional nuances related to digital waste management through fifteen semi-structured interviews with policymakers, municipal officers, technology providers, and community representatives in India, Sri Lanka, Bangladesh, and Pakistan. This approach seeks to move beyond technological determinism and offers a more comprehensive view of how data-driven waste systems translate into practice. Ultimately, the chapter aims to provide empirically grounded insights that can inform policy, promote inclusive adoption of technological innovations, and support a transition towards more sustainable and innovative waste management systems in the region.

2 Conceptual Foundations

The waste intelligence concept signifies a shift in waste management approaches, evolving from traditional collection and disposal methods to integrated, data-driven, and proactive systems. It involves strategically applying digital technologies such as the Internet of Things (IoT), artificial intelligence (AI), machine learning (ML), and geographic information systems (GIS) to record, analyse, and interpret real-time data on waste production, structure, and movement (Kumar et al. 2024). Waste intelligence assists decision-makers in optimising collection routes, reducing operational costs, identifying recycling opportunities, and predicting waste surges before they happen by converting raw data into actionable insights. This approach enhances conventional waste management by adding intelligence, adaptability, and foresight within urban and industrial ecosystems. The concept also closely aligns with the circular economy paradigm, which promotes the active use of resources and waste reduction through reuse, recycling, and recovery (Singh et al. 2025). This transition is enabled by real-time data analytics, allowing policymakers and practitioners to track material flows, evaluate system efficiency, and recycle resources effectively. Waste intelligence can transform inert waste systems into lively circular ecosystems through digital dashboards, smart bins, and predictive modelling. It provides the empirical foundation for advancing sustainability goals, such as responsible production and consumption (SDG12) and climate action (SDG13), by linking environmental performance with economic and social indicators (Dewasiri 2025; Jayarathne et al. 2025). As a result, waste intelligence functions both as a technological enabler and a governance tool for strengthening circularity and sustainability paradigms.

The importance of real-time analytics lies in its ability to replace static, retrospective reporting with dynamic, adaptable, and forward-looking decision-making. Traditional waste management systems often rely on periodic data collection, which yields delayed information that fails to capture the evolving nature of waste generation accurately. In contrast, real-time data streams from sensors and digital monitoring platforms enable quick identification of operational inefficiencies or environmental hazards, allowing for swift intervention. This level of informational agility makes waste management a predictive and preventive process. For example, municipalities can quickly detect bin overflows, illegal dumping, or hazardous waste leaks, ensuring timely responses to minimise environmental damage and operational costs.

Theoretically, this chapter presents the adoption of waste intelligence from three complementary perspectives: Diffusion of Innovation Theory, Socio-technical Systems Theory, and Stakeholder Theory. The Diffusion of Innovation (DOI) model (Rogers 2003) describes how social systems adopt innovations over time, emphasising key factors such as relative advantage, compatibility, and observability. Implementing waste intelligence in developing regions can therefore be seen as a process influenced not only by technological benefits but also by institutional readiness and user perceptions. This idea is supported by the Socio-technical Systems Theory, which states that technological infrastructure and social structures are interconnected, requiring synergies between technology, people, and processes for effective digital waste systems (Sony and Naik 2020). Lastly, Stakeholder Theory emphasises the importance of inclusive engagement among governments, technology companies, communities, and waste workers in assessing the legitimacy and sustainability of introducing innovations. When combined, these frameworks offer a cohesive conceptual basis for examining the real-world challenges of deploying waste intelligence. They help to appreciate how digital technologies intersect with human behaviour, institutional capacity, and governance within the complex socio-economic environments of South Asia.

3 Methodology

In this study, the designed approach used to investigate the subtle aspects of waste intelligence and real-time analytics adoption in South Asia was a qualitative exploratory design. This method was chosen because it enables revealing the complex interaction between technology, institutions, and society. This aspect cannot be fully addressed through the quantitative evaluation of waste management systems. The qualitative inquiry allows the researcher to engage directly with participants' experiences, perceptions, and contextual meanings, thereby enabling the derivation of in-depth, grounded knowledge about the social and organisational processes involved in technological innovation.

Semi-structured interviews were the primary method for collecting data, supplemented by focus group discussions and document analysis. These interviews were intentionally flexible, enabling the exploration of emerging themes while keeping a clear structure aligned with the research aims. The interview protocol covered awareness and understanding of waste intelligence, perceived opinions and challenges of adoption, institutional readiness, stakeholder partnerships, and community engagement. Fifteen interviews were carried out in India, Sri Lanka, Bangladesh, and Pakistan, involving a diverse range of stakeholders such as policymakers, municipal waste managers, technology service providers, non-governmental organisations, and community leaders. This purposive sampling strategy ensured that insights from both strategic and operational levels were obtained, offering a comprehensive view of waste intelligence implementation across different governance and socio-economic settings.

In addition to interviews, some selected policy documents, municipal reports, and project evaluations were also examined to triangulate and verify findings. Such materials further helped provide contextual insights into the waste management systems in the country, digital transformation projects, and sustainability strategies in the area. Various data sources contributed to enhancing the credibility and transferability of the findings, which aligns with best practices in qualitative research.

Thematic analysis was employed to examine the data collected, following the six-phase model outlined by Braun and Clarke (2006). It involved an iterative process of familiarising oneself with the data, generating initial codes, identifying themes, refining and redefining theme patterns, and establishing the final analytical categories. Themes were inductively identified from participants' narratives, guided by the theoretical frameworks of diffusion of innovation, socio-technical systems, and stakeholder engagement. The main themes identified, such as drivers, barriers, enablers, and contextual dynamics, underscore the complex nature of adopting waste intelligence in the South Asian context.

The qualitative design was chosen for this study because it enables the researcher to explore lived experiences and socio-cultural aspects that are often not captured by quantitative or model-based research. Through open-ended conversations and interpretative analysis, the study uncovers the what, why, and how of waste intelligence within specific institutional and cultural contexts. Ultimately, this approach deepens understanding of the human and policy dimensions of waste intelligence, providing a more detailed and context-aware perspective that can guide future research, policy development, and technological adaptations across emerging economies.

4 Findings and Thematic Analysis

The qualitative study offered deep insights into perceptions, implementation, and sustainability of waste intelligence and analytics in various South Asian settings. Thematic analysis of fifteen semi-structured interviews with policymakers, municipal officials, technology innovators, and community representatives revealed four main themes: perceived benefits, barriers, enablers, and contextual realities. The following subsections present these themes along with some examples.

4.1 Perceived Benefits of Waste Intelligence

In the four countries, respondents highlighted that digital waste monitoring enhanced operational efficiency, transparency, and environmental outcomes. An Indian senior municipal engineer stated:

> *Previously, our waste collection lorries followed fixed routes regardless of waste levels. Now, sensor data allows us to optimise routes daily - reducing fuel costs and improving response times.*

The feeling is universal across the world because novel waste management methods increase efficiency and lower emissions (Citton et al. 2020; Kaza et al. (2018). The respondents perceived real-time analytics as a tool for informed decision-making and accountability. One of the policymakers of Sri Lanka noted:

> *We can no longer afford to work by guesswork. Dashboards display the real-time situation, which increases public trust and encourages performance-based budgeting.*

The technology vendors in Bangladesh and Pakistan have also reported using predictive analytics as a proactive management tool. As one of the respondents from Lahore observed:

Data enables the prediction of peak waste periods in advance, such as during festivals or weekends, allowing for more effective resource allocation.

Other participants linked these digital systems to broader sustainability agendas. As one Bangladeshi environmental officer explained:

When we learn that real-time reports on waste reduction have been achieved, it is not only a sign that operations are successful but also a step in the right direction towards the circular economy.

These results align with the arguments of Guerrero et al. (2013) and Ahmed et al. (2023), who assert that the integration of technologies in waste management is necessary to promote environmental sustainability and transparency.

4.2 Barriers to Adoption

Although optimism exists, participants pointed out ongoing barriers that hinder the widespread adoption of waste intelligence systems. Common themes emerged, often related to technical limitations and infrastructural deficiencies. One of the Bangladeshi IT specialists remarked:

Sensors, the technology, are not as stable as the environment can tolerate; heat, humidity, or ad hoc maintenance are known to cause failures.

These findings are in line with the results of Wang et al. (2021), who observe that climatic and maintenance-related issues tend to decrease the reliability of the IoT in developing countries. There also appeared economic and institutional restraints. One of the urban planners in Sri Lanka described it as follows:

Smaller municipalities cannot afford the initial capital for smart bins and data servers. The absence of special budgets paralyses these projects along the way.

Social and behavioural aspects were also eminent. One of the NGO workers in Pakistan remarked:

People believe that sensors and cameras are used to spy on them rather than to improve waste management. To develop trust, we must have awareness campaigns.

To this, a municipal officer of Chennai remarked:

We have occasionally encountered resistance from our staff to new systems; they view data tracking as a challenge to their usual authority.

All these obstacles highlight the multidimensional nature of sustainability transitions—not just technical, but highly social and institutional. This aligns with Schot and Geels (2013), who note that any shift to an innovative sustainability system involves institutionalising, aligning stakeholders, and gaining cultural acceptance alongside technological readiness.

4.3 Drivers and Enablers

Conversely, the research identified some significant enablers that facilitate progress in the adoption of waste intelligence. Policy requirements, foreign alliances, and public–private partnerships (PPPs) were recognised as key drivers by interviewees. The Swachh Bharat Mission, with its innovative city structures in India, provided momentum for local innovation. As one of the interviewees from Pune mentioned:

> *They were financed and legitimised through government schemes. Once the pilot was successful, private players joined the project.*

In Sri Lanka and Bangladesh, PPPs played a vital role in mobilising resources and expertise. A waste management director based in Colombo stated:

> *Collaborations with ICT companies have helped us to deploy IoT gadgets without significant capital expenditure, as they manage the technology, while we handle the logistics.*

Education and gamification proved to be effective instances of behavioural change at the community level. One of the leaders of a Bangladeshi community commented:

> *The residents became motivated to clean up their bins after seeing the data show improvement in their neighbourhoods on the app, which sparked a competition among them.*

A Pakistani municipal coordinator made another observation, and he observed:

> *The city requires champions in each of the cities - one leader who believes in technology and encourages others to follow him.*

The examples show that waste intelligence can succeed when technological innovation aligns with institutional support and citizen involvement. The significance of collaboration highlighted here mirrors the findings of Kirchherr et al. (2017) and Gaur et al. (2024), emphasising that cooperation among multiple actors and stakeholder engagement are vital for the success of transitions to a circular economy (Singh et al. 2024b).

4.4 Regional Contexts and Variations

The unique social, political, and infrastructural conditions shaped the development of waste intelligence across different countries. India had well-organised national programmes with private sector involvement, while Sri Lanka, despite its emerging technology, demonstrated strong potential for citizen-generated data systems in local councils.

Bangladesh was heavily reliant on donor-funded pilot projects centred on innovation through social enterprise. However, in comparison to Pakistan, it made use of localisation dashboards and mobile-based systems at the provincial levels. According to a policy analyst in Islamabad, she summarised:

Every nation is experimenting with new approaches. It is not the inertia of technology but the inertia of continuity, so that projects must have a longer life than a political administration.

Likewise, one of the Bangladeshi policymakers said:

Donor projects begin with hope but collapse with the termination of funding. Sustainability implies local ownership.

This result aligns with Kemmerling (2023), who emphasises policy continuity and institutional learning as essential for maintaining environmental innovations in the Global South. The above contextual differences highlight that, although waste intelligence shares a common technological foundation, its success ultimately depends on governance systems, financing structures, and country-specific public participation (Table 1).

5 Discussion

These results highlight that waste intelligence is not merely a technological shift, but a socio-institutional change. The findings show that there is an active interaction between human agency, institutional maturity, and digital infrastructure—confirming that the effectiveness of waste intelligence systems depends not only on technology but also on governance coherence and culturally tailored approaches. This aligns with the Socio-technical Systems Theory, which states that co-evolution between social systems and technical systems is essential whenever implementing new technology (Geels 2004). Such co-evolution has yet to develop in South Asia, where policy fragmentation and resource constraints are widespread.

The results also support the Diffusion of Innovation Theory (Rogers 2003), particularly in relation to perceived usefulness, compatibility, and observability. Interviewees also observed that visible outcomes, such as reduced waste overflow or increased citizen satisfaction, encouraged adoption in municipal councils. However, complexity and low trialability hindered diffusion in smaller municipalities, which aligns with the findings of Hannan et al. (2015) on innovative city technologies in an emerging economy.

Additionally, the paper supports Stakeholder Theory (Freeman and Phillips 2002), because the effective operation of waste intelligence programmes relies on cooperation between governments, technological companies, and local communities. Those that were most successful involved citizens not as passive data subjects, but as partners in developing sustainable practices. As one environmental officer in Sri Lanka stated:

According to the technology, local behaviour is what should be designed, not the local people, who should be asked to learn imported systems.

This observation is similar to that of Mishra et al. (2021), who emphasise the significance of community participation and indigenous knowledge in driving circular economy solutions in South Asia.

Finally, the paper concludes by highlighting how global literature praises the effectiveness of the new waste system (Gupta et al. 2019; Iqbal et al. 2024), whereas this paper reveals the discrepancy between high-tech standards and on-the-ground realities. South Asian cities are on the path to digitisation, yet success in this process depends on long-term policy commitment, inter-agency cooperation, and social trust. As one Indian policy advisor summed up:

> *Waste intelligence is easy to start but hard to sustain. This information is practical, provided that individuals take action.*

Collectively, these results contribute to the conceptual knowledge base by demonstrating that waste intelligence adoption is a multilayered activity influenced by technology, institutions, and human behaviour. Interregional cooperation and exchange of experience among South Asian countries could enhance not only the efficiency but also the equity of waste analytics implementation, ensuring that digital changes make a positive contribution towards sustainability and inclusive development.

6 Implications and Recommendations

The conclusions of this qualitative investigation hold significant policy, management, and academic implications for the future. Waste intelligence and real-time analytics are transformative in promoting sustainability in South Asia, but their implementation requires systemic adjustments in governance, technology, and social alignment (Islam 2025). This chapter details how the lessons learned during the study can be utilised to create a more coherent and resilient data-driven waste management framework.

Regarding policy, the governments of South Asia should move beyond piecemeal, project-based efforts in favour of institutionalised digital waste management systems. The research revealed that the absence of policy continuity and specific funding often halts promising pilot projects when donor or political funding is withdrawn. Waste intelligence must be integrated into long-term municipal strategies by national urban development ministries with stable budget allocations, performance indicators, and incentives through regulation. Harmonising policies across ministries such as the environment, technology, and local governance is crucial to prevent overlap and redundancy. Inspired by Rogers' Diffusion of Innovation theory, governments are encouraged to create supportive ecosystems that foster observability and trialability, enabling municipalities to test and refine digital solutions before large-scale implementation. It may also be beneficial to promote cross-learning, common standards, and joint investments in open-data infrastructure for waste analytics through regional cooperation, such as the South Asian Association for Regional Cooperation (SAARC) or the Bay of Bengal Initiative for Multi-Sectoral Technical and Economic Cooperation (BIMSTEC) platforms.

Practically, municipal governments and non-governmental technology providers need to shift their focus from solely technology-driven models to participatory and context-specific implementation approaches. The findings showed that engaging citizens, providing education, and building trust are even more crucial than deploying smart bins or sensors. Municipal managers should incorporate feedback systems within communities, utilise waste dashboards, promote transparency through open data portals, and adopt gamified mobile applications to encourage behavioural change. Public–private partnerships should emphasise long-term service contracts and capacity-building rather than short-term installations. Moreover, adaptive maintenance systems, local technician training, and designing sensors with climate resilience can enhance operational resilience, which is especially important in tropical areas like Bangladesh and Sri Lanka (Dewasiri et al. 2025). Technology companies should develop comprehensive business models that make digital waste services affordable for small and mid-sized cities, potentially through leasing or shared-service arrangements.

The study also highlights several implications for future research. Firstly, there is a need for mixed-methods studies to evaluate the economic and environmental advantages of qualitative findings, helping to link perceptions with measurable outcomes. Longitudinal research could examine how municipal and community behaviours change over time following exposure to data-driven systems, offering a better understanding of technology adoption patterns. A comparative analysis of South Asian countries and other developing nations would help identify models and policy lessons that can be adapted and scaled in similar socio-economic contexts. Furthermore, incorporating insights from behavioural economics and cultural psychology could provide valuable insights into how social norms, trust, and motivation affect household waste management practices. Finally, researchers should investigate how waste intelligence relates to broader sustainability efforts, such as the adoption of renewable energy, carbon accounting, and supply chain management within a circular economy (Mishra et al. 2024).

Overall, the study's implications emphasise that waste intelligence is not just a technological breakthrough but signifies a paradigm shift in how societies should conceptualise, regulate, and participate in sustainable waste management. The case of South Asia, where demographic and environmental pressures are considerable, offers both an opportunity and a pressing need to embed intelligence into waste systems. Through a combination of technological innovation, institutional reform, and civic engagement, the region can make a significant step towards a data-driven, circular, and inclusive model of urban sustainability.

7 Conclusion

This chapter discusses how waste intelligence and real-time analytics are transforming the South Asian region and provides a qualitative overview of how digital innovations are changing existing waste management practices. Based on 15 detailed interviews with policymakers, municipal officers, technology providers, and community leaders across India, Sri Lanka, Bangladesh, and Pakistan, the study finds that implementing waste intelligence is a multidimensional process influenced by technological capability, institutional maturity, and social involvement.

The results indicated that real-time waste analytics are crucial for operational efficiency, transparency, and sustainability, aligning with global circular economy goals. However, the research also revealed persistent obstacles—such as technological fragility, financial constraints, broken policies, and mistrust among people—that hinder implementation in developing countries. Positively, the presence of policy requirements, public–private partnerships, community involvement, and global engagement as enablers demonstrates that, even amid these barriers, adaptive and inclusive models can succeed when grounded in local realities.

The chapter emphasises that the usefulness of digital transformation in waste management depends on the co-evolution of technology and society, framing its analysis within the context of Diffusion of Innovation, Socio-technical Systems, and Stakeholder theories. The findings advocate for unified governance structures, collaborative implementation, and international knowledge sharing as key strategies for achieving sustainable development. Finally, the chapter concludes by asserting that waste intelligence is not merely a technological intervention but a transformative approach—one that can foster more innovative, resilient, and circular waste ecosystems in South Asia and beyond.

8 Future Directions and Research Agenda

The growing use of digital technologies in waste management offers a promising area for future research and innovation. Based on the qualitative results of the current study, several new pathways are worth academic and policy consideration. Firstly, advanced analytics and artificial intelligence (AI) could help expand existing waste intelligence systems beyond mere monitoring to include prescriptive and predictive functions. Future research should explore how artificial intelligence-based decision-support systems and digital twins can be utilised to simulate waste streams, optimise resource distribution, and assess the carbon reduction impact within the circular economy context (Singh et al. 2024a, 2024c).

Second, integrating traceability systems based on blockchains can enhance transparency and accountability in waste value chains, particularly in industrial and e-waste contexts (Centobelli et al. 2022). Comparative studies of blockchain-based traceability in South Asian supply chains may provide valuable insights into governance reforms and ethical waste trading practices.

Third, the social aspect of digital waste management has not been adequately explored. Researchers are encouraged to investigate behavioural, cultural, and ethical factors that influence citizens' participation in innovative waste initiatives. There could be interdisciplinary collaboration among environmental engineers, behavioural economists, and sociologists to develop a more comprehensive understanding of how technology is adopted.

Finally, future research should focus on cross-regional partnerships and policy-making based on information. By developing a shared South Asian waste intelligence platform that incorporates open data, artificial intelligence, and policy interoperability, it is possible to foster collaborative innovation and strengthen the sustainability agenda across the region. Essentially, future research must move beyond technology adoption towards systemic change, where data intelligence forms the basis for sustainable, inclusive, and resilient urban futures.

Table 1 Summary of emergent themes from the qualitative analysis

Theme	Description	Illustrative quotations	Supporting literature
1. Perceived benefits of waste intelligence	Waste intelligence enhances operational efficiency, transparency, and environmental outcomes by enabling real-time decision-making, optimised collection, and predictive analysis	"Earlier, our collection trucks moved on fixed routes regardless of waste levels. Now, with sensor data, we optimise routes daily—fuel costs are down, and response time is faster." (Municipal Engineer, India)	Guerrero et al. (2013), Islam et al. (2025), Kaza et al. (2018)
		"Dashboards show what is happening in real time, which strengthens public trust." (Policymaker, Sri Lanka)	
2. Barriers to adoption	Technical, financial, and social challenges limit implementation, including unreliable sensors, high costs, fragmented policy, and public mistrust	"Sensors often fail due to heat, humidity, or irregular maintenance—the technology is only as reliable as the environment allows." (IT Specialist, Bangladesh)	Schot and Geels (2013), Hannan et al. (2015)
		"People think sensors and cameras are for surveillance, not for better waste management." (NGO Worker, Pakistan)	
3. Drivers and enablers	Strong policy mandates, international collaborations, PPPs, and citizen participation promote adoption and scalability	"Government schemes gave us both legitimacy and funding. Once the pilot succeeded, private players came on board." (Official, India)	Kirchherr et al. (2018),s Azevedo et al. (2022)
		"When residents saw their neighbourhood waste levels improving on the app, they started competing to keep bins cleaner." (Community Leader, Bangladesh)	

(continued)

Table 1 (continued)

Theme	Description	Illustrative quotations	Supporting literature
4. Regional contexts and variations	Differences in governance, finance, and institutional maturity influence adoption patterns across South Asia	"Each country is experimenting in its own way. The real challenge is not the lack of technology but the lack of continuity." (Policy Analyst, Pakistan)	Geels (2004), Freeman and Phillips (2002)
		"Donor projects start with promise but end when funding ends. Sustainability means local ownership." (Policymaker, Bangladesh)	

Source Authors' qualitative fieldwork (2025)

References

Ahmed Z, Hasan MK, Islam MJ (2023) Urban waste management and circular economy in Bangladesh: a systematic review. J Int Environ Appl Sci 18(2):55–63

Braun V, Clarke V (2006) Using thematic analysis in psychology. Qual Res Psychol 3(2):77–101

Centobelli P, Cerchione R, Del Vecchio P, Oropallo E, Secundo G (2022) Blockchain technology for bridging trust, traceability and transparency in the circular supply chain. Inf Manag 59(7):103508

Citton M, Croonenberg S, Shami AE, Chammas G, Kayed S, Aoun Saliba N, Abou Najm M, Tamim H, Zeineldine S, Makki M, Kalot M, Al-Hindi M (2020) Multisource groundwater contamination under data scarcity: the case study of six municipalities in the proximity of the Naameh landfill, Lebanon. Water 12(5):e1358

Dewasiri NJ (2025) Highlights of circular economy actions in the climate change policies: a call for action for the Asia-Pacific region. In: Climate change and social responsibility, pp 23–41.https://doi.org/10.1108/S2043-052320250000025002

Dewasiri NJ, Choudhary M (2025) Harnessing artificial intelligence (AI) for sustainable innovation: case studies and practical applications. In: Innovation ecosystems and sustainable technologies: enhancing competitiveness and sustainability. Emerald Publishing Limited, pp 169–188. https://doi.org/10.1108/978-1-83662-368-720251008

Dewasiri NJ, Gangodawilage D, Singh R (2025) Pathways to climate neutrality: integrating sustainability in transport, energy, and communication for the indo-pacific region. In: Climate neutrality and sustainable ecosystems: practical approaches to addressing global warming. Singapore: Springer Nature Singapore, pp 95–114. https://doi.org/10.1007/978-981-96-7181-6_6

Freeman RE, Phillips RA (2002) Stakeholder theory: a libertarian defence. Bus Ethics Q 12(3):331–349

Gaur TS, Yadav V, Mittal S, Sharma MK (2024) A systematic review on sustainable E-waste management: challenges, circular economy practices, and a conceptual framework. Manag Environ Qual: Int J 35(4):858–884

Geels FW (2004) From sectoral systems of innovation to socio-technical systems: insights about dynamics and change from sociology and institutional theory. Res Policy 33(6–7):897–920

Guerrero LA, Maas G, Hogland W (2013) Solid waste management challenges for cities in developing countries. Waste Manag 33(1):220–232

Gupta PK, Shree V, Hiremath L, Rajendran S (2019) The use of modern technology in innovative waste management and recycling: artificial intelligence and machine learning. Recent advances in computational intelligence. Springer International Publishing, Cham, pp 173–188

Hannan MA, Al Mamun MA, Hussain A, Basri H, Begum RA (2015) A review on technologies and their usage in solid waste monitoring and management systems: issues and challenges. Waste Manag 43:509–523

Iqbal A, Yasar A, Nizami AS, Sultan IA, Haider R, Tabinda AB, Kedwii A, Chaudary MM, Ghori MU (2024) Modelling the Nexus of municipal solid waste sector for climate resilience and adaptation to nature-based solutions: a case study of Pakistan. Heliyon 10(11):e31235

Islam FS (2025) Artificial intelligence-driven intelligent waste-to-energy networks for climate-resilient circular resource management in vulnerable megacities. Int J Environ Clim Change 15:381–415

Jayarathne PGSA, Dewasiri NJ, Khan S (2025) Challenges of integrating social responsibility and climate change for the sustainable development goals: experience from the south Asian context. In: Climate change and social responsibility. Emerald Publishing Limited, pp 205–226. https://doi.org/10.1108/S2043-052320250000025011

Kaza S, Yao L, Bhada-Tata P, Van Woerden F (2018) What a waste 2.0: a global snapshot of solid waste management to 2050. World Bank Publications

Kemmerling A (2023) Policy innovation in the global south and south-north policy learning. J Comp Policy Anal: Res Pract 25(5):475–486

Khan SAR, Tabish M, Yu Z (2023) Mapping and visualising research output on waste management and green technology: a bibliometric review of literature. Waste Manag Res 41(7):1203–1218

Kirchherr J, Reike D, Hekkert M (2017) Conceptualising the circular economy: an analysis of 114 definitions. Resour Conserv Recycl 127:221–232

Kumar A, Singh R, Khan S, Kumar V (2024) Artificial intelligence-enabled management: an emerging economy perspective. De Gruyter. Accessed 23 Dec 2024

Maalouf A, Mavropoulos A (2023) Re-assessing global municipal solid waste generation. Waste Manag Res 41(4):936–947

Mishra JL, Chiwenga KD, Ali K (2021) Collaboration as an enabler for the circular economy: a case study of a developing country. Manag Decis 59(8):1784–1800

Mishra U, Rathnasiri MSH, Dewasiri NJ (2024) Net-zero economy and energy sustainability: a Nepalese perspective. In: Singh R, Khan S, Kumar A, Luthra S, Chokshi H (eds) Net zero economy, corporate social responsibility and sustainable value creation. CSR, sustainability, ethics & governance. Springer, Cham. https://doi.org/10.1007/978-3-031-55779-8_9

Rogers EM (2003) Diffusion of innovations (5th ed.). Free Press

Schot J, Geels FW (2013) Strategic niche management and sustainable innovation journeys: theory, findings, research agenda, and policy. Dyn Sustain Innov Journeys 17–34

Singh R, Ravesangar K, Narayanan S, Singh KKH (2024a) HRM practices in creating net zero: supporting organisational sustainability. In: Singh R, Khan S, Kumar A, Luthra S, Chokshi H (eds) Net zero economy, corporate social responsibility and sustainable value creation. CSR, sustainability, ethics & governance. Springer, Cham. https://doi.org/10.1007/978-3-031-55779-8_1

Singh R, Gupta A, Kandpal V (2025) ESG excellence in the digital world: circular economy e-business solutions. In: Fintech for ESG and the circular economy, pp 193–204. https://doi.org/10.1002/9781394234035.ch10

Singh R, Gupta A, Bajpai A, Kandpal V (2024b) Exploring challenges of circular economy initiatives for smart cities. In: Smart cities and circular economy: the future of sustainable urban development. Emerald Publishing Limited, pp 59–69. https://doi.org/10.1108/978-1-83797-957-820241005

Singh R, Crowther D, Ravesangar K (2024c) Harnessing human capital for a sustainable future: HR's pivotal role toward net zero policies. In Singh R, Crowther D (eds) Transition towards a sustainable future. Approaches to global sustainability, markets, and governance. Springer, Singapore. https://doi.org/10.1007/978-981-97-5756-5_2

Sony M, Naik S (2020) Industry 4.0 integration with socio-technical systems theory: a systematic review and proposed theoretical model. Technol Soc 61:101248

Wang C, Qin J, Qu C, Ran X, Liu C, Chen B (2021) An innovative municipal waste management system based on deep learning and the internet of things. Waste Manag 135:20–29

Circular Economy, Blockchain, and Future Sustainability Pathways

Artificial Intelligence for Sustainable Water, Energy, and Food Security (WEF)

Yogeeswari Subramaniam, Manuel A. Zambrano-Monserrate, Paulina Permatasari, and Liu Guangling

Abstract To promote sustainable development and ensure economic growth, Malaysia is currently making significant efforts to become a hub for artificial intelligence innovations. Therefore, this examines relationships between artificial intelligence and water, food, and energy resources, especially within the case of Malaysia, since such relationships are often overlooked. In being able to explore linkages of artificial intelligence in Malaysia, the current study examines data from the period 2019–2023 using a methodology of a direct impact assessment. The findings show that there is a positive correlation between artificial intelligence and food, energy, and water resources and thus ensure resource sustainability. Additionally, the study displays frameworks outlining an artificial intelligence-based resource security model and emphasizes AI's sustainability contributions to food, energy, and water resources and its role in achieving the Sustainable Development Goals (SDGs). Consequently, it is therefore recommended that policymakers should be supportive of incorporating artificial intelligence technologies to ensure sustainability activities for Malaysia.

Keywords Artificial intelligence · Water · Food · Energy · Malaysia · Sustainability

Y. Subramaniam (✉)
Universiti Teknologi Malaysia, Johor Bahru, Malaysia
e-mail: yogeeswari.s@utm.my

M. A. Zambrano-Monserrate
Universidad Espíritu Santo, Samborondón, Ecuador
e-mail: manuelzambranom@uees.edu.ec

P. Permatasari
Parahyangan Catholic University, Bandung, Indonesia
e-mail: paulina@unpar.ac.id

L. Guangling
Northwest University of Political Science and Law, Xi'an, China
e-mail: lgl1008@sina.com

R. Singh et al. (eds.), *Data-Driven Sustainability: Harnessing Technology for A Greener Future and Environmental Resilience*, World Sustainability Series,
https://doi.org/10.1007/978-3-032-13412-7_22

1 Introduction

WEF are key to social progress, human well-being, and long-term economic growth (Biggs et al. 2015; Zahedi et al. 2024). As the global population grows and urban areas expand, the pressure on these resources keeps rising. Climate change and rapid tech developments are also making it harder to strike a balance between people's needs, environmental protection, and economic priorities. Water is necessary to sustain life on Earth because it facilitates industrial processes, drinking, agriculture, sanitation, and ecosystem support, which has long been acknowledged. Energy is essential for modern agriculture, food processing, and human well-being. It also powers transportation, industries, and other critical services, all of which contribute to economic growth (Jaiswal et al. 2022). Food is a fundamental human necessity that guarantees nutritional security and is closely associated with both social stability and economic development. The management of one resource has a direct impact on the management of the others. For instance, producing food and energy depends heavily on water, while treating and moving water around takes a lot of energy. That's why it's so important to make sure there's enough of all three, food, energy, and water not just for today, but for the future too. Keeping these resources available is key to supporting human life and driving both social and economic progress.

As Malaysia continues to grow economically, the demand for food, energy, and water is rising, making it more important than ever to manage these resources sustainably. Malaysia's population has grown from 32.5 million in 2019 to 34.1 million in 2024, and projections suggest it could reach 41 million by 2050 (Department of Statistics Malaysia 2024). This steady rise is placing increasing strain on the country's already interconnected food, energy, and water systems, especially as demand climbs for both present and future needs. Water plays a crucial role in nearly 90% of the country's economic processes, supporting everything from power generation and industrial production to large-scale agriculture (Sophocleous 2004). Major crops like oil palm, paddy, and rubber are highly reliant on irrigation, and industries and thermal power plants need immense amounts of water to run with optimum efficiency (Ahmed et al. 2014; Hanafiah et al. 2019).

Food and energy not only constitute the key to the health of the population; they are also at the center of keeping Malaysia's manufacturing and industrial supply chains in good condition (Ali et al. 2012; Osabohien et al. 2024). Controlling the three critical resources of food, energy, and water, therefore has become a priority in ensuring the long-term sustainability and economic stability of the nation. Against this background, the government of Malaysia has stepped up its endeavors to reign over these crucial resources in a more sustainable and visionary way. With increasing concerns about climate change, population growth, and resource limitation, policymakers have promulgated a range of bills and policies aimed at safeguarding the national interests in the long term. For water, initiatives like the Water Supply Services Industry Act 2006, the National Water Resources Policy, and the River of Life project in Kuala Lumpur are all designed to improve water quality, enhance supply efficiency, and build a stronger water services sector to support agriculture, industry, and urban communities (Afroz et al. 2014). The goal is to transform Malaysia into a high-income, knowledge-based economy. In fact, Malaysia's score on the Government AI Readiness Index increased from 67.37 points in 2022 to 68.71 points in 2023, moving it from 29th to 23rd place (Ministry of Science, Technology and Innovation 2023). A better ranking reflects improved AI-related infrastructure, policies, and strategies, demonstrating Malaysia's advancement and dedication to strengthening its AI capabilities and enhancing its standing in the global AI readiness arena. Additionally, the government's MyDIGITAL Malaysia initiative aims to advance AI integration in the public services, manufacturing, finance, and healthcare sectors (Mahusin et al. 2024; Geetha et al. 2024). Through these efforts, the country seeks to foster a national mindset, expand Malaysia's digital economy, and create a digitally inclusive society. All these pose concerns about whether the innovation and development of AI will impact the sustainability of food, water, and energy resources. Since the development of AI in Malaysia will impact the entire ecosystem, it is essential to determine the "right way" to handle and support the SDGs of eradicating hunger (SDG 2), providing access to clean, affordable energy (SDG 7), and maintaining clean water and sanitation (SDG 6). With Malaysia moving quickly towards developing a robust ecosystem through the adoption of AI development and innovations, this study aims to examine the relationship between AI and water, energy, and food resources. By thoroughly analysing AI's associations on these resources, Malaysia can formulate suitable strategies for AI advancement that will ensure sustainable and integrated management of these resources while addressing present and future challenges.

1.1 Stylized Facts

Many scholars have explored the connection between economic growth and water, food and energy nexus. Salim et al. (2014) for the OECD, Zezza and Tasciotti (2010) for developing nations, Saladini et al. (2018) for Mediterranean region Sarkodie et al. (2019) for Australia, Raihan et al. (2023) for Egypt, Han et al. (2022), Ringler et al. (2016) for the global and so on examined the significance of economic growth on water, food, and energy, showing that an increase in economic growth results in an increase in the demand for these resources. The reason is that as the economy expands, resources for industrialization, urbanization, agriculture, and transportation, which depend on water, energy, and food, are needed more frequently, leading to over exploitation and pollution. On the other hand, Salim et al. (2014), Zezza and Tasciotti (2010), Raihan et al. (2023), and Long et al. (2006), argue that economic growth improves the sustainability of food, water, and energy resources by allocating resources towards advanced and efficient infrastructure, technologies, and techniques. As an example, Zhang et al.'s recent study from 2021 shows that as economies develop, investments in and adoption of cutting-edge, environmentally friendly technologies rise. This, in turn, drives resource consumption that is efficient and sustainable, resulting in more effective management of water, food, and energy resources. As a result, water, food, and energy are significantly impacted by economic growth, but the exact direction remains uncertain.

In addition, climate change affect water, food and energy, as revealed by Eriksson et al. (2009), Rockström et al. (2012), Han et al. (2022), Said et al. (2019), Herrera-Franco et al. (2023), Mirón et al. (2023), Mperejekumana et al. (2024), Awandu et al. (2024), Shrestha and Aryal (2011). Because of greenhouse gas emissions, all three of these resources are extremely vulnerable to climate change. As specified by Eriksson et al. (2009), Rockström, et al. (2009) and Shrestha and Aryal (2011), climate change drop the productivity of agricultural, alter water availability and production of energy, leading to water and energy shortages and food insecurity. For example, crop yields and production are extremely sensitive to greenhouse gas emissions; high temperatures and extreme weather can lower crop yields and jeopardize the productivity of agriculture. This could result in decreased food security and an increase in the likelihood of food insecurity. Aside from that, Rockström et al. (2012) and Shrestha and Aryal (2011) clarify that decreased snowpack and altered precipitation patterns lead to droughts and floods, which reduce the amount of water available for daily use and economic activity. Water resource scarcity and temperature changes subsequently impact energy resources as well, decreasing the output of energy and reducing the efficiency of thermoelectric and hydropower plants. Therefore, the availability of food, energy, and water resources is harmed by climate change, which has a negative impact on these resources.

Imasiku and Ntagwirumugara (2019), Rosegrant et al. (2002), Perrone and Hornberger (2014), Hellegers et al. (2008), Brears (2023), Morales-García and García-Rubio (2024), Zahedi et al. (2024) and so on analyzed the impact of population growth on the Water-Energy-Food nexus. Growing populations lead to increased demand for resources, which in turn leads to the overuse of those resources. This causes scarcities of food, water, and energy resources, endangering the population's ability to meet its present and future needs. In Rwanda, Imasiku and Ntagwirumugara (2019) discovered that increased population growth puts a strain on the country's food, energy, and water supplies due to increased demand for these resources. This impacts resource sustainability and availability. As Malaysia's population grows, so does the demand for water, especially for daily use. But industries and agriculture are also pulling heavily from water supplies, often leading to overuse and shortages. Godfray et al. (2010) looked into how food production is tied to population growth, showing that as more people need to be fed, the demand for land, water, and energy goes up. This can lead to deforestation and soil erosion, making food, water, and energy systems even more vulnerable.

With Malaysia's growing population, there is the added energy requirement in residential, transport, and industrial sectors. Additional demands for resources are likely to add to dependence on fossil fuels, consequently causing increasing emissions of greenhouse gases. Essentially, population growth places ever-higher demands on already strained food, energy, and water systems of nations. In such a rising scenario of challenges, Artificial Intelligence (AI) can be a lifesaver. Even as economic growth and global warming have transformed patters of using resources, AI seeks to change such systems in a cleaner and more efficient way. In a generic sense, AI comprises an array of technologies such as robotics, machine learning, predictive analytics, and computer systems developed specifically to simulate human thinking and decision-making processes. Such technologies are being deployed ever-more to make utilizing needed resources more efficient, reduce waste, improve operating efficiencies, and aid sustainability efforts.

Recent research shows how AI can be used. Scholars like Uen et al. (2018), Liu et al. (2018), Oviroh et al. (2023), Zhang et al. (2020), Zhou et al. (2024), Shao et al. (2023), D'Amore et al. (2022), and López-Flores et al. (2023) studied how data analysis, IoT, cloud systems, and automation change resource management. Their work finds that these technologies help improve water use, support precise farming, and assist in using renewable energy with existing power grids. These tools help save resources and reduce costs. Table 1 shows some projects where AI helps manage food, water, and energy. In water management, AI predicts demand, keeps water clean, and improves delivery systems. In agriculture, AI-based systems control water use in real time. They help save money and increase crop growth. The works of Shao et al. (2023), and Zhou et al. (2024) show how machine learning and remote sensing are used to forecast crop harvests and detect pests early. These tools help farmers make better decisions, reduce chemical use, and grow more food with less pollution. In energy, the studies by D'Amore et al. (2022) and López-Flores et al. (2023) show how AI manages smart grids. They help predict energy use and balance supply and demand. These innovations, collectively, facilitate the accelerating of the transition to renewables through the optimization of the efficiency of systems and a lesser reliance on fossil fuels, a necessary step towards a cleaner and less polluting energy infrastructure.

As outlined earlier and illustrated in Table 1, a notable research gap exists: none of the previous studies have explored how Artificial Intelligence (AI) intersects with food, water, and energy specifically within the Malaysian context. Given Malaysia's current leadership in applying AI to support sustainability, manage resources, and enhance environmental protection efforts, this investigation is both timely and essential. This study sets itself apart by concentrating exclusively on Malaysia, analyzing how AI relates to food, water, and energy resources in this national setting. To the best of the author's knowledge, this is the first empirical examination of the connections between AI and these three critical resources in Malaysia.

Moreover, this research proposes an AI-Driven Resource Security Framework tailored to the Malaysian context. The framework represents an original contribution designed to improve the sustainability and security of food, energy, and water resources through AI-powered strategies. The chapter presents new evidence illustrating how AI can be used to support sustainability in these areas and offering lessons for policymakers in Malaysia to develop successful AI-informed resource management policies. The study also addresses broader global challenges. These food, energy, and water challenges are not unique to Malaysia; these same challenges are confronting most developing countries owing to the rapid pace of urbanization, population increase, and climate change pressures. The Malaysian experience serves as a case study through which this research illustrates an exemplar use of how emerging economies can leverage AI to balance their sustainability challenges

Table 1 Studies on AI and water, energy, food resources

Author(s)	Country	Result(s)
Uen et al. (2018)	Taiwan	• Water shortage rates reduced • Increased food production rate • Increase hydropower output
Liu et al. (2018)	Globe	• Promote hybrid renewable energy production • Without adequate water, food production is compromised
Zhang et al. (2020)	China	• Enhances water management • Enhances crop yields and supply chains • Improves efficiency and grid management
D'Amore et al. (2022)	Global	• Increasing importance in water, energy, and food sectors
Oviroh et al. (2023)	Uganda	• Optimizing Water, Energy, and Food Decision-Making
Shao et al. (2023)	China	• Enhances the integration of water, food, and land resources for better crop yield
López-Flores et al. (2023)	Mexico	• Improving predictions and decision-making for water, food and energy management
Zhou et al. (2024)	Taiwan	• Enhancing water, food and energy synergies and urban sustainability

Source Author's own illustration

and achieve the United Nations Sustainable Development Goals (SDGs). Malaysia's socio-economic and environmental context poses the proper environment to test AI initiatives aimed at solving local problems and facilitating worldwide action towards sustainability.

The second relevance of the study is its potential for policy implication. While governments are building AI strategies, not very many have explicitly connected these strategies to the management of valuable assets like energy, water, and food. By laying out empirical data regarding the application of AI in these industries, the report offers Malaysian policy makers practical recommendations for establishing sustainable and creative policies. The report identifies the use of AI, not just for economic development, but also for building resilience and achieving sustainability targets.

Finally, the AI-Driven Resource Security Framework in this chapter goes beyond theory. It is a practical, actionable guide that encourages synergistic collaboration among governments, private sector companies, and research institutions. With the illustration of how AI can be utilized in agriculture, water infrastructure, and energy infrastructure, the model promotes an integrated strategy towards resource security. With its holistic perspective, the research is a timely and useful addition to the literature and adds merit to Malaysia's vision of being a regional head in AI-driven sustainable development.

2 The Connection Between Artificial Intelligence and Water, Energy and Food

This chapter employs annual data for Malaysia from 2019 until 2023, with data selected based on availability. Artificial intelligence (AI) as the core independent variable is measured by the AI readiness index. Additionally, the dependent variables such as water (WS), energy (ES), and food (FS) are measured by water productivity, energy trilemma index, and food security index, respectively. Table 2 summarizes the definitions, measurements, and sources of the dependent and independent variables. Hence, in this present study, a direct impact analysis was conducted to examine the impact of AI on water, food, and energy resources in Malaysia.

The scatter plot indicates a high correlation between AI and the sectors of food, energy, and water in Malaysia, to show how AI can make positive headway in sustainability (see Fig. 1). The agriculture and food sector begins with AI applications such as precision farming, crop monitoring, and supply chain predictive analytics. Precision agriculture utilizes technology like satellite images, sensors, and artificial intelligence to enable farmers to use fertilizers, water, and pesticides more efficiently. Not only does this enhance the volume of crops harvested but also decreases costs and minimizes environmental disruption. Logistically, predictive analytics helps manage the supply chain by forecasting demand, locating potential disruptions, and optimizing food distribution. This means lower post-harvest loss and an efficient system overall. All these innovations add up meaningfully to Malaysia, especially as the country remains very reliant on imports of such staple foods as rice. By improving methods of food production and also delivering it, AI is strengthening the food system and making it less vulnerable to global shocks, a large step towards more resilient national food security.

Using AI in Malaysia's energy sector is increasing. This trend is similar to what is happening worldwide, as shown in studies by Uen et al. (2018), Zhang et al. (2020), and Oviroh et al. (2023). This study shows that AI is becoming more important. AI helps make the energy system more efficient, flexible, and sustainable. In Malaysia, the demand for energy is rising. This is due to more industry, faster city growth, and a growing population. Because of this, better energy management is needed. This makes advanced solutions necessary. One example is smart grids powered by AI. These grids monitor how much energy people use, balance the electrical load,

Table 2 Data sources

Variables	Symbols	Definition/Measurement	Source
AI	*AI*	AI Readiness Index	Oxfordinsights (2024)
FS	*FS*	Food security index	FAOSTAT (2024)
ES	*ES*	Energy Trilemma Index	World Energy council (2024)
WS	*WS*	Water productivity, total (constant 2015 US$ GDP per cubic meter of total freshwater withdrawal)	World Bank (2024)

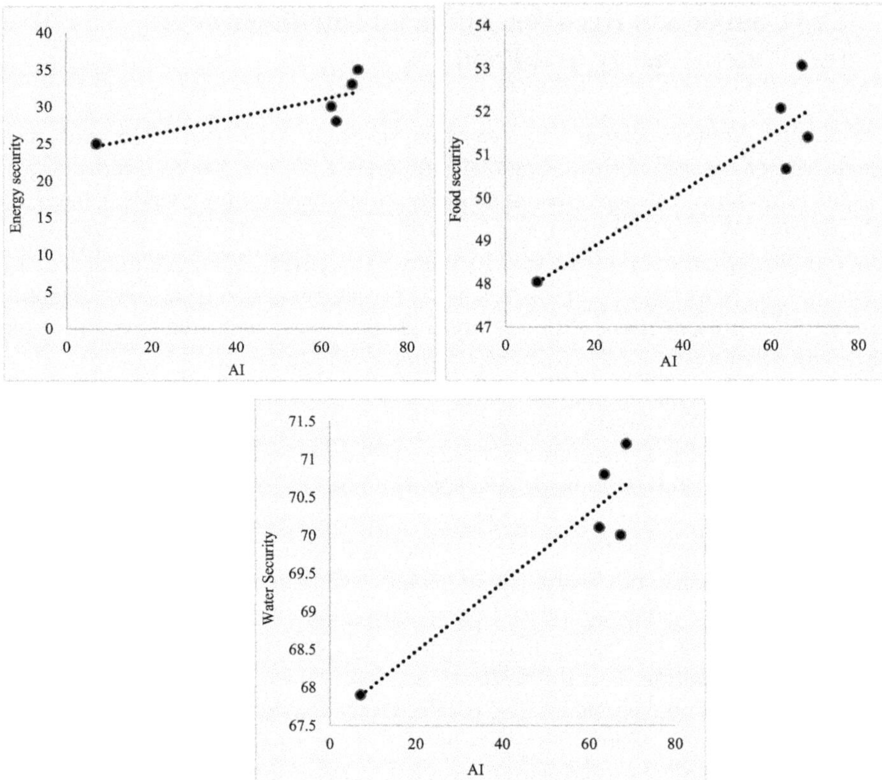

Fig. 1 Artificial intelligence correlation with Malaysia's WEF

and reduce energy loss. Machine learning algorithms are used to predict how much energy will be generated from renewable sources like wind and solar. This helps connect renewable energy to the national grid. AI also helps with maintenance. It predicts when equipment might fail. This reduces downtime and costs and makes the grid more reliable.

While Malaysia aims to be more dependent on green energy, AI could be a driving force as an enabler to facilitate the transition at lower costs and better. While water data were not obviously presented in the scatter plots of the study, the future for AI in the water industry is also monumental. AI-based monitoring systems can identify pipe leakages, water quality monitoring, and irrigation optimization, which is particularly practical in agricultural or rural areas. With increasing problems in water shortage and water pollution, these technologies present effective solutions in conserving water and relieving pressure on existing resources.

AI also has an increasing role in disaster risk reduction. Machine learning systems can send early alerts for disasters like floods and droughts. These alerts help communities prepare and respond better. In Malaysia, where the weather changes often, these technologies support national efforts for safety. Artificial intelligence helps manage food, energy, and water supplies. It plays a key role in Malaysia's Water–Energy–Food (WEF) plan. This plan considers how these sectors are connected and work together. Using AI in this plan helps save resources, reduce climate risk, and improve efficiency. Research shows that AI also helps Malaysia meet its development goals. These include zero hunger, clean water, affordable energy, and responsible use of resources. In this way, AI adoption puts Malaysia on the right trajectory toward a more sustainable and resilient future.

3 AI-Driven Resource Security Framework

The current study provides a clear and direct analysis of the relationship between artificial intelligence (AI) and water, food, and energy resources, despite not employing advanced estimation techniques to investigate the effects of AI on these resources. The analysis shows a positive correlation between AI and food, energy, and water resources, demonstrating that Malaysia's efforts to develop and apply AI promote resource security. Considering this, the study offers a framework for AI-driven resource security that encompasses food, energy, and water and highlights Malaysia's paths to sustainable development (see Fig. 2).

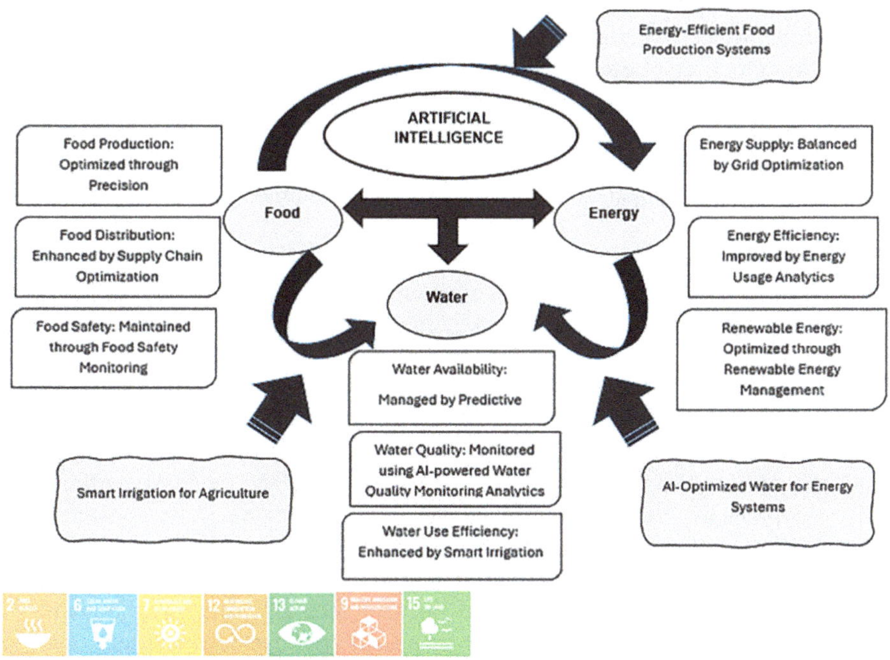

Fig. 2 AI-driven resource security framework. *Source* Author's own illustration

Water resource highlights the role of AI technologies by examining the aspects of water availability, quality, and efficiency. (i) The capacity to guarantee a sufficient and consistent supply of water to meet the needs of various economic sectors, including households, industries, ecosystems, and agriculture, is emphasized in water availability; (ii) water quality refers to the preservation of clean water and protection from pollution, contamination, and chemical levels for human consumption and environmental health; and (iii) water efficiency denotes the efficient and effective management of water resources by reducing water waste and consumption while encouraging sustainable water management practices. AI predictive analytics improve water resource by helping to forecast water availability through the analysis of weather trends, river flows, and drought risk. This facilitates the allocation of water resources. Similarly, AI-driven IoT sensors and intelligent irrigation systems enhance water quality through continuous monitoring and provide real-time information about pollution levels in water supplies. For instance, to reduce non-revenue water losses, which previously accounted for about 36% of the total water supply, Air Selangor, the largest water service provider in Malaysia, implemented AI-powered water leak detection technologies. Smart irrigation systems, which minimize water waste and ensure sustainable water management, are especially beneficial for Malaysia's agricultural sectors, including rice and palm oil production, helping to optimize water efficiency. AI is helping Malaysia make real progress toward SDG 6 and SDG 12 by boosting how efficiently the country manages its water resources.

Artificial intelligence has a great impact on three major areas in the food industry: distribution, production, and safety. In the area of production, AI maximizes farm output through efficient use of resources and at the same time minimizes environmental degradation. In terms of distribution, AI facilitates efficient food supply chains and gets food to the necessary places on time, increasing delivery drastically and waste minimally. In terms of safety, AI keeps a tight check on the whole continuum from farm to table so that health standards are maintained and contamination is eliminated.

Precision farming in Malaysia uses AI technology. Farmers get data from drones and sensors. They use this data to make better decisions about fertilizing, watering, and planting. MARDI has created tools with AI to watch over crops. These tools help manage water, fertilizer, and other inputs. The focus is on major crops like rice. AI also supports food delivery by reducing delays and waste in the supply chain. This helps move food more efficiently from farms to consumers across the country. Lastly, AI technology is also streamlining food safety monitoring in the supply chain, reducing opportunities for contamination and allowing Malaysia to stay within food safety compliances. Thus, AI contributes to SDG 2 and SDG 12 by assisting food supply chains, reducing food waste, and ensuring that needs are met steadily and on time. Besides that, energy emphasizes the use of efficient and renewable energy sources, combined with dependable and sustainable energy, to meet the country's energy needs. The availability and supply of energy are referred to as energy supply, while energy efficiency focuses on the efficiency of energy consumption. Renewable energy refers to the integration of renewable energy sources, which ultimately contributes to sustainability and lowers emissions. AI improves the steady supply and demand of energy and prevents energy blackouts by balancing conventional energy sources with renewable energy sources and grid optimization. Reduced energy waste and optimized power usage are among the benefits of using AI-driven analytics to monitor and enhance energy efficiency in both the residential and industrial sectors. AI supplements solar and hydropower, the renewable energy sources, and minimizes the use of fossil fuels, aside from enhancing energy supply and efficiency. Tenaga Nasional Berhad, for one, is leading in Malaysia in harnessing AI technologies to optimize the country's electrical grid, making it easier to transition to renewable energy sources and balance energy supply and demand. This helps Malaysia attain its objectives within the Malaysia Renewable Energy Roadmap of generating 35 percent of its energy from renewable sources by 2025. As AI improves energy supply, efficiency, and renewable sources of energy, Malaysia is thus in a position to attain SDG 7 (affordable and clean energy) and energy security.

Lastly, the AI integration across these sectors facilitates the governance of Malaysia's water, food, and energy resources, making resource utilization more efficient and unlocking sector synergies. For instance, AI-optimized agronomic water management improves crop yields while conserving water, progressing both food and water. Similarly, AI aids energy-efficient food production systems by reducing the energy consumed in operations such as supply chains and food storage. These AI-driven innovations contribute to the Sustainable Development Goals (SDGs) of zero hunger, clean water and sanitation, affordable and clean energy, and climate action. Thus, the framework addresses resource challenges and promotes long-term sustainability in line with Malaysia's Vision 2030.

4 Conclusion

As existing studies have focused on developed countries and overlooked developing countries like Malaysia, the present study addresses this critical gap by examining the associations of artificial intelligence (AI) on water, food, and energy, particularly in Malaysia. In this study, we employed direct impact analysis using data from 2019 to s2023, due to limited data availability on AI as a core variable. Using direct impact analysis, this study confirms that AI significantly influences water, food, and energy resources in Malaysia by optimizing the sustainable use of natural resources. AI shows a positive association with water, food, and energy, enhancing their security. Additionally, this study provides an AI-driven resource framework tailored to Malaysia. It demonstrates that AI can improve the sustainability of food, energy, and water resources, all of which are directly related to the Sustainable Development Goals (SDGs). Given that Malaysia's ongoing efforts and initiatives to leverage AI tools and technologies will help the country achieve its sustainability goals.

Based on the study outcomes, this study recommends that policymakers actively promote the adoption of AI technologies across the food, water, and energy sectors, not only to optimize resource efficiency but also to enhance sustainability. In this regard, it is recommended that the following policies be strengthened and encouraged: AI-driven water management systems that optimize irrigation and provide real-time water quality monitoring; AI technologies that enhance productivity in food security through precision agriculture and improve supply chain efficiency, minimizing food waste and ensuring timely delivery; and AI-integrated policies for energy grid optimization, renewable energy usage, and energy efficiency. In addition, the government should support AI innovation through research and development (R&D) and offer financial incentives or subsidies to businesses and households that adopt AI technology. By promoting collaboration between the public and private sectors, Malaysia can ensure the long-term sustainability of its natural resources and accelerate progress toward achieving the Sustainable Development Goals (SDGs).

To determine the impacts of AI on water, energy, and food, the current study employed a simple correlation analysis rather than more advanced techniques due to limited data availability on the core variable. Therefore, once more data becomes available in the future, researchers are encouraged to apply advanced estimation techniques to examine the full extent of artificial intelligence on water, food, and energy.

References

Abas MA, Abdull RMNA, Muhammad SAS (2021) The impact of climate change disaster to the socio-economic of paddy farmer in Malaysia. Disaster Adv

Abughlelesha SM, Lateh HB (2013) A review and analysis of the impact of population growth on water resources in Libya. World Appl Sci J 23(7):965–971

Abd Aziz AJ, Baharuddin NA, Khalid RM, Kamarudin SK (2024) Review of the policies and development programs for renewable energy in Malaysia: progress, achievements and challenges. Energy Explor Exploit 01445987241227509

Afroz R, Masud MM, Akhtar R, Duasa JB (2014) Water pollution: challenges and future direction for water resource management policies in Malaysia. Environ Urban ASIA 5(1):63–81

Ahmed F, Siwar C, Begum RA (2014) Water resources in Malaysia: issues and challenges. J Food Agric Environ 12(2):1100–1104

Ali R, Daut I, Taib S (2012) A review on existing and future energy sources for electrical power generation in Malaysia. Renew Sustain Energy Rev 16(6):4047–4055

Awandu W, Kanda EK, Kimokoti SN (2024) The water-energy-food nexus in Kenya: climate change impacts and adaptation strategies—a review. In: The water, climate, and food nexus: linkages, challenges and emerging solutions, pp 59–70

Biggs EM, Bruce E, Boruff B, Duncan JM, Horsley J, Pauli N, McNeil K, Neef A, Van Ogtrop F, Curnow J, Haworth B, Imanari Y (2015) Sustainable development and the water–energy–food nexus: a perspective on livelihoods. Environ Sci Policy 54:389–397

Brears RC (2023) The green economy and the water-energy-food nexus. The green economy and the water-energy-food nexus. Springer International Publishing, Cham, pp 31–61

D'Amore G, Di Vaio A, Balsalobre-Lorente D, Boccia F (2022) Artificial intelligence in the water–energy–food model: a holistic approach towards sustainable development goals. Sustainability 14(2):867

Department of Statistics Malaysia (2024) The population of Malaysia. https://open.dosm.gov.my/dashboard/population. Accessed on July 2024

Ericsson K, Rosenqvist H, Nilsson LJ (2009) Energy crop production costs in the EU. Biomass Bioenerg 33(11):1577–1586
Geetha C, Ayub MS, Chandran EVV (2024) The influence of adopting artificial intelligence (AI) on Malaysia's economic environment. Malays J Bus Econ (MJBE) 11(1)
Godfray HCJ, Beddington JR, Crute IR, Haddad L, Lawrence D, Muir JF,... Toulmin C (2010) Food security: the challenge of feeding 9 billion people. Sci 327(5967):812–818
Han X, Hua E, Engel BA, Guan J, Yin J, Wu N, Sun S, Wang Y (2022) Understanding implications of climate change and socio-economic development for the water-energy-food nexus: A meta-regression analysis. Agric Water Manag 269:107693
Hanafiah MM, Ghazali NF, Harun SN, Abdulaali HS, AbdulHasan MJ, Kamarudin MKA (2019) Assessing water scarcity in Malaysia: a case study of rice production. Desalin Water Treat 149:274–287
Hellegers P, Zilberman D, Steduto P, McCornick P (2008) Interactions between water, energy, food and environment: evolving perspectives and policy issues. Water Policy 10(S1):1–10
Herrera-Franco G, Bollmann HA, Lofhagen JCP, Bravo-Montero L, Carrión-Mero P (2023) Approach on water-energy-food (WEF) nexus and climate change: a tool in decision-making processes. Environ Dev 46:100858
Imasiku K, Ntagwirumugara E (2020) An impact analysis of population growth on energy-water-food-land nexus for ecological sustainable development in Rwanda. Food Energy Secur 9(1):e185
Jaiswal KK, Chowdhury CR, Yadav D, Verma R, Dutta S, Jaiswal KS, Karuppasamy KSK (2022) Renewable and sustainable clean energy development and impact on social, economic, and environmental health. Energy Nexus 7:100118
Khan I, Hou F, Irfan M, Zakari A, Le HP (2021) Does energy Trilemma a driver of economic growth? The roles of energy use, population growth, and financial development. Renew Sustain Energy Rev 146:111157
Liow JY (2023) A comparative study: exploring the food security in Malaysia and Singapore (Doctoral dissertation, UTAR)
Liu J, Mao G, Hoekstra AY, Wang H, Wang J, Zheng C, van Vilet MT, Wu M, Ruddell B, Yan J (2018) Managing the energy-water-food nexus for sustainable development. Appl Energy 210:377–381
Long SP, Ainsworth EA, Leakey AD, Nosberger J, Ort DR (2006) Food for thought: lower-than-expected crop yield stimulation with rising CO2 concentrations. Science 312(5782):1918–1921
López-Flores FJ, Sánchez-Zarco XG, Rubio-Castro E, Ponce-Ortega JM (2023) A machine learning approach for optimizing the water-energy-food-ecosystem nexus: a resilience perspective for sustainability. Environ Dev Sustain 1–29
Mahusin N, Sallehudin H, Satar NSM (2024) Malaysia public sector challenges of implementation of artificial intelligence (AI). IEEE Access
Ministry of Science, Technology and Innovation (2023) Malaysia's ranking climbed to 23 from 29. In: The goverment AI readiness index compared to the previous year. Retrieved on Oct 2025 from https://www.mosti.gov.my/en/siaran-kenyataan-media/malaysias-ranking-climbed-to-23-from-29-in-the-govermentai-readiness-index-compared-to-the-previous-year/
Mirón IJ, Linares C, Díaz J (2023) The influence of climate change on food production and food safety. Environ Res 216:114674
Morales-García M, García-Rubio MA (2024) Security vs efficiency of the water-energy-food nexus. A study of the economies of the regions of Spain. Environ Sustain Indic 24:100543
Mperejekumana P, Shen L, Zhong S, Muhirwa F, Gaballah MS, Nsigayehe JMV (2024) Integrating climate change adaptation into water-energy-food-environment nexus for sustainable development in East African Community. J Clean Prod 434:140026
Osabohien R, Aderemi TA, Jaaffar AH, Oloke E, Bassey R, Yusoff NYBM, Ifekwem NE (2024). Electricity consumption and food production in Malaysia: implication for the sustainable development goal 2. Int J Energy Econ Policy 14(3):119–126

Oviroh PO, Austin-Breneman J, Chien CC, Chakravarthula PN, Harikumar V, Shiva P, Kimbowa AB, Lutnz J, Miyingo EW, Papalambros PY (2023) Micro water-energy-food (MicroWEF) Nexus: a system design optimization framework for integrated natural resource conservation and development (INRCD) projects at community scale. Appl Energy 333:120583

Perrone D, Hornberger GM (2014) Water, food, and energy security: scrambling for resources or solutions? Wiley Interdiscip Rev Water 1(1):49–68

Poumanyvong P, Kaneko S (2010) Does urbanization lead to less energy use and lower CO2 emissions? A cross-country analysis. Ecol Econ 70(2):434–444

Raihan A, Ibrahim S, Muhtasim DA (2023) Dynamic impacts of economic growth, energy use, tourism, and agricultural productivity on carbon dioxide emissions in Egypt. World dev sustain 2:100059

Ringler C, Willenbockel D, Perez N, Rosegrant M, Zhu T, Matthews N (2016) Global linkages among energy, food and water: an economic assessment. J Environ Stud Sci 6

Rockström J, Steffen W, Noone K, Persson Å, Chapin III FS, Lambin E,... Foley J (2009) Planetary boundaries: exploring the safe operating space for humanity. Ecol Soc 14(2)

Rockström J, Falkenmark M, Lannerstad M, Karlberg L (2012) The planetary water drama: dual task of feeding humanity and curbing climate change. Geophys Res Lett 39(15)

Rosegrant MW, Cai X, Cline SA, Nakagawa N (2002) The role of rainfed agriculture in the future of global food production. World Dev Indic

Rusli NM, Noor ZZ, Taib SM, Sabli NSM (2022) Water footprint of rice production in Malaysia: a review of evapotranspiration and factors of climate change for rice and food security in Malaysia. Malays J Fundam Appl Sci 18(4):497–510

Said M, Komakech HC, Munishi LK, Muzuka ANN (2019) Evidence of climate change impacts on water, food and energy resources around Kilimanjaro, Tanzania. Reg Environ Change 19(8):2521–2534

Saladini F, Betti G, Ferragina E, Bouraoui F, Cupertino S, Canitano G, Gogliotti M, Aution A, Pulselli FM, Riccaboni A, Bidoglio G, Bastianoni S (2018) Linking the water-energy-food nexus and sustainable development indicators for the Mediterranean region. Ecol Indic 91:689–697

Salim RA, Hassan K, Shafiei S (2014) Renewable and non-renewable energy consumption and economic activities: Further evidence from OECD countries. Energy Econ 44:350–360

Sarkodie SA, Strezov V, Weldekidan H, Asamoah EF, Owusu PA, Doyi INY (2019) Environmental sustainability assessment using dynamic autoregressive-distributed lag simulations—nexus between greenhouse gas emissions, biomass energy, food and economic growth. Sci Total Environ 668:318–332

Shao W, Ding Y, Wen J, Zhu P, Ou L (2023) Optimal decision-making in the water, land and food nexus using artificial intelligence and extreme machine learning. Water Supply 23(10):4166–4177

Shrestha AB, Aryal R (2011) Climate change in Nepal and its impact on Himalayan glaciers. Reg Environ Change 11:65–77

Sophocleous M (2004) Global and regional water availability and demand: prospects for the future. Nat Resour Res 13:61–75

Sovacool BK (2014) Environmental issues, climate changes, and energy security in developing Asia. Asian Dev Bank Econ Work Pap Ser (399):17–14

Uen TS, Chang FJ, Zhou Y, Tsai WP (2018) Exploring synergistic benefits of water-food-energy nexus through multi-objective reservoir optimization schemes. Sci Total Environ 633:341–351

Vo DH, Vo AT (2021) Renewable energy and population growth for sustainable development in the Southeast Asian countries. Energy Sustain Soc 11(1):30

Vorosmarty CJ, Green P, Salisbury J, Lammers RB (2000) Global water resources: vulnerability from climate change and population growth. Science 289(5477):284–288

Zahedi R, Yousefi H, Aslani A, Ahmadi R (2024) Water, energy, food and environment nexus (WEFEN): sustainable transition, gaps and covering approaches. Energ Strat Rev 54:101496

Zezza A, Tasciotti L (2010) Urban agriculture, poverty, and food security: empirical evidence from a sample of developing countries. Food Policy 35(4):265–273

Zhang X, Cao Z, Dong W (2020) Overview of edge computing in the agricultural internet of things: key technologies, applications, challenges. Ieee Access, 8:141748–141761

Zhou Y, Chang FJ, Chang LC, Herricks E (2024) Elevating urban sustainability: an intelligent framework for optimizing water-energy-food nexus synergies in metabolic landscapes. Appl Energy 360:122849

Industry 5.0 for a Climate-Resilient South Asia: A Human-Centric Approach to Net-Zero Futures

Narayanage Jayantha Dewasiri, Rubee Singh, Asanga Ranasinghe, Jacinta Dsilva, and Mohamed Irfan Shaikh

Abstract This chapter examines how Industry 5.0, emphasising human—technology cooperation, can boost climate resilience and net-zero goals in South Asia. Through qualitative interviews with policymakers, industry leaders, and experts from Sri Lanka, India, Bangladesh, Nepal, and the Maldives, the study identifies enablers, obstacles, and strategies for adopting Industry 5.0. Four themes emerge: human-technology synergy in green innovation; policy coherence and institutional readiness; technological adaptation for circular, resilient production; and socio-cultural inclusive change. The study highlights collaborative governance, ethical artificial intelligence (AI), and capacity-building as vital for aligning tech innovation with climate aims. It also proposes a Human-Centred Climate Resilience Framework linking Industry 5.0 tools (digital twins, robotics, AI) with adaptive, regenerative

N. J. Dewasiri
Department of Accountancy and Finance, Faculty of Management Studies, Sabaragamuwa University of Sri Lanka, Belihuloya, Sri Lanka

Qasim Ibrahim School of Business, Villa College, Maldives

N. J. Dewasiri
e-mail: jayantha@mgt.sab.ac.lk

R. Singh (✉)
Institute of Business Management, GLA University, Mathura, India
e-mail: dr.rubeerajput@gmail.com

Department of Management Studies, Kumaun University, Nainital, Uttarakhand, India

A. Ranasinghe
Postgraduate Institute of Management, University of Sri Jayewardenepura, Nugegoda, Sri Lanka
e-mail: director@pim.sjp.ac.lk

J. Dsilva · M. I. Shaikh
SEE Institute, The Sustainable City, Dubai, United Arab Emirates
e-mail: jacintadsilva2010@gmail.com

M. I. Shaikh
e-mail: irfan@seeinstitute.ae

R. Singh et al. (eds.), *Data-Driven Sustainability: Harnessing Technology for A Greener Future and Environmental Resilience*, World Sustainability Series,
https://doi.org/10.1007/978-3-032-13412-7_23

industrial systems. Concluding with policy recommendations, it advocates leveraging Industry 5.0 for climate resilience, decarbonisation, and sustainable equity in South Asia.

Keywords Industry 5.0 · Climate resilience · Net-zero transition · Human-centric innovation · South Asia

1 Introduction

The twenty-first century is paradoxical: as global economic growth gains momentum due to technological advances, it also heightens risks to the environment and climate. This tension is extreme in South Asia, a region inhabited by almost a quarter of the world's population, yet simultaneously suffering some of the worst effects of climate change. With populated coastal areas, agrarian economies, and rapidly urbanising city centres, South Asian countries are highly vulnerable to rising sea levels, unpredictable monsoons, resource shortages, and severe heatwaves. As the Intergovernmental Panel on Climate Change (IPCC) points out, not only ecosystems but also economic stability, social cohesion, and human wellbeing are threatened; therefore, climate resilience and a net-zero future must now be central to the development agenda rather than optional aspects of environmental protection (Change 2007; Jayarathne et al. 2025).

Although policies have progressed, South Asia's industrial sectors still rely heavily on carbon-based production. The need to decouple growth from greenhouse gases has highlighted new paradigms redefining human, technology, and planet relations. Industry 5.0 emerges as a distinct approach, moving beyond Industry 4.0's automation focus. It emphasises human collaboration with intelligent tech like AI, robotics, and IoT. Its core principles are co-creation, ethics, and sustainability, prioritising well-being and environmental health over mere economic efficiency.

Climate resilience involves an economy and ecosystem's ability to anticipate, absorb, and adapt to environmental shocks while maintaining core functions. Industries must create adaptive, low-carbon, inclusive systems to withstand shocks. Net-zero transitions require reducing emissions and restructuring resource use, energy, and industry (Singh et al. 2024c). Industry 5.0 bridges these by fostering sustainable, technology-driven, people-oriented innovation (Barata and Kayser 2023).

South Asia's industrial sector faces a crisis and opportunities. Key industries like textiles, energy, construction, and manufacturing are major emitters and vital for livelihoods and exports. The challenge is balancing environmental goals with socio-economic needs. Diversity affects readiness for Industry 5.0: India and Sri Lanka progress in digitalisation and renewables, but Bangladesh, Nepal, and the Maldives face infrastructure, skills, and institutional hurdles. Tailored, human-centred models are needed, integrating technology, governance, education, and culture. The chapter includes both theoretical and practical inputs. On a conceptual level, it situates Industry 5.0 within a broader discussion on sustainability, the circular economy, and resilience, and extends these ideas to developing economies. Empirically, it is based on insights gathered from fifteen in-depth interviews with policymakers, practitioners, and sustainability experts who have backgrounds in Sri Lanka, India, Bangladesh, Nepal, and the Maldives. These perspectives offer a practical understanding of how Industry 5.0 is perceived and operationalised in different institutional contexts.

In this connection, the chapter aims at three main goals:

1. To investigate how key stakeholders perceive the role of Industry 5.0 in supporting climate resilience and net-zero transitions.
2. To identify the enabling factors and institutional obstacles that influence the adoption of human-centric industrial innovation.
3. To propose a Human-Centric Climate Resilience Framework (HCCRF) that links the ideas of Industry 5.0 and local climate strategies.

The following sections review the theoretical framework of Industry 5.0, describe the research methodology, and present thematic findings based on qualitative analysis. Policymaking and managerial implications are discussed, culminating in an outline of policy and management considerations for integrating human-focused innovation into South Asia's sustainable industrial future.

2 Theoretical and Conceptual Background

Through ongoing industrial revolutions, humans have continuously transformed their relationship with technology, production, and nature. The first industrial revolution mechanised labour by using steam power, while the second increased reliance on fossil fuels through electricity and engines. The third revolution computerised production with electronics and computing, leading to the fourth revolution, Industry 4.0, which introduced automation, robotics, and cyber-physical systems. Although Industry 4.0 has boosted productivity through artificial intelligence (AI), big data, and the Internet of Things (IoT), concerns have been raised that it exploits social and environmental sustainability (Kumar et al. 2024). Optimisation has tended to worsen inequalities, increase energy consumption, and contribute to ecological degradation.

As a corrective paradigm, Industry 5.0 reintroduces human values into industrial innovation. It aims to make technology meaningful rather than solely productive by positioning human beings as co-creators rather than mere operators. The European Commission (2021) defines Industry 5.0 as a human-centric, sustainable, and resilient model where the purpose of industry is redefined to enable society to enjoy wellbeing within planetary limits. Its three main pillars—human-centricity, sustainability, and resilience—align with the global climate agenda.

- Human-centricity emphasises collaboration between humans and smart machines, ensuring that automation enhances creativity, ethics, and empathy.

- Sustainability promotes the circular use of resources, renewable energy, and decarbonisation.
- Resilience emphasises adaptable systems that can withstand shocks from climate change, pandemics, or geopolitical upheavals.

The principles position Industry 5.0 within an industrial philosophy focused on achieving prosperity without damaging the environment.

2.1 Key Theoretical Foundations

The connection between Industry 5.0, climate resilience, and net-zero transitions is based on three main theoretical perspectives. Circular Economy theory advocates for the development of production systems that mimic nature by reducing waste and recycling resources (Korhonen et al. 2018; Dewasiri 2025). It moves away from the linear 'take-make-dispose' model towards a 'reduce-reuse-recycle-recover' approach. Industry 5.0 technologies, such as IoT monitoring, digital twins, and predictive analytics, support circularity by tracking resources and optimising the life cycle. These practices foster restorative, low-carbon production ecosystems essential for climate resilience.

Resilience Theory, initially used in ecology (Holling 1973; Van-Breda 2018), explains a system's ability to withstand shocks and maintain functions. In Industry 5.0, resilience includes social adaptability and institutional flexibility. The COVID-19 crisis revealed weaknesses in global supply chains, highlighting the need for decentralised, tech-enabled, human-guided production systems. In climate-vulnerable South Asia, integrating resilience into industrial systems ensures continuity, flexibility, and recovery.

The Human-Centric Innovation Theory prioritises human inventiveness, morals, and diversity as key aspects of technological progress (Zhang et al. 2013). It challenges the dominance of the all-technocratic innovation approach by advocating for participatory design, meaningful stakeholder involvement, and co-creation. These principles are reflected in the context of Industry 5.0 through ethical artificial intelligence, responsible automation, and collaborative interaction between humans and robotic systems. In South Asian economies, characterised by population heterogeneity and institutional complexity, this approach will serve as a guide to promote technological development that balances equity, employment, and community welfare while pursuing environmental objectives. Consequently, these perspectives argue that Industry 5.0 is not merely a technological shift but a socio-technical one rooted in human and environmental sustainability.

2.2 Linkages with Global Sustainability Agendas

Industry 5.0 directly supports the Sustainable Development Goals (SDGs), including SDG 7 (Clean Energy), SDG 9 (Industry, Innovation, and Infrastructure), SDG 12 (Responsible Consumption and Production), and SDG 13 (Climate Action). Industrial innovation that aligns with these global aims strengthens Industry 5.0, based on the belief that industrial progress should coincide with environmental preservation and social fairness.

2.3 Research Gaps and Regional Relevance

Although the sustainability potential of Industry 5.0 is increasingly recognised in global scholarship, there is minimal empirical data, especially in developing regions. Existing research has focused on its technological and operational aspects, but none address its human-focused contribution to the sustainability transition. The predominant literature focuses on Europe and East Asia. However, it is underrepresented in South Asia, a region marked by resource scarcity, poor policy coordination, and socio-economic diversity, presenting unique challenges. South Asian literature has so far examined renewable energy, green supply chains, and sector-specific circular strategies (Mishra et al. 2024). However, such efforts rarely address the comprehensive Industry 5.0 paradigm that integrates technology, policy, and culture into a unified sustainability model. This lack of qualitative, stakeholder-led evidence hampers the region's ability to achieve Industry 5.0 that is climate resilient.

3 Methodology

This research paper employs a qualitative, interpretivist research design to explore how Industry 5.0 can enhance South Asia's resilience to climate change and accelerate net-zero transitions. Because Industry 5.0 is relatively new and there is currently limited empirical data linking it to climate action, a qualitative approach was deemed the most suitable to understand stakeholders' perspectives, their lived experiences, and contextual factors. The study aimed to investigate how policymakers, industry leaders, scholars, and sustainability professionals conceptualise and implement human-centric Industry 5.0 principles within climate-related industrial policy. Two research questions guided the inquiry:

1. What is the perceived role of Industry 5.0 by South Asian stakeholders in achieving climate-resilient and net-zero futures?
2. What enablers, barriers, and strategic pathways can be used to incorporate these principles into regional climate agendas?

The study was intended to be based on the constructivist paradigm because of the recognition that knowledge on sustainability and technological transformation is socially constructed through interaction and interpretation.

The population of interest comprised professionals and experts working at the intersection of industry, sustainability, and policy in five South Asian countries: Sri Lanka, India, Bangladesh, Nepal, and the Maldives. Participants were selected using purposive and criterion-based sampling, requiring a minimum of five years' experience in technology, policy, or environmental management. The sample pool was also expanded through snowball sampling, where initial participants referred additional individuals.

Overall, 15 semi-structured interviews were conducted: four in Sri Lanka (SL), four in India (IN), three in Bangladesh (BD), two in Nepal (NP), and two in the Maldives (MV), representing government agencies, the industry, and academia, respectively. The twelfth interview marked data saturation, and three additional interviews were conducted to ensure cross-country balance.

The interviews took place between March and June 2025 and lasted between 45 and 60 min, either in person or on an online platform such as Zoom or Microsoft Teams. Open-ended questions were used to encourage participants to reflect on topics such as human–machine collaboration, barriers to the adoption of green technologies, and the institutional readiness for sustainable transformation (Singh et al. 2024b). All interviews were recorded with audio, conducted with informed consent, transcribed verbatim, and anonymised (using country-coded identifiers, e.g., P3-IN: Participant 3 from India).

NVivo 14 was used for data management and thematic analysis to conduct the analysis (Braun and Clarke 2006). Six iterative steps were followed, including familiarisation with the data, coding, theme development, review, defining, and synthesising. Inductive coding produced primary categories, which were further refined into four main themes.

1. Human–machine synergy for green innovation
2. Policy alignment and institutional readiness
3. Technological adaptation for circular and resilient production
4. Socio-cultural integration for inclusive transformation

4 Findings and Discussion

The fifteen detailed interviews conducted across five countries in South Asia uncovered a nuanced understanding of how Industry 5.0 could foster a climate-resilient transformation, as seen by stakeholders. The interviewees, including policymakers, industrial experts, scholars, and sustainability practitioners, provided insights into the opportunities, challenges, and socio-political intricacies involved in adopting a human-centric industrial approach. The thematic analysis identified four interconnected themes: machine synergy for green innovation; policy alignment and institutional readiness; technological adaptation towards circular and resilient production; and socio-cultural inclusiveness for an inclusive transformation. These themes collectively highlight different yet related aspects of how Industry 5.0 might support the realisation of net-zero and climate-resilient futures in South Asia (Mishra et al. 2024).

4.1 Human–Machine Synergy for Green Innovation

In all five countries, participants emphasised that Industry 5.0 goes beyond mechanised efficiency to include intentional collaboration between humans and intelligent technologies. Rather than seeing automation as a threat to jobs or an innovation challenge, most stakeholders viewed it as an opportunity to boost human problem-solving, especially in sustainable innovation (Dewasiri and Choudhary 2025). One Indian participant added that machines should not be perceived as substitutes but as assistants capable of thinking sustainably and making decisions more rapidly and effectively (P3-IN). Similarly, a Sri Lankan policymaker highlighted that "*with AI-inspired systems, it is possible to monitor emissions or energy use in real-time. Yet, it remains human judgment that interprets these findings into ethical and contextually relevant outcomes*" (P1-SL).

The human–machine synergy concept reflects the theory of human-centred innovation, where technology is considered a facilitator of human creativity rather than its substitute. In this research, participants highlighted that combining artificial intelligence, robotics, and the Internet of Things (IoT) can help optimise resource use, reduce waste, and support low-carbon production. They also stressed that these technologies should be implemented and regulated carefully to uphold ethical values and social justice. A Bangladeshi scholar remarked, "*Technology can lead to fewer emissions, but, at the cost of more inequality or joblessness, we will lose the human component of sustainability* (P2-BD)."

This human–machine alliance aligns with the model of the circular economy, where intelligent systems support closed-loop production using predictive analytics and real-time monitoring (Singh et al. 2024b). For example, participants cited instances of AI-powered predictive maintenance systems that saved energy and reduced machine failure time, making operational processes and the environment more sustainable (Dewasiri et al. 2025). In India and Sri Lanka, participants highlighted pilot programmes where robotics could be employed to handle hazardous substances, thereby reducing occupational risks and enhancing environmental safety.

The findings support the idea that Industry 5.0, when applied with a human-centric focus, can promote innovations that are both technologically advanced and ethically sound. However, participants also cautioned that this potential might stay unrealised unless there are sufficient efforts to enhance digital literacy and reskill the workforce. It is evident what is needed: to leverage Industry 5.0 for climate resilience, South Asia must cultivate human capital capable of working effectively with intelligent systems. Therefore, educational institutions and governments play a vital role in designing interdisciplinary curricula that integrate technology, sustainability, and ethics.

4.2 Policy Alignment and Institutional Readiness

Although the respondents recognised the potential of Industry 5.0, they all agreed that there is a lack of policy frameworks to support such transitions. In all five countries, there was a strong perception that industrial and climate policies are not well coordinated; instead, they operate in silos and lack harmony. A Bangladeshi respondent noted that *while they have national climate policies, these do not address technology or Industry 5.0. The sustainability agenda and industrial innovation are still not linked* (P2-BD). Similarly, a Nepalese policymaker noted that their institutions remain focused on Industry 2.0 or 3.0, with regulation lagging behind innovation, which in turn hinders industries from making *ambitious decisions* (P4-NP).

The two perspectives highlight the core idea of resilience theory: that institutions, much like ecosystems, should possess adaptive capacity to incorporate new ideas and technologies in response to a changing environment. Respondents on several occasions emphasised the importance of inter-ministerial coordination and regional cooperation as a means of easing policy alignment. For example, one Sri Lankan sustainability professional stated, "W*e can never have green transformation without aligning technology, environment, and education policies. Fragmentation wastes both time and opportunity*" (P3-SL).

Simultaneously, positive institutional changes were identified. India's focus on the Digital India campaign and the integration of renewable energy, Sri Lanka's current efforts in green financing, and Bangladesh's emerging emphasis on circular manufacturing can serve as starting points for embedding Industry 5.0 principles. However, participants warned that these initiatives remain largely project-based rather than systematic. One Maldivian respondent noted that *while they have climate-sensitive interventions, these are isolated islands of innovation, not integrated into a nationwide system* (P1-MV).

The results suggest that achieving institutional readiness for Industry 5.0 in South Asia requires three mutually compatible changes: the alignment of industrial and climate policies, strengthening the capacity of regulatory bodies to assess new technologies, and fostering cross-border cooperation for knowledge sharing and standardisation. This approach is a best practice observed globally, particularly in the European Union and Japan, where policy frameworks clearly integrate human-centric and sustainable innovation principles. Such policy alignment in the South Asian context could foster regional resilience, as industries would gain access to similar standards, digital infrastructure, and mechanisms for green financing (Rathnasiri et al. 2024).

Simply put, this theme emphasises that technological evolution alone is not enough; the success of Industry 5.0 depends on institutional ecosystems that can support the adaptation of governance. Without these, Industry 5.0 will merely become another imported policy slogan, rather than a locally grounded approach to climate resilience.

4.3 Technological Adaptation for Circular and Resilient Production

All countries expressed hope regarding the opportunities that Industry 5.0 technologies offer in fostering a circular and resilient production system. However, they also recognised the need for context-specific adaptation rather than directly transplanting models from developed economies to achieve such transformation. According to a Sri Lankan manufacturing manager, the sensors introduced to monitor water and energy consumption are a small step. However, they are gradually shifting the psychological mindset towards efficiency and responsibility (P5-SL). Similarly, an environmental expert in the Maldives highlighted *that circular production requires both technology and culture. It is not sufficient to rely solely on imported models; we must develop our own version of sustainability that is adaptable* (P1-MV).

These lessons reflect the circular economy theory, which views resource efficiency as a combination of technological and behavioural processes. The interviewees mentioned new practices such as waste valorisation, resource recovery, and the integration of renewable energy in sectors like textiles and tourism. One of the Bangladeshi respondents said, "*In our garment factories, we are conducting experiments with fabric recycling and solar-powered dyeing machines. These technologies will help us become less dependent on fossil fuels and reduce waste at the same time.*" (P3-BD).

A key theme here is the resilience of production processes, which also ties into resilience theory, especially its focus on adaptive systems that can absorb shocks and reorganise themselves. One or two participants linked their Industry 5.0 ambitions to what they learned during the COVID-19 pandemic, where supply chain shocks exposed the weaknesses of the old manufacturing paradigms. One of the policy analysts in Nepal noted that "*as our imports ceased during the pandemic, we realised that digitalisation and automation are not just about efficiency, but about survival*" (P2-NP).

Nonetheless, participants also recognised other significant barriers to technological adoption. The most frequently mentioned issues included limited funds for upgrading technology, inadequate digital infrastructure, and a shortage of skilled technicians to operate the new and advanced systems. One Sri Lankan industrialist remarked, "*We would like to go to Industry 5.0, but we do not have the capital, nor the expertise in our SMEs. They will not be supported financially and in training; they will be left behind*" (P2-SL). This point emphasises why technological progress must be complemented by financial inclusion and skills development.

The policy implications of this theme involve establishing green innovation centres and regional knowledge-sharing platforms to assist SMEs in adopting sustainable technologies. However, the introduction of green financing tools, including concessional loans, tax breaks, and carbon credits, can be advantageous for investing in digital sustainability. Technological adaptation and circular production systems are among the most significant ways in which Industry 5.0 can be utilised to foster a more climate-resilient South Asian context.

4.4 Socio-Cultural Integration for Inclusive Transformation

The final theme is that implementing Industry 5.0 in South Asia relies on cultural and social acceptance as well as technological and policy preparedness. Participants consistently argued that the transition towards sustainability should be inclusive, ethically grounded, and locally appealing. One of the Indian sustainability directors stated that Industry 5.0 is not solely about machines; it is about thinking. "*Sustainability should become a human aspect rather than merely a policy goal*" (P3/IN). Similarly, a Nepali policymaker remarked that *"By involving local communities, technology will benefit everyone, not just large corporations. Communities will become partners in change when they understand the significance of green innovation"* (P2-NP).

Such inclusiveness also reflects a human-centric approach to innovation and social sustainability, which adheres to the principles of co-creation, empathy, and equity. The participation of both genders, along with the roles of youth and local knowledge in technological transitions, was highlighted by participants across all countries. A sustainability consultant in Bangladesh stated that *"Our female labour force in the garment industry would exemplify green manufacturing, provided they receive adequate digital training and recognition"* (P4-BD).

Moral considerations were also integral to what participants believed. Another issue raised was data privacy, labour displacement, and the potential for increasing inequalities if automation is implemented without adequate protections. According to one Maldivian expert, "*Industry 5.0, when it neglects ethics, will repeat the mistakes of industrialisation under a different guise*" (P2-MV). These points suggest that human-centricity should move beyond mere rhetoric and be integrated into institutional processes of fair governance and accountability.

The respondents proposed several culturally rooted approaches for inclusive change. These included integrating sustainability and education within schools, launching awareness programmes by communities, and promoting local innovation projects. Connecting local artisans, cooperatives, and rural enterprises was seen as vital to embedding the principles of Industry 5.0 into traditional economic frameworks. A Sri Lankan scholar remarked that "I*nclusivity gives sustainability a soul. It will transform Industry 5.0 from a purely technocratic endeavour into a genuine social movement without it*" (P4-SL).

In theory, this theme supports the idea that the human-centric vision of Industry 5.0 is also consistent with the resilience perspective. In this place, adaptability is rooted in diversity and inclusion. In a practical sense, it highlights the need for social innovation policies that democratise access to technology and enable local communities to be co-creators of change.

4.5 *Integrative Discussion: Proposed Human-Centric Climate Resilience Framework (HCCRF)*

Figure 1 helps explain the relationship between Industry 5.0, potentially acting as a strategic facilitator of climate-resilient change in South Asia through the integration of technological intelligence, institutional capacity, and socio-cultural adaptability into a cohesive system: the Human-Centric Climate Resilience Framework (HCCRF). Its core principle is human–machine synergy to promote green innovation, viewing it as a collaborative effort between human ingenuity and intelligent technologies. In this setup, automation, robotics, and artificial intelligence are not seen as replacements for human labour but as tools that enhance human capabilities. This synergy enables real-time data analysis, predictive maintenance, and process optimisation, leading to reductions in emissions and waste. It also restores the ethical and cognitive dimensions of industrial innovation, ensuring that values rather than solely data guide decision-making.

Surrounding the core is the area of institutional readiness and policy alignment, which serves as the structural backbone of resilience. Strong institutions, supported by coherent policies, regulatory tools, and financial instruments, establish the necessary environment for implementing Industry 5.0. This layer reflects the concepts of

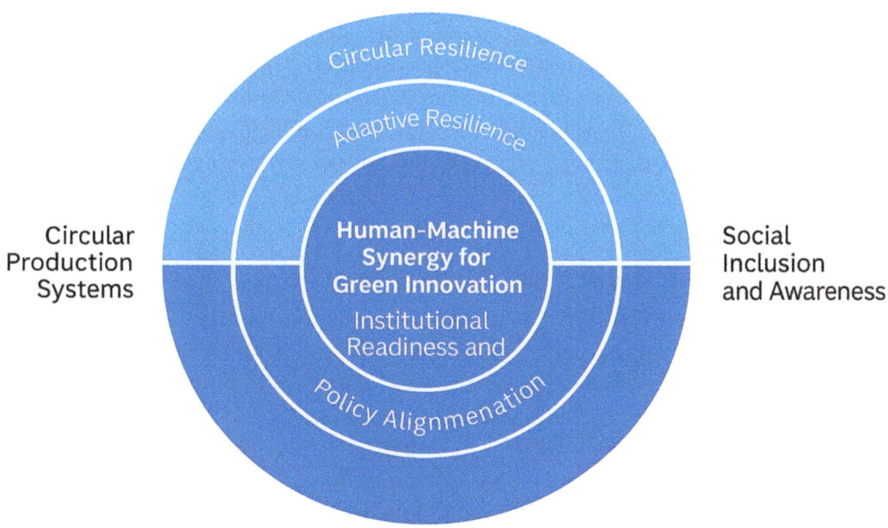

Fig. 1 Human-centric climate resilience framework (HCCRF). *Source* Authors' own

resilience theory, whereby adaptive governance enables systems to restructure and adapt to shocks. In the context of South Asia, policy alignment is characterised by the integration of industrial, environmental, and educational policies, strengthening relations between the state and corporations, and promoting regional cooperation through organisations like the South Asian Association for Regional Cooperation (SAARC) and Bay of Bengal Initiative for Multi-Sectoral Technical and Economic Cooperation (BIMSTEC).

The outermost layer pertains to socio-cultural integration and inclusivity, where technological and policy development align with local contexts. This element reflects the human-centred innovation approach, emphasising that cultural acceptance, equity, and ethical governance are vital for sustainable change. Technological advancement within society is grounded in the community's lived experiences, achieved through social participation of women, engagement of the youth, and community programmes. Trust in the long-term outcomes is fostered by ethical awareness and a shared sense of ownership regarding sustainability.

These three layers comprise a dynamic, interdependent system. The tools of innovation are supplied through technological intelligence, the regulatory and financial framework through institutions, and moral legitimacy and inclusivity through socio-cultural systems. The presence of feedback loops between the layers indicates continuous learning and adaptation, where data generated by Industry 5.0 systems guide policy decisions, and the community can influence the design of technology.

The outcome of this encounter is the development of climate-resilient systems, such as adaptive, low-carbon production methods that can withstand environmental shocks, and the shift towards net-zero, supported by circular economy principles and green innovation. The framework promotes the sustainable development of competitive and environmentally friendly industries, ensuring inclusive well-being and making the benefits of industrial changes accessible to all population segments.

The HCCRF offers a conceptual framework and strategic roadmap to policymakers, practitioners, and researchers by integrating the ideas of the circular economy, the theory of resilience, and human-driven innovation. It positions Industry 5.0 not as a technological revolution but as a societal project that redefines industrial success as sustainable, equitable, and rooted in human flourishing.

5 Practical and Managerial Implications

The current research paper identifies several key implications for policymakers, industry leaders, and development institutions aiming to adopt Industry 5.0 to foster sustainable transformation in the South Asian context.

Governments at the policy level need to shift their focus towards integrated country strategies rather than fragmented efforts when incorporating Industry 5.0 principles into industrial, technological, and environmental policies. This approach requires policy frameworks that prioritise aligning industrial competitiveness with environmental sustainability, digitalisation, and human development. Ministries of industry, environment, and education should collaborate in developing national roadmaps that connect the advancement of digital skills, transitions to renewable energy, and circular economy practices. Human-centric metrics of innovation and carbon-reduction targets should be embedded into industrial policies and established as the fundamental principles guiding sustainable development.

Regional cooperation is vital where the region shares common weaknesses. Platforms like SAARC and BIMSTEC can be utilised to promote cross-border collaboration by developing shared research infrastructure, technology transfer, renewable energy systems, and standardised regulatory practices for ethical AI and sustainable manufacturing. A collective development ecosystem would rely on a cooperative approach among South Asian countries to jointly pursue goals like net-zero emissions and reduce dependence on imported technologies.

Monetary systems must also be established to support the green transformation. Small and medium-sized enterprises (SMEs), often the industrial backbone of the region, frequently lack the resources to invest in long-term technologies. The risks linked to innovation can be reduced by expanding green finance options, such as concessional funding, tax credits, blended finance, and sustainability-linked bonds, which may slow down the pace of Industry 5.0 implementation. Regional carbon markets could also serve as additional motivation for trading emissions and offer opportunities to reinvest in low-carbon technologies.

The propagation of a human-centric approach cannot occur without education and capacity building. Interdisciplinary programmes that integrate sustainability, technology, and ethics should be developed in universities and vocational institutes. The Industry 5.0 learning hubs established through collaboration between the public and private sectors can connect academia, entrepreneurs, and industry to ensure the workforce is equipped with both digital and environmental skills. Upskilling programmes should ensure that automation complements human labour rather than replacing it.

As a manager, organisations should view Industry 5.0 as a revolutionary breakthrough rather than merely a technological upgrade aimed at sustainability. Business leaders must integrate human and environmental principles into corporate strategies, incorporate sustainability metrics into performance evaluations, adopt transparent ESG reporting, and foster creativity and engagement among staff (Singh et al. 2025). This approach is crucial because the effective blending of human knowledge and digital intelligence enhances organisational flexibility, creativity, and sustainability over the long term.

Operationally, industries should adopt circular business model strategies that minimise waste and extend product lifespan. Digital technologies (such as the Internet of Things and simulating a digital twin) could enhance the efficiency and resilience of the energy system, but developed ethical frameworks are necessary to ensure responsibility in automated systems and data usage. Most importantly, the principle of inclusiveness must underpin all managerial and policy interventions. Through proactive engagement of women, youth, and marginalised populations in developing green technologies, the creative reservoir is expanded and social equity is fostered.

These implications emphasise that the transition to Industry 5.0 in South Asia will require coordinated efforts across various levels of governance, including the government, industry stakeholders, educational institutions, and civil society. Through collaborative governance and a shared commitment to human-centred innovation, the region can utilise technology to promote inclusive, climate-resilient, and sustainable development.

6 Conclusion and Future Directions

This chapter examined how Industry 5.0, understood through human centricity, sustainability, and resilience, can serve as a transformative process towards climate-resilient and net-zero visions in South Asia. Based on fifteen in-depth interviews with policymakers, industrial practitioners, sustainability experts, and academics in Sri Lanka, India, Bangladesh, Nepal, and the Maldives, this study demonstrates how this new paradigm is being conceptualised and implemented within the diverse industrial and policy contexts of the region.

The results indicate that South Asia is at a crossroads where rapid industrialisation clashes with an emerging climatic threat. Industry 5.0 offers a vision of redesigning production systems centred on sustainability, low-carbon methods, and inclusivity. Four key themes emerge: the human–machine partnership in green innovation; policy coherence and institutional preparedness; technical adaptation through circularity and resilient production; and social-cultural inclusiveness to foster a comprehensive transformation. These themes highlight that the shift to Industry 5.0 is not solely technological but also fundamentally human, requiring a balance between innovation, government action, and cultural considerations.

The paper explores Industry 5.0's potential to personalise technology, positioning human beings as creative partners rather than passive consumers. Artificial intelligence, robotics, and the Internet of Things, guided by ethical considerations, can be utilised to make technologies more efficient and environmentally responsible. However, decentralised government, policy stagnation, and uneven technological readiness still hinder this transformation. Climatic-resilient industrialisation must therefore be achieved through integrated policies that connect technology, sustainability, and education, while promoting inclusivity across gender, community, and socioeconomic lines.

One of the theoretical contributions of the chapter is the Human-Centric Climate Resilience Framework (HCCRF), which combines Industry 5.0 principles with the circular economy and resilience theory. This framework conceptualises sustainable industrial transformation as a combination of three pillars (technological intelligence, institutional readiness, and socio-cultural adaptability) that support each other in a process of continuous learning. Human–machine synergy facilitates innovations, while policy coherence and inclusive governance provide stability and legitimacy.

In the future, several research directions are possible. Sector-specific studies could focus on the applications of Industry 5.0 technologies across major industries such as textiles, agriculture, construction, and renewable energy. Long-term research might evaluate how changes brought by Industry 5.0 impact emissions, productivity, and employment, while cross-country comparative analyses could identify adaptable models suitable for the Global South. Additional studies could also explore the micro-dynamics of human–machine cooperation, the ethics of AI governance, and higher-level regulations that prevent environmental injustices. To develop comprehensive approaches to sustainable industrial development, interdisciplinary teams comprising management, engineering, environmental science, and behavioural studies would be essential.

Finally, Industry 5.0 represents not only a technological revolution but also a moral and developmental one, aiming to reimagine industry as a means of promoting human and planetary wellbeing. In South Asia, it offers a rare opportunity to combine digital transformation with climate action, whereby economic growth complements rather than undermines environmental and social resilience.

References

Barata J, Kayser I (2023) Industry 5.0–past, present, and near future. Procedia Comput Sci 219:778–788

Braun V, Clarke V (2006) Using thematic analysis in psychology. Qual Res Psychol 3(2):77–101

Change OC (2007) Intergovernmental panel on climate change. World Meteorol Organ 52(1–43):1–10

Dewasiri NJ, Choudhary M (2025) Harnessing artificial intelligence (AI) for sustainable innovation: case studies and practical applications. In: Innovation ecosystems and sustainable technologies: enhancing competitiveness and sustainability. Emerald Publishing Limited, pp 169–188. https://doi.org/10.1108/978-1-83662-368-720251008

Dewasiri NJ, Gangodawilage D, Singh R (2025) Pathways to climate neutrality: integrating sustainability in transport, energy, and communication for the indo-pacific region. In: Climate neutrality and sustainable ecosystems: practical approaches to addressing global warming. Singapore: Springer Nature Singapore, pp 95–114. https://doi.org/10.1007/978-981-96-7181-6_6

Dewasiri NJ (2025) Highlights of circular economy actions in the climate change policies: a call for action for the Asia-pacific region. Clim Chang Soc Responsib 23–41. https://doi.org/10.1108/S2043-052320250000025002

European Commission (2021) Industry 5.0—towards a sustainable, human-centric and resilient European industry. Publications Office of the European Union. Research and innovation+2Horizon-europe.gouv.fr+2

Holling CS (1973) Resilience and stability of ecological systems. Annu Rev Ecol Syst 4:1–23

Jayarathne PGSA, Dewasiri NJ, Khan S (2025) Challenges of integrating social responsibility and climate change for the sustainable development goals: experience from the South Asian context. In: Climate change and social responsibility. Emerald Publishing Limited, pp 205–226. https://doi.org/10.1108/S2043-052320250000025011

Korhonen J, Honkasalo A, Seppälä J (2018) Circular economy: the concept and its limitations. Ecol Econ 143:37–46

Kumar A, Singh R, Khan S, Kumar V (2024). Artificial intelligence-enabled management: an emerging economy perspective. De Gruyter. Retrieved December 23, 2024

Mishra U, Rathnasiri MSH, Dewasiri NJ (2024) Net-zero economy and energy sustainability: a Nepalese perspective. In: Singh R, Khan S, Kumar A, Luthra S, Chokshi H (eds) Net zero

economy, corporate social responsibility and sustainable value creation. CSR, Sustainability, Ethics & Governance. Springer, Cham. https://doi.org/10.1007/978-3-031-55779-8_9
Rathnasiri MSH, Dewasiri NJ, Kumar AA (2024) Policy framework and implementation strategies for Sri Lanka's transition to a net-zero economy. In: Singh R, Khan S, Kumar A, Luthra S, Chokshi H (eds) Net zero economy, corporate social responsibility and sustainable value creation. CSR, Sustainability, Ethics & Governance. Springer, Cham. https://doi.org/10.1007/978-3-031-55779-8_3
Singh R, Gupta A, Bajpai A, Kandpal V (2024b) Exploring challenges of circular economy initiatives for smart cities. In: Smart cities and circular economy: the future of sustainable urban development. Emerald Publishing Limited, pp 59–69. https://doi.org/10.1108/978-1-83797-957-820241005
Singh R, Crowther D, Ravesangar K (2024b) Harnessing human capital for a sustainable future: HR's pivotal role toward net zero policies. In: Singh R, Crowther D (eds) Transition towards a sustainable future. Approaches to Global Sustainability, Markets, and Governance. Springer, Singapore. https://doi.org/10.1007/978-981-97-5756-5_2
Singh R, Ravesangar K, Narayanan S, Singh KKH (2024c) HRM practices in creating net zero: supporting organisational sustainability. In: Singh R, Khan S, Kumar A, Luthra S, Chokshi H (eds) Net zero economy, corporate social responsibility and sustainable value creation. CSR, Sustainability, Ethics & Governance. Springer, Cham. https://doi.org/10.1007/978-3-031-55779-8_1
Singh R, Gupta A, Kandpal V (2025) ESG excellence in the digital world: circular economy e-business solutions. Fintech ESG Circ Econ 193–204. https://doi.org/10.1002/9781394234035.ch10
Van Breda AD (2018) A critical review of resilience theory and its relevance for social work. Social Work/maatskaplike Werk 54(1):1–18
Zhang Y, Zhou Y, McKenzie J (2013) A humanistic approach to knowledge-creation: People-centric innovation. Towards organisational knowledge: the pioneering work of Ikujiro Nonaka. Palgrave Macmillan UK, London, pp 167–189

Blockchain in Resource Recovery for Securing Traceability in Circular Systems

Wasswa Shafik, Hassan Tumwiine, and S. Mojtaba Matinkhah

Abstract The increasing concern for environmental pollution and the depletion of natural resources has directed global attention to an alternative approach that considers such issues, the circular economy. Circular systems are favorable from both an economic and environmental point of view, reducing the need for virgin materials and non-renewable energy sources. Creating closed supply chains using the principles of a circular economy is complex. It requires high levels of coordination between stakeholders, from raw material origin to end user destination. Because of the large number of transactions in this system, it is important to maintain the integrity of information in the supply chain. Blockchain provides advantages for supply chains, such as transparency, accountability, and supporting collaboration. The significant benefits that blockchain services provide for resource recovery in circular systems are now recognized by governments and industry leaders. Although the contributions to circular economy principles are intangible, their help in enabling companies to reduce GHG emissions can be quantified. With the data chain approach, buildings can incorporate advanced material management and digital service models such as leasing, which are essential for the circular economy. The discussion then extends to how blockchain can transform building management systems by supporting renewable energy trading, automated billing services, and the incorporation of distributed supply chains, delivering real benefits to circularity.

Keywords Blockchain traceability · Circular economy · Data immutability · Digital ledger · Resource recovery · Smart contracts · Supply transparency · Sustainable tracking · Token incentives · Waste valorization

---

W. Shafik (✉)
Dig Connectivity Research Laboratory (DCRLab), Kampala, Uganda
e-mail: wasswashafik@ieee.org

School of Digital Science, Universiti Brunei Darussalam, Gadong, Brunei

H. Tumwiine
Department of Civil and Environmental Engineering, Monash University, Clayton, Australia

S. M. Matinkhah
Computer Engineering Department, Yazd University, Yazd, Yazd Province, Iran

R. Singh et al. (eds.), *Data-Driven Sustainability: Harnessing Technology for A Greener Future and Environmental Resilience*, World Sustainability Series,
https://doi.org/10.1007/978-3-032-13412-7_24

1 Introduction

The pressing need to reduce GHG emissions and cope with resource depletion majorly drives the adoption of the circular economy (CE). Preparing waste for resource recovery and consequent reuse or recycling is a vital phase of any circular system. The objective is to minimize the generation of residual waste during the CE operation. It is not enough for the final user to collect the product post reaching the end of life so that material can be recovered; nevertheless also assures that the material is prepared for resource recovery. This relates to the concept of product stewardship, in which all stakeholders in the loop participate and collaborate to make the operation successful (Zhang et al. 2024). This compliance requires transparency in the product's journey, which blockchain technology can help with.

1.1 Definition of Circular Economy

Resource efficiency and circular economy are widely accepted concepts at the core of sustainable development. Strategies and policies have been prepared to enhance reuse and recyclability, reduce waste and resource use, and incentivize the use of secondary raw materials. A well-designed circular economy can consider the environment optimally and generate substantial economic benefits such as employment creation, economic growth, reduction of dependence on scarce and expensive natural resources, and reduction of import prices. Circular economy principles comprise repair, reuse, remanufacturing, and recycling, ultimately preventing the accumulation of waste through resource recovery. A circular economy model consolidates repair, reuse, remanufacturing, and recycling from life cycle and material flow perspectives (Chatterjee et al. 2023). Resource recovery aims to bring back energy and material resources to the productive economy without any further sorting or processing. Traceability is required to preserve the quality of raw materials and to prove transparency in these processes within an economic model based on resource recovery.

Blockchain can secure traceability in resource recovery-based circular systems through transparency, accountability, and collaboration, as the key features of blockchain address the traceability challenges in resource recovery arising from the system's complex information flows, stakeholder involvement, conflicting interests, and regulatory measures. Blockchain applications in resource recovery through the circular economy lens shed light on the economic, environmental, and social impacts of blockchain itself and provide an implementation perspective on detailed case studies, required technologies, and involved stakeholders. Resource recovery minimizes the impact of waste on the environment by utilizing secondary raw materials, reducing the depletion of natural resources by cutting down primary raw materials, and reducing the vulnerability of raw material supply chains by reducing dependency on import centers (Wazid et al. 2021). Resource recovery also generates new sources of income by contributing to employment creation, firms' competitiveness, and GDP growth.

1.2 Principles of Resource Recovery

The concept of the circular economy can be operationalized by using resource recovery from waste streams. Resource recovery means obtaining resource equivalents from those materials whose economic value is lower than the economic value of the achieved product or service. Waste generation is an intermediary step in the value chain. It is a resource that includes prolific feedstock for the circular economy if losses can be minimized and the material loop can be perpetuated along the value chain (lakhan et al. 2023). Resource recovery calls for substituting virgin resources with recovered ones, closing the gaps between waste and resources, and reducing the environmental impact. Four principles must be adhered to:

(a) avoid resource loss,
(b) avoid contamination of resources,
(c) protect natural resources, and
(d) Use the recovered resources.

Resource loss can be conceptualized on three tiers: dynamic, physical, and economic. Dynamic loss relates to low recovery rates, excessive landfilling, and recycling techniques that are below possible kinetics. Contamination refers to the dissolution of harmful entities into the recovered resources, often arising from inefficient recovery processes or poor management. The encroachment of natural resources like water and soil is also categorized under the contamination of resources. Physical losses denote destruction of materials and energy as well as the creation of by-products. Economic loss entails a negative net benefit concerning economic allocation, embodied energy, and capital. The protection of natural resources requires avoiding biodiversity loss, soil erosion, ecological imbalance, and environmental degradation. Recovered resources should be accounted for in the market and policy planning (Ahmed and MacCarthy 2023). The continuous reintroduction of those types of suboptimal recovered resources that tend to deteriorate the quality of production of specific products should be avoided.

2 Benefits of Circular Systems

Integrating circular system features and resource-recovery strategies into business development paves the way towards sustainable development. Circular economy–derived development safeguards the environment by reducing waste generation and mitigating ecosystem contamination through resource-recovery strategies. It also conserves natural resources by lowering product demand through the reuse of certain components and product life–extension strategies, reduces transportation costs by enabling product- or component-level reuse, conserves bundle-services resources by implementing function-life maximization, reduces manufacturing costs through production optimization, and safeguards human life by reducing the content of potentially harmful materials in recycled products. The reuse and remanufacturing of components minimizes raw material extraction and the manufacturing process, thus providing economic as well as environmental benefits by diverting waste from landfill and minimizing carbon emissions (Alkatheiri and Alghamdi 2023). However, several barriers exist in pursuing these strategies. Not all phases of the product life cycle provide the opportunity for circularity, as the time cycle for different products varies considerably. For time-sensitive products, such as smartphones, the lifecycle of products is very short. The lack of consumer awareness and limited innovation in product design is also detrimental to promoting resource efficiency for these products. The infrastructure available to support a circular economy, such as reverse logistics, is also currently limited. Although companies perceive circular economy–based activities to be worthy, only a few can successfully implement them. The lack of enabling government regulations for circular processes and operations significantly influences circular economy adoption among actors. Although several organizations have a circular economy business model, most of these have a low circular uptake (Kumar et al. 2023).

3 Blockchain Technology Overview

Circular systems and resource recovery aim at a more sustainable approach to the management of resources by considering utilizing secondary materials as the feedstock for new production. Although circular operations tend to be more environmentally friendly in comparison with traditional linear supply chain activities, collaboration among the supply chain stakeholders is more challenging. Circular operations require constant tracing, i.e., to identify the provenance, conditions, and location of any residual materials when being processed or circulated. Blockchain technology can address the traceability challenges by increasing the collaborative transparency requirements among the supply chain processes and stakeholders. Blockchain is a distributed and decentralized system where shared information is stored in the ledgers of multiple organizations within a network (Jouini and Sethom 2023). Within a blockchain network, a transaction represents the objects of the protocol that will trigger a state change.

3.1 What is Blockchain?

The term "blockchain" was first introduced in 1991 by Stuart Haber and W. Scott Stornetta, in the scope of a cryptographically secured chain of blocks concept. Two decades later, in 2009, Bitcoin was released by an unknown person or group of people under the pseudonym Satoshi Nakamoto. Bitcoin was the first implementation of a digital currency system that used the idea of a cryptographically secured blockchain. Transaction records have a timestamp and are placed in a block that is linked to the previous blocks in a linear and chronological order, built on peer-to-peer network principles, and no central administrator or trusted authority regulates the system (Nguyen et al. 2024). Although Bitcoin achieved success, the scalability of its blockchain was a major challenge to its use in new business cases. Ethereum was launched in 2014 as a new blockchain model that introduced smart contracts, which can replicate business logic and provide additional features to distributed applications (DApps) built on a public blockchain network. The public Ethereum blockchain was launched on 30 July 2015 with 11.9 million Ether mined, comprising a deflationary currency (a cryptocurrency) that could be used to trade on the execution of smart contracts. Smart contracts are self-executing computer programs that can be applied in different business models and change the dimensions of blockchain beyond just a transaction blockchain (Ghayvat et al. 2024).

3.2 Types of Blockchain

Blockchains can be grouped according to different features of their design, such as access rights, operation, and consensus algorithms. Regarding access rights, public blockchains allow any user to join the network and have equal privileges. Private blockchains restrict participation to a closed set of verified users who can change the ledger's design and configuration. Hybrid (or federated/consortium) blockchains consist of a group of participants governing the network, providing additional control and reducing computing power requirements. Blockchain technology offers several key features relevant to resource recovery in circular systems. Decentralization, or distributed data storage, enables a common set of source data to be accessed and amended independently by a heterogeneous range of users (Huang and Liu 2024). This feature creates opportunities to address trust issues in complex, globalized supply chains, as demonstrated in the De Beers platform, Tracr, which tracks gemstones using a private, permissioned blockchain. Immutability, achieved through hashing and digital signatures, enhances accountability and credibility by securing data on the blockchain from retroactive alteration or censorship. Smart contracts formalize a series of conditional statements or business rules in executable code residing on the blockchain, operating automatically and autonomously within the network. These functions underpin transparency and traceability objectives within circular economy activities (Santhuja and Anbarasu 2023).

3.3 Key Features of Blockchain Technology

A blockchain is an append-only data structure consisting of a sequence of blocks. Each block contains the cryptographic hash of the previous block, a timestamp, and transaction data represented by a Merkle tree. The first phase of data verification during block creation, known as mining or forging, relies on the consensus algorithm used by the network. The mining procedure involves the current block miner calculating the nonce—a number that, when hashed with the other block information, produces a hash value with a specific format. This format may be, for example, a certain number of zeros found at the beginning of the hash. Miners repeatedly adjust the nonce and rehash the block contents until the hash meets the required format. The difficulty target is dynamically adjusted to ensure new blocks are appended at a consistent rate, such as one every 10 min. Several variants of blockchains exist, including public, consortium (or federated), and private. Public blockchains are open to everyone and are governed solely by the rules encoded in their protocols. Consortium blockchains have predefined sets of nodes that govern the network. Private blockchains are centralized entities governed by very limited members (Rahmadika et al. 2023). In consortium and private blockchains, permissioned access controls ensure that only authorized parties can participate in consensus and view transactions.

3.4 Role of Blockchain in Resource Recovery

Resource recovery is an activity aimed at reducing resource consumption by feeding materials back into the economy instead of landfilling or disposing of them. It is essential for closing material and information loops in circular systems. The connection between resource recovery and the circular economy has been studied in the literature, with Apple cited as an example of a company that is trying to recover resources. Several studies have also established theories explaining how the idea of resource recovery supports the development of a circular economy. Several key features of blockchain, including decentralization, visibility, trust, and immutability, contribute to secure traceability and accountability in circular systems. Recent literature has demonstrated that traceability in circular systems can be implemented using blockchain technology (Bag et al. 2024). However, implementing blockchain requires thoroughly understanding the processes in a specific circular system that can be secured through traceability.

The role of blockchain in resource recovery within circular systems is to implement secure traceability that contributes to accountability and transparency. Several challenges related to organizing data and executing component/product backflows must be addressed before implementing blockchain. Resource recovery challenges include determining the nature of the flow to be established, managing large and multifaceted physical backflows, and ensuring funds are available despite the high costs of executing these flows. Proper organization of data thus becomes crucial for managing the reverse flow. Circular systems face difficulty in organizing and securely maintaining transaction data, along with ensuring collaboration among stakeholders to support the process. Key traceability challenges include maintaining immutable details about resource recovery processes, embodying reconciliation rules in smart contracts for automated reconciliation, and embedding transition rules in smart contracts for automatic initiation of backflows (Shafik 2024). Appropriate organization and secure maintenance of transactional details can be realized through blockchain implementation.

3.5 *Enhancing Transparency*

The circular economy approach emphasizes collaboration among economic actors to extend loop lifetimes and generate economic value. These objectives cannot be realized without transparency, which increases accountability and responsibility among stakeholders as they contribute to loop closure. Transparency promotes cooperation across supply chains and circular systems: participants often refrain from sharing information due to concerns about misuse, manipulation, or loss of future benefit. Therefore, resource recovery efforts require higher transparency levels to support efficient collaboration throughout the loop. The nature of resource recovery demands shared ownership, as it involves multiple actors and value-generating phases and starts with waste. Bitcoin and Ethereum solved the trust problem; however, the blockchain-powered supply chain—the key enabler of the circular economy—is poorly known and not fully adopted. The OECD emphasizes commitment and responsibility in performing supply chains: all actors must consider the impact of their decisions on social, environmental, and economic issues and take required actions. Transparency supports responsible sourcing practices; it motivates companies to improve social and environmental behaviors and enhances consumers' knowledge of production practices, products, and companies. Transparency also improves product safety and quality. Traceability technologies have made the concept of transparency a reality (Shafik 2025). Digital technology, specifically blockchain, can enable transparency and ensure sustainable behaviors by identifying superior choices. Transparency can also build trust between countries in relation to the international trade of waste materials, thereby fostering circularity.

3.6 *Improving Accountability*

Accountability is crucial for sustainable development. Circular economy stakeholders need to demonstrate that they only use recirculated materials. Blockchain offers an attractive option because of its inherent design, which meets transparency requirements for tracking the safety of recovered materials. The World Economic Forum noted in 2019 that blockchains have the potential to chart the complex route of metals in the supply chain. When embedded in circular business models, transparency can be leveraged as a motivator—the pride factor—to support market development and collaboration. Recall that members of a permissioned blockchain collaborate and replicate the ledger. Members can examine and verify the behavior of each party, which enforces accountability. As the European Commission emphasized, the digital product passport will contribute to guaranteeing the traceability of recycled content between operators and throughout the supply chain (Zhang et al. 2024). Transparency in the lead recovery ecosystem has the potential to reshape the industry in three important ways.

3.7 Facilitating Collaboration

The circular economy model is based on an inter-enterprise collaboration due to its nature, which requires an operative linkage with value chains from several sectors, resource recovery, reintroduction into the production cycle, and the adoption by consumers. A large number of circular economy initiatives have tried to involve collaboration among the different actors, normally in a local context. Different initiatives and policies regarding the implementation of a circular economy require a sharing of information and plans. This information sharing can be presented as a SWOT analysis with respect to nurture interaction, exchange of knowledge, and, finally, collaboration (Chatterjee et al. 2023). However, access to confidential information is limited by regulations related to business secrets, privacy, or tax data, not always consistent for different associations or countries. Blockchain technology improves collaboration by secure sharing information and transaction sharing in the circular economy. On one hand, the dimension of sharing information, such as best practices for waste management, can be improved using an independent permission blockchain in which information is registered anonymously, so users can get the information and have established trust that the information is correct, but without disclosing confidential data. This is particularly useful when regulatory requirements are not in agreement. On the other hand, blockchain is generating international initiatives (Wazid et al. 2021). Various countries, like the Netherlands, the United States, Denmark, China, and the Russian Federation, have issued white papers that establish blockchain-led strategies in line with their national objectives, as summarized in Table 1.

3.8 Traceability Challenges in Circular Systems

Traceability is a key element of a circular system because it equips all stakeholders with the critical information required to verify that the principles of the circular economy are honored at all scales. However, managing and securing traceability is also one of the pivotal challenges of the circular economy, being the foundation that

Table 1 Blockchain in resource recovery and circular systems

Dimension	Focus area	Blockchain role	Traceability benefit	Circular impact	Reference
Material flows	Plastic recycling	Immutable digital records	Proof of origin and quality	Enhance recycling markets	(Zhang et al. 2024)
Energy systems	Waste-to-energy	Smart metering with blockchain	Transparent energy credits	Incentivizes clean energy recovery	(Chatterjee et al. 2023)
Industrial symbiosis	Byproduct exchange	Distributed ledger for trading	Verified cross-industry transactions	Promotes resource efficiency	(Wazid et al. 2021)
Supply chain	E-waste management	Token-based collection system	Rewards for compliance	Reduces landfill burden	(lakhan et al. 2023)
Construction sector	Demolition waste	Blockchain-enabled BIM	Authentic material reuse tracking	Encourages green construction	(Ahmed and MacCarthy 2023)
Agriculture	Organic waste recovery	Smart contracts for compost sales	Verifies nutrient value	Boosts soil regeneration	(Alkatheiri and Alghamdi 2023)
Mining sector	Rare earth recovery	Blockchain mineral passports	Certification of recovered minerals	Reduces primary mining	(Kumar et al. 2023)
Textile industry	Fabric recycling	QR-coded blockchain tags	Ensures authenticity in reuse	Lowers textile waste	(Jouini and Sethom 2023)
Food industry	Food waste traceability	Blockchain food surplus tracking	Records redistribution	Reduces food insecurity	(Nguyen et al. 2024)
Packaging sector	Circular packaging	Digital twin of packaging materials	Lifecycle traceability	Drives packaging reuse	(Ghayvat et al. 2024)
Water systems	Wastewater reuse	Blockchain-enabled water credit system	Secures transaction data	Promotes water circularity	(Huang and Liu 2024)
Healthcare sector	Medical waste recovery	Smart ledger compliance system	Tracks hazardous disposal	Reduces environmental hazards	(Santhuja and Anbarasu 2023)

(continued)

Table 1 (continued)

Dimension	Focus area	Blockchain role	Traceability benefit	Circular impact	Reference
Transportation	Battery recycling	Blockchain-powered EV battery passport	Verifies second-life batteries	Extends the lifecycle of materials	(Rahmadika et al. 2023)
Finance & policy	Green incentives	Token-based subsidy distribution	Prevents fraud and misuse	Strengthens sustainability finance	(Bag et al. 2024)
Governance	Regulatory compliance	Blockchain-based audit trails	Transparent verification	Improve policy enforcement	(Nasr Esfahani et al. 2024)

enables scaling and implementing a circular model across a wide range of sectors. Traceability in a circular system is required to build public trust and acceptance, offering verifiable information, and ultimately confirming the actual sustainability of products and services (Zhang et al. 2024). A key step that requires traceability information—whether it is with regard to the composition of the different parts, or the location of each material—is the resource recovery and/or product reprocessing stage. This information is currently seldom easily attainable and often is the responsibility of market actors within the circular system, in addition to their main role(s). This presents a number of challenges, particularly in achieving the source-separation of different waste and end-of-use products; optimizing the logistics processes involved; and building and maintaining enabling partnerships with the rest of the stakeholders, especially the municipalities. In addition to ensuring transparent traceability, resource recovery organizations can benefit from blockchain technology because it facilitates the management and control of information and data and makes processes more efficient through the automation of smart contract-driven agreements between parties (Chatterjee et al. 2023).

3.9 Data Management Issues

Preparing resource and waste flows for reuse requires several processes, many often performed by resource recovery companies and facilities (e.g., cutting and sizing, shredding, and cleaning). The inclusion of these actors in the reverse supply chain is an important progress toward securing traceability in circular systems. Many other applications, such as maintenance and development of material passports and databases, can similarly contribute. Nevertheless, although utilization of these data may be limited at the current stage of the circular economy, active and proper maintenance may make them progressively more useful and even critical as the focus reverses from downstream flows (i.e., former waste flows) to resource flows with assessed material properties and guaranteed provenance (lakhan et al. 2023). Related to the big data problem in resource recovery is also the curvature of the supply chain. While the supply chain for newly produced goods is normally relatively simple, the one responsible for the collection, transport, treatment, remanufacturing, and reassembly of resource flows is usually much larger and more complex. When many different resource streams remain mixed through the different steps of upstream operations, available information regarding material quality becomes less reliable. Furthermore, during the utilization phase, information on maintenance, ownership, and displacement is lost. One of the major reasons for the current low rate of reuse of recovered materials, therefore, is the lack of proper information management (Alkatheiri and Alghamdi 2023). In addition to the volume/quantity of data, internal and external quality assurance is a relevant aspect to guarantee the overall quality of data, maintain reliability, and prevent the circulation of misleading information.

3.10 Supply Chain Complexity

Supply chain complexity increasingly represents a major deterrent to resource recovery. Circular supply networks can incorporate different players from various circular economies. Likewise, involved companies can be assigned to different tiers (raw materials suppliers, parts and components suppliers, manufacturing, assembling, distributors, etc.). This level of circularity has numerous benefits, such as reduction of waste, more efficient use of natural resources, and stimulation of new business opportunities. Nevertheless, the increasing number of actors within a circular supply chain represents a major challenge for organizational structures. The wider and more complex the supply chain, the greater the difficulty in managing the production phase, the flow of goods, and the communication process (Jouini and Sethom 2023). Resource recovery paths often consist of a variety of stages, each of which requires the preparation and sharing of documents and information between different actors to guarantee the correct transfer and subsequent recovery. When recovery activities take place at a transnational level, the exchange and sharing of technical information becomes even more important, as several countries with different rules and codes are involved. Several initiatives promoting the introduction of traceability and resource recovery for waste management have been established (Kumar et al. 2023). Nonetheless, satisfaction levels related to these systems continue to be mediocre because of not only the poor informatization of the involved actors, but also the levels of confidence concerning the correctness and completeness of the exchanged information.

3.11 Regulatory Compliance

Achieving regulatory compliance is a key consideration for successful blockchain implementation in resource recovery. Regulations and policies concerning resource recovery vary across regions, making it challenging for circular systems to comply with all requirements. This regulatory mismatch may discourage participation in resource recovery activities and hinder the effective implementation of blockchain technology. Traceability and auditability throughout the resource recovery process enable circular businesses to demonstrate compliance effortlessly, thereby fostering broader adoption of circular principles. The application of blockchain to resource recovery is observed across various sectors, including waste management, reverse logistics, and reasoned hoarding. Within these domains, the regulatory landscape imposes constraints that can impede sustainability efforts. By enhancing transparency and accountability, blockchain-assisted traceability streamlines compliance procedures and supports the overcoming of regulatory barriers (Nguyen et al. 2024). Consequently, regulators may be empowered and incentivized to advocate for the adoption of circular economy practices, amplifying the positive impacts of blockchain integration.

3.12 Implementing Blockchain for Traceability

Resource recovery processes can benefit from the traceability of data related to resources and waste materials. Blockchain provides a shared data infrastructure and secure data management solution that can increase traceability and sustainability in resource recovery. A framework built upon blockchain technology can integrate key elements of traceability in resource recovery, identifying relevant use cases and assessing impacts. Blockchain is a distributed ledger underlying the cryptocurrency Bitcoin that permutes transactional data, enabling cheaper services without requiring an intermediary. The management of transactional data records and smart contracts enables an enterprise with a defendable competitive advantage to exchange goods and services directly with a partner located anywhere in the world, without requiring a trusted third-party organization. Traceability processes for goods and waste materials, supported by the distributed and immutable nature of blockchain data, align with the goals of the circular economy. Connecting different supply and value chains offers a wide range of new business areas, along with economic and environmental benefits. Combining a traceability framework with other technologies improves circular economy solutions across various industries (Ghayvat et al. 2024). The economic, environmental, and social impacts of blockchain in circular economy use cases are also highlighted.

4 Technological Requirements

Resource recovery bases its efficiency on traceability, transparency, and the exchange of accurate, timely information on material flows between actors over the lifecycle of materials and products. Circular activities thrive on transparency because they are also run to increase transparency. Secure storage of information over the lifecycle provides accountability and traceability for circular activities. The protected, accessible shared information platform increases trust among the circular groups and their suppliers. The collaborative interaction, with visibility of each other's operations, optimizes the functioning of activities and therefore leads to social networking growth and triggers more business environment development. It is important to note that circular manufacturing creates its own operational activities and broadens supply-and-demand networks (Huang and Liu 2024). Expansion of the operational matrix tropically enhances circularity, apart from the benefits of social networking and business environment development.

4.1 *Stakeholder Engagement Strategies*

Circular economic systems require multiple parties to cooperate. The current linear system encourages short-term benefits through immediate profits, resulting in an uneven power play among its actors, especially when it comes to primary resource extraction and trade. It disregards the responsibility of taking care of waste, which leads to considerable externalities but is offloaded on communities and the Earth. Achieving circularity requires users to become stewards of the Earth and think of the community in the short, middle, and long term. When these powers are redistributed, thereby shifting the cost burden from future communities to the current ones, an equitable future could be built. Blockchain technology can facilitate such a fundamental power shift due to its inherent properties. However, blockchain systems have their limitations when it comes to working with untrusted parties (Rahmadika et al. 2023). Misaligned incentives, stakeholder disengagement, and delays can result in the system becoming unviable. Thus, stakeholder engagement strategies are needed to help untrusted parties work cohesively on the system.

Stakeholder engagement strategies boil down to three questions: "Who needs to work together?" Scheduling conflict-based delays that occur because some important counterparties are unable to commit their approvals in time can be repaid with an equitable distribution of benefits within the network. Economics-based waning interest, where counterparties discover that committing full effort into the network does not generate enough benefits in the medium term, can be repaid with other positive externalities that the network can generate. Finally, utility-based misalignment, which happens when counterparties find certain steps of the workflow to be against their mission or values, can be repaid by holding the greater cause behind the network in the forefront throughout (Bag et al. 2024). Ultimately, parties need to build trust towards one another to work on an untrusted network and shape an equitable future together.

4.2 *Impact Assessment of Blockchain on Resource Recovery*

The first-order economic outcome of integrating blockchain technology in resource recovery is securing traceability. Stellar performance in this area results in a second-order contribution to all other economic objectives: transparency, accountability, and empowerment of stakeholders. Such empowerment enables trade and collaboration beyond the traditional supply chain boundaries, exhibiting true functionality as a circular enabler. In the absence of a robust traceability system, supply chain optimization cannot produce a truly circular outcome (Nasr Esfahani et al. 2024). Empowerment of stakeholders is not only related to transactional issues but also to envisioning the system beyond the natural scope of vertical organizations. A comprehensive assessment reveals a circular economy synergy enabled by blockchain: first, traceability contributes to a closed-loop flow of recovered resources; second, it allows accountability to the community and environment throughout the product life cycle; and third, it enables transparency of the circular economy scheme to support the broader recovery of other materials. A broader look at the comprehensive benefits of resource recovery further highlights the crucial nature of ensuring accountability and transparency within the life cycle (Zhang et al. 2024). Ultimately, the positive impacts achieved by adopting blockchain in resource recovery provide robust support for the implementation of a circular system.

4.3 Economic Implications

A careful analysis of the economic implications is a key step in a blockchain implementation for traceability support in circular systems (see also the discussion of the traceability challenges). The applications of blockchain in the field of economic activities related to the environment, waste, and resources, designated here as a smart green economy, are considered in several recent studies. Traceability features provide added value to economic activities related to resources and nature, such as green energy, circular materials, sustainable agriculture and food, clean water, or sustainable urbanization. Saving in transaction costs and the creation of green jobs, which mostly occur during the construction and implementation phases, can be mentioned in this respect (Chatterjee et al. 2023). In addition, blockchain potentially enables the involvement of companies that have never been active in the field of environment, resources, or sustainable development before. Blockchain technology applied to circular-system-related activities reduces information asymmetry and increases the reliability of information. Although in the initial phase, it seems to be most suitable for service and marketing activities, in the future, production activities related to the circular economy will certainly be operated in this way. With the spread of the tourism service of chain traceability, understanding consumer demands and offering targeted sales plans can bring increased profits (Wazid et al. 2021). Of course, empirical evidence is also needed for measuring the impact of using blockchain on the various economic activities of the circular economy.

4.4 Environmental Outcomes

The environmental benefits of blockchain-based platforms operating with physical resource recovery processes depend on the total emission reduction, energy-saving performance, and potential environmental risks caused by blockchain. The environmental impact assessment of recovering resources for a PET bottle deposit-refund system, supported by blockchain, was evaluated. Empirical LCA results indicate that the application of blockchain for deposit-refund systems virtually reduces the GHG emissions embodied in the material recovery stage. A considerable number of emissions is saved because it is possible to create a reliable tracking database that facilitates more resource-efficient and environmentally friendly solutions, such as human reminders and certified shipments (Ahmed and MacCarthy 2023). More concretely, the decentralized control enforced by the blockchain makes the transportation of empty bottles for recycling convenient and legal (like, with a proper transaction vehicle ID, trip route, and real cost). Battery energy consumption has been recognized as a crucial barrier to the large-scale adoption of blockchain-based platforms. Electricity demand arising from the decentralized platform can greatly influence environmental performance, and the battery demand for blockchain-powered applications is significantly affected by rewards information, transaction information, and task volume. Furthermore, it has been found that the negative effect of blockchain energy consumption dominates the environmental benefits of the recycling process when the virtual shared battery economy exchange on the blockchain is below a certain threshold, pointing to the fact that the development of task volume and reward policies deserves further attention (Jouini and Sethom 2023).

4.5 Social Benefits

Resource recovery has significant social implications in addition to economic advantages. As a job-creating industrial process, it not only reduces landfill requirements but also lowers the demand for resource mining in natural environments, which is linked with high accident risks. Resource recovery supplies raw materials for different production processes in various industries. Moreover, reducing landfilling waste through resource recovery decreases the formation of gas emissions and acidic leachate, which can seep into water bodies and produce foul odors, thereby improving the quality of life near landfill sites. The social benefits of successful implementation are waste-to-resource conversion, as well as improvements in society and the environment (Zhang et al. 2024). Blockchain technology enables transparency and accountability across all aspects of supply chain management. It can be utilized to underpin collaboration among stakeholders by providing transparent transaction processes and a robust environment for contractual agreements. Technology employs cryptography alongside distributed consensus to ensure that data written to the blockchain cannot be manipulated or destroyed. Its inherent features—decentralization, immutability, and smart contract capabilities—make it suitable for tracking the as-generated, as-collected, and as-processed stages in waste-resource value chains. Several real-world projects have applied blockchain technology to support resource recovery processes within the circular economy (Chatterjee et al. 2023). Economic, environmental, and social indicators assess the impact of technology's implementation for traceability in circular systems.

5 Future Trends in Blockchain and Circular Economy

Emerging technologies and trends appear poised to accelerate blockchain solutions for enhanced traceability in circular systems. The Internet of Things (IoT) will provide large volumes of real-time information needed for in-depth supply chains with numerous nodes and inventories. Artificial intelligence will allow blockchain applications to monitor manufacturing production, inventory, sales, and supply chains to autonomously forecast and minimize supply–demand imbalances. 5G communications and communication satellites will enable blockchain access in remote locations and from a variety of devices to include drones, robots, autonomous vehicles, and surveillance satellites (Wazid et al. 2021). Besides technology, future progress will require international cooperation and coordination of policies and implementation strategies. Furthermore, adjustments in cross-border intellectual property rights and antitrust are needed to maximize benefits for users and society. In this respect, it is particularly important to observe how the differing views on privacy protection are handled at the international level. Obtaining benefits from convergences while preserving respect for human rights associated with the development of blockchain requires strong international cooperation in policymaking, standardization, and technology and application development (Ahmed and MacCarthy 2023).

5.1 Emerging Technologies

The rapid evolution of blockchain capabilities continues to define current technological frontiers. The properties of Distributed Ledger Technologies (DLTs), notably decentralization, immutability, and transparency, position them at the heart of initiatives aimed at empowering participants involved in Industry 4.0, data management, and international transactions through traceability, accountability, logging, and accessibility. Recent DLT advancements expand the scope of blockchain applications in circular economy analysis, addressing metasearch, complex asset management, and the plethora of Environmental, Social, and Corporate Governance (ESG) principles (Jouini and Sethom 2023; Nguyen et al. 2024). Despite these advances, dependencies remain on both centralized resources and insecure parameters. The standards for defining smart contract semantics with higher abstraction levels are yet to be released. · The trustworthiness of oracles—essential for feeds beyond cyclic data management—is not guaranteed. · The conflict between scalability and security suggests potential improvements through sharding or distributing the blockchain across multiple organizations. · Miner centralization remains an inherent inefficiency despite diversification efforts. Furthermore, the variety of consensus algorithms introduces complexity in the choice and implementation of suitable frameworks for circular economy assessment (Ghayvat et al. 2024).

5.2 Policy Developments

Several recent developments within the regulatory landscape are expected to support the implementation of Circular Economy models, i.e., related to Extended Producers' responsibility, Product passports, and Carbon Border Adjustment Mechanisms. The Extended Producers' Responsibility Framework Operating Model provides guidance to help establish a level playing field for producer responsibility organizations (PROs) within the framework of the Circular Economy Action Plan of the European Commission. It provides guidance on key operational aspects of PROs, such as their governance, (joint) liability, performance standards, and transparency. A consistent framework on these matters contributes to heightened quality and transparency and hence increases trust that waste targets will be met (Huang and Liu 2024). This, in turn, contributes to a level playing field between PROs by clarifying the conditions under which garbage collection, treatment, and recycling activities should be organized.

Product passports can be supplemented by location tracking to enable manufacturers or distributors to trace the movement of materials and products throughout the supply chain, even after sale to the end user. Tracking different attributes of materials as they move along and between supply chains will increasingly require the coupling of digital tracking technologies with blockchain networks—providing an immutable, shared, and distributed ledger for an auditable trail of materials. Public and private sector organizations need to plan carefully for the combination of a product passport, location tracking, digital product passports, and digital supply chains so that technologies and standards are designed jointly to support the long-term circular economy vision. carbon border adjustment mechanisms implicitly de-risk investments in a circular economy by systematically pricing in the pollution externalities that are created today by linear supply chains (Rahmadika et al. 2023). Carbon border taxes work towards levelling the playing field for producers, preventing so-called carbon leakage and reducing global emissions.

6 Global Collaboration Efforts

The integration of blockchain technology into circular economy initiatives depends strongly on policies, regulations, and public funding. Presently, regulatory and governmental restrictions hinder the full adoption of blockchain in resource recovery, even if the economic benefits appear compelling. Early development of new regulations and policies inspired by recent trends can foster blockchain adoption. In this context, national and international cooperative research projects or programs have a catalytic effect. Such cooperation also facilitates the sharing of knowledge, experiences, and lessons learned between countries that are at different stages of circular economic implementation (Bag et al. 2024). The European Union (EU), for example, launched Horizon 2020, its research and innovation framework program, through the

Circular Economy in Europe, Accelerating Transition (CircE) project. This initiative supports the implementation of the circular economy in Europe by promoting collaboration and the exchange of information between European industry and other partners. Countries support cooperation between firms in different sectors to develop innovative products and services within the circular economy (Chatterjee et al. 2023; lakhan et al. 2023). Ongoing initiatives in the United States, Australia, and New Zealand demonstrate the extended commitment of countries to investing research resources in this field.

7 Conclusion

Blockchain technology transforms how data is recorded and shared in circular systems. It establishes a coding template to ensure data uniformity and creates secure communication channels among participants. Technology's intrinsic traceability helps link companies with social and environmental objectives, fostering controlled and environmentally responsible material recovery. Furthermore, blockchain enables the integration of stakeholders across different segments and geographic locations, supporting supply chain transparency and enhancing consumer awareness of circular products. Technology suits activities entailing material extraction, transformation, and consumer disposal, essential steps for transitioning towards a resource management economy. Despite its potential, blockchain's application in resource recovery remains limited. Current implementations rely on smart contracts, consensus mechanisms, non-fungible tokens, and decentralized applications. Underpinned by four principles, transparency, accountability, collaboration, and cost savings, blockchain in resource recovery improves costs, revenues, transparency, trust, and collaboration among participants. It also enables functionalities such as transaction logging/forwarding, transaction management, sharing of information and documentation, data management, and tokenization.

References

Ahmed WAH, MacCarthy BL (2023) Blockchain-enabled supply chain traceability—How wide? How deep? Int J Prod Econ 263. https://doi.org/10.1016/j.ijpe.2023.108963

Alkatheiri MS, Alghamdi AS (2023) Blockchain-assisted cybersecurity for the internet of medical things in the healthcare industry. Electronics (Switzerland) 12(8). https://doi.org/10.3390/electronics12081801

Bag S, Rahman MS, Routray S, Shrivastav SK, Agrawal S (2024) Exploring the potential of blockchain-enabled smart contracts for achieving net-zero emissions: an empirical study. Bus Strat Environ 33(5). https://doi.org/10.1002/bse.3691

Chatterjee K, Singh A, Neha (2023) A blockchain-enabled security framework for smart agriculture. Comput Electr Eng 106. https://doi.org/10.1016/j.compeleceng.2023.108594

Ghayvat H, Zuhair M, Shukla N, Kumar N (2024) Healthcare-CT: solid PoD and blockchain-enabled cyber twin approach for healthcare 5.0 ecosystems. IEEE Internet Things J 11(4). https://doi.org/10.1109/JIOT.2023.3312448

Huang C, Liu S (2024) Securing the future of industrial operations: a blockchain-enhanced trust mechanism for digital twins in the industrial Internet of Things. Int J Comput Appl 46(5). https://doi.org/10.1080/1206212X.2024.2318821

Jouini O, Sethom K (2023) AgriBIoT: a blockchain-based IoT architecture for crop insurance. Lect Notes Netw Syst 655 LNNS. https://doi.org/10.1007/978-3-031-28694-0_32

Kumar P, Kumar R, Gupta GP, Tripathi R, Jolfaei A, Najmul Islam AKM (2023) A blockchain-orchestrated deep learning approach for secure data transmission in IoT-enabled healthcare system. J Parallel Distrib Comput 172. https://doi.org/10.1016/j.jpdc.2022.10.002

lakhan A, Mohammed MA, Ibrahim DA, Abdulkareem KH (2023) Bio-inspired robotics enabled schemes in blockchain-fog-cloud assisted IoMT environment. J King Saud Univ-Comput Inf Sci 35(1). https://doi.org/10.1016/j.jksuci.2021.11.009

Nasr Esfahani M, Ghahfarokhi BS, Etemadi Borujeni S (2024) Blockchain-based end-to-end privacy-preserving scheme for IoT-based healthcare systems. J Supercomput 80(2). https://doi.org/10.1007/s11227-023-05522-7

Nguyen TV, Le LS, Shah SA, Hameed S, Draheim D (2024) PenChain: a blockchain-based platform for penalty-aware service provisioning. IEEE Access 12. https://doi.org/10.1109/ACCESS.2023.3344038

Rahmadika S, Astillo PV, Choudhary G, Duguma DG, Sharma V, You I (2023) Blockchain-based privacy preservation scheme for misbehavior detection in lightweight IoMT devices. IEEE J Biomed Health Inform 27(2). https://doi.org/10.1109/JBHI.2022.3187037

Santhuja, P., & Anbarasu, V. (2023). Blockchain-Enabled IoT Solution for e-Waste Management and Environmental Sustainability through Tracking and Tracing. *International Journal of Engineering Trends and Technology*, *71*(12). https://doi.org/10.14445/22315381/IJETT-V71I12P216

Shafik W (2024) Blockchain-based internet of things (B-IoT): challenges, solutions, opportunities, open research questions, and future trends. In: Blockchain-based Internet of Things. Chapman and Hall/CRC, pp 35–58. https://doi.org/10.1201/9781003407096-3

Shafik W (2025) Leveraging emerging technologies for smart and secure software development: Blockchain, IoT, and Beyond. Mod Insights Smart Secur Softw Dev 41–76

Wazid M, Das AK, Park Y (2021) Blockchain-envisioned secure authentication approach in AIoT: applications, challenges, and future research. In: Wireless communications and mobile computing, vol 2021. https://doi.org/10.1155/2021/3866006

Zhang Z, Zeng K, Yi Y (2024) Blockchain-empowered secure aerial edge computing for AIoT devices. IEEE Internet Things J 11(1). https://doi.org/10.1109/JIOT.2023.3294222

Data-Driven Change: The Role of Focal Firms in Advancing Sustainability Standards in Global Supply Chains

Kristina Encinas Bartos, Julia Schwarzkopf, and Martin Müller

Abstract This study leverages data-driven analysis to explore the institutional pressure from focal firms in embedding sustainability standards within global supplier networks. Utilizing a comprehensive dataset of over 5,000 suppliers from the automotive industry, we examine the temporal evolution of sustainability standard implementation as mandated by focal firms. Our analysis highlights significant differences across supplier groups, shedding light on how data-driven insights can inform the adoption of more sustainable practices. The findings underscore the role of focal firms in driving environmental resilience and offering a technological roadmap for addressing global challenges like climate change and resource management. We conclude with strategic recommendations for leveraging technology and data analytics to enhance sustainability across supply chains, presenting valuable insights for policymakers and multinational enterprises aiming to create a greener future.

Keywords Sustainability standards · Supply chains · Sustainable supply chain management · Panel data

1 Introduction

In recent decades, there has been an increasing focus on sustainable supply chain management (SSCM) in academic research (Ahi and Searcy 2015; Schaltegger and Burritt 2014). In light of the mounting challenges afflicting the Global South, including climate change, migration, and food insecurity, multinational enterprises (MNEs) are poised to assume a pivotal role in the pursuit of effective solutions. Their global reach and market power give them the chance to improve working conditions

K. Encinas Bartos · M. Müller
Institute of Sustainable Corporate Management, Ulm University, Ulm, Germany
e-mail: martin.mueller@uni-ulm.de

J. Schwarzkopf (✉)
HTW Berlin Business School, Berlin, Germany
e-mail: Julia.Schwarzkopf@htw-berlin.de

R. Singh et al. (eds.), *Data-Driven Sustainability: Harnessing Technology for A Greener Future and Environmental Resilience*, World Sustainability Series,
https://doi.org/10.1007/978-3-032-13412-7_25

and environmental management in their supply chains (Wettstein et al. 2019). In management literature, these firms are called focal firms, defined as companies that "[...] govern the supply chain, provide the direct contact to the customer, and design the product or service offered" (Seuring and Müller 2008, p. 1699). Research in the management field has focused on focal firms' SSCM, developing conceptual frameworks, assessment and management strategies as well as enablers for sustainable supply chains and the role of transparency across multi-tier supply chains, among other aspects (e.g., Beske-Janssen et al. 2015; Fraser et al. 2020b; Gualandris and Kalchschmidt 2016; Schöggl et al. 2016; Seuring et al. 2022; Seuring and Müller 2008). A relevant body of literature uses institutional theory to explain developments, practices and strategies in the field at hand (e.g., Glover et al. 2014; Sauer and Seuring 2018; Zeng et al. 2017). In this context, institutional pressure from focal firms is said to be an incentive for sustainability in supply chains. Literature mentions sustainability requirements set by the focal firm towards its suppliers to drive this development (Rao 2005; Roberts 2003; Sarkis 2001). Such criteria are often linked to internationally recognized sustainability standards that have the purpose to improve supply chain sustainability (Beske et al. 2008; Mueller et al. 2009; Sarkis 2001) and thus contribute to solving the above-mentioned global challenges.

In the following article, we raise the question to what extent the described institutional pressure from focal firms leads to the institutionalization of sustainability standards in global supplier networks and how this development is reflected in company data over time. In particular, we analyze the following research questions:

- How does the institutionalization of global sustainability standards evolve over time when a focal firm sets them as a mandatory requirement for its direct suppliers?
- What differences are there between supplier groups with different characteristics?

Our research aims to understand how coercive institutional pressure drives the use of sustainability standards in global supplier networks. Section Two discusses the term "sustainability standard" in the context of global supply chains. It also covers institutionalization and institutional theory in the context of SSCM, followed by a literature review. Section Three presents our data analysis procedure and results. This section contains a quantitative analysis of a data set of over 5,000 suppliers from a focal firm in the automotive industry. The data analysis offers insights into the implementation of sustainability standards at suppliers globally. Section Four discusses the results. The article ends with conclusions and suggestions for future research in Section Five.

2 Background

2.1 Sustainability Standards in the Supply Chain

The term "sustainability standard" appears in various contexts in management literature. It is used in connection with management certification schemes, such as ISO14001, to reduce a company's environmental impact. Meanwhile, standards for reporting and transparency are also labelled environmental, social, or sustainability standards (Larrinaga et al. 2002; Momin and Parker, 2013). Examples include the standards set by the Global Reporting Initiative (GRI 2023; Halkos and Nomikos 2021; Vigneau et al. 2015) or legislative requirements, such as the Corporate Sustainability Reporting Directive (DIRECTIVE (EU) 2022/2464, 2022).

The term is often used in the context of voluntary certifications such as Fairtrade or standards developed by industry collectives. Sustainability standards help focal firms evaluate their suppliers' and/or supply chain's sustainability performance. According to Darnall et al. (2023), sustainability standards "[…] internationally standardize firms' internal sustainability procedures, which generally promote greater environmental (and/or social) responsibility through risk management, assessment, performance expectations, and third-party audits." The term can thus be classified as a behavioural standard (performance standard) following the differentiation made by McIntosh et al. (2003). Performance standards "[…] define clear instructions for actions in the internal company processes, defining precisely what a company is allowed and forbidden to do." This definition can be applied to suppliers as well, who obtain instructions on what practices they should implement.

Standards are rules suppliers must comply with to help companies evaluate their sustainability performance. While concepts to evaluate the whole supply chain exist, they are difficult to operationalize. Standards for evaluating direct suppliers can be implemented. Examples of these standards include a supplier code of conduct, life-cycle assessments, and certification programs. More industry standards have been developed in recent decades to set a shared understanding of sustainable business principles and clarify expectations within supply chains. However, these standards address supplier management for risk and performance, not supply chain management for sustainable products. Examples include the chemical industry's TfS (Together for Sustainability) Assessment and the Automotive Industry Guiding Principles (Drive Sustainability). These standards are integrated into the SSCM practices of the focal companies and suppliers must follow them through contractual obligations.

From the use of the term *sustainability standard* in literature and practice as well as the above-mentioned definition of performance standards by Mueller et al. (2009), we derive the following broad definition of the term for the remainder of this article.

Def. 1: A sustainability standard is a set of rules with the purpose of minimizing negative environmental or social impacts.

When firms (or suppliers) hence follow a sustainability standard, it is supposed to bring an improvement in terms of environmental and/or social sustainability. We furthermore assume that the degree of compliance with a sustainability standard correlate positively with a company's (or supplier's) sustainability performance. I.e., in our context, we assume that suppliers' sustainability performance is better if they comply with a sustainability standard to a higher degree.

To analyze how focal firms integrate sustainability standards into supplier evaluation processes, we must first understand how these standards are used in SSCM. In this context, the degree to which a supplier complies with a standard indicates the supplier's sustainability performance. For example, in the automotive industry, a standardized tool is largely used for such an assessment by focal companies (Fraser et al. 2020a). This standard—also known as the *Self-Assessment Questionnaire on CSR / Sustainability for Automotive Sector Suppliers (SAQ)*—provides information on a suppliers' performance in terms of social sustainability, environmental performance, business ethics and supplier management. The SAQ asks for site-specific policies and certificates validated by an external party, aiming to control for compliance with a set of requirements. The degree to which a supplier complies with the requirements is indicated by a score between 0 and 100.

The rest of this article will assume the SAQ score reflects the degree of institutionalization of sustainability standards at a supplier level. This assumption is based on the SAQ's broad use by several original equipment manufacturers (OEMs) and its creation by the industry initiative *Drive Sustainability,* comprising global automotive manufacturers (Drive Sustainability 2019; Encinas Bartos et al. 2023).

Suppliers can improve their performance by disclosing additional policies, procedures, or certificates, as the level of SAQ compliance may change. Figure 1 shows the process: (1) The focal firm requests information from the supplier as needed. (2) The supplier provides all relevant documents and information. (3) A third party validates the answers. (4) The supplier is given a score and feedback on the answers. (5) The focal firm receives the score and feedback. (6) The supplier implements feedback and improves its answers (e.g., developing new policies or implementing an environmental management system). If more information is provided to comply with the industry standard, the process re-starts from step (2).

In our further analysis, we focus on one automotive MNE with a global supply base. The MNE uses the SAQ to assess its suppliers' sustainability performance on a regular basis and as a prerequisite for going into a business relation. We hence use an already existing and institutionalized assessment standard to evaluate the degree of institutionalization.

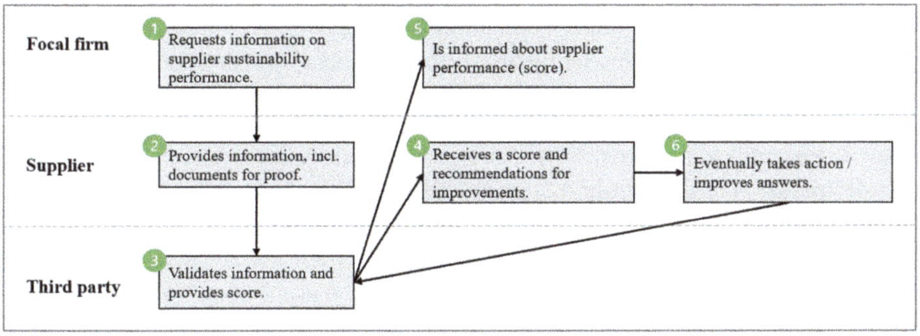

Fig. 1 SAQ process. *Source* Authors own work

2.2 *The Concept of Institutionalization and Institutional Theory in SSCM*

Institutionalization in the context of sustainability standards can be described as *"[…] a process turning new values or practices into broadly accepted rules, norms and standards"* (Hofstetter and Grimm 2019, p. 529). This definition and the previous considerations show the following definition for the use of sustainability standards in a supply chain.

> Def. 2: Sustainability standards are institutionalized when the actors of the supply chain adopt internationally recognized frameworks and rules on sustainability by implementing policies and procedures to comply with social, environmental and business ethics standards.

Our focus is on direct suppliers and their focal firm, which is the buying company. Literature describes these actors as the organizational or institutional field (DiMaggio and Powell 1983; Sauer and Seuring 2018). Policy implementation is driven by different factors within this field. For example, in Hamprecht and Schwarzkopf's (2014) analysis of why MNE subsidiaries implement environmental practices, it is assumed that subsidiaries face institutional pressure from two sides. This is also described as institutional duality, which the authors extend by a third factor, leading to institutional trinity. They argue that the subsidiary's norms and values influence the extent to which environmental practices are implemented and hence become institutionalized. This concept can be translated to a supply chain context, where suppliers are influenced by three factors: (1) the institutional context of the country they are located in, (2) the expectations and institutional pressure from the buying company, and (3) their own values and norms. These factors form the institutional pressure that determines the extent to which suppliers implement sustainability practices.

This inference is in line with the management literature, where institutional pressure is found to be a major driver for SSCM (Seuring and Müller 2008). Taking a closer look at institutional theory, institutional pressure has different origins (Cavusoglu et al. 2015; DiMaggio and Powell 2000; Hirsch 1975; Zeng et al. 2017).

Institutional pressure may come from laws and regulations, values and norms or expectations and culture, which influence firm activities. According to DiMaggio and Powell (1983), institutional pressure leads to organizations becoming more similar to each other. The authors distinguish between coercive, mimetic and normative isomorphism. Isomorphism is defined as *"[...] the convergence of firm characteristics within an institutional field"* (Sauer and Seuring 2018, p. 563). Coercive isomorphism has its origin in pressure from *"[...] those in power"* (Sarkis et al. 2011, p. 7). This can for instance be governments that promote new policies, leading to changes in firm activity. Considering global value chains, *those in power* may however also be buying companies (i.e., focal firms) that set criteria for their suppliers. Mimetic institutional isomorphism on the other hand occurs when actors mime the actions of other organizations such as successful activities from competitors. And lastly, normative institutional pressure leads to organizations conforming in order to legitimize their activities. Environmental awareness in a society and private customers' ethical principles, for instance, are expected to lead to a higher normative pressure on companies for more sustainable business activities (Ball and Craig 2010; Sarkis et al. 2011).

In our further analysis, we argue that the institutionalization of sustainability standards in global supply chains is a result of institutional pressure. In particular, we focus on coercive pressure from focal firms because *"[w]hen the focal company is pressured, it usually passes this pressure on to suppliers"* (Seuring and Müller 2008, p. 1703). We argue that pressure through mandatory sustainability requirements from focal firms lead to the implementation of sustainability standards at direct suppliers at a global scale. Over time, we expect this development to be reflected in respective practices at supplier sites and thus lead to the institutionalization of sustainability standards at suppliers according to the definition given in the beginning of this section that new practices and values become the broadly accepted norm.

2.3 *Institutionalization of Sustainability Standards in Global Supply Chains*

We identified the following five main factors that are mentioned to influence the institutionalization of sustainability standards in global supply chains.

(1) **Focal firms' requirements**, inducing coercive pressure on suppliers (Caniëls et al. 2013; Ehrgott et al. 2011; Gualandris and Kalchschmidt 2014; Kauppi and Hannibal 2017; Mani and Gunasekaran 2018).
(2) **National institutional conditions**, inducing regulatory pressure (Carbone and Moatti 2011; Kauppi and Hannibal 2017; Kuruvilla and Li 2021; Mani and Gunasekaran 2018; Manning et al. 2012).
(3) **Requirements from external stakeholders** such as NGOs, social organizations, media etc., inducing normative pressure (Mani and Gunasekaran 2018; Sancha et al. 2015).

(4) **Voluntary sustainability standards and global initiatives**, offering frameworks for a common understanding of good business practice (Derkx and Glasbergen 2014; Dietz and Grabs 2022; Grabs 2020; Reinecke et al. 2012).
(5) **Firm culture and management**, being induced by institutional pressure and at the same time determine the way firms (both focal firms and suppliers) react to external pressure (Caniëls et al. 2013; Carbone and Moatti, 2011; Grimm et al. 2011, 2022; Mani and Gunasekaran, 2018).

This text focuses on the first factor, but other factors were considered as well, particularly those associated with pressure from focal firms. In analyzing social sustainability standards, Mani and Gunasekaran (2018) analyzed the role of multiple factors driving their adoption in Portugal and India. They found social pressure from customers to significantly and positively influence the adoption of social sustainability practices in supply chains. This pressure can be defined as pressure from buying companies for socially sustainable business practices at the supplier. Mani and Gunasekaran (2018) also found a link between social sustainability adoption, regulation compliance, and external stakeholder pressure. Implementation of social sustainability practices improves suppliers' performance in terms of social aspects like human rights and safety compliance. They also found a connection between organizations' sustainability culture and the adoption of sustainability practices that go beyond regulatory pressures. This is in line with the role of norms and values of subsidiaries in determining the degree of institutionalization of environmental practices at a firm level (Hamprecht and Schwarzkopf 2014).

Kuruvilla and Li (2021) focused on the implementation of freedom of association and collective bargaining in global supply chains. They pointed out that the implementation of these two is highly dependent on the institutional environment and they determined ways of institutionalizing them as a major future research field.

On social sustainability, Kelling et al. (2021) analyzed social sustainability practices in South Africa's mining sector and found that institutional uncertainty is a major factor hindering the institutionalization of social sustainability standards.

Kauppi and Hannibal (2017) examined the role of assessment initiatives in promoting social sustainability standards in global supply chains. They found that coercive pressure due to consumer and compliance requirements has the largest effect. Sancha et al. (2015) found that mimetic pressure has a positive effect on the adoption of sustainable supplier development practices. They did not find a positive effect from coercive or normative pressures. Ehrgott et al. (2011) also tested different types of pressure and their impact on focal firms' social supplier selection. They found that regulatory pressure had no positive or significant effect. However, the mid-level management of buying companies was found to play a relevant role. A cultural factor seemed to drive the development towards more sustainable supply chain management (Mani and Gunasekaran 2018).

Grimm et al. (2011) examined how a focal company can succeed in institutionalizing sustainability standards. The authors identified the capabilities needed for this, such as interfirm dialogue and risk management. Grimm et al. (2022) expanded the study to buying companies' roles in institutionalizing sustainability in multi-tier supply chains, emphasizing the same capabilities.

Also Caniëls et al. (2013) analyzed factors leading suppliers to apply more sustainable business practices. They found supplier readiness (i.e., a supplier's capability to internally implement sustainable processes and change) as well as customer (i.e., focal firms') requirements to have a significant influence on supplier participation. For larger suppliers, the investment from customers as well as a cooperative relationship between the buying company and supplier were found to function as additional drivers to implement sustainable practices.

The literature review shows that different factors influence the institutionalization of sustainability standards and the role of focal firms in their supply chains. Most studies find that focal firms play a relevant role in this process. Particularly, their sustainability requirements as customers are found to be a driving factor (e.g., Caniëls et al. 2013; Mani and Gunasekaran 2018). However, the review also shows that there is a scarcity of quantitative analysis with supplier data, especially large-scale panel data analysis. This article aims to contribute to this field by analyzing changes over time. This is particularly important when establishing a new rule (i.e., sustainability standards) as accepted standards, according to the definition of institutionalization. Changes over time provide an indication of the implementation of sustainability standards (Delmas 2002; Grabs 2020). As the institutionalization of standards requires time, our data analysis provides unique insights into these developments.

3 Quantitative Analyses

3.1 Data Set

The data used in our panel analysis comes from the above-mentioned automotive MNE. It includes 5,543 supplier sites that directly deliver parts or services to the focal firm, the MNE. The data set provides information on a supplier site's location (country), headcount, and type of service (manufacturer or service provider). This descriptive information remains stable over time. The supplier sustainability performance (SSP) (proxied by the result a supplier obtains (site-specific) in the SAQ described in Sect. 2.1) changes over time. The data set includes the time frame from January 2019 to January 2021 and is divided into five points of time (data measurement every six months). The data set provides information on the SAQ score the suppliers obtained at that given date. We refer to these points in time as periods in the description of our results below. The first point in time is January 2019, the second point is June 2019, the third, fourth, and fifth points in time are January 2020, June 2020, and January 2021, respectively.

Table 1 shows summary statistics for our sample, including the number of suppliers for each supplier group, the mean performance score, the maximum and minimum score, and the standard deviation. The sample includes characteristics

Table 1 Summary statistics

Supplier group	Count	Mean	Max	Min	Sd
Total	5543	70.9	100	1	19.0
Manufacturing	4536	73.2	100	1	17.4
Services	1007	60.5	97	1	22.0
SE	847	52.8	98	1	23.8
ME	1765	67.6	98	1	19.0
LE	2931	78.1	100	2	12.2
Africa	65	75.0	98	14	15.2
Americas	691	73.7	100	6	15.3
Asia	576	77.8	98	1	12.4
Europe	4210	69.4	99	1	20.0
Oceania	1	83	83	83	0

such as business category, site headcount, and region. The headcount is grouped into small enterprises (SE) with less than 50 employees, medium-sized enterprises (ME) with 50–249 employees and large enterprises (LE) with 250 or more employees. Most suppliers deliver manufactured parts. More than half of the suppliers are large enterprises, and around one quarter are located outside of Europe. The average sustainability performance score differs as well. Manufacturing and large suppliers perform better, while suppliers in Europe perform worse than those outside of Europe. Asia-based suppliers reach the highest average score, followed by those in Africa and the Americas. The summary statistics provide an overview of which types of suppliers have implemented the requested SAQ policies and internationally recognized sustainability standards.

3.2 *Analysis Over Time*

The analysis over time is insightful because the focal firm required suppliers to implement and communicate mandatory assessment criteria in July 2019. Therefore, analyzing differences before and after the assessment criteria became mandatory may show how much focal firm requirements influence suppliers. A fixed-effects approach would be suitable, but the criteria became mandatory for different groups of suppliers at different times. Because there was no fixed point in time for a renewal of contracts or new contracts with suppliers, the exact point in time when the criteria became mandatory for single suppliers cannot be derived from the data set. This means it may have been different for all sites in the sample.

Therefore, we needed to take a less specific approach and looked into the development over time in general. We divided the data set into the different groups shown in the summary statistics, because certain characteristics were found to have a statistically significant effect on supplier sustainability performance. Results are shown in Figs. 2, 3, 4 and 5 for the different supplier groups in terms of business category, headcount and continent.

To better understand regional differences, we examined compliance with the sustainability standard in the ten countries with the most suppliers. See Fig. 6 for details.

Figure 2 shows the overall compliance trend over time. The average score increased until June 2020, and then the curve flattened. This means that the overall improvement of scores happened at a faster rate until June 2020 and that the improvement slowed down in the remaining periods of the sample.

In Fig. 3, the development is shown for the two groups of manufacturing suppliers as well as service providers. Reflecting the difference in average score presented in the summary statistics, the group of service providers starts at a much lower average score than manufacturing suppliers. Both groups improve their performance over

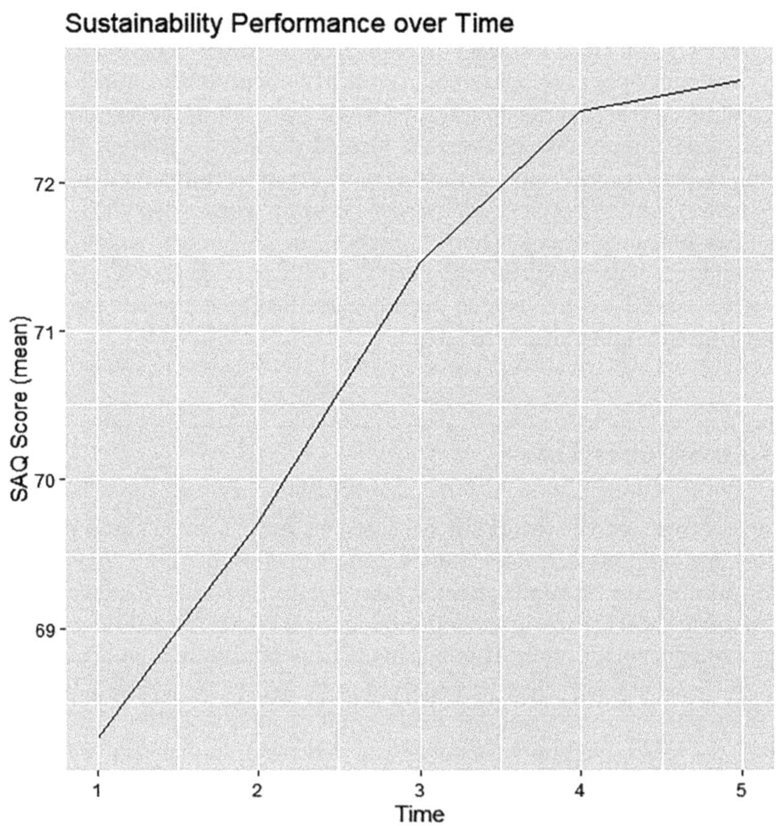

Fig. 2 Sustainability performance over time. *Source* Authors own work

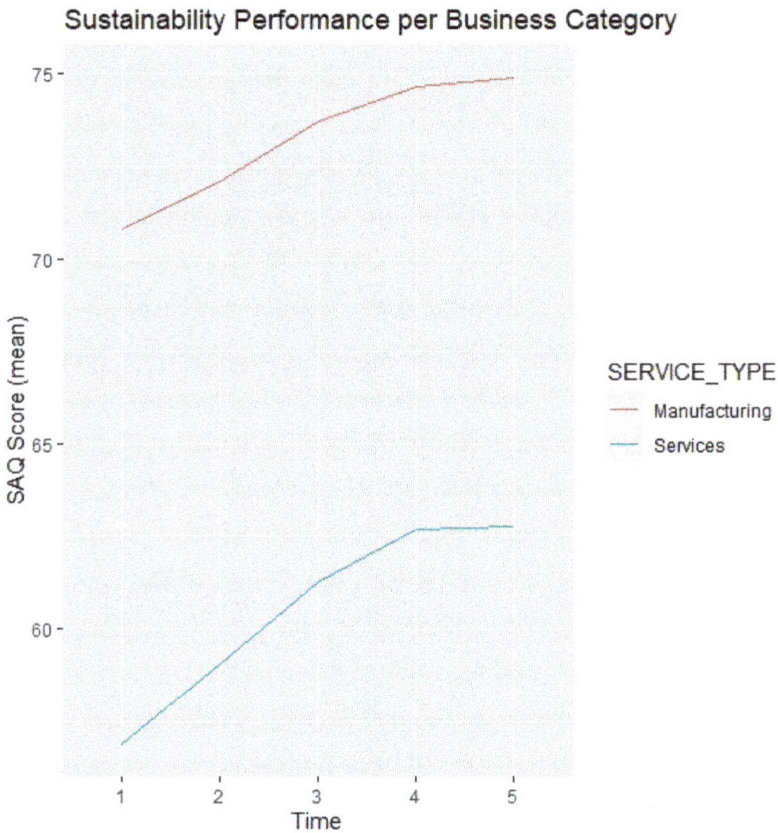

Fig. 3 Sustainability performance per business category. *Source* Authors own work

time. The curve for the service providers is steeper until period four, indicating a larger increase relative to the group of manufacturers. After period four—just like in the figure for the whole sample—the curve flattens. The difference in average score between the two groups, however, remains large with more than twelve percentage points in difference after the last period.

In Fig. 4 the performance change is shown for the different groups divided by headcount. Here as well, the starting points of average scores reflect the differences shown in the summary statistics with the group of large enterprises showing the highest score, i.e., the highest degree of compliance with the sustainability standard. The performance of the large and medium-sized suppliers seems to converge slightly towards the last periods, however, while the difference was more than ten percentage points in the first period, there is still a difference of around eight percentage points in the two groups' average performance in the last period. The difference to the group of small suppliers remains similar over time. The development in general shows the same trend as the other supplier groups, i.e., over time, suppliers in all groups

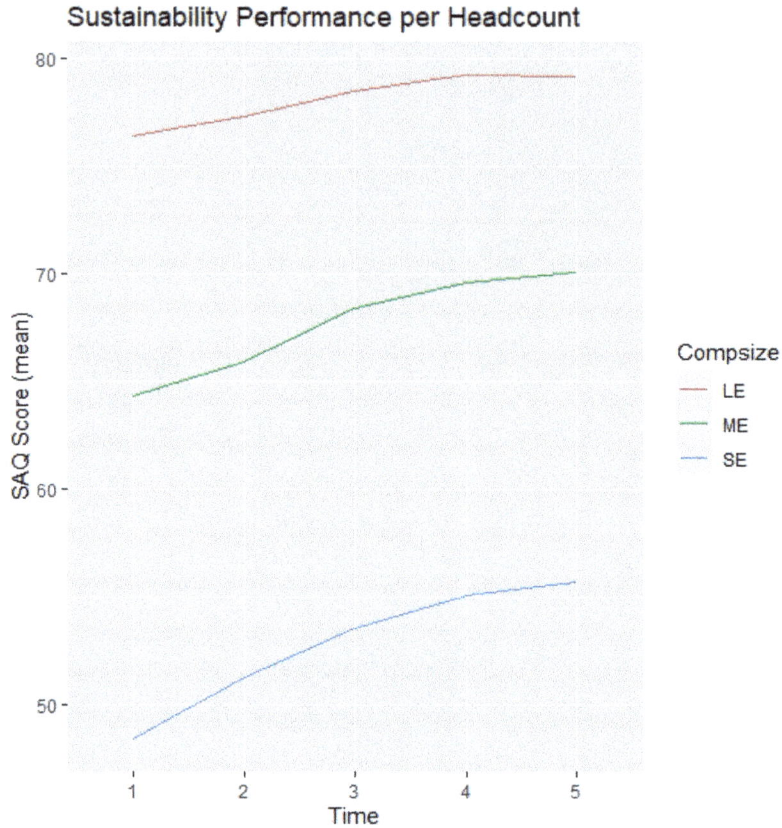

Fig. 4 Sustainability performance per headcount. *Source* Authors own work

improve their performance and hence comply with the sustainability standard to a higher degree.

Figure 5 shows supplier sites grouped by location. European suppliers are the largest group and have the lowest average scores, improving over time with steeper improvement in the early periods. Suppliers in the Americas are the second-largest group, with almost linear performance improvement until period four, followed by a flat period. African suppliers, the smallest group (except Oceania, one supplier), follow a different improvement pattern. Asia has the highest average performance and lowest increase, with scores over ten percentage points higher than Europe in January 2019. While the difference narrows over time, it remains comparably large in the final period of the sample.

The focal firm's assessment criteria appear to have induced coercive pressure, leading to heightened sustainability standard compliance within a global supplier network. However, this development seems to stagnate over time, as suppliers reach

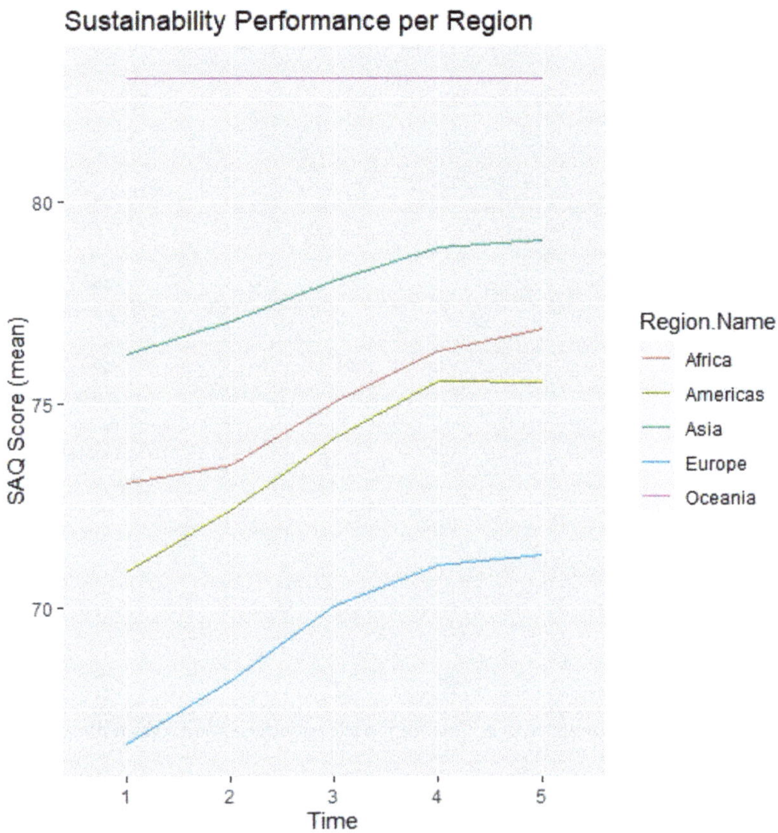

Fig. 5 Sustainability performance per region. *Source* Authors own work

a point where they are not further incentivized to improve their performance. Performance differences among supplier groups persist. The institutional pressure argument by (DiMaggio and Powell 1983) does not fully explain our sample, but seems applicable within supplier groups.

Figure 6 shows the development over time for the ten countries with the highest number of supplier sites. Most supplier sites are located in Germany, followed by China, Mexico, Spain, Czech Republic, Italy, Poland, Sweden, Brazil and the United States of America.

Suppliers in China demonstrate the highest performance with an average score of nearly 77% and an improvement to 80% over time. Polish suppliers show steady improvement, followed by those in the USA, with an average of around 73.5%. Czech suppliers also improve, reaching higher compliance with the sustainability standard by the end of the fifth period. Mexican and Spanish suppliers have similar developments over time, reaching approximately 75% in the fifth period. German and Italian suppliers have the lowest scores initially but show improvement over

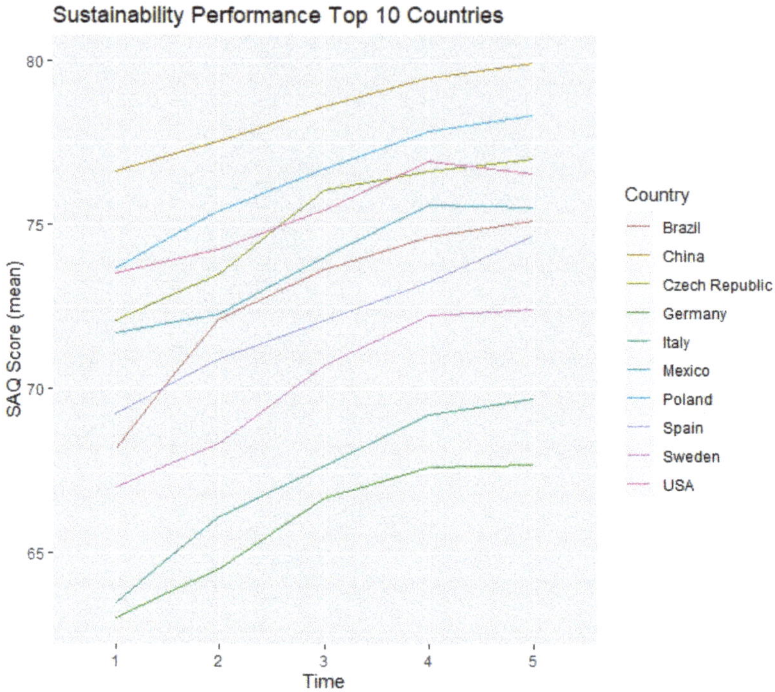

Fig. 6 Sustainability performance over time for top ten countries. *Source* Own illustration

time, with Italy's development stronger towards the end. The analysis indicates that pressure from the focal firm leads to SSP improvement, with distinct country-related differences.

4 Discussion

The overarching result of an improvement in sustainability performance of suppliers over time is in line with Hofstetter and Grimm (2019) who argue that the institutionalization of sustainability standards from focal firms in their supply chain is about making suppliers aware of their requirements. The authors furthermore find that these requirements are a "*[…] communication and education challenge*" (Hofstetter and Grimm 2019, p. 530) to many suppliers. It seems plausible that suppliers gradually comply with the sustainability standard as they learn about the focal firms' requirements. However, the lack of a specific improvement point deserves attention. As section three notes, the focal firm's mandatory assessment criteria for suppliers become relevant at specific times. This could explain differences in supplier development. Over time, this may reflect differences in the focal firms' expectations of different groups. Manufacturing suppliers, for instance, may face harder criteria to maintain business relationships or due to other pressures, such as obtaining an ISO 14001 certification. Lowered criteria for service providers may reflect their lower performance. The same could apply to country-specific assessment criteria. The analysis shows that differences in supplier sustainability performance persist between supplier groups, even with mandatory requirements. An objective comparison of development over time is difficult. This may be due to differences in the focal firm's requirements and supplier characteristics. For example, Encinas Bartos et al. (2022) found that suppliers' headcounts and business category significantly influence supplier sustainability performance.

Mimetic pressure might explain supplier group development. Suppliers in a group improve performance because competitors are doing so. This means improvement is induced by both focal firm pressure and mimetic pressure from competitors (Latif et al. 2020; Martínez-Ferrero and García-Sánchez 2017).

Further types of pressure may be influenced by a country's institutional environment, i.e., the stringency of policy and legislation. For instance, Jackson et al. (2020) found that firms respond less flexibly to certain stakeholder interests in countries with more stringent policies. Gulenko (2018) also found that mandatory reporting regulations increase the quantity of topics covered by firms, but without improving disclosure quality. This may explain lower performance in countries like Germany, Italy, and Sweden, which have stricter environmental policies than other top ten analyzed countries (Kruse et al. 2022; OECD. Stat 2022). Research shows contradictory results regarding mandatory reporting regulations, discussed by Kuruvilla and Li (2021). The role of the institutional environment and its impact on sustainability performance in terms of compliance with global standards may warrant further analysis. Cultural differences may explain discrepancies between suppliers in different countries.

More generally, Literature questions the use of coercion as a strategy (Caniëls et al. 2013). Analysis shows that suppliers comply with sustainability standards, but efforts wane at certain thresholds. This could be due to firm expectations, but other factors may be at play. Carbone and Moatti (2011) showed that coercion tends to lead to cost-reduction SCM practices, not innovation and differentiation. This suggests that an excellent sustainability performance is based on innovation and differentiation, not on coercion. Thus, coercion might improve supplier sustainability up to the point of maintaining a business relationship with the firm, or to the point where more sustainable practices lead to cost savings, like energy efficiency.

The improvement may stagnate later, but it may continue eventually; it might just require time for certain policies and management systems to be implemented, and for that, the timeframe of the analysis may not have captured this.

The results indicate that focal firms need to proactively communicate and educate specific groups of suppliers on their requirements. Service providers or groups of suppliers in regions with less-than-perfect sustainability standards implementation may need more support. Also, focal firms may consider pushing their suppliers more after they reach a certain threshold as performance stagnates. In general, focal firms should analyze the reasons for the varying degrees of standard implementation so that cultural factors or organizational influences do not skew the overall picture.

5 Conclusions

In this article, we examined how leading companies encourage the use of global sustainability standards in their supply chains. Specifically, we assessed automotive suppliers' compliance with an internationally recognized standard. Our data, from over 5,000 sites, revealed suppliers' progress. Our findings suggest that leading companies' push for sustainability does indeed drive its adoption.

The data analysis showed a general improvement over time, i.e., the mandatory assessment criteria of the focal firm analyzed led to a higher level of institutionalization of the given sustainability standard at the suppliers included in the sample. It also showed that supplier characteristics, such as business category or headcount, play a relevant role in explaining differences in the level of compliance with the sustainability standard that are not overcome over time. Country-specific factors also seem to play an important role, causing differences in the level of compliance. For the different supplier groups, it furthermore became evident that once a certain threshold in compliance with the standard was reached, the improvement in average performance became weaker, i.e., the incentive to improve further seemed to decrease. We derive two main results from these findings: (1) MNEs do play a relevant role when it comes to institutionalizing sustainability standards in their supply chains; (2) the assessment requirements from focal firms determine the degree of institutionalization of sustainability standards at their suppliers.

These findings raise new research questions, but our analysis has limitations. The two-year time frame may not show the full evolution of compliance with the sustainability standard. The SAQ, a widely used automotive industry sustainability standard, includes internationally recognized standards like ISO certification, but it may not fully measure suppliers' commitment to sustainability. The question of actual change, whether policies and processes are genuinely embraced, remains unanswered. Other studies, like Mani and Gunasekaran's (2018), found higher compliance with global standards, but Kuruvilla and Li (2021) raised questions about the translation of these standards to local contexts. Further research could verify our supplier performance data with additional on-site information from, for example, sustainability audits.

To measure how well sustainability standards have spread to global supply chains, future research should consider all tiers of the supply chain, not just the top ones. Our data set could not do that. But for real institutional change, these tiers need to be examined too. One key to spreading institutional change across many tiers is spreading it across the first tier. Therefore, our article is an important contribution to this field. It shows that Multinational Enterprises (MNEs) play an essential role in driving compliance with international standards worldwide and managing global challenges like we mentioned earlier.

References

Ahi P, Searcy C (2015) An analysis of metrics used to measure performance in green and sustainable supply chains. J Clean Prod 86:360–377. https://doi.org/10.1016/j.jclepro.2014.08.005

Ball A, Craig R (2010) Using neo-institutionalism to advance social and environmental accounting. Crit Perspect Account 21(4):283–293. https://doi.org/10.1016/j.cpa.2009.11.006

Beske P, Koplin J, Seuring S (2008) The use of environmental and social standards by German first-tier suppliers of the Volkswagen AG. Corp Soc Responsib Environ Manag 15(2):63–75. https://doi.org/10.1002/csr.136

Beske-Janssen P, Johnson MP, Schaltegger S (2015) 20 years of performance measurement in sustainable supply chain management-what has been achieved? Supply Chain Manag: Int J 20(6):664–680. https://doi.org/10.1108/SCM-06-2015-0216

Caniëls MCJ, Gehrsitz MH, Semeijn J (2013) Participation of suppliers in greening supply chains: an empirical analysis of German automotive suppliers. J Purch Supply Manag 19(3):134–143. https://doi.org/10.1016/j.pursup.2013.02.005

Carbone V, Moatti V (2011) Towards greener supply chains: an institutional perspective. Int J Log Res Appl 14(3):179–197. https://doi.org/10.1080/13675567.2011.609160

Cavusoglu H, Cavusoglu H, Son J-Y, Benbasat I (2015) Institutional pressures in security management: direct and indirect influences on organizational investment in information security control resources. Inf Manag 52(4):385–400. https://doi.org/10.1016/j.im.2014.12.004

Darnall N, Iatridis K, Kesidou E, Snelson-Powell A (2023) Penalty zones in international sustainability standards: where improved sustainability doesn't pay. J Manage Stud 61(6):2373–2405. https://doi.org/10.1111/JOMS.12975

Delmas MA (2002) The diffusion of environmental management standards in Europe and in the United States: an institutional perspective. Policy Sci 35:91–119

Derkx B, Glasbergen P (2014) Elaborating global private meta-governance: an inventory in the realm of voluntary sustainability standards. Glob Environ Chang 27:41–50. https://doi.org/10.1016/j.gloenvcha.2014.04.016

Dietz T, Grabs J (2022) Additionality and implementation gaps in voluntary sustainability standards. New Political Economy 27(2):203–224. https://doi.org/10.1080/13563467.2021.1881473

DiMaggio PJ, Powell WW (2000) The iron cage revisited: institutional isomorphism and collective rationality in organizational fields. In: Baum JAC, Dobbin F (eds) Economics meets sociology in strategic management. Advances in Strategic Management, vol 17. Emerald Group Publishing Limited, pp 143–166. https://doi.org/10.1016/S0742-3322(00)17011-1

DiMaggio PJ, Powell WW (1983) The iron cage revisited: institutional isomorphism and collective rationality in organizational fields. Am Sociol Rev 48(2):147–160. https://doi.org/10.2307/2095101

DIRECTIVE (EU) 2022/2464 (2022). https://eur-lex.europa.eu/legal-content/EN/TXT/?uri=CELEX:32022L2464

Ehrgott M, Reimann F, Kaufmann L, Carter CR (2011) Social sustainability in selecting emerging economy suppliers. J Bus Ethics 98(1):99–119. https://doi.org/10.1007/s10551-010-0537-7

Encinas Bartos K, Schwarzkopf J, Mueller M, Hofmann-Stoelting C (2022) Explanatory factors for variation in supplier sustainability performance in the automotive sector—a quantitative analysis. Clean Logist Supply Chain 5:10068. https://doi.org/10.1016/j.clscn.2022.100068

Encinas Bartos K, Schwarzkopf J, Mueller M (2023) The role of trainings in improving supplier sustainability performance. World Dev 175:106482. https://doi.org/10.1016/j.worlddev.2023.106482

Fraser IJ, Müller M, Schwarzkopf J (2020b) Transparency for multi-tier sustainable supply chain management: a case study of a multi-tier transparency approach for SSCM in the automotive industry. Sustainability 12(5):1814. https://doi.org/10.3390/su12051814

Fraser IJ, Müller M, Schwarzkopf J (2020a) Dear supplier, how sustainable are you? Sustainability Management Forum | Nachhaltigkeits Management Forum 28(3–4):127–149. https://doi.org/10.1007/s00550-020-00507-z

Glover JL, Champion D, Daniels KJ, Dainty AJD (2014) An institutional theory perspective on sustainable practices across the dairy supply chain. Int J Prod Econ 152:102–111. https://doi.org/10.1016/j.ijpe.2013.12.027

Grabs J (2020) Assessing the institutionalization of private sustainability governance in a changing coffee sector. Regul GovAnce 14(2):362–387. https://doi.org/10.1111/rego.12212

GRI (2023) GRI—home. https://www.globalreporting.org/ (accessed 09 April 2025)

Grimm JH, Hofstetter JS, Sarkis J (2022) Corporate sustainability standards in multi-tier supply chains—an institutional entrepreneurship perspective. Int J Prod Res 61(14):4702–4724. https://doi.org/10.1080/00207543.2021.2017053

Grimm JH, Hofstetter S, Müggler M, Peters NJ (2011) Institutionalizing proactive sustainability standards in supply chains: which institutional entrepreneurship capabilities matter? In: Marcus A, Shrivastava P, Sharma S, Pogutz S (eds.) Cross-sector leadership for the green economy. Palgrave Macmillan, pp. 177–193. https://doi.org/10.1057/9781137015891_10

Gualandris J, Kalchschmidt M (2014) Customer pressure and innovativeness: their role in sustainable supply chain management. J Purch Supply Manag 20(2):92–103. https://doi.org/10.1016/j.pursup.2014.03.001

Gualandris J, Kalchschmidt M (2016) Developing environmental and social performance: the role of suppliers sustainability and buyer-supplier trust. Int J Prod Res 54(8):2470–2486. https://doi.org/10.1080/00207543.2015.1106018

Gulenko M (2018) Mandatory CSR reporting-literature review and future developments in Germany. Nachhalt Manag Forum 26:3–17. https://doi.org/10.1007/s00550-018-0476-9

Halkos G, Nomikos S (2021) Corporate social responsibility: trends in global reporting initiative standards. Econ Anal Policy 69:106–117. https://doi.org/10.1016/J.EAP.2020.11.008

Hamprecht J, Schwarzkopf J (2014) Subsidiary initiatives in the institutional environment. Manag Int Rev 54(5):757–778. https://doi.org/10.1007/s11575-013-0191-2

Hirsch PM (1975) Organizational effectiveness and the institutional environment. Adm Sci Q 20(3):327–344. https://www.jstor.org/stable/2391994%0AJSTOR

Hofstetter JS, Grimm JH (2019) Multi-tier sustainable supply chain management. In: Sarkis J (ed) Handbook on the sustainable supply chain. Edward Elgar Publishing Limited, pp. 526–541. https://doi.org/10.4337/9781786434272

Jackson G, Bartosch J, Avetisyan E, Kinderman D, Knudsen JS (2020) Mandatory non-financial disclosure and its influence on CSR: an international comparison. J Bus Ethics 162:323–342. https://doi.org/10.1007/s10551-019-04200-0

Kauppi K, Hannibal C (2017) Institutional pressures and sustainability assessment in supply chains. Supply Chain Manag: Int J 22(5):458–472. https://doi.org/10.1108/SCM-01-2017-0004

Kelling NK, Sauer PC, Gold S, Seuring S (2021) The role of institutional uncertainty for social sustainability of companies and supply chains. J Bus Ethics 173:813–833. https://doi.org/10.1007/s10551-020-04423-6

Kruse T, Dechezleprêtre A, Saffar R, Robert L (2022) Measuring environmental policy stringency in OECD countries: An update of the OECD composite EPS indicator. In: OECD economics department working papers No. 1703, OECD Publishing. https://doi.org/10.1787/90ab82e8-en

Kuruvilla S, Li C (2021) Freedom of association and collective bargaining in global supply chains: a research agenda. J Supply Chain Manag 57(2):43–57. https://doi.org/10.1111/jscm.12259

Larrinaga C, Carrasco F, Correa C, Llena F, Moneva J (2002) Accountability and accounting regulation: the case of the Spanish environmental disclosure standard. Eur Account Rev 11(4):723–740. https://doi.org/10.1080/0963818022000001000

Latif B, Mahmood Z, San OT, Said RM, Bakhsh A (2020) Coercive, normative and mimetic pressures as drivers of environmental management accounting adoption. Sustainability 12(11):4506. https://doi.org/10.3390/su12114506

Mani V, Gunasekaran A (2018) Four forces of supply chain social sustainability adoption in emerging economies. Int J Prod Econ 199:150–161. https://doi.org/10.1016/J.IJPE.2018.02.015

Manning S, Boons F, von Hagen O, Reinecke J (2012) National contexts matter: the co-evolution of sustainability standards in global value chains. Ecol Econ 83:197–209. https://doi.org/10.1016/j.ecolecon.2011.08.029
Martínez-Ferrero J, García-Sánchez I-M (2017) Coercive, normative and mimetic isomorphism as determinants of the voluntary assurance of sustainability reports. Int Bus Rev 26(1):102–118. https://doi.org/10.1016/j.ibusrev.2016.05.009
McIntosh M, Thomas R, Leipziger D, Coleman G (2003) International standards for corporate responsibility. Ethical Corp Mag 13:22–29
Momin MA, Parker LD (2013) Motivations for corporate social responsibility reporting by MNC subsidiaries in an emerging country: the case of Bangladesh. Br Account Rev 45(3):215–228. https://doi.org/10.1016/j.bar.2013.06.007
Mueller M, Gomes dos Santos V, Seuring S (2009) The contribution of environmental and social standards towards ensuring legitimacy in supply chain governance. J Bus Ethics 89:509–523. https://doi.org/10.1007/s10551-008-0013-9
Rao P (2005) The greening of suppliers—in the south East Asian context". J Clean Prod 13(9):935–945. https://doi.org/10.1016/j.jclepro.2004.04.010
Reinecke J, Manning S, von Hagen O (2012) The emergence of a standards market: multiplicity of sustainability standards in the global coffee industry. Organ Stud 33(5–6):791–814. https://doi.org/10.1177/0170840612443629
Roberts S (2003) Supply chain specific? understanding the patchy success of ethical sourcing initiatives. J Bus Ethics 44(2–3):159–170. https://doi.org/10.1023/A:1023395631811
Sancha C, Longoni A, Giménez C (2015) Sustainable supplier development practices: drivers and enablers in a global context. J Purch Supply Manag 21(2):95–102. https://doi.org/10.1016/j.pursup.2014.12.004
Sarkis J (2001) Manufacturing's role in corporate environmental sustainability concerns for the new millennium. Int J Oper Prod Manag 21(5/6):666–686. https://doi.org/10.1108/01443570110390390
Sarkis J, Zhu Q, Lai KH (2011) An organizational theoretic review of green supply chain management literature. Int J Prod Econ 130(1):1–15. https://doi.org/10.1016/j.ijpe.2010.11.010
Sauer PC, Seuring S (2018) A three-dimensional framework for multi-tier sustainable supply chain management. Supply Chain Manag 23(6):560–572. https://doi.org/10.1108/SCM-06-2018-0233
Schaltegger S, Burritt R (2014) Measuring and managing sustainability performance of supply chains: review and sustainability supply chain management framework. Supply Chain Manag 19(3):232–241. https://doi.org/10.1108/SCM-02-2014-0061
Schöggl J-P, Fritz MMC, Baumgartner RJ (2016) Toward supply chain-wide sustainability assessment: a conceptual framework and an aggregation method to assess supply chain performance. J Clean Prod 131:822–835. https://doi.org/10.1016/j.jclepro.2016.04.035
Seuring S, Müller M (2008) From a literature review to a conceptual framework for sustainable supply chain management. J Clean Prod 16(15):1699–1710. https://doi.org/10.1016/j.jclepro.2008.04.020
Seuring S, Aman S, Hettiarachchi BD, de Lima FA, Schilling L, Sudusinghe JI (2022) Reflecting on theory development in sustainable supply chain management. Clean Logist Supply Chain 3:100016. https://doi.org/10.1016/j.clscn.2021.100016
OECD.Stat (2022) Environmental policy stringency index, OECD. https://stats.oecd.org/viewhtml.aspx?datasetcode=EPS&lang=en (accessed 06 September 2023)
Drive Sustainability (2019) Progress report driving change since 2013. https://drivesustainability.org/wp-content/uploads/2019/05/20190507-CSR-DS_ProgressREPORT-A4_V10_web96dpi.pdf (accessed 09 April 2025)
Drive Sustainability (2023) Guiding principles 4.0 translations—drive sustainability. https://www.drivesustainability.org/guiding-principles-4-0-translations/ (accessed 31 May 2023)
Together for Sustainability (2023) Assessment—TfS initiative. https://www.tfs-initiative.com/how-we-do-it/assessment (accessed 09 April 2025)

Vigneau L, Humphreys M, Moon J (2015) How do firms comply with international sustainability standards? processes and consequences of adopting the global reporting initiative. J Bus Ethics 131:469–486. https://doi.org/10.1007/s10551-014-2278-5

Wettstein F, Giuliani E, Santangelo GD, Stahl GK (2019) International business and human rights: a research agenda. J World Bus 54(1):54–65. https://doi.org/10.1016/J.JWB.2018.10.004

Wiengarten F, Pagell M, Fynes B (2013) ISO 14000 certification and investments in environmental supply chain management practices: identifying differences in motivation and adoption levels between Western European and North American companies. J Clean Prod 56:18–28. https://doi.org/10.1016/j.jclepro.2012.01.021

Zeng H, Chen X, Xiao X, Zhou Z (2017) Institutional pressures, sustainable supply chain management, and circular economy capability: empirical evidence from Chinese eco-industrial park firms. J Clean Prod 155:54–65. https://doi.org/10.1016/j.jclepro.2016.10.093

AI and Climate Change Prediction: Trends, Challenges, and Future Directions

Nilantha Randeniya, Richard Haigh, and Dilanthi Amaratunga

Abstract Climate change involves complex interactions of atmospheric conditions, ocean currents, human activities, and environmental factors that challenge traditional modelling approaches. With vast, expanding datasets from satellites, sensors, and weather stations, conventional methods struggle with real-time analysis and accurate predictions. Artificial Intelligence has emerged as a transformative data-driven technology for forecasting climate scenarios through enhanced analysis of dynamic climate data. However, current AI-CC interconnections remain unclear, with fragmented knowledge hindering climate adaptation strategies. This study systematically evaluates AI and Climate Change Prediction through bibliometric analysis of Scopus literature following PRISMA criteria. Using Vos Viewer, Biblioshiny, and R Statistical Software, the research analyses publication trends, sources, and countries across 1990–2024 (historical trends) and 2020–2024 (next-generation technologies). Results demonstrate exponential publication growth ($R^2 = 0.8205$), particularly post-pandemic. While the USA dominated until 2015, China has emerged as a major research hub. Thematic mapping identifies climate modelling and predictive analytics as pivotal for environmental resilience. Critically, overall CC-AI integration remains limited, marked by disciplinary disconnects. Future priorities include advancing scenario prediction, adaptive techniques, predictive modelling accuracy, and clarifying data-driven AI-CCP relationships to maximize next-generation technologies' potential in climate adaptation.

Present Address:
N. Randeniya (✉) · R. Haigh · D. Amaratunga
School of Applied Sciences, University of Huddersfield, Huddersfield, UK
e-mail: Nilantha.Randeniyaa@hud.ac.uk; Rnilandeniya@gmail.com

R. Haigh
e-mail: R.Haigh3@hud.ac.uk

D. Amaratunga
e-mail: D.Amaratunga@lboro.ac.uk; D.Amaratunga2@hud.ac.uk

D. Amaratunga
School of Architecture, Building and Civil Environment, Loughborough University, Loughborough, UK

R. Singh et al. (eds.), *Data-Driven Sustainability: Harnessing Technology for A Greener Future and Environmental Resilience*, World Sustainability Series,
https://doi.org/10.1007/978-3-032-13412-7_26

Keywords Climate change · Artificial intelligence · AI · State of arts

1 Introduction

Human actions, urban expansion, and the construction of infrastructure have reduced natural coverage, including green spaces (Nyuk and Chen 2015). Climate Change (here onwards CC), resulting from human activities like fossil fuel combustion, triggers enduring shifts in Earth's climate patterns (IPCC 2021). The heightened presence of greenhouse gases, particularly carbon dioxide, causes global warming and diverse environmental impacts (NASA 2021). These repercussions involve extreme weather events, rising temperatures, and ice melting (IPCC 2021). Mitigation strategies prioritize emission reduction, while ongoing changes are addressed through adaptation measures (UNFCC 2015). Global cooperation, exemplified by agreements like the Paris Agreement (UNFCC 2015) imperative to confront this pressing challenge and secure a sustainable future. These observed trends align with the scientific consensus presented in reports from authoritative bodies like the Intergovernmental Panel on Climate Change. According to the IPCC (2021), human activities notably the combustion of fossil fuels and deforestation, stand as the principal catalysts for CC. Addressing this challenge involves international endeavours, policies, and agreements dedicated to mitigating greenhouse gas emissions, adapting to impacts, and fostering sustainable practices (UNFCC 2015; Adger 2014). Climate Change Predictions (here onwards CCP) are essential for assorted reasons, providing insights into potential impacts on the environment, including weather patterns, sea levels, and ecosystems (Peiris, et al. 2025) (Randeniya, et al. 2026). This information is crucial for individuals, communities, and governments to make informed decisions about adaptation and mitigation strategies (Abatzoglou and Williams 2016). Understanding these predictions helps people anticipate and prepare for changes in local climates, reducing vulnerability to extreme weather events (Greg Garrard 2019). In decision-making processes, CCP inform policymakers about potential risks and challenges, aiding in the formulation of effective policies (Nisbet 2017). Furthermore, collaboration is essential in addressing the global nature of CC and Predictions create a mutual understanding, fostering global collaboration among nations, organizations, and communities (UNEP 2023).

Artificial intelligence (here onwards AI) is accelerating the development of alternative computing models stimulated by the abilities and efficiency of the human brain. AI plays a pivotal role in advancing CCP, leveraging its capacity to analyse extensive datasets, discern intricate patterns, and generate valuable insights (Greg Garrard 2019). AI's proficiency in processing diverse data sources, including satellite imagery and climate models, contributes to more accurate outputs. Predictions compared to traditional methods with an aptitude for recognizing complex relationships within data, AI enhances climate models and simulations, fostering a deeper understanding of the complex dynamics inducing CC (Goodfellow et al. 2016; Markus et al. 2019). The speed and adaptability of AI algorithms further revolutionize climate prediction by enabling faster data processing and continuous learning from current information (Nyuk and Chen 2015). Integrating AI into climate research not only increases prediction accuracy but also facilitates the adoptability and development of early warning systems via optimizes resource allocation, valuable insights for decision-makers and researchers (Salas 2022). The incorporation of AI into CCP holds the promise of transforming the field, providing more reliable and efficient tools for addressing the challenges of a changing climate (Nisbet 2017; Randeniya et al. 2026).

It is important to use AI in CCP, but the current trend is unclear, and there is a lack of understanding of the connection between these subject areas (Salas 2022). Additionally, fragmented theoretical knowledge is also major problems in these areas (UNEP 2023). To comprehensively understand the current landscape of the two study areas, it is essential to conduct a thorough review of the state of the art (SOA). This review will interpret the latest advancements and emerging trends within the domains of CCP and AI. By identifying the SOA, can gain an understanding of the prevailing methodologies and technological innovations, thereby establishing benchmarks for performance and quality through the analysis of current best practices. Moreover, this approach will provide a clear overview of future research directions by pinpointing research areas and aligning R&D efforts accordingly. It will also facilitate the identification of gaps and unmet needs, thereby highlighting potential opportunities for novel innovations.

To identify the SOA the Bibliometric analysis is as an important quantitative analysis tool (Berkhout et al. 2023). It can effectively describe the overall trend of the development of a subject or field, and it has been widely used in various fields (Pritchard 1969; Berkhout et al. 2023). The method utilized by (Zhaohua et al. 2018) for analysing CC adaptation relies on extensive research literature data. The study reveals that the complexity inherent in examining studies from various countries, regions, and organizations stances significant challenges to statistical analysis.

1.1 Research Background

The study utilizes two primary timelines: the Extended Timeline and the Recent Timeline. The Extended Timeline spans from 1990 to 2024, covering the last three decades. This timeline is used to identify historical trends and developments over an extended period, capturing trends that may not be apparent in shorter intervals. It provides insight into long-term patterns and shifts in the field. The Recent Timeline covers the last three years, from 2020 to 2023. Its primary purpose is to highlight current and focused trends and developments. This shorter period allows for a detailed examination of recent advancements and areas that require immediate attention. Additionally, some topics may not need extended analysis or may lack sufficient data for a longer timeline. The Recent Timeline enables the study to concentrate on areas that demand more in-depth exploration and current relevance.

The paper aims to identify and highlight current state of the art in bridging the gap between AI and CCP. The bibliometric analysis will systematically review the scientific literature documented in the Scopus database over the two primary timelines.

The study aims to achieve several objectives, to identifying the current state of research by analysing publication trends, sources, subject areas, and the countries most involved in this research over a specified period. It will also evaluate key studies, notable authors, and shifting research trends across different periods. Additionally, the analysis will cover the geographical distribution of studies, emphasize significant aspects of CC, AI, and climate change prediction (CCP) that have garnered notable attention, and suggest potential future research directions in this evolving field. The study will address key questions, what are the emerging trends and research directions in AI-based CCP? What are the major research gaps and challenges in applying AI to CCP?

2 Research Methodology

The assessment of the literature was conducted according to the Preferred Reporting Items for Systematic Reviews and Meta-Analyses; PRISMA (Annexes 01). The research has adopted the research methodology into PRISMA and These criteria, shown in Fig. 1, were used to create a knowledge base of research on CC, AI, and CC prediction. The methodology consists of the following steps: When reviewing the literature in the Scopus database, keywords were selected to ensure that all relevant publications were included. The Search Criteria is as follows: *(TITLE-ABS-KEY ("artificial intelligent" OR "AI" OR "ML (Machine Learning)" OR "Machin Learning" OR "Deep Learning" OR "Neural Network" OR "CNN" OR "Conventional Neural Network") AND TITLE-ABS-KEY ("Climate Change" OR "Climate" OR "Climate Impact" OR "Climate change Impact") AND TITLE-ABS-KEY ("Climate Change Prediction" OR "Climate Impact" OR "Climate Prediction")).* N = 4490. The second step were screening and sub step of S1 is to select documents after 2020 to 2023 and Number of Records after Remove is N—940. The Next step is for Select Only English Language document, and the excluded languages are mentioned in the step S2 in Fig. 1 (remaining Number of documents are 480). Third step of screening is to Exclude Subject Areas which are not relevant and there are remining around 321 documents left and excluded Subject Areas are mentioned in the S3 in Fig. 1.

The database was analysed using several software programmes. Vos Viewer and Biblioshiny were used for network analysis of keywords and authors, respectively. In addition, SciMAT algorithms were used to create an overlap graph and a map of thematic development. Bibliometric Software is a web interface that is integrated into the R statistical software. Bibliometric offers a range of functions and visualisations to conduct these analyses efficiently. The Biblioshiny package starts by downloading the data from various bibliographic databases and, once the data has been uploaded, it is possible to carry out the analysis, including with appropriate and dynamic graphics. VOS viewer (Visualisation of Similarities viewer) is a software tool developed for the creation and visualisation of networks. It is often used for the

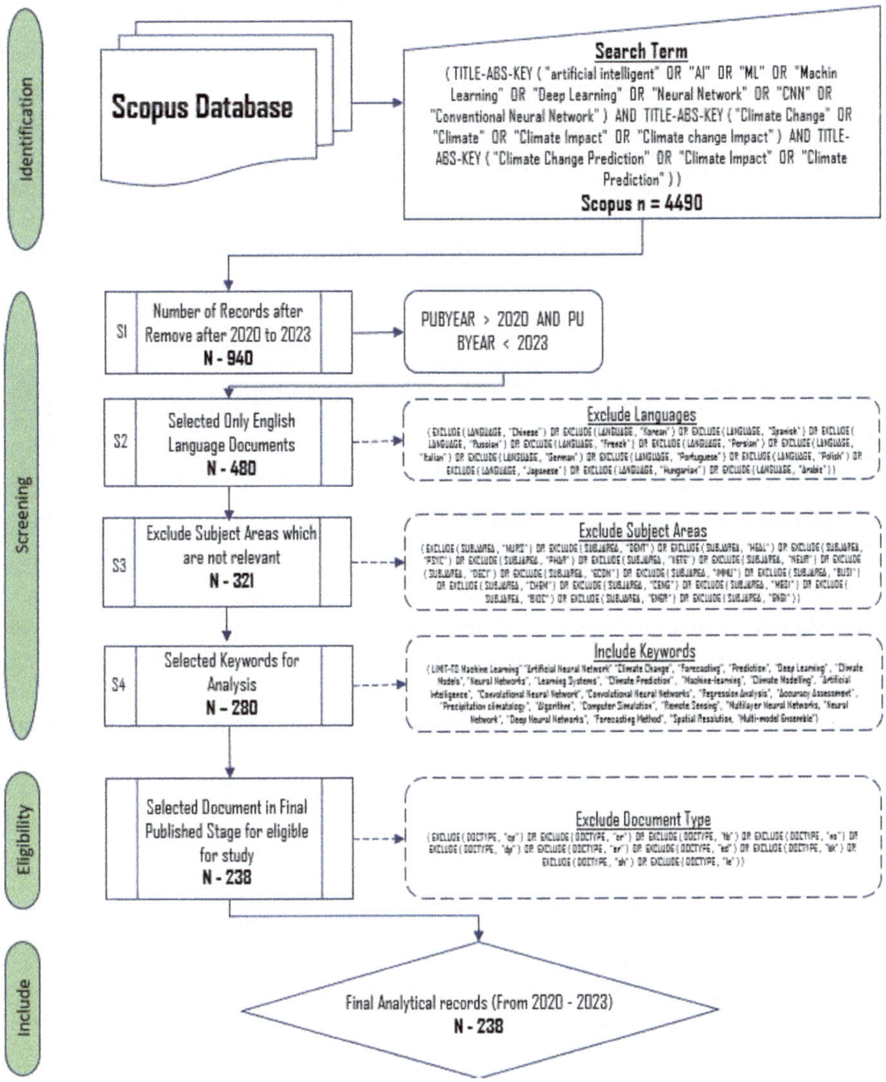

Fig. 1 Research methodology. *Source* Compiled by Author, 2025

analysis and visualisation of patterns and relationships in scientific literature based on bibliographic data.

3 Data Sources

The relevant data for the bibliometric analysis can be selected from various databases such as Web of Science, Scopus, PubMed, and ProQuest. Scopus was selected as the primary database for this study as it is the largest collection of abstracts and citations from peer-reviewed literature. Scopus includes high quality databases covering academic papers in the fields of engineering, natural sciences, social sciences, education, arts and humanities, making it a widely used resource for researchers. The following details refer to the database used in this study.

Table 1 Data sources table

Timespan 2020:2023	Sources 94	Documents 238	Annual Growth Rate 31.52 %
Authors 905	Authors of single-authored docs 11	International Co-Authorship 28.57 %	Co-Authors per Doc 4.59
Author's Keywords (DE) 755	References 0	Document Average Age 2.15	Average citations per doc 8.714

As shown in Table 1, the data covers 94 diverse sources (Journals) from the Scopus database throughout a three-year period, from 2020 to 2023. There were 238 documents after filtered documents over this time, representing an annual growth rate of 31.52%, suggesting a significant increase in studies about AI and CC. Only 11 of the 905 authors that contributed to these articles were single authors, which is a small percentage of all the publications. The average number of authors per document is about 4.59, and the overall co-authorship percentage is 28.57%. In all documents, the author's keywords total 755.

4 Analysis and Results

4.1 Identification of Current State of Art in CC Prediction via AI

The Publications in Last 3 Years and Decades

AI and CC have evolved substantially over the last 33 years and have become significant topics for discussion. Through evaluating both fields' present conditions, this analysis looks to bring insight on their historical growth and present situation.

According to Fig. 2, There were two papers that discussed AI and CC in the early 1990s, which suggests that the topic was not incredibly significant at the time. Nonetheless, here has been a significant increase in these fields' attention and research over time. Publications increased over 100% by 2000, and in the years that followed,

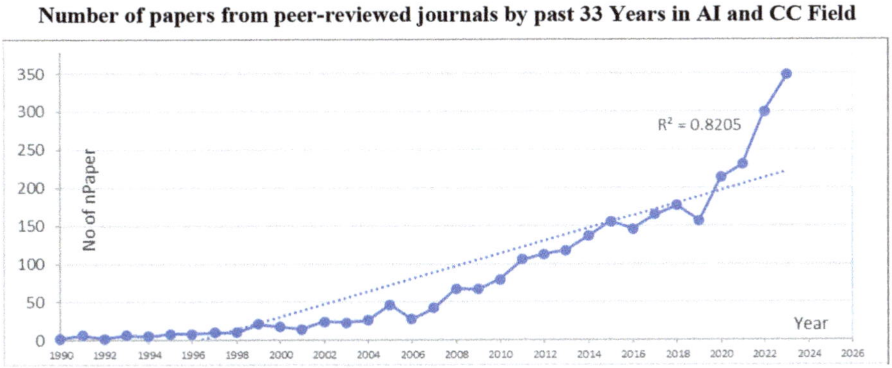

Fig. 2 Number of papers from peer-reviewed journals by past 33 Years in AI and CC Field. *Source* Compiled by Author, 2025

annual growth rates ranged from 50 to 70%. According to the study "Global Publication Trend and Forecast" (2016), there has been a 50% increase in global databases covering papers from peer-reviewed journals across various disciplines (see Fig. 2). This notable growth reflects a comparative divergence between fields such as Artificial Intelligence (AI) and Climate Change (CC) and other disciplines, marking an atypical occurrence, as such significant expansion is uncommon relative to trends seen in other areas.

AI and CC have experienced a global attention boom in the previous three years (from 2020 to 2023), especially after the pandemic, which has made these subjects extremely popular. There has been exponential growth in publications, as evidenced by a high growth rate with an R-squared value of 0.8205, underscoring the substantial expansion of the dataset. According to the "Global Publication Trend and Forecast" (2023), this growth is atypical compared to trends in other fields, which have shown relatively low growth rates with an R-squared value of 0.505 in recent years (before 2020) due to the global pandemic and economic crises (see Fig. 3).

4.2 Most Relevant Sources and Subject Areas

The cornerstone of this study's analytical framework lies in the identification of the most pertinent sources. As represented in Fig. 4, the tabulation succinctly captures the volume of documents produced by key sources over the preceding 3 years. Leading the pack is Atmosphere, which notably generated 21 documents within this period. Following closely is the Journal of Advances in Modelling Earth Systems, making a substantial contribution with 16 documents. Remote Sensing appears as another significant source, boasting 14 documents over the past three years. Additionally, Geophysical Research Letters garners attention as a noteworthy source, producing a total of 12 documents. These findings underscore the influential role that specific

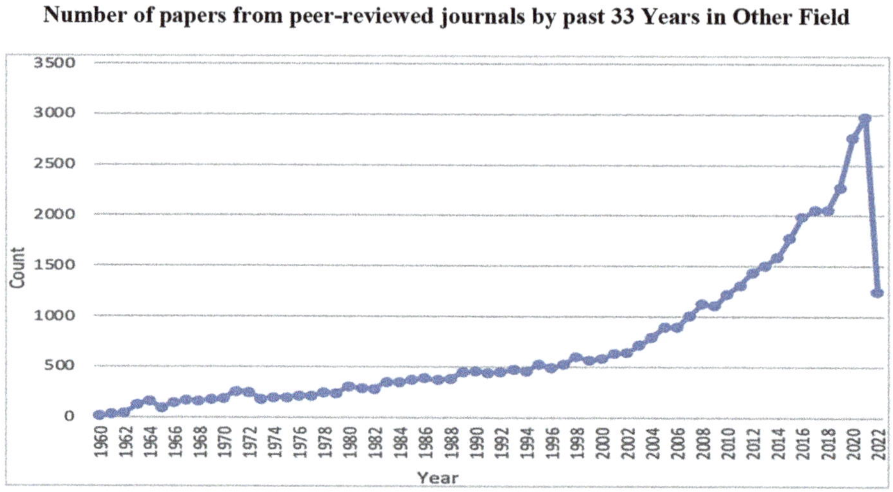

Fig. 3 Number of papers from peer-reviewed journals by past 33 Years in Other Field. *Source* Compiled by Author, 2025

sources play in shaping contemporary discourse within the field, highlighting the diverse and scholarly depth inherent in the academic landscape.

The publications mostly focus on environmental and physical science fields. This suggests that research is still mostly contained within narrow academic boundaries even though this field of study has evolved over decades and the complexity of climate challenges is enormous. Furthermore, rather than AI or Computer Science related journals, environmental and physical science-based journals are where AI is most often used. This will be confirmed by the Distribution of publications across

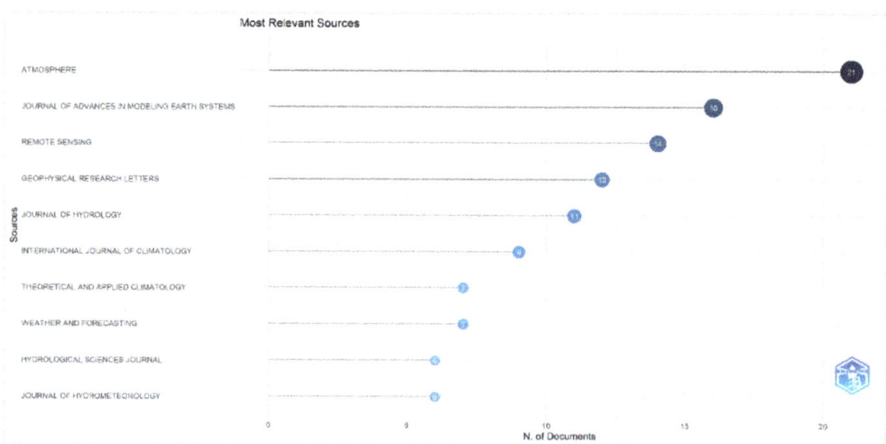

Fig. 4 Most relevant sources. *Source* Compiled by Author, 2025

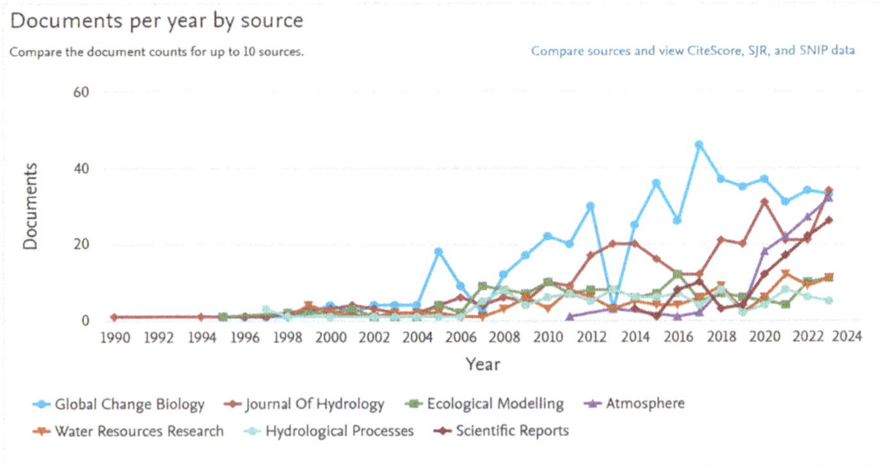

Fig. 5 Distribution of publications across subject categorisations and journals. *Source* Compiled by Author, 2025

subject categorisations and journals over the past 33 years Fig. 5. Initially, the Journal of Hydrology held prominence in the early 1990s as a foundational journal in this area of study. The establishment of the Journal of Ecological Modelling in 1995 further contributed to publishing research findings in the field. The Journal of Global Change Biology depicts significant fluctuations in publication numbers over time; for instance, there were around 20 publications in 2004–2005 but only 3 in 2007, reflecting dynamic changes within the journal. By 2018, this journal was publishing more than 45 articles annually. Similar trends are seen in the Journal of Hydrology, which saw a notable increase in publications after 2010, solidifying its position as a leading journal in CC and AI research.

The distribution of study areas (in Fig. 6) shows where the research areas of CC, and AI are studied in the subjects. The main subject is environmental science and about 69% of the research papers are from this subject area. Then after the multidisciplinary research area, which accounts for 13.3% of the research documents, social sciences standing for 12% of publications. The humanities and arts and cultural studies together account for 5% of the share.

In addition, CC is a multidisciplinary phenomenon with profound impacts on societal, economic, and ethical dimensions. This analysis reveals a significant gap over the past three decades: CC has predominantly been viewed through the lenses of environmental and physical sciences. While there has been some input from social sciences, it has not been sufficient to fully understand the breadth of its societal implications. Moreover, considerations from economic and ethical perspectives have been largely absent. In contrast, AI stands for a cutting-edge technological advancement on a global scale, driving revolutionary developments across various fields. However, this analysis underscores a critical gap: the insufficient integration of technological studies in understanding CC. Bridging this gap is essential to foster comprehensive

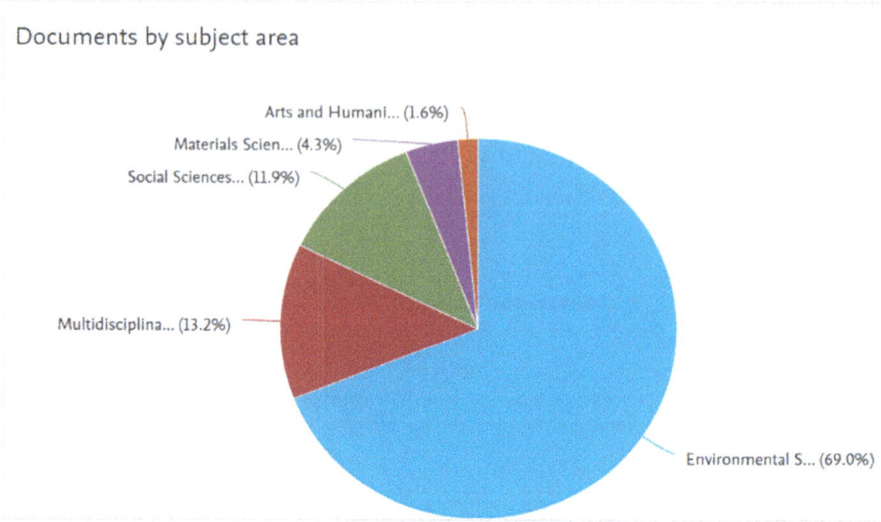

Fig. 6 Distribution of subject area. *Source* Compiled by Author, 2025

insights into the intersection of AI and CC, ensuring a comprehensive approach to addressing its multifaceted challenges.

4.3 Countries Are Actively Engaged in Research

The countries of origin showed by the authors (Fig. 7) were used to analyse the origin of the scientific works. China and the United States of America (USA) were the countries with the highest production of scientific articles in the research areas of CC, AI, and CC prediction. India is in third place. Over the span of 33 years, the USA appeared as a leading contributor to scientific research in this field, with a predominant focus on qualitative methodologies. However, a notable shift occurred after 2015 towards quantitative analysis techniques, particularly driven by the integration of AI and ML initiatives. Subsequently, research efforts became more pronounced within Chinese based universities and among scholars. Presently, there is a discernible preference for quantitative research, with a growing trend post-2019 towards the integration of AI into this subject area. However, there is an absence of multidisciplinary research and insufficient joints effort. Moreover, the examination reveals an absence of researchers in fields like AI and computer science. Professionals in Disaster Risk Reduction (DRR) are conducting most of the AI-based research on CC.

The world map of country collaboration (Figs. 8 and 9) is the minimum of 5 research publications in the last 3 years, based on the number of publications. According to the map, the thickness of the line is the weight of the collaboration. The figure shows that collaboration between China and the USA has the greatest influence

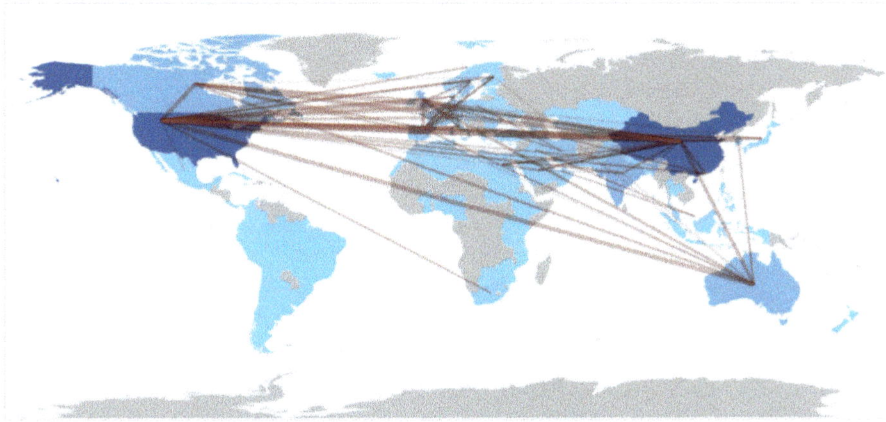

Fig. 7 Countries' collaboration world map. *Source* Compiled by Author, 2025

on collaboration. According to Fig. 8 the Network Diagram of Country Collaboration, USA collaborates with the European countries, which have quite influential collaboration, as well as with Japan, Brazil, Norway, the Netherlands, Mexico, and Canada. There are also collaborations in the European region, such as those between the UK and Italy, the UK and France and Germany and France and Italy, i.e. cluster collaborations. There is also another cluster co-operation in India with Vietnam, Iran, Malaysia, Indonesia, and Egypt, which is quite small but growing.

The analysis shows that, USA has been at the leading edge of climate-based research, proving itself as the main force in the international scientific community in the early years. Following 2019, China became another hub for the advancement of research worldwide, producing a substantial number of studies and publications. The previous three years have seen the rise to importance of the USA and China as global leaders, separating the world into two primary centres for AI and CC research. Nevertheless, recent patterns show the emergence of more epicentres than these two. New research groups are being set up in Europe, the United Kingdom, India, and Australia. These semi-clusters continue to grow, with each group focusing on unique research topics associated with their unique geographic challenges. Instead of only considering CC to be a worldwide occurrence, these clusters often place a higher priority on recognizing and resolving regional environmental challenges.

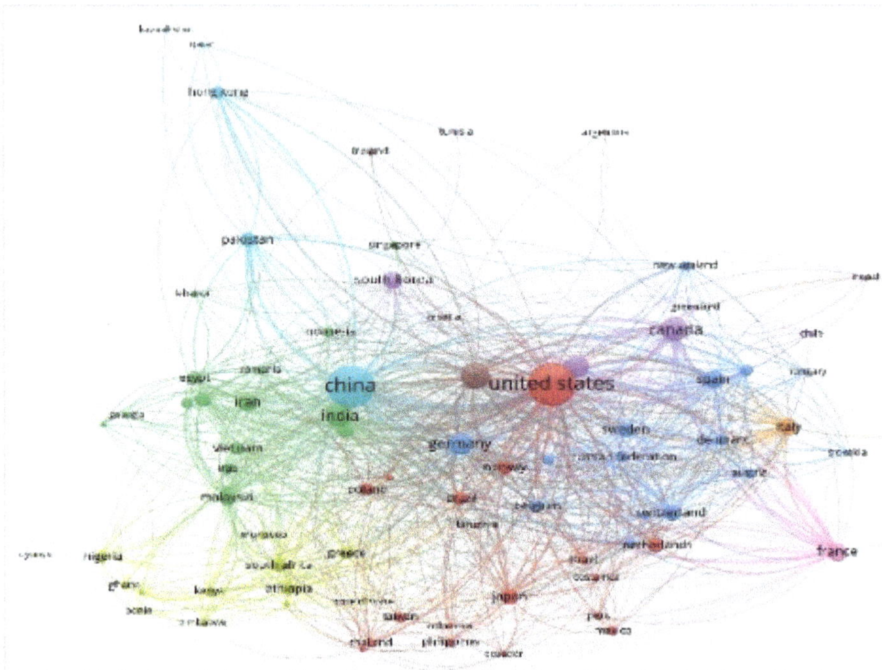

Fig. 8 Network diagram of country collaboration. *Source* Compiled by Author, 2025

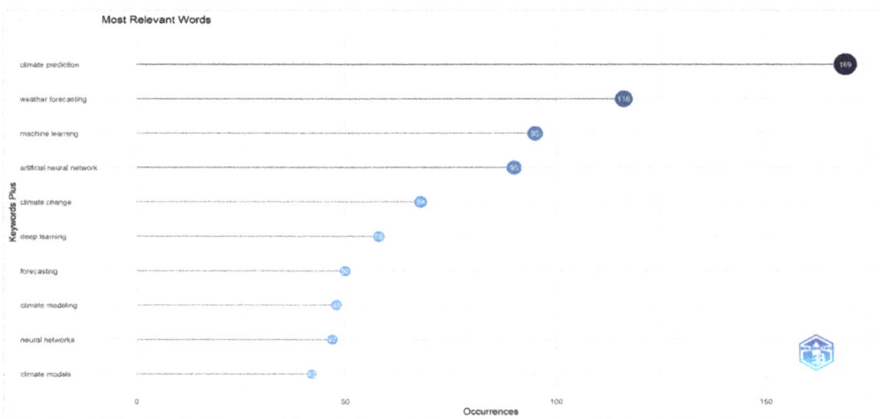

Fig. 9 The most relevant key words. *Source* Compiled by Author, 2025

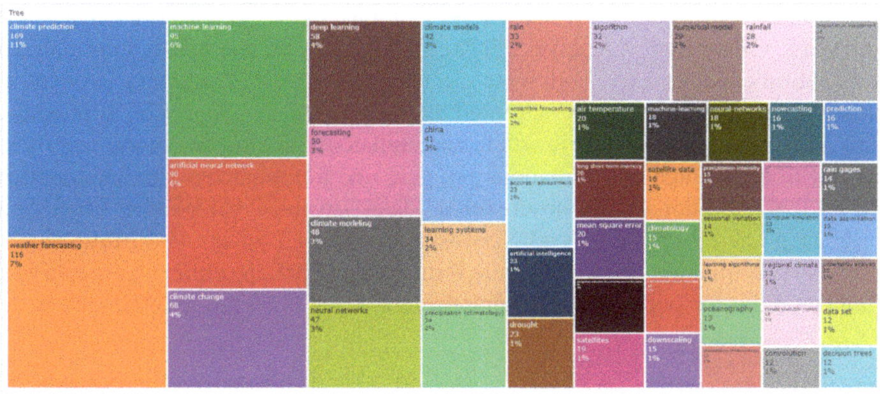

Fig. 10 The tree map of the most frequent words. *Source* Compiled by Author, 2025

5 Identification of Trends, Emerging Areas, and Research Directions in the Subject Area

5.1 The Key Words

The keywords appearing in research publications between 2021 and 2023 were collected and analysed to discover the latest trends, describe the current focus areas, and decide the direction of future development of the field. The main keywords are presented in Fig. 9 by accumulating the frequency of vocabulary found in the individual research papers focussing on the analysis. Climate Prediction is the most often used keyword with a frequency of 169. The second most often mentioned keyword is Weather forecasting, which is mentioned 116 times in the research. Machine learning (ML) is also mentioned around 95 times, AI 90 times and CC 68 times.

The tree map of the most frequent words is shown in the Fig. 10. It shows the number of frequent keywords and the proportion of keywords in the frequency of mentions. Climate Prediction is the most frequent keyword with 11% of mentions. The second most often mentioned keyword is Weather forecasting with a frequency of 7% and ML (6%), AI (6%), CC (4%), Deep learning (4%) and Forecasting (3%), Climate modelling (3%) and Nural Networks (3%) each receive 6% to 3% of mentions.

5.2 Most Citied Articles and Documents

Table 2 presents a compilation of the most cited articles and documents. Topping the list is the article titled 'A Benchmark Data Set for Data-Driven Weather Forecasting' by Rasp S., Dueben P.D., Scher S., Weyn JA, Mouatadid S., and Thuerey N. This article garnered 136 citations in 2020 alone, making it the most cited article in the past three years. On average, it receives 27.20 citations per year and boasts a normalized total citation count of 5.01. Following closely is the article 'Predicting Drought in Pakistan using ML Algorithms' authored by Sachindra D.A., Shahid S., Ahmed K., Shiru M.S., and Nawaz N. This article accumulated 125 citations in the last three years, with an average of 25 citations per year and a normalized citation count of 4.61 in 2020. Additionally, the article titled 'Data-Driven Global Weather Prediction using Deep Convolutional Neural Networks on a Cubed Sphere and Multi-Model Ensemble precipitation and temperature forecast using MLalgorithms' garnered 113 and 99 citations, respectively, with annual averages of 22.60 and 19.80 citations.

Table 2 Most citied articles and documents

Paper	DOI	Total Citations	TC per Year	Normalized TC
RASP S, 2020, J ADV MODEL EARTH SYST	https://doi.org/10.1029/2020MS002203	136	27.20	5.01
KHAN N, 2020, ADV WATER RESOUR	https://doi.org/10.1016/j.advwatres.2020.103562	125	25.00	4.61
WEYN JA, 2020, J ADV MODEL EARTH SYST	https://doi.org/10.1029/2020MS002109	113	22.60	4.17
AHMED K, 2020, ATMOS RES	https://doi.org/10.1016/j.atmosres.2019.104806	99	19.80	3.65
BONAVITA M, 2020, J ADV MODEL EARTH SYST	https://doi.org/10.1029/2020MS002232	51	10.20	1.88
ISLAM MA, 2020, J HYDROL	https://doi.org/10.1016/j.jhydrol.2020.125474	50	10.00	1.84
FANG Y, 2021, SCI TOTAL ENVIRON	https://doi.org/10.1016/j.scitotenv.2020.143841	46	11.50	4.25
KANG J, 2020, ATMOSPHERE	https://doi.org/10.3390/atmos11030246	35	7.00	1.29
CHATTOPADHYAY A, 2020, J ADV MODEL EARTH SYST	https://doi.org/10.1029/2020MS002084	35	7.00	1.29
BI K, 2023, NATURE	https://doi.org/10.1038/s41586-023-06,185-3	33	16.50	18.77

Source Compiled by Author, 2025

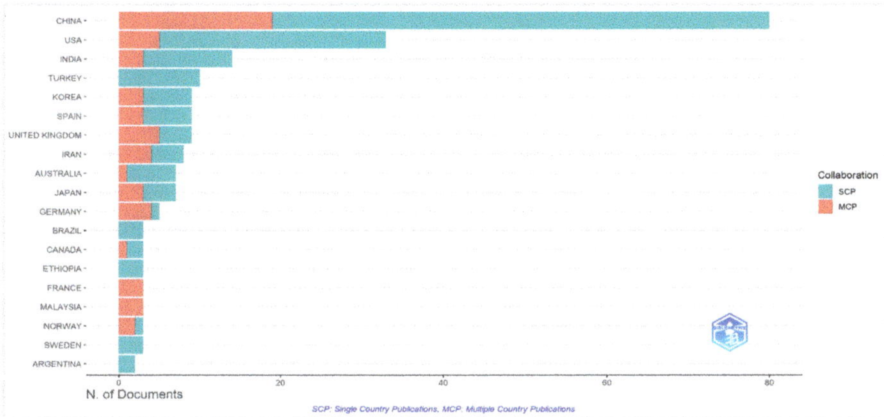

Fig. 11 Country collaboration and publication. *Source* Compiled by Author, 2025

5.3 Country Collaboration and Publication

The Fig. 11 illustrates the collaborative efforts and research publications in the field of CC and AI over the past three years. The green colour denotes Single Country Publications (SCP), while the red colour stands for Multi-Country Publications (MCP). China, USA, and India appear as the top three countries with the highest number of publications. China leads with the highest number of publications, predominantly SCPs (Single Country Publications), accounting for about 80% of the total, with the remaining 20% being MCPs (Multi Country Publications). In the case of the USA and India, SCPs form approximately 90%, with the remaining 10% being MCPs. This pattern continues by the Korea and Span while Turkey is the 99% of SCPs. However, the United Kingdom stands out with a different pattern where most publications are MCPs, forming 70%, while SCPs are only 30%. This shows an elevated level of international collaboration in the study of CC and AI in the UK.

6 Identification of Research Gaps in the Subject Area

6.1 Thematic Map

The thematic map presented in Figure 12 delineates the evolving trends and thematic framework within the research domain. It employs a graph where themes are positioned along the x-axis based on their centrality (degree of relevance) and along the y-axis based on their impact (degree of development). The size of each bubble correlates with its influence within the field.

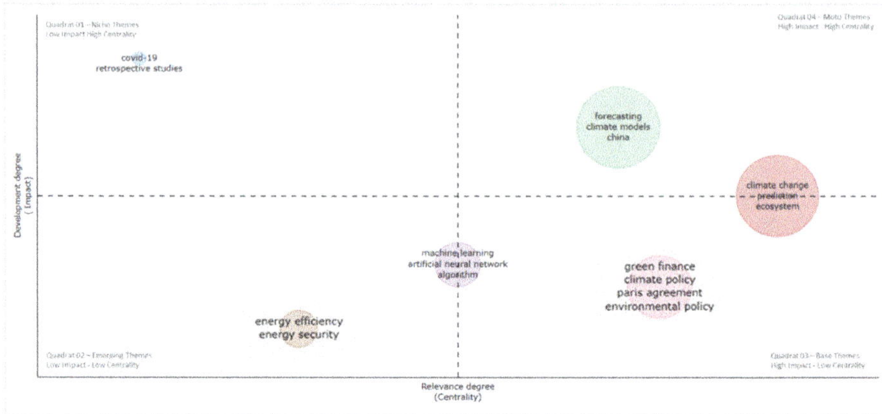

Fig. 12 Thematic map. *Source* Compiled by Author, 2024. *Source* Compiled by Author, 2025

Quadrant 01—Niche Themes: These themes cater to specific, often niche audiences with specialized interests. They typically show high centrality but moderate impact. Notable examples include studies focused on Covid-19 impacts and restorative practices.

Quadrant 02—Emerging Themes: This quadrant encompasses topics currently in development, such as advancements in energy efficiency, security considerations, and the emerging roles of ML and AI in CC research.

Quadrant 02—Emerging Themes: This quadrant encompasses topics currently in development, such as advancements in energy efficiency, security considerations, and the emerging roles of ML and AI in CC research.

Quadrant —Emerging Themes: This quadrant encompasses topics currently in development, such as advancements in energy efficiency, security considerations, and the emerging roles of ML and AI in CC research.

Quadrant 02—Emerging Themes: This quadrant encompasses topics currently in development, such as advancements in energy efficiency, security considerations, and the emerging roles of ML and AI in CC research.

Quadrant 02—Emerging Themes: This quadrant encompasses topics currently in development, such as advancements in energy efficiency, security considerations, and the emerging roles of ML and AI in CC research.

Quadrant 02—Emerging Themes: This quadrant encompasses topics currently in development, such as advancements in energy efficiency, security considerations, and the emerging roles of ML and AI in CC research.

Quadrant 03—Motor Themes: Fundamental and cross-cutting topics that influence multiple sides of the field exist in this quadrant. Topics such as green finance, climate and environmental policy frameworks, and the implementation dynamics of the Paris Agreement are pivotal in shaping discourse within the CC and AI research landscape.

Quadrant 04—Moto Themes: Themes occupying this quadrant show both high impact and centrality within the study of CC and AI. They encapsulate guiding principles or foundational ideals that direct organizational and individual research endeavours. Examples include climate modelling methodologies and focused research initiatives, which are critical for shaping the research agenda and methodologies within the field. The largest bubbles within this quadrant signify themes of utmost influence and strategic importance.

This thematic mapping not only provides a nuanced depiction of the thematic diversity within the research area but also underscores the strategic significance of different thematic categories in delineating the contours of CC and AI research.

6.2 *Network Visualisation Keyword Co-Occurrence.*

The network visualization of keyword co-occurrence (Figure 13) and nodes stand for data points, and links are the connections between them. Among the clusters, the red-green colour cluster is the main cluster being CC, climate, change, ecosystems, projections, forecasts scenarios and responses. CC has become a core of this cluster. The next one is the Blue-Yellow Cluster which is AI, machine learning, and applications for CC are becoming a trending topic in this cluster. The yellow colour is the accuracy, performance and parameters of the studies and is also a colouring research area for AI and CC.

About the ultramodern, this analysis reveals that these two subject areas are working separately towards the same vision. Although there are connections between these areas, when necessary, the links are not strong, as showed by the comparatively

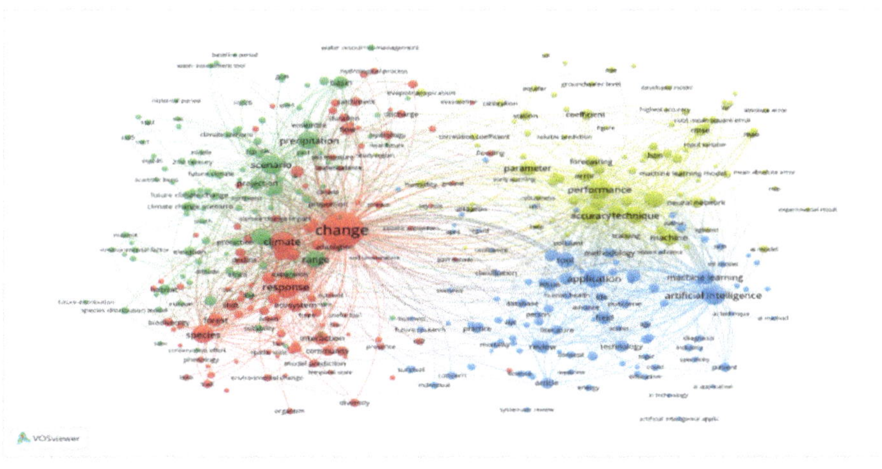

Fig. 13 Network visualisation keyword co-occurrence. *Source* Compiled by Author. *Source* Compiled by Author, 2025

thin linking lines between the two clusters. Specifically, there are very few connections between CCP and AI predictions, reflecting the weak linkage between the green cluster and the yellow cluster. Also, these connections will be more elaborate as follows.

The network visualization displayed strong links between CC and AI, with almost 80% of the linkages being started (All the percentages derived from VOS viewer). However, deeper analysis reveals gaps and insufficient linkages between, and Fig. 13**.** Depicts the link between CC and AI. CC is related to the scenarios, CCP, and forecasts cluster, proving its importance as a research area. An analysis proves a strong association between the green segment of the illustration and keywords related to CC and climate, proving their importance in the study. Furthermore, connections with AI tend to be disturbed, with current links being independent or restricted in scope.

Figure 14 depicts the linkage between AI and CC, with around 20% of the linkages being created. AI makes significant connections to performance, accuracy, and parameters. However, the primary connections are not entirely consistent with CC and Scenario Predictions. There is just one strong correlation noticed, and it has not been shown in more than five studies conducted in the last three years. Furthermore, the themes of Prediction, CC Forecast, and CC Adaptation have low linkage and are underrepresented in the research produced during the last three years.

Figure 14 depicts the relationship between Accuracy within the domain of CC around AI, with approximately 15% of the linkages found. In this context, accuracy is inextricably linked to AI and ML, as well as other relevant areas, and it also has a direct relationship with scenario planning, given the importance of precise modelling and forecasting. Nonetheless, concerns about accuracy at the nexus of CC and AI

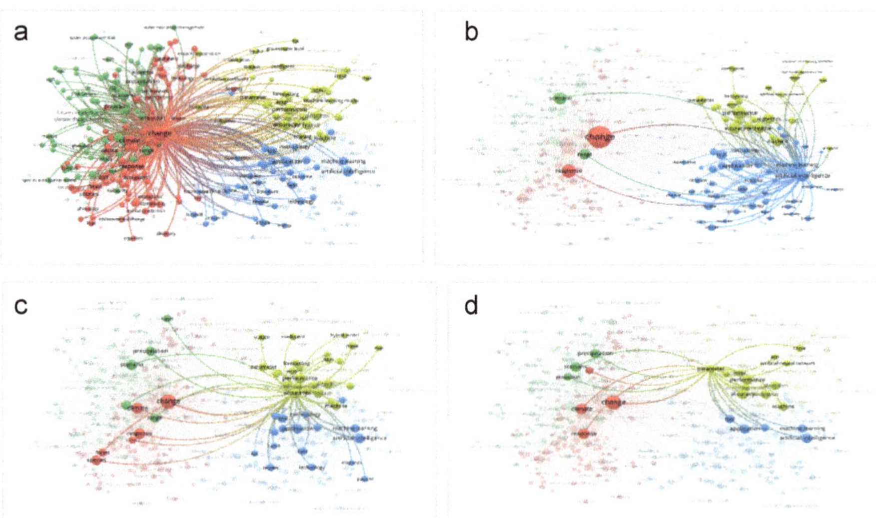

Fig. 14 **a** The linkage of CC with AI. **b** The linkage of AI with CC. **c** The linkage of accuracy in CC & AI. **d** The linkage of predictions in CCP and AI. *Source* Compiled by Author, 2025

are not heavily stressed, highlighting a distinct research gap identified in this study. It is important to note that the performance of AI models relies mainly on achieving accuracy before verifying it. Figure 14 depicts the prediction linkage, with major linkages to AI, Performance, Accuracy, Parameters, and Response. CC is not the primary connection within the domain of prediction; at this point within the field of CCP assume a significant part. Predictions, on the other hand, are inextricably linked to AI and ML applications and tools due to their reliance on these technologies.

Finally, it has been determined that there is limited integration, thin coverage, disconnections and Neglected connections. The limited integration of CC and AI highlights the need for more focused efforts to close this gap. While some variables, such as performance and accuracy, exhibit substantial correlations, total integration is lacking, impeding progress in tackling climate concerns with AI-driven solutions. Critical subjects such as scenario prediction, adaptive techniques, and forecasting have been underrepresented in current AI research, stressing the need to investigate AI's potential in these areas. Concerns about accuracy and performance in AI models for climate applications are not sufficiently addressed, compromising the dependability of AI-powered solutions. Furthermore, the relationship between CCP and AI is barely specified, despite the potential for AI to improve prediction abilities in climate science. Addressing these limitations is critical for increasing the effectiveness of AI in addressing climate concerns.

7 Conclusions

This study undertook a bibliometric analysis to explore the evolution of research at the intersection of CC and AI, using a dataset of 238 papers retrieved from Scopus spanning the years 2020 to 2023. While some aspects considered data from the past 33 years due to the limitations in finding contemporary trends in research publications over the last three years, the study provides a comprehensive overview of these fields. It includes insights into annual paper production, publication types, global distribution, research categories, influential journals, languages, highly referenced articles, current trends, and thematic connections within the subject area. Methodological tools employed included VOS viewer, Biblioshiny, and SciMAT.

Key findings show a significant rise in research attention and output in CC and AI since the 1990s, with a notable increase of 100% in publications by 2005. Subsequently, annual growth rates ranged from 50 to 70%. The global prominence of these topics has surged in the past three years, particularly post-pandemic, contributing to exponential publication growth clear from a high growth rate (R-squared value of 0.8205), underscoring the robust expansion of the dataset. Analysis reveals a predominant disciplinary focus within the field, highlighting gaps in interdisciplinary collaboration, learning, action, and study. Despite the historical dominance of the USA in this field up until 2015, China has appeared as a significant research hub, fostering many clusters dedicated to regional environmental challenges. This trend underscores a global shift towards localized approaches in addressing CC through AI solutions.

China's increasing prominence is shown by its high number in citations, affiliations, and international collaborations, introducing novel movement to the global scientific landscape. However, thematic mapping finds critical areas such as climate modelling and focused research initiatives as pivotal for shaping future research agendas within the field.

Nevertheless, integration between CC and AI is still limited, marked by sparse coverage, disconnects, and neglected connections. This gap needs concerted efforts to foster greater constructive collaboration between disciplines. While certain variables show strong correlations, overall integration stays insufficient, hindering advancements in AI-driven solutions for climate-related challenges. Key areas such as scenario prediction, adaptive techniques, and forecasting warrant further exploration to use AI's potential effectively in climate science. Moreover, addressing concerns about the accuracy and performance of AI models in climate applications is imperative for enhancing the reliability of AI-powered solutions. The relationship between CCP and AI capabilities requires clearer delineation to maximize AI's predictive abilities in climate science.

7.1 Future Research and Study Limitations

Future research in the intersection of AI and CC should focus on several critical areas to advance knowledge and impact. Firstly, fostering enhanced interdisciplinary collaboration between climate scientists and AI experts is crucial for developing integrated solutions that address complex environmental challenges effectively. Secondly, exploring advanced AI applications in climate modelling, scenario forecasting, adaptive strategies, and risk assessment will unlock new potentials for improving prediction accuracy and decision-making in climate science. Thirdly, investigating the ethical implications of AI, including issues of bias, transparency, and accountability, is essential to ensure responsible deployment of AI technologies in climate-related contexts. Additionally, addressing challenges related to data quality, availability, and privacy in AI applications for climate research will be pivotal in harnessing the full potential of data-driven approaches.

Limitations of this study should be carefully considered in interpreting its findings. Firstly, the dataset used was derived from a subset of papers sourced from Scopus, potentially limiting the study's ability to fully encompass the breadth of global research in CC and AI. Secondly, the temporal scope, while extending to 2023, may benefit from longer-term analysis to capture evolving trends and sustained impacts of AI in climate research. Additionally, the reliance on bibliometric analysis, while valuable, may overlook qualitative nuances and insights that could be captured through qualitative research methods. The study's focus on regions with higher research output could introduce bias, potentially overlooking perspectives and contributions from underrepresented regions or countries.

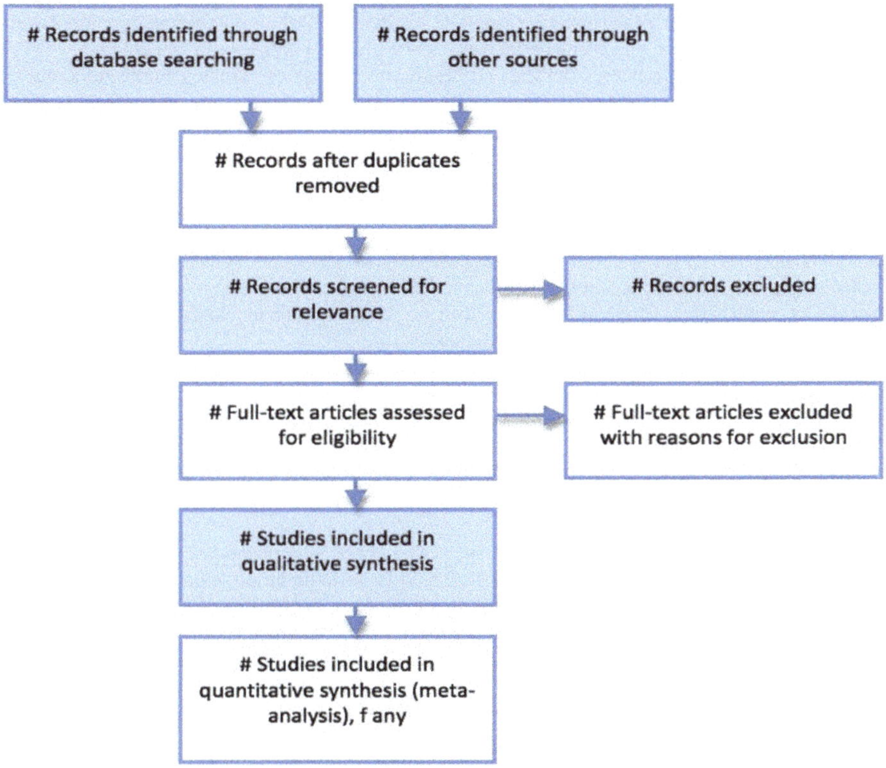

Fig. 15 Preferred reporting items for systematic reviews and meta-analyses; PRISMA. *Source* (PRISMA 2024)

8 Annexes 01

See Figure 15.

References

Abatzoglou JT, Williams (2016) Impact of anthropogenic climate change on wildfire across western US forests. In: Proceedings of the national academy of sciences of the United States of America

Adams J (2023) Global research report from the institute for scientific information. Clarivate. https://clarivate.com/blog/exploring-chinas-research-landscape-and-its-future-direction/

Adger WA (2014) Assessment of adaptation practices, options, constraints and capacity %U. In:Parry ML, Canziani OF, Palutikof JP, Hanson CE, vander Linden PJ (Eds) In: Climate Change 2007: impacts, adaptation and vulnerability. Contribution of working group II to the fourth assessm..https://ueaeprints.uea.ac.uk/ 25215.: University Press, Cambridge, pp 719e743

Berkhout F, van den Hurk B, Bessembinder J, Bregman B, van Drunen M (2023) Framing climate uncertainty: Socio-economic and climate scenarios in vulnerability and adaptation assessments

Cobo MJ, López-Herrera AG, Herrera-Viedma E, Herrera F (2011). Science mapping software tools: Review, analysis, and cooperative study among tools. J Am Soc Inf Sci Technol
Giovanni Di Matteo, Pierfrancesco N (2018) Bibliometric analysis of climate change vulnerability assessment research
Greg Garrard (2019) Never too soon, always too late: Reflections on climate temporality. Edited by Mike Hulme, Editor-in-Chief
Goodfellow, Bengio Y, Courville, A (2016) Deep learning. MIT Press
IPCC I (2021) Climate change 2021: the physical science basis. https://www.ipcc.ch/report/ar6/wg1/
Joaquin Salas GP (2022) A systematic mapping of artificial intelligence solutions for sustainability challenges in latin america and the caribbean. IEEE Lat Am Trans 20:11
Kroll H, Frietsch R (2022) China's changing role in global science and innovation, discussion papers innovation and policy analysis No. 73. Fraunhofer ISI
Kuglitsch M, Pelivan I, Ceola S, Menon M (2022) Facilitating adoption of AI in natural disaster management through collaboration. Nature Communications, 05
Markus R, Gustau C-V, Bjorn S, Martin J (2019) Deep learning and process understanding for data-driven Earth system science. Nature volume
NASA N (2021) Climate change: how do we know? https://climate.nasa.gov/evidence/
Matthew C, Nisbet SS (2017) The oxford encyclopedia of climate change communication. Oxford University Press
Nyuk, H., & Chen, Y. (2015). *Study of green areas and urban heat island in a tropical city.*
Peiris MTOV, Randeniya N, Bopitiyegedara N (2025) How do anthropogenic factors define flood risk perception of vulnerable communities? Evidence from Kelani River Lower Basin Colombo Sri Lanka. Int J Disaster Risk Reduct 120(105352). https://doi.org/10.1016/j.ijdrr.2025.105352.
Prisma (2024) The PRISMA 2020 statement. PRISMA extensions. https://www.prisma-statement.org/
Pritchard A (1969) Statistical bibliography or bibliometrics?
SR Weiskopf et al. (2020). Climate change effects on biodiversity, ecosystems, ecosystem services, and natural resource management in the United States. Sci Total Environ
Randeniya JN, Haigh R, Amaratunga D (2026) Responsible intelligence: ethical AI governance for climate prediction in the Australian context Abstract AI Ethics 6(1). https://doi.org/10.1007/s43681-025-00909-z
Ruben, H, Hector P-M, David F-A, Francisco J, Enrique H-V (2015) A bibliometric analysis of 20 years of research on software product lines. Inf Softw Technol
Salama M, Bahsoon R, Bencomo N (2017) Managing trade-offs in self-adaptive software architectures: a systematic mapping study. Manag Trade-Offs Adapt Softw Arch
National AI Strategy (2021) Secretary of state for digital, culture, media and sport by command of her majesty
Taqia R, Muhammad Zudhy Irawan Z, Anissa NM, Rizka F (2023) Knowledge mapping of cool pavement technologies for urban heat island Mitigation
UNEP U (2023) STORY As shortages mount, countries hunt for novel sources of water. UNEP—https://www.unep.org/
UNFCC U (2015) The Paris agreement. https://unfccc.int/process-and-meetings/the-paris-agreement/the-paris-agreement
Zhaohua W, Yuandong Z, Bo W (2018) A bibliometric analysis of climate change adaptation based on massive research literature data

GPSR Compliance
The European Union's (EU) General Product Safety Regulation (GPSR) is a set of rules that requires consumer products to be safe and our obligations to ensure this.

If you have any concerns about our products, you can contact us on

ProductSafety@springernature.com

In case Publisher is established outside the EU, the EU authorized representative is:

Springer Nature Customer Service Center GmbH
Europaplatz 3
69115 Heidelberg, Germany

www.ingramcontent.com/pod-product-compliance
Ingram Content Group UK Ltd.
Pitfield, Milton Keynes, MK11 3LW, UK
UKHW020320200926
475648UK00006B/18

9783032134110